The series Lecture Notes in Physics (LNP), founded in 1969, reports new developments in physics research and teaching - quickly and informally, but with a high quality and the explicit aim to summarize and communicate current knowledge in an accessible way. Books published in this series are conceived as bridging material between advanced graduate textbooks and the forefront of research and to serve three purposes:

- to be a compact and modern up-to-date source of reference on a well-defined topic;
- to serve as an accessible introduction to the field to postgraduate students and non-specialist researchers from related areas;
- to be a source of advanced teaching material for specialized seminars, courses and schools.

Both monographs and multi-author volumes will be considered for publication. Edited volumes should however consist of a very limited number of contributions only. Proceedings will not be considered for LNP.

Volumes published in LNP are disseminated both in print and in electronic formats, the electronic archive being available at springerlink.com. The series content is indexed, abstracted and referenced by many abstracting and information services, bibliographic networks, subscription agencies, library networks, and consortia.

Proposals should be sent to a member of the Editorial Board, or directly to the responsible editor at Springer:

Dr Lisa Scalone
lisa.scalone@springernature.com

Maxim Kagan

Lecture Notes
of Professor Yuri Kagan
in Theoretical Solid-State
Physics

 Springer

Maxim Kagan
Donostia International Physics Center
San Sebastián-Donostia, Guipúzcoa, Spain

ISSN 0075-8450   ISSN 1616-6361   (electronic)
Lecture Notes in Physics
ISBN 978-3-032-14620-5   ISBN 978-3-032-14621-2   (eBook)
https://doi.org/10.1007/978-3-032-14621-2

This Springer imprint is published by the registered company Springer Nature Switzerland AG
The registered company address is: Gewerbestrasse 11, 6330 Cham, Switzerland

If disposing of this product, please recycle the paper.

Preface

This book is based on the Lecture course of my father Prof. Yuri Kagan (1928–2019) who was one of the first direct pupils of Prof. Landau. He was among the founding fathers of the Modern Solid-State Theoretical Physics in Soviet Union and worldwide together with other famous Russians including Abrikosov, Andreev, Bogoliubov, Dzyaloshinskii, Eliashberg, Ginzburg, Gor'kov, Keldysh, Khalatnikov, Larkin, Lifshitz, Migdal, Pitaevskii, Tolmachev, Zel'dovich, Americans including Anderson, Ashcroft, Bardeen, Cohen, Fisher, Kadanoff, Kittel, Kohn, Leggett, Luttinger, Mermin, Pines, Schrieffer, Slater, Thouless, Europeans De Gennes, Friedel, Frohlich, Fulde, Hubbard, Kosterlitz, Mott, Nozieres, Peirls, Rice, Ziman, as well as Kasuya, Kondo, Maki, Nagaoka, Yosida, and others in Japan.

He read to me this course in 1981–1982 in my senior years in the university. I conserved the notes and now update them checking all the calculations and adding the recent results in the field such as Kosterlitz–Thouless theory of two-dimensional superfluidity.

My father started to read lectures in the beginning of the 1960s on the Chair of Theoretical Nuclear Physics in Moscow Engineering Physical Institute (famous MEPHI) and later on he proceeded on the Chair of the Solid State and Laser Physics of MEPHI.

In those years, the Solid-State Theoretical Physics was rapidly developing both in our country and abroad first of all due to the pioneering efforts of the leading scientific schools of the Soviet Union and the USA with an active participation of the famous professors of the UK, France, Germany, and Japan.

Fundamental contribution in the Condensed Matter Theory in Russia belongs to Prof. L. D. Landau and his leading pupils. Many important results in this field were obtained also by the theorists of Profs. I. E. Tamm and N. N. Bogoliubov schools. In the development of the semiconductors physics in Russia, a pioneering contribution is connected with the Saint-Petersburg School of Prof. A. F. Ioffe.

My father Academician Yuri Kagan belongs to the glorious generation of the first pupils of Academician Landau and stood on the grounds of the Modern Solid-State Physics in the beginning of the 1960s. During many years, he was the Head of the Theoretical Department of Kurchatov Institute in Moscow and read the basic course on the Solid-State Theoretical Physics in MEPHI. In his young years, he made an important contribution to the development of the atomic industry in our country.

In those years, it was a lack of good textbooks and the leading theorists of the Soviet Union, the United States, and Europe constructed their lecture courses starting from the first principles (how it is fashionable to say now) such as Coulomb laws and Slater determinants for the interactions of ions and electrons in crystals.

By doing that, Russian theorists as usual were fond of exact analytical solutions for the low-density (small values of the gas parameter) expansion in Fermi–Bose gases of neutral particles or the dense Coulomb plasma in metals. At the same time, Americans paid more attention to the physically transparent pictures and models, as well as symmetry considerations and numerical calculations in the actual (intermediate) parameter regions where exact diagrammatic expansions often proved to be impossible.

Both schools possessed excellent physical intuition, wonderful knowledge of the grounds, and broadness of the scientific interests, which allow them to read such extensive and multi-facet courses as the theory of condensed matter state selecting very carefully at the same time the most substantial material. All these qualities were demonstrated by my father while the preparation and reading of the large two-semester course on the Solid-State Theoretical Physics in MEPHI.

In fact, the approaches of the American and Russian schools were complimentary to each other. Both schools based their research on the Landau Fermi–Bose theory with the transition from strongly interacting real particles to the gas of weakly interacting quasiparticles while describing the observable quantities such as specific heat in the solid state. In this approach the most important was the careful investigation of the different bosonic and fermionic branches of the quasiparticle spectrum and damping of the elementary excitations.

My father did not retreat from this fundamental principle in his lecture course as well, devoting the first semester to the bosonic branches of the elementary excitations (ideal and slightly non-ideal (Bogoliubov) Bose gas, phonons and rotons in helium, sound and phonons in crystals, and magnons in the ferromagnets).

At the same time, the second semester was devoted to the electronic branches of the spectrum in normal metals and superconductors. His course started from the second half of the third year in the university and assumed the good knowledge of the audience of the grounds of Quantum Mechanics and Statistical Physics.

Later on, my father began to read lectures using my notes. Unfortunately, he had not enough time to work on the notes and prepare them for the publication as the monograph. Till the last years of his long career in physics, he was concentrated on writing research articles and reviews. This book is an attempt to compensate this drawback.

Ideas, lectures, and classical papers of my father together with the ideas and classical papers of my scientific supervisor Prof. A. F. Andreev exerted a strong influence on my scientific development during my young years as the theoretical physicist representing Landau School and young professor. I permanently returned to this set of ideas during all my research and professorial career in Kapitza Institute, MEPHI and MIEM as well as during extended courses on Condensed Matter Theory for Ph.D. students and postdocs of the leading universities in Europe.

Modern students in their senior years in leading Russian and western universities have in the front of them the whole bunch of the excellent textbooks and monographs in the Solid-State Theoretical Physics including classical books of Landau, Lifshitz, Pitaevskii on *Statistical Physics Part II Theory of the Condensed State, Quantum Theory of Solids* by Prof. Kittel, J. M. Ziman book *Electrons and Phonons*, A. A. Abrikosov monography *Fundamentals of the Theory of Metals*, Ashcroft–Mermin book on Solid-State Physics, Madelung book on the Solid-State Physics, and so on.

Most of these books were published in our country or translated into Russian in the 1980s—just in my late years in the university. I have all of these books on my bookshelf and often look in them with great pleasure while preparing lectures for Ph.D. students, majors, and bachelors of MIEM and MEPHI.

It is very pity that my father did not publish his Lecture course in these years. However, looking on my lecture notes I see that they do not lose their actuality even now.

Indeed, several interesting aspects of the Condensed Matter Theory connected, e.g., with Mossbauer effect and phonon degrees of freedom, phonon density of states and phonon damping due to anharmonic interactions, weakly non-ideal Bose gas and spin-polarized atomic hydrogen, effects of coherent and incoherent neutron scattering on the different crystalline structures, and of course metastable states of the metallic hydrogen are often rather vaguely described in the modern literature.

Thus, I hope that it is not late to correct this evident drawback and to publish the Lecture course of my father together with the citation of his several classical papers which conserve their actuality in our times.

Note that the recent experimental discovery of metallic state of hydrogen under high external pressure $P = 5\,\mathrm{Mbar}$ together with the observation of Eliashberg room-temperature superconductivity in metallic hydrides makes the thorough investigation of the phonon branches of the excitation spectrum and the electron–phonon interaction in metals an actual problem again.

The course of my father will serve as a valuable starting point to prepare the young specialists working in Material Science Divisions to put their teeth deeply in metallic hydrogen problem.

Working on the course of my father I tried where it is possible to conserve the original style of expressing his thoughts and physical considerations making additional clarifications only when it is necessary.

I hope that the content of this book will be interesting to the broad physical community both in Russia and abroad conserving the memory about my father—an outstanding theoretical physicist and multi-facet intellectual belonging to the golden age of Russian Science.

San Sebastian-Donostia, Spain   Maxim Kagan
September 2023–September 2025

About Professor Yuri Kagan

Professor Yuri Kagan

- 1992 Jacob Wallenberg Professor, Uppsala
- 1992 Emil Warburg Professor
- 1990 Honorary van der Waals Professor Amsterdam
- 1988 and 1996 Morris Loeb lecturer Harvard University, Boston
- 1962 Full professor of physics, Physical Engineering Institute, Moscow
- Head theory division, Kurchatov Institute, since 1962

Honours and Awards

- 1994 Award by the Alexander von Humboldt Foundation, Germany
- 1994 Karpinskii Award Toepfer Foundation, Germany
- Member of the American Physical Society, Russian Academy of Sciences, Academy Europaea, Hungarian Physical Society, Hungarian Academy of Sciences (foreign member)

Acknowledgements I am grateful to my wife Tatiana for the idea of this book, to my son Alexander Kagan, Mirumid Mirakhbarov and Victor Naumov who helped me to transfer the Word file into LATEX and improved the quality of the figures. The stimulating discussions with Vladimir Glushkov, Renat Ikhsanov, Andrey Kleev, Kliment Kugel, Alexey Kuznetsov, Eugene Mazur, Alexey Menushenkov, Vyacheslav Silkin, Alexey Vagov, and Andrey Vasenko helped me a lot in preparing the book.

I acknowledge the support and hospitality of Ricardo Diez Muino the Director of DIPC, Pedro Echenique the DIPC President, and all my colleagues during my regular visits to San Sebastian.

Competing Interests The author has no competing interests to declare that are relevant to the content of this manuscript.

About This Book

The 2-year basic course of Prof. Yuri Kagan consists of two parts. Part I of the course Bosonic branches of elementary excitations in condensed matter physics including Bogoliubov Bose gas and u-v transformation, second quantization technique, Bose–Einstein condensation (BEC) in spin-polarized atomic hydrogen, superfluid helium, sounds and vortices, crystal lattices, phonons in crystals, acoustic and optical branches, Debye approximation and Debye density of states, Van Hove singularities, phonon-specific heat, Lindemann criterion and melting of crystal lattices, quantum kinetic equation for phonons and thermal conductivity, anharmonisms and lifetime of phonons, Mossbauer effect, neutron scattering in crystals, exchange interactions, magnons in ferromagnetic materials, Curie temperature, and second-order phase transitions.

Part II Electronic excitations in normal metals and superconductors including ion–ion, electron–ion, and electron–phonon interactions, effective electron–electron interaction and screening of electron charge in metals, dielectric function and polarization operator, plasmons, Kohn anomaly and Fridel oscillations, correlation radius, energy functional and structure of elementary cell, metallic hydrogen, electronic bands in metals, magnetic properties: Landau diamagnetism and Pauli paramagnetism, De Haas–Van Alphen effect, BCS superconductivity, Tolmachev logarithm, superconducting current, London equations and Meissner effect, superconductors of the first and second orders, and superconducting alloys.

An approach of Prof. Yuri Kagan in the Part II of his lecture course (Electronic excitations in normal metals and superconductors) was ideologically close to David Pines–Philippe Nozieres books on the Theory of Quantum Fluids and Elementary Eexcitations in Solid State published in the middle of the 1960s. It is different from Alexey Abrikosov approach in Fundamentals of the metal theory and Lifshitz–Azbel–Kaganov approach in Electron Theory of Metals since it is based on the deep microscopic investigations of the basic interactions in metals, while nice books of Prof. Abrikosov and Prof. Lifshitz et al. mostly use purely phenomenological one-particle description in the spirit of Landau Fermi liquid theory.

Part I of the book on bosonic branches of elementary excitations is unique since it covers many aspects of phonon physics (anharmonisms, quantum kinetic equation, different mechanisms of thermal conductivity in dielectric crystals, Mossbauer effect) and Bose condensation theory (Bogoliubov Bose gas, Bose condensation in spin-polarized atomic hydrogen) which are rather vaguely represented or are totally absent in the other books such as a famous book on Electrons and Phonons of Prof. Ziman.

The bosonic branches become very important nowadays in connection with Eliashberg theory of high-temperature superconductivity (based on strong electron–phonon interaction) in metallic hydrogen and metallic hydrides, the search of the Pines demon (acoustic plasmon mode) in strongly correlated electron systems as well as in connection with the discovery of the Bose–Einstein condensation in the ultracold quantum gases in magnetic traps.

The Concluding Remarks briefly summarizing modern aspects in the theory of superconductivity and strongly correlated electron systems.

Contents

Part I
Bosonic Branches of Elementary Excitations

Lecture 1. Elementary Excitations in Solid State Physics

Abstract

In the first Lecture we analyze the main types of bosonic and fermionic elementary excitations emerging in the condensed matter physics. They are ranging from phonons in dielectrics and superfluid liquids and gases, electrons and plasmons in metals till excitons in semiconductors and magnons in magnetic systems. We describe different types of the ground state in stable and metastable (amorphous) systems.

1.1 Introduction

At temperature equal zero the system is in the ground state. For temperatures $T > 0$ the collective motions appear in the system (the excitations are created in it see Fig. 1.1). These excitations [1] in the first approximation are undamped. Besides that, we should introduce a weak interaction between the excitations.

In Landau theory the excitations are called quasiparticles [2–5]. If the quasiparticles still strongly interact with each other, we should construct the new quasiparticles from the initial ones, that is to reconstruct the vacuum (as in the case of interacting Bose gas or electrons in superconductor [6,7]).

In fact, the excitations are the fluctuations of the homogeneous quantum liquid [3,3–5,8–12,14]. Due to the homogeneity, there is a translational symmetry in the system. Hence, the wave function of the system will look like $e^{i\mathbf{kr}}$, i.e., the propagation of the excitations usually takes place in the form of the plane waves [15–17].

We can think about the propagation of the excitations on the language of the secondly quantized quasiparticles with the momentum $\mathbf{p} = \hbar\mathbf{k}$ (as e.g., for photons and phonons in solid state [13,18–21]).

In crystal the momentum is not conserved. The conserved quantity is a quasimomentum [13,18–23].

Because of that the wave function of electron in crystal besides a plane wave contains also the Bloch amplitude [24,25]: $\Psi = e^{i\mathbf{kr}}u(\mathbf{r})$, where $\mathbf{k}$ is quasimomentum.

© The Author(s), under exclusive license to Springer Nature Switzerland AG 2026

M. Kagan, *Lecture Notes of Professor Yuri Kagan in Theoretical Solid-State Physics*, Lecture Notes in Physics 1048, https://doi.org/10.1007/978-3-032-14621-2_1

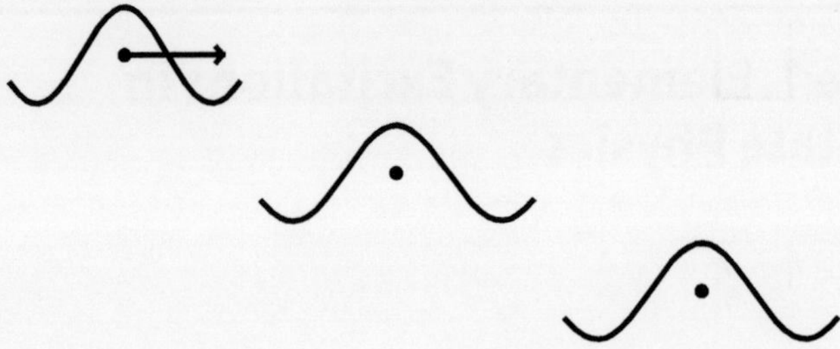

Fig. 1.1 Emergence of collective excitations in the system for non-zero temperature $T \neq 0$ [1]

Precisely in the same manner we can also quantize the problem in vacuum. As a result, we find the quanta- and quasiparticles with the momentum $\mathbf{p} = \hbar\mathbf{k}$ and spectrum (dispersion law) $\varepsilon(\mathbf{p})$. The wave function of quasiparticles is proportional to $e^{i\mathbf{kr} - i\varepsilon(\mathbf{k})t}$. Quasiparticles have the finite lifetime (similar to the resonance, peak of the finite width corresponds to them). At the same time, they can have a complicated dispersion law consisting of cosine and sine functions in crystal [21–28].

1.2   Bosonic and Fermionic Branches of Elementary Excitations

Let us count the main bosonic and fermionic branches of excitations in metals and semiconductors.

(1) Quanta of the vibrational field the phonons [1,8,11,18–31]. Phonons exist in crystals, amorphous systems, liquids and so on. It is a bosonic branch of the excitation spectrum. We will detailly investigate the phonon spectrum in the first part of the course.

(2) The main fermionic branch corresponds to electrons [1,7–9,12–14,18–36]. From electron in vacuum remains the charge (-e). It is not renormalized due to gradient invariance. Spin of electron $S = 1/2$ is conserved as well. At the same time, the kinematic properties of the electron in crystal are totally different from the electron in vacuum: an effective mass emerges instead of the bare one, the motion of electron parallel and antiparallel to electric field is possible and so on. We will study the electron spectrum in the second part of our course.

(3) One more fermionic branch is polaron [11,37–40]. The polaronic branch describes the electron slowly moving in substance, which in the process of motion strongly polarizes the media and as a result is dressed in a coat arising from this interaction. It moves together with the displacements of the media atoms.

(4) Spin waves (magnons) [1,11,12,41–47] in principle can represent both bosonic as well as fermionic branch of excitations. In ferromagnet (e.g., in one-dimensional ferromagnetic chain) the magnon corresponds to the flip of one spin in the collinear ground state due to the exchange interaction with neighboring spins (see Fig. 1.2).

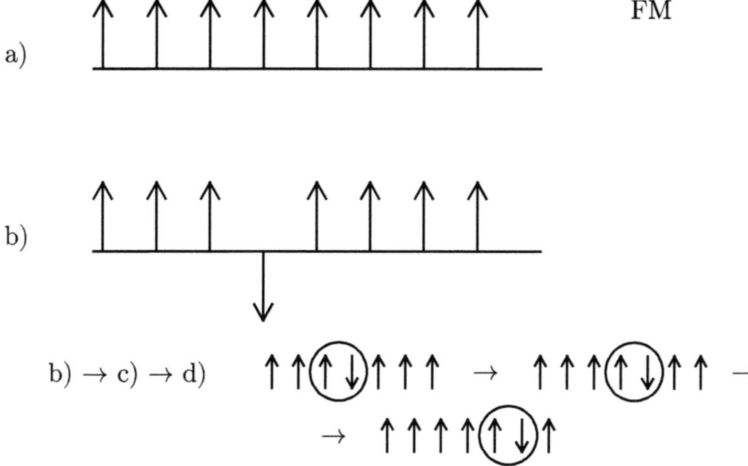

Fig. 1.2 Qualitative description of the spin wave in ferromagnet (FM). On the upper Fig. **a** the collinear ferromagnetic state is described. On the **b** we illustrate one flipped spin due to the exchange interaction between the spins. Further on the **b c d** the propagation along the FM chain from the left to the right of the pair of antiparallel spins on the neighboring sites is described. The equal energies of the states **b c d** makes the propagation of this "singlet" pair possible. As a result of this propagation a spin wave runs along the chain [1]

As a result of the flip process the pair of the antiparallel spins appears on the neighboring sites. In the spin waves this pair propagates ("runs" along the chain from the left to the right).

(5) <u>Excitons</u> [48–51] represent one more bosonic branch of the spectrum. Excitons appear as a result of the excitation by light of the electronic state of one atom. This excited state described, e.g., by the two-level system afterwards runs along the crystal (see Fig. 1.3).

In other words, the excited state effectively «collapses»on the given site of the crystal lattice (the electron on his site returns back to the ground state), but at the same time an excited state emerges on the neighboring site and so on. As a result, the state excited by light in a 1D chain effectively propagates (runs along the chain from the left to the right) together with the wave propagation.

(6) One more bosonic branch of excitations in metals corresponds to <u>plasmons</u> [1,9, 10,27,52]. Plasmonic branch describes collective oscillations of electron density in metals. We will also thoroughly consider it in the second part of the lecture course.

Note that the correlated pair of fermions is a boson. Let us remind that the phenomenon of the BCS superconductivity is based on the formation of the Cooper pairs consisting of two electrons in the momentum space. In the bi-polaron superconduc-

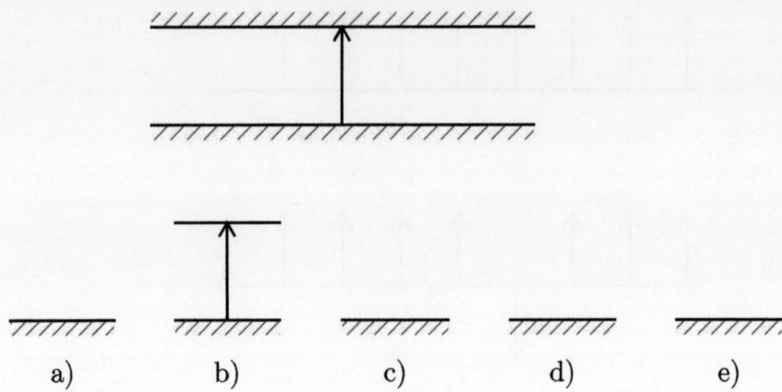

Fig. 1.3 Qualitative description of the optical exciton. On the upper figure we show the two-level system, e.g., two-level atom in which electron can be in a ground and in an excited state. On the lower figure we show the wave propagation along the crystal which is accompanied by the processes of the collapse on the site (**a**) and emergence on the neighboring site (**b**) of the excited electronic[1] state

tivity of Alexandrov–Ranninger [53] the local pair of two polarons (bi-polaron) emerges in the real space.

In nuclei we can have the pair of two protons or two neutrons. As a result, the compound boson (bi-neutron or bi-proton), i.e., the Bose particle with the total integer spin $S = 0$ (or $S = 1$ as in neutron stars [26,28,54–57]) can be formed leading to the superfluidity.

At low temperatures $T \sim 1\,\mathrm{mK}$ two helium-3 atoms form Cooper pairs and He-3 becomes superfluid [54]. Moreover, the Cooper pairs in superfluid He-3 are formed again in a state with total spin of the pair $S = 1$ [26,28,54–56,58].

1.3 The Ground State of the System

From Quantum mechanics [15] we know that the correction to the energy of the ground state arising in the second order of the perturbation theory is always negative.

As a result, we have attraction at large distances. At the same time, at small distances between the atoms we have repulsion due to Coulombic interactions between the nuclei.

If we denote the volume on one electron as Ω_0, then this volume will be proportional to the third power of the mean distance between electrons in metal $\Omega_0 \sim r_0^3$.

Let us stress, that the kinetic energy of electrons estimated from the uncertainty principle by the order of magnitude equals to

$$T_{\mathrm{el.}} \sim \frac{\hbar^2}{mr_0^2}. \tag{1.1}$$

At the same time the Coulomb interaction of electrons with nuclei:

$$V \sim -\frac{1}{r_0}.$$

(1.2)

Thus, for large densities $r_0 \to 0$ kinetic energy grows faster than potential energy. Hence the total energy should be positive: $E = T + U > 0$, and electrons have the tendency to delocalize on large densities to increase r_0 and in this way to decrease the total energy.

As a result, the <u>metallization takes</u> place in the system since $E \sim T \to \min$ [1,59,60].

Note, that for heavy particles (for atoms in helium or for ions in crystals) $T_{\text{at.}} \sim \frac{\hbar^2}{Mr_0^2}$. Moreover, the ion mass $M \gg m$ is large. If the kinetic energy of ions is small in comparison with the absolute value of the electron–ion potential energy, then the minimum of the total energy of the ions is determined by the potential energy and is realized on the periodic structure (or on the crystal).

In fact, the fixed distance r_0 appears in the system which equals to the period of the crystalline lattice (or the distance between ions in crystal). It corresponds to the realization of the minimum of the total ionic energy and, as a consequence, of that to the total energy of the full system of ions and delocalized electrons (see Fig. 1.4).

Let us stress that this is not a universal situation. There exist for the example amorphous systems without the crystalline structure. However, for amorphous systems [61–68] we always have a <u>metastable</u> and not a ground state (see Fig. 1.5).

It is interesting, that if we reduce the ion mass (considering hydrogen or helium), then the kinetic energy of heavy particles $T_{\text{at.}} \sim \frac{\hbar^2}{Mr_0^2}$ increases and begins to play an important role. As a result, the <u>quantum crystals</u> appear [69–72].

Note that in helium and hydrogen usually the interactions are weak and potential wells are shallow [26,28,65,69–75]. This happens in He-3 and He-4. It becomes

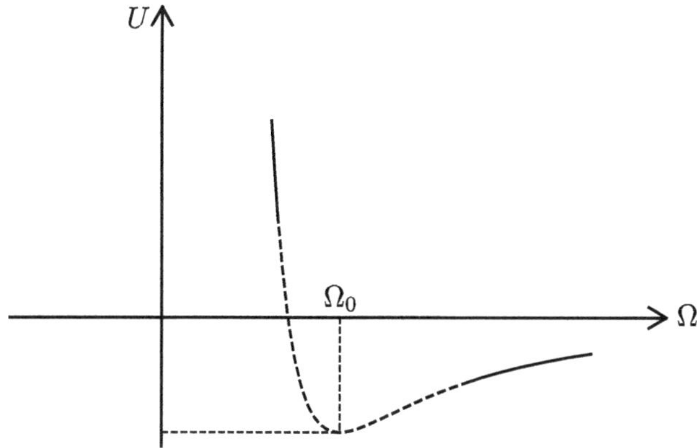

Fig. 1.4 Dependence of the energy of metal in the crystalline state from the volume of the elementary cell [1]

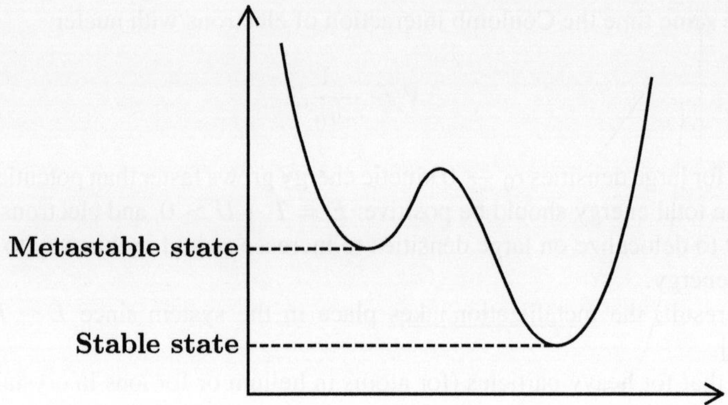

Fig. 1.5 Metastable state of the amorphous system. For $T = 0$ the system can practically infinitely long time live in the metastable state [1]

beneficial to delocalize for the system of atoms. The total energy should remain negative meanwhile.

In fact, for He-3 and He-4 the energy of the "smeared" state is lower, than the energy of the ordered state. Let us emphasize that helium has a small mass and hence a large kinetic energy. It has a filled electron shell and thus the small absolute value of the attractive potential energy. In fact, the potential energy in helium has a van der Waals character [1,2,26,28,69,73].

To answer the question what is the ground state of helium-liquid or solid we should perform the analysis with respect to the atomic mass and the magnitude of the interaction between the atoms (to determine the De Boer [76] parameter in today terminology of the low temperature physics).

1.4   The Spin-Polarized Hydrogen

For hydrogen the interaction between molecules is larger than for helium. That is why hydrogen wants, but cannot be a liquid [1,74–78].

It is important to note, that (in the absence of external pressure) the pressure in minimum $P = -\frac{\partial E}{\partial \Omega} = 0$.

We know from the Heitler–London theory [15,79], that molecular phase of hydrogen is realized in the spin-singlet state. At the same time spin-polarized hydrogen remains in the atomic state (see Fig. 1.6), though even for the spin-triplet term there is a shallow van der Waals potential well with the depth of 6 K [1,2,74–78]. This well, however, is too shallow and the bound state does not arise in it.

Suppose the magnetic field H (which creates the spin polarization) is very large ($H \sim 10^5$ Oe, $\frac{\mu_B H}{k_B} \sim 6.5$ K), while the temperature $T \leq 1$ K is small. Then the ratio $\frac{\mu_B H}{k_B T} \gg 1$, and hydrogen atoms are spin-polarized [74,75].

However, we know that the effective interaction between two particles should be expressed in terms of the scattering amplitude $U_{\text{eff.}} \sim a$ [2,15]. It occurs, that the

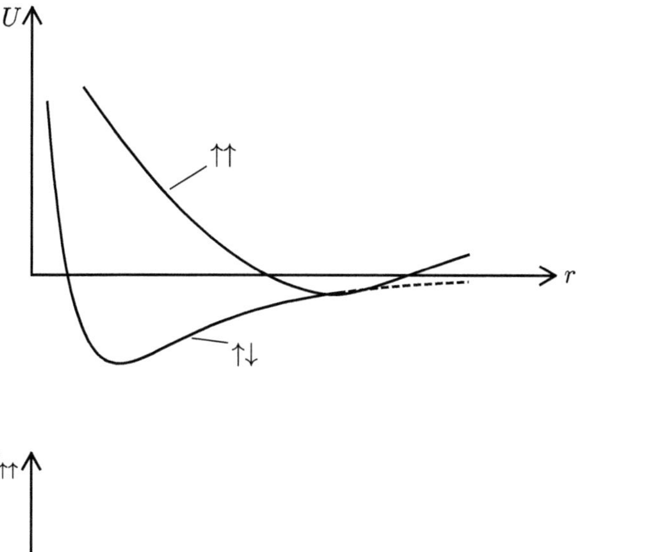

Fig. 1.6 Spin-singlet and spin-triplet terms for the interaction between two hydrogen atoms. For singlet term a deep enough potential well arises which leads to the bound state of two atoms (the hydrogen molecule is formed). For the spin-triplet term we also have a shallow van der Waals potential well of the depth of the order of 6 K. However, the bound state is not formed in this well [1]

scattering amplitude is positive $a > 0$ and corresponds to repulsion in spite of the fact that the van der Waals potential energy corresponds to attraction.

Then the actual dependence of the total energy on the volume of the elementary cell looks as follows (see Fig. 1.7) and the energy minimum on this Figure is realized for the density equal to zero. Thus, the spin-polarized atomic hydrogen remains in the quantum gas state for $T = 0$.

This fact manifests the difference between quantum and classical gas. The classical gas would always condense even in the shallow potential well.

Concluding this section, note that hydrogen atom is the composed boson. It consists of proton and electron and has a total spin $S = 0$ or 1.

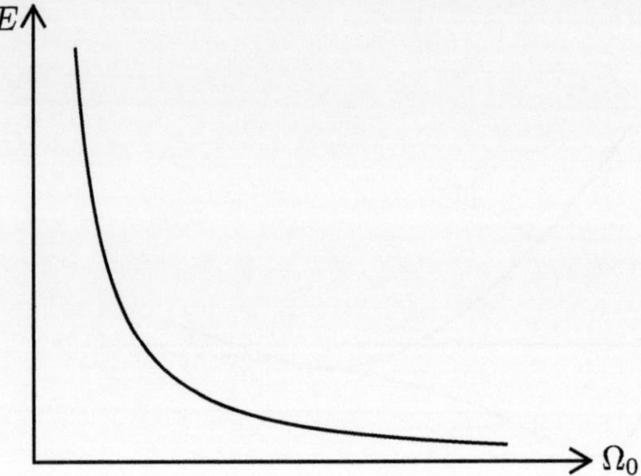

Fig. 1.7 Dependence of the energy of the spin-polarized atomic hydrogen on the average volume per one atom Ω_0. The energy is minimal for the density of the gas equal to zero [1]

At the same time He-3 atom has 2 protons, 1 neutron and 2 electrons on the closed shell with total electron spin which is equal to 0. Thus He-3 is a composed fermion. Analogously He-4 atom consists of 2 protons, 2 neutrons and 2 electrons and hence is a composed boson.

1.5   The Types of the Ground State

For the closed electron shells the van der Waals ground state is typical. It is characterized by the small binding energies and low melting temperatures.

For the non-closed shells (in the covalent crystal) the ground state is described by the two-well potential (see Fig. 1.8). This potential is often called «the Lifshitz trousers» [1, 15].

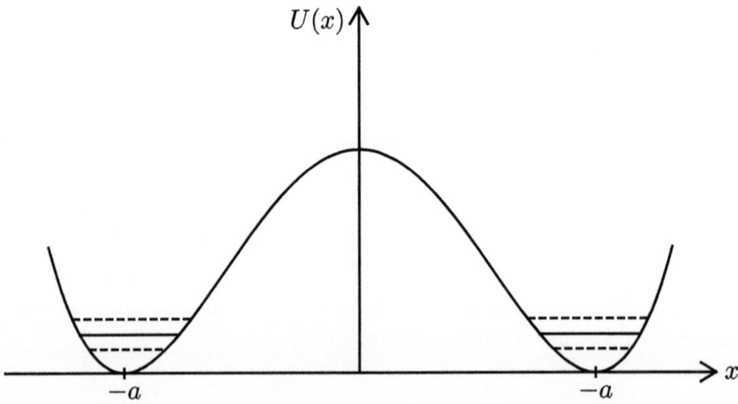

Fig. 1.8 Two-well potential and the splitting of the ground state level in it. The electron is delocalized between the two wells in this potential [1]

If we bring two systems with identical levels close to each other, then we get the ground state shown in Fig. 1.8. It has lower energy due to the fact that a particle living in the first well spends part of the time in the second well and thus is delocalized between the wells.

Suppose that we have one electron in each potential well. In this case we get the competition between the gain in kinetic energy due to delocalization ($\Delta E < 0$) and the loss of energy due to Coulomb repulsion ($\Delta E > 0$).

If the potential wells are located on the substantially large distance from each other, then the Coulomb locking takes place promoting the dielectric state of the system [80–86].

If we bring the potential wells very close to each other, then the energy gain due to delocalization prevails over the loss due to Coulomb energy. In this case the electron starts to move freely between the wells and the delocalization in the real space takes place. It leads to the metallization of the ground state.

As a result, as a function of the change of the density we have the metal–insulator transition in the system. On a modern language it is often called Mott–Hubbard localization transition [80–84]. Moreover, if the minimum of the energy corresponds to the close location of the wells in the two-well Lifshitz potential [15], then the system is emerged in the metallic state. If the separation between the wells is large enough, then the dielectric ground state is realized.

On the border between metals and dielectrics there are semiconductors such as germanium (Ge) or silicium (Si) [87–90].

References

1. Y. Kagan, *Lectures on the Solid-state Theoretical Physics* (MEPHI, Moscow, 1981–1982). Part I. Phonons, Part II Electrons - unpublished
2. L.D. Landau, E.M. Lifshitz, *Statistical Physics, Part I* (Butterworth-Heinemann, Oxford, 1999)
3. E.M. Lifshitz, L.P. Pitaevskii, *Statistical Physics, Part II: Theory of the Condensed State* (Elsevier, 2013)
4. L.D. Landau, Sov. Phys. JETP **3**, 920 (1957)
5. L.D. Landau, Sov. Phys. JETP **5**, 101 (1957)
6. N.N. Bogoliubov, V.V. Tolmachev, D.V. Shirkov, *New Method in the Theory of Superconductivity* (Consultants Bureau, New York, London, Chapman and Hall, 1959)
7. J.R. Schrieffer, *Theory of Superconductivity* (W.A. Benjamin Inc., New York, 1964)
8. D. Pines, *Elementary Excitations in Solids* (W.A. Benjamin Inc., New York-Amsterdam, 1963)
9. D. Pines, P. Nozieres, *The Theory of Quantum Liquids*, vol. 1. Normal Fermi Liquids (W.A. Benjamin Inc. New York-Amsterdam, 1966)
10. P. Nozieres, D. Pines, *The Theory of Quantum Liquids: Superfluid Bose Liquids*, vol. 2 (Addison Wesley Publishing Company, USA, Canada, 1990)
11. R.P. Feynman, *Statistical Mechanics, A Set of Lectures* (Advanced Books Classics) (Westview Press, Boulder, 1998)
12. P. Fulde, *Electron Correlations in Molecules and Solids*. Springer Series in Solid State Sciences (Springer, Berlin, Heidelberg, New York, 1993)
13. C. Kittel, *Quantum Theory of Solids* (Wiley, New York-London, 1963)
14. A.A. Abrikosov, L.P. Gor'kov, I.E. Dzyaloshinskii, *Methods of Quantum Field Theory in Statistical Physics* (Prentice Hall, Englewood Cliffs, New Jersey, 1963)

15. L.D. Landau, E.M. Lifshitz, *Quantum Mechanics: Non-Relativistic Theory* (Pergamon Press, New York, 1977)
16. R.P. Feynman, A. Hibbs, *Quantum Mechanics and Path Integrals* (McGraw-Hill Companies, New York, 1965)
17. R.P. Feynman, *The Feynman Lectures on Physics* (Addison Wesley Longman, 1970)
18. J.M. Ziman, *Electrons and Phonons, The Theory of Transport Phenomena in Solids* (Clarendon Press, Oxford, 1960)
19. N.W. Ashcroft, N.D. Mermin, *Solid State Physics* (Holt, Rinehart and Winston, 1976)
20. O. Madelung, *Introduction to Solid-State Theory*. Springer Series in Solid-State Sciences (1995)
21. A.A. Abrikosov, *Fundamentals of the Theory of Metals* (Elsevier Science Publishers, North-Holland, Amsterdam, 1988)
22. I.M. Lifshitz, M.Y. Azbel', M.I. Kaganov, *Electron Theory of Metals* (Consultants Bureau, New York, 1973)
23. J.M. Ziman (ed.), *The Physics of Metals* (Cambridge University Press, 1960)
24. J.M. Ziman, *The Calculation of Bloch Functions*. Solid State Physics. Advances in Research and Applications (Academic, New York and London, 1971)
25. A.H. Wilson, *Quantum Theory of Metals* (Cambridge University Press, New York, 1953)
26. M.Y. Kagan, *Modern Trends in Superconductivity and Superfluidity*. Lecture Notes in Physics, vol. 874. (Springer, Dordrecht, 2013)
27. M. Yu. Kagan, K.I. Kugel, A.L. Rakhmanov, A.O. Sboychakov, *Electronic Phase Separation in Magnetic and Superconducting Materials Recent Advances*. Springer Series in Solid-State Sciences, vol. 201 (Springer International Publishing, 2024)
28. M.Y. Kagan, *Physics of the Macroscopic Quantum Systems* (Publishing House of the Moscow Energetic Institute (MEI), Moscow, 2014). (in Russian)
29. N.B. Brandt, V.A. Kul'bachinskii, *Quasiparticles in Condensed Matter Physics* (Fizmatlit Publishing House, Moscow, 2005). ((in Russian))
30. Y.I. Frenkel, *Introduction to the Theory of Metals* (State Technical and Theoretical Publishing House, Leningrad-Moscow, 1948). ((in Russian))
31. Y.G. Dorfman, I.K. Kikoin, *Physics of Metals*. Optical and Magnetic Properties, State Technical and Theoretical Publishing House (Leningrad-Moscow, Electric, 1933). (in Russian)
32. S.W. Lovesey, *Condensed Matter Physics: Dynamic Correlations (Frontiers in Physics)* (The Benjamin Cummings Publishing Company, 1986)
33. S.V. Vonsovskii, M.I. Kaznelson, *Quantum Solid-State Physics* (Springer Series in Solid-State Sciences, Berlin-Heidelberg- New York, 1989)
34. W.A. Harrison, *Electronic Structure and the Properties of Solids* (W.H. Freeman and Company, San Francisco, 1980)
35. W.A. Harrison, *Solid State Theory* (McGraw-Hill Book Company, New-York-London-Toronto, 1970)
36. J.W. Negele, H. Orland, *Frontiers in Physics* (Addison-Wesley Publishing Company, Quantum Many-Particle Physics, 1988)
37. L.D. Landau, Phys. Z. Sowjetunion **3**, 664 (1933). ((in German))
38. S.I. Pekar, J. Phys. USSR **10**, 341 (1946)
39. A.S. Alexandrov, N.F. Mott, *Polarons and Bi-polarons* (World Scientific, Singapore, 1995)
40. J.T.L. Devreese, *Polarons in Ionic Crystals and Polar Semiconductors* (North-Holland, Amsterdam, 1972)
41. A.I. Ahiezer, V.G. Bar'yachtar, S.V. Peletminskii, *Spin Waves* (Nauka Publishing House, Moscow, 1967). ((in Russian))
42. S.V. Vonsovskii, *Magnetism* (Wiley, 1974)
43. S.V. Tyablikov, *Methods in Quantum Theory of Magnetism* (Springer Science + Business Media, 1967)
44. D.C. Mattis, *The Theory of Magnetism: An Introduction to the Study of Cooperative Phenomena* (Harper and Row Publishers, New York, Evanston, London, 1965)
45. Y.A. Izumov, Y.N. Skryabin, *Basic Models in Quantum Theory of Magnetism* (Ekaterinburg, 2002) (in Russian)

46. A.M. Tsvelik, *Quantum Field Theory in Condensed Matter* (Cambridge University Press, 1995)
47. G. Parisi, *Statistical Field Theory (Frontiers in Physics)* (Addison-Wesley Publishing Company, 1988)
48. V.M. Agranovich, V.L. Ginzburg, *Crystal Optics with an account of Spatial Dispersion, and Theory of Excitons* (Springer, Berlin-Heidelberg-New York, 1984)
49. V.M. Agranovich, *Theory of Excitons* (Nauka Publishing House, Moscow, 1968). ((in Russian))
50. R.S. Knox, *Theory of Excitons* (Academic, New York-London, 1973)
51. L.D. Landau, E.M. Lifshitz, *Electrodynamics of Continuous Media* (Pergamon Press, New York, 1984)
52. E.M. Lifshitz, L.P. Pitaevskii, *Physical Kinetics* (Butterworth-Heinemann, 2012)
53. A. Alexandrov, J. Ranninger, Phys. Rev. B **23**, 1796 (1981)
54. D. Vollhardt, P. Wolfle, *The Superfluid Phases of Helium 3* (Taylor and Francis, London, New York, Philadelphia, 1990)
55. B. Svistunov, E. Babaev, N. Prokof'ev, *Superfluid States of Matter* (CRS Press, Taylor and Francis group, London, New York, 2015)
56. G.E. Volovik, *The Universe in Helium Droplet* (Oxford University Press, Oxford, 2002)
57. G. Baym, C. Pethick, D. Pines, Nature **224**, 673 (1969)
58. D.D. Osheroff, R.C. Richardson, D.M. Lee, Phys. Rev. Lett. **28**, 885 (1972)
59. E. Wigner, H.B. Huntington, J. Chem. Phys. **3**, 764 (1935)
60. E.G. Brovman, Yu. Kagan, A. Kholas, Structure of the metallic hydrogen at zero pressure. Sov. Phys. JETP **34**, 1300 (1972)
61. P.W. Anderson, B.I. Halperin, C.M. Varma, Phil. Mag. **25**, 1 (1972)
62. W.A. Phillips, J. Low Temp. Phys. **7**, 351 (1972)
63. N.F. Mott, E.A. Davis, *Electronic processes in Non-Crystalline Materials* (Oxford University Press, 1979)
64. I.M. Lifshitz, S.A. Gredeskul, L.A. Pastur, *Introduction to the Theory of Disordered Systems* (Wiley, New York, 1988)
65. J.M. Ziman, *Models of Disorder: The Theoretical Physics of Homogeneously Disordered Systems* (Cambridge University Press, 1979)
66. B.I. Shklovskii, A.A. Efros, *Electronic Properties of Doped Semiconductors* Springer Series in Solid-State Sciences (Berlin-Heidelberg-New York, 1984)
67. M.H. Brodsky (ed.), Amorphous semiconductors, in *Topics in Applied Physics* (Springer, Berlin-Heidelberg-New York, 1979)
68. P. Esquinazi, (ed.), *Tunneling Systems in Amorphous and Crystalline Solids* (Springer, 1998)
69. A.F. Andreev, I.M. Lifshitz, Sov. Phys. JETP **29**, 1107 (1969)
70. A.J. Leggett, Phys. Rev. Lett. **25**, 1543 (1970)
71. E. Kim, M.H.W. Chan, Science **305**, 1941 (2004)
72. E. Kim, M.H.W. Chan, Nature **427**, 225 (2004)
73. K.H. Benneman, J. Ketterson (eds.), *The Physics of Liquid and Solid Helium* (Wiley, New York, London, Sydney, Toronto, 1975)
74. A. Griffin, D.W. Snoke, S. Stringari (eds.), *Bose–Einstein Condensation* (Cambridge University Press, UK, 1995)
75. T.J. Greytak, D. Kleppner, in *New Trends in Atomic Physics*, ed. by G. Grynberg, R. Stora vol. II (North-Holland, New York, 1984), p.1127
76. J. De Boer, Physica **14**, 139 (1948)
77. I.F. Silvera, J.T.M. Walraven, Phys. Rev. Lett. **44**, 164 (1980)
78. Yu. Kagan, I.A. Vartanyanz, G.V. Shlyapnikov, Sov. Phys. JETP **54**, 590 (1981)
79. W. Heitler, F. London, Zeit. fur Physik **44**, 455 (1927)
80. N.F. Mott, *Metal–Insulator Transitions* (Taylor and Francis LTD, London, 1974)
81. J. Hubbard, Proc. Royal Soc. London A **276**, 238 (1963)
82. J. Hubbard, Proc. Roy. Soc. London A **277**, 237 (1964)
83. J. Hubbard, Proc. Roy. Soc. London A **281**, 401 (1964)
84. J. Hubbard, Proc. Roy. Soc. London A **285**, 542 (1965)
85. H. Frohlich, *Theory of Dielectrics* (Clarendon Press, Oxford, 1958)

86. W. Braun, *Dielectrics* (Springer, Berlin-Gottingen-Heidelberg, 1956)
87. P.Y. Yu, M. Cardona, *Fundamentals of Semiconductors: Physics and Materials Properties* (Springer, 2002)
88. A.I. Ansel'm, *Introduction to the Theory of Semiconductors, State Publishing House of Physical and Mathematical Literature* (Moscow-Leningrad, 1962)
89. V.L. Bonch-Bruevich, S.G. Kalashnikov, *Semiconductor Physics* (Nauka Publishing House, Moscow, 1977). ((in Russian))
90. K. Seeger, *Semiconductor Physics* (Springer, Wien-New York, 1973)

Lecture 2. Ideal Bose Gas

2

Abstract

In this Lecture we analyze the thermodynamic properties of the ideal Bose gas. We introduce Bose–Einstein distribution function and calculate Bose condensation (BEC) temperature. We consider the number of condensed and non-condensed particles and investigate carefully the structure of the Bose condensate. We find temperature dependence of the specific heat at low temperatures and near the critical λ-point in the ideal Bose gas.

2.1 Bose–Einstein Distribution for the Ideal Bose Gas

In ideal Bose gas [1–16] each level can be occupied by the arbitrary number of particles. For the derivation of the Bose–Einstein distribution [2–4] we need the particle identity [1,17,25] which means the absence of the exchanges between the particles which occupy the same level in combinatorics. Note that the particles which occupy the different levels do not interact in the ideal Bose gas.

Let n_k to be the occupation number (the number of particles in the state with momentum $\mathbf{k}$).

For bosons $n_k = 0, 1, 2, \ldots$ is an arbitrary integer number.

We should start from the Gibbs Grand canonical distribution [1,2] for large alternating number of particles ($k_B = 1$):

$$w_{n_k k} = \exp\left(\frac{\Omega_k + \mu n_k - n_k \varepsilon_k}{T}\right), \tag{2.1}$$

where $\Omega = \sum_k \Omega_k$ is the (total) thermodynamic potential of the system, Ω_k is the thermodynamic potential of the particles occupying the state with momentum $\mathbf{k}$, $E = \sum_k E_k$ is the total energy of the system, $E_k = n_k \varepsilon_k$ is the energy of the state with momentum $\mathbf{k}$ in the ideal (non-interacting) Bose gas.

© The Author(s), under exclusive license to Springer Nature Switzerland AG 2026

M. Kagan, *Lecture Notes of Professor Yuri Kagan in Theoretical Solid-State Physics*, Lecture Notes in Physics 1048, https://doi.org/10.1007/978-3-032-14621-2_2

The normalization condition for the probability

$$\sum_{n_k=0,1,2,\ldots} w_{n_k k} = 1 \tag{2.2}$$

can be represented as follows:

$$\exp\left(\frac{\Omega_k}{T}\right) \sum_{n_k=0}^{\infty} \left(\exp\frac{\mu-\varepsilon_k}{T}\right)^{n_k} = 1. \tag{2.3}$$

Summing up the infinite geometrical series in (2.3), we get

$$\frac{\Omega_k}{T} = -\ln\frac{1}{\left(1-\exp\frac{\mu-\varepsilon_k}{T}\right)}. \tag{2.4}$$

Thus, the thermodynamic potential of the **k** state [1,2]:

$$\Omega_k = T\ln\left(1-\exp\frac{\mu-\varepsilon_k}{T}\right). \tag{2.5}$$

As a result, the average occupation number of the **k** state:

$$\bar{n}_k = -\left(\frac{\partial\Omega_k}{\partial\mu}\right)_T = \frac{1}{\left(\exp\frac{\varepsilon_k-\mu}{T}-1\right)}. \tag{2.6}$$

At the same time the full thermodynamic potential of the system:

$$\Omega = \sum_k \Omega_k = T\sum_k \ln\left(1-\exp\frac{\mu-\varepsilon_k}{T}\right) = -PV. \tag{2.7}$$

From the positiveness of $\bar{n}_k$ the chemical potential should be negative or zero: $\mu \leq 0$. In the classical (Boltzmann [1,2]) statistics $\mu < 0$, and its absolute value is very large.

2.2 Bose Condensation (BEC) Temperature

Let us write an expression for the total number of particles in ideal Bose gas via the average occupation numbers [2]:

$$N = \sum_k \bar{n}_k = V\int\frac{d^3\mathbf{k}}{(2\pi)^3}\frac{1}{\left(\exp\frac{\varepsilon_k-\mu}{T}-1\right)}. \tag{2.8}$$

Now let us proceed to the integration over the energy $\varepsilon = \frac{\hbar^2 k^2}{2m}$. Then with an account of simple expressions $k = \frac{1}{\hbar}\sqrt{2m\varepsilon}$, $d\varepsilon = \frac{\hbar^2}{m}k\,dk$, we get

$$d^3\mathbf{k} = 4\pi k(k\,dk) = 4\pi \frac{1}{\hbar}\sqrt{2m}\,\frac{m}{\hbar^2}\sqrt{\varepsilon}\,d\varepsilon. \tag{2.9}$$

As a result (2.8) is written as follows:

$$N = V\frac{m^{3/2}}{\hbar^3}\frac{4\pi\sqrt{2}}{(2\pi)^3}\int_0^\infty d\varepsilon \frac{\sqrt{\varepsilon}}{\left(\exp\frac{\varepsilon-\mu}{T} - 1\right)}. \tag{2.10}$$

But the gas density $n = \frac{N}{V}$ is fixed. Moreover, the chemical potential of the Bose gas is not positive $\mu = -|\mu| \le 0$. Hence:

$$\frac{N}{V} = \frac{m^{3/2}}{\hbar^3}\frac{\sqrt{2}}{2\pi^2}\int_0^\infty d\varepsilon \frac{\sqrt{\varepsilon}}{\left(\exp\frac{\varepsilon+|\mu|}{T} - 1\right)} = \text{const.} \tag{2.11}$$

Let us lower the temperature T. We should find μ as a function of T: $|\mu| = f(T)$. Since the temperature T becomes smaller, the exponent $\exp\frac{\varepsilon+|\mu|}{T}$ increases. To compensate this increase we should lower $|\mu|$. Finally for $T = T_0$: $|\mu| = 0$.
Introducing dimensionless variable $x = \frac{\varepsilon}{T}$ for the energy at $T = T_0$, we obtain

$$\frac{N}{V} = n = \frac{m^{3/2}T_0^{3/2}}{\hbar^3}\frac{\sqrt{2}}{2\pi^2}\int_0^\infty dx \frac{\sqrt{x}}{(\exp(x) - 1)}. \tag{2.12}$$

The integral which enters in Eq. (2.12) is tabulated. As a result, we get the famous Einstein formula [4]:

$$T_0 = \frac{\hbar^2}{m}n^{2/3} \cdot 3.31. \tag{2.13}$$

2.3 The Number of Condensed and Non-condensed Particles

For temperatures $T < T_0$ we have the contradiction, connected with the impossibility to satisfy the condition on the conservation of the total number of particles. The contradiction can be resolved in the following way: in the equation for total number of particles we proceed from the sum to the integral according to

$$N = \sum_k n_k \rightarrow V \int \frac{d^3\mathbf{k}}{(2\pi)^3}\bar{n}_k. \tag{2.14}$$

This replacement is valid when there are no specially selected (distinguished) states (that is when occupation numbers of each state are very small in comparison with the total number of particles).

We have a distinguished state with momentum $k = 0$, since

$$\bar{n}_k = \frac{1}{\exp\left(\frac{\frac{\hbar^2 k^2}{2m} + |\mu|}{T}\right) - 1}$$

and as a result, for $k = 0$: $|\mu| \to 0$, and hence, $\bar{n}_{k=0} \to \infty$. That is why we have to select the contribution of zero energy in the sum and write for the total number of particles:

$$N_{\text{tot.}} = N_\varepsilon = 0 + N_{\varepsilon > 0}. \tag{2.15}$$

For $T \leq T_0$ the number of the non-condensed particles:

$$N_{\varepsilon > 0} = \left(\frac{T}{T_0}\right)^{3/2} N_{\text{tot.}}. \tag{2.16}$$

For $T = T_0$ all the particles are washed out from the condensate $N_{\varepsilon > 0} = N_{\text{tot.}}$. (they occupy the states with $\varepsilon > 0$).

It is important that for $T < T_0$ the macroscopic number of particles occupy the level with $k = 0$:

$$N_{\varepsilon = 0} = N_0 = N_{\text{tot.}} \left(1 - \left(\frac{T}{T_0}\right)^{3/2}\right). \tag{2.17}$$

2.4   Structure of the Bose Condensate

Let us demonstrate that Bose condensation takes place not in the narrow interval of levels close to $k = 0$, but precisely at this level [1].

Suppose that for low temperatures $T \leq T_0$ the chemical potential of the Bose gas μ is very small ($|\mu| \ll T$), but is not equal to zero.

We would like to show that the chemical potential is much less, than the level spacing between zeroth and next level. By this demonstration we will prove that it is valid to put the chemical potential equal to zero.

We will show also that the number of particles living on the next level, though still much larger than 1, at the same time is much smaller than the number of particles on the zeroth level.

Rigorously speaking, on the zeroth level (for $\varepsilon = 0$) the number of particles

$$N_0 = \frac{1}{\exp\left(\frac{|\mu|}{T}\right) - 1} \approx \frac{T}{|\mu|}. \tag{2.18}$$

Thus, the chemical potential for non-zero temperatures:

$$|\mu| = \frac{T}{N_0}, \tag{2.19}$$

and hence $|\mu|$ is really very small, because $N_0 \sim N \sim 10^{24}$ is macroscopically large quantity.

Let us find now the occupation of the next level (which follows after the one with $k = 0$). If we put our system in the cubic box with the volume $V = L^3$, then the inter-level spacing in the box

$$\Delta\varepsilon \sim \frac{\hbar^2}{m}\frac{\pi^2}{L^2} \tag{2.20}$$

(since the minimal wave vector in this case $k_{\min} \sim \frac{\pi}{L}$). At the same time the gas density $n = \frac{N}{V}$, and correspondingly:

$$L^2 = V^{2/3} = \left(\frac{N}{n}\right)^{2/3}. \tag{2.21}$$

Then the level spacing:

$$\Delta\varepsilon \sim \frac{\hbar^2}{m} n^{2/3} \frac{1}{N^{2/3}} \sim T_0 \frac{1}{N^{2/3}}. \tag{2.22}$$

Now since $T \sim T_0$ and $N_0 \sim N$, the estimates for the chemical potential and level spacing can be written as follows: $|\mu| \sim \frac{T_0}{N_0}$ and $\Delta\varepsilon \sim \frac{T_0}{N_0^{2/3}}$. Then for the ratio between the chemical potential and the level spacing we get

$$\frac{|\mu|}{\Delta\varepsilon} \sim \frac{1}{N_0^{1/3}} \to 0. \tag{2.23}$$

If $N_0 \sim 10^{24}$, then $N_0^{1/3} \sim 10^8$. That is why for the next level with the energy $(0 + \Delta\varepsilon)$ the chemical potential $|\mu| = 0$ with a fantastic accuracy.

Moreover, the average occupation number of the next (the first) level:

$$N_1 \sim \frac{1}{\exp\left(\frac{0+\Delta\varepsilon+|\mu|}{T}\right) - 1} \approx \frac{1}{\exp\left(\frac{\Delta\varepsilon}{T}\right) - 1}$$

$$\approx \frac{T}{\Delta\varepsilon} \sim \frac{T}{|\mu|}\frac{|\mu|}{\Delta\varepsilon} \sim N_0 \frac{1}{N_0^{1/3}} \sim 10^{-8} N_0 \sim 10^{16} \tag{2.24}$$

is much smaller, than the occupation of the zeroth level ($\frac{N_1}{N_0} \sim 10^{-8}$) and is not macroscopical already. Thus, we demonstrated that the macroscopic number of particles $N_0 \sim N$ lives only on the zeroth level.

That is why, the particle accumulation (Bose condensation) takes place at <u>one level</u> indeed.

Note, that in the three-dimensional Bose gas at high temperatures as well as in the classical Boltzmann gas the density of states $g(\varepsilon)$ (the number of states in the

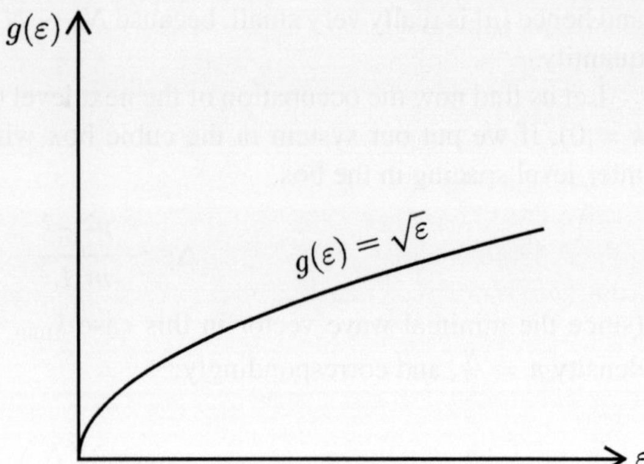

Fig. 2.1 Density of states in the three-dimensional Bose gas at high temperatures [1]

energy interval $d\varepsilon$) is defined from the condition $\frac{d^3\mathbf{k}}{(2\pi)^3} = g(\varepsilon)d\varepsilon$, and behaves in the square root fashion (see Fig. 2.1):

$$g(\varepsilon) \sim \sqrt{\varepsilon}. \tag{2.25}$$

At the same time, when the Bose condensate appears, the density of states for $\varepsilon = 0$ is proportional to the delta function: $g(\varepsilon) \sim \frac{N_0}{V}\delta(\varepsilon)$ since

$$\int g(\varepsilon)d\varepsilon \sim \frac{N_0}{V}\int \delta(\varepsilon)d\varepsilon = \frac{N_0}{V}. \tag{2.26}$$

As a result, the density of states in the 3D Bose gas at low temperatures acquires the form illustrated in Fig. 2.2.

The valid question appears meanwhile: how this particle accumulation agrees with entropy increase? The answer on this question is the following: the phase transition takes place in the system. As a result, the distinguished level appears and the symmetry of the system reduces (with respect to the system with "equal rights" between the levels).

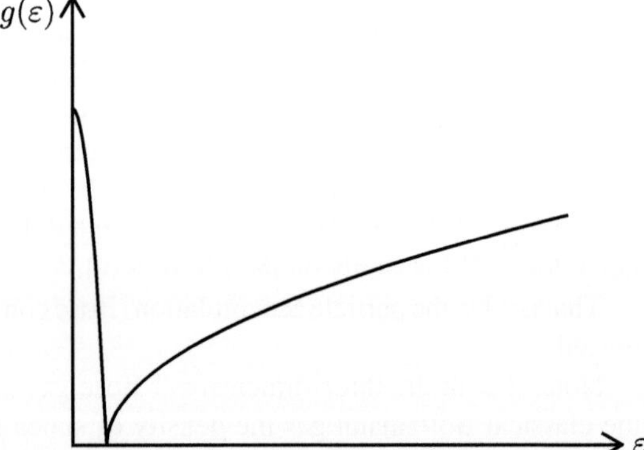

Fig. 2.2 Density of states in the 3D Bose gas at low temperatures [1]

We can say that when we make the temperature lower (reduce the ratio $\frac{T}{T_0}$), the larger number of particles is in Bose condensate (see Eq. (2.17)) and the less is entropy.

Thus, there is no violation of the principle of the entropy increase. The entropy is reduced only with the temperature reduction.

We have here the analogy with the electric field: if the electric field corresponds to very large occupation numbers (of photons), then it can be considered classically (the commutation relations for the field are not bosonic already). The level with $k = 0$ is an analog of the external field. It can be also considered classically.

2.5 The Origin of the Bose Condensation Temperature T_0

The Bose condensation temperature is connected with the uncertainty principle. The total density $n \sim \frac{1}{r_0^3}$, where r_0 is an average distance between the particles. Then $n^{2/3} \sim \frac{1}{r_0^2}$, and the energy per one particle $T_0 \sim \frac{\hbar^2}{mr_0^2}$. Moreover, for the temperature $T \sim T_0$ the average distance between the particles in the Bose gas by order of magnitude equals to the De Broglie wave length: $\lambda_T \sim \frac{\hbar}{\sqrt{mT_0}} \sim r_0$.

For high temperatures $T \gg T_0$ the classical statistics works. The occupation numbers are very small in this case and we cannot distinguish between fermionic and bosonic particles (they behave identically).

However, for $T \sim T_0$ their behavior is very different. If the particles are fermionic [1,2,17,27], then the energy per one particle is nevertheless of the order of $T_0 \sim \frac{\hbar^2}{mr_0^2}$, because of the Pauli principle which does not allow for two particles to be in one state.

Thus, the average distance between fermions will be still of the order of r_0 (the antisymmetric properties of the many-particle fermionic wave function with respect to the inter-changes between particles will promote that). Moreover, in the fermionic system the energy per one particle or the chemical potential becomes constant at low temperatures (as in degenerate Fermi gas, where $\mu = E_F$).

If the particles are bosonic, then for the temperatures $T \leq T_0$ they will be continuously accumulated in the state with $k = 0$, and thus the energy of the state will be continuously decreasing.

The total energy of the system for $T \leq T_0$ is only due to the non-condensed particles since the particles in Bose condensate have zero energy. The chemical potential for the non-condensed particles is still $\mu = 0$, and they are described by the Planck distribution with the density of states $g(\varepsilon) \sim \sqrt{\varepsilon}$. As a result, the total energy of the system:

$$E = V \int \frac{d^3 \mathbf{k}}{(2\pi)^3} \frac{\varepsilon_k}{\exp\left(\frac{\varepsilon_k}{T}\right) - 1} = V \int g(\varepsilon) d\varepsilon \frac{\varepsilon}{\exp\left(\frac{\varepsilon}{T}\right) - 1}. \tag{2.27}$$

Further on introducing the dimensionless variable for the energy $x = \frac{\varepsilon}{T}$ and making the integral dimensionless, we get

$$E = \text{const } VT^{5/2} \int_0^\infty \frac{x^{3/2}dx}{\exp(x) - 1} = AVT^{5/2}. \tag{2.28}$$

We can understand the temperature dependence of the energy $E \propto T^{5/2}$ as the product of the average number of non-condensed particles $N_{\text{non-cond.}} = \left(\frac{T}{T_0}\right)^{3/2} N_{\text{tot.}}$ and the average temporal energy of each particle $\bar{\varepsilon} \sim T$ which is proportional to

$$T^{3/2} \cdot T = T^{5/2}.$$

Specific heat of the ideal Bose gas at fixed volume of the system ($V = \text{const}$), for low temperatures $T \leq T_0$ and fixed number of particles $N = \text{const}$:

$$C_V = \left(\frac{\partial E}{\partial T}\right)_{V,N} = \frac{5}{2}\frac{E}{T} \propto T^{3/2}. \tag{2.29}$$

We can find now the entropy of the system utilizing the condition:

$$C_V = T\frac{dS}{dT} = \frac{5}{2}\frac{E}{T}. \tag{2.30}$$

Then $\frac{dS}{dT} = \frac{5}{2}\frac{E}{T^2} \propto T^{1/2}$, and taking into account the Nernst theorem ($S \to 0$ for $T \to 0$) we obtain for the entropy:

$$S = \frac{2}{3} \cdot \frac{5}{2} \cdot \frac{E}{T} = \frac{5}{3} \cdot \frac{E}{T} \propto T^{3/2}. \tag{2.31}$$

Now we can evaluate the Free energy:

$$F = E - TS = E - \frac{5}{3}E = -\frac{2}{3}E = -\frac{2}{3}AVT^{5/2}. \tag{2.32}$$

As a result, we get for the pressure:

$$P = -\left(\frac{\partial F}{\partial V}\right)_{T,N} = \frac{2}{3}\frac{E}{V} = \frac{2}{3}AT^{5/2}. \tag{2.33}$$

We derived a very interesting result, namely: we reduce the volume, but the pressure is not changed because a fraction of particles goes to the Bose condensate (the pressure is defined by non-condensed particles).

Indeed, the total number of the non-condensed particles can be written via the volume of the system using the relation: $T_0^{3/2} \sim n = \frac{N}{V}$. Then $N_{\text{non-cond.}} = \left(\frac{T}{T_0}\right)^{3/2} N_{\text{tot.}} \sim \frac{NT^{3/2}}{n} = VT^{3/2}$, and the number of the non-condensed particles

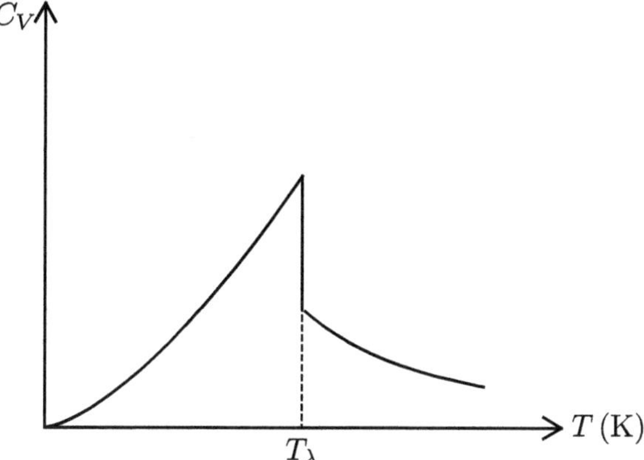

Fig. 2.3 Typical temperature behavior of the specific heat for the second order phase transition. In the critical point specific heat diverges. In the vicinity of the phase transition specific heat behaves similar to the Greek letter λ. This behavior of specific heat is observed in the non-ideal low density Bogoliubov Bose gas [28] with repulsive interaction between particles as well as in dense superfluid He-4 in the so-called roton region of temperatures (see Lecture 4 and [8]). In superfluid He-4 the critical temperature $T_\lambda \approx 2.2\,\mathrm{K}$ [1,34]

is proportional to the volume. Hence, the density of the non-condensed particles: $n_{\text{non-cond.}} = \frac{N_{\text{non-cond.}}}{V} = \text{const}$ for the given temperature. Correspondingly, the pressure $P \sim n_{\text{non-cond.}}\,T = \text{const}$ for the given temperature.

2.6   λ-Point in the Specific Heat and Phase Transition

Why we have phase transition in the λ-point?

All the thermodynamic potentials are continuous (not divergent) in the λ-point. However, the derivatives are already divergent. In the non-ideal

Bose gas [1,10,11,28–32] (contrary to the ideal gas) specific heat is divergent in the λ-point (see Fig. 2.3). The interaction in the gas meanwhile guarantees the superfluidity [12–16,33–37].

2.7   Specific Heat of the Ideal (Non-interacting) Bose Gas Close to the Critical Point

Thus, for low temperatures $T \ll T_0$ specific heat of the ideal Bose gas $C_V \propto T^{3/2}$ and the derivative of the specific heat with respect to temperature is positive $\left(\frac{dC_V}{dT} > 0\right)$.

Note that for high temperatures $T \gg T_0$ we have Boltzmann gas with N particles. From the theorem of equal energy distribution between the degrees of freedom this monoatomic gas has specific heat

$$C_V \approx \frac{3}{2} k_B N = \text{const}.\tag{2.34}$$

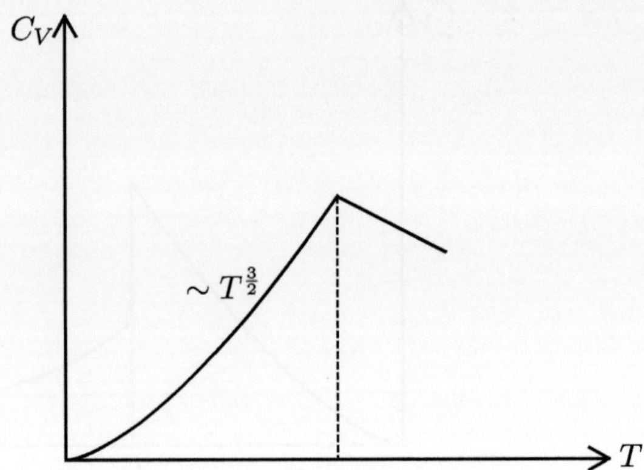

Fig. 2.4 Qualitative behavior of the ideal Bose gas close to the critical temperature $T = T_C^{\mathrm{BEC}}$ (from the monograph The Theory of the Bose gas by V.V. Tolmachev) [1, 10]

When we approach the Bose condensation temperature T_0 from above, then, according to the monograph of V.V. Tolmachev [10], specific heat gains (additionally to the constant) a small term, linearly decreasing with respect to temperature:

$$C_V \approx \frac{3}{2} k_B N - B k_B N \left(\frac{T - T_0}{T_0} \right) \qquad (2.35)$$

for $\frac{T-T_0}{T_0} \ll 1$. This term goes to zero for $T = T_0$.

Thus, for the temperatures larger than a critical one the derivative of the specific heat with respect to temperature becomes negative ($\frac{dC_V}{dT} < 0$).

As a result, the temperature behavior of the specific heat in the ideal Bose gas in the vicinity of the critical point can be qualitatively described in Fig. 2.4 taken from the Tolmachev monograph [10]. From Fig. 2.4 we can see, that specific heat of the ideal Bose gas, though has a kink, is a continuous function in the critical point and only the temperature derivative of the specific heat experiences a jump for $T = T_0$.

At the same time for the second order phase transitions already the second derivative of the thermodynamic potentials (that is specific heat itself $C_V(T) = T \frac{dS}{dT} = -T \frac{\partial^2 F}{\partial T^2}$.) should have a jump.

In fact, in this terminology the ideal Bose gas corresponds to the "third-order" phase transition. In the critical point only the third derivative experiences a jump

$$\Delta \left(\frac{\partial C_V}{\partial T} \right)_V = -3.66 \frac{k_B N}{T_0},$$

where $V = \mathrm{const}$, N is the total number of particles, while the specific heat has only a kink.

2.8 Supplementary Material: The Evaluation of the Jump in the Temperature Derivative of Specific Heat in the Ideal (Non-interacting) Bose Gas

Let us utilize Eq. (2.11) for the total number of particles for the temperatures slightly higher than a critical one $\left(\frac{T-T_0}{T_0} \ll 1\right)$. Then:

$$\frac{N}{V} = \frac{m^{3/2}}{\hbar^3} \frac{\sqrt{2}}{2\pi^2} \int_0^\infty d\varepsilon \frac{\sqrt{\varepsilon}}{\left(\exp \frac{\varepsilon + |\mu|}{T} - 1\right)} = \text{const}. \tag{2.36}$$

Further on let us introduce the notation $N_0(T) = N(\mu = 0)$. As a result:

$$\frac{N_0(T)}{V} = \frac{m^{3/2}}{\hbar^3} \frac{\sqrt{2}}{2\pi^2} \int_0^\infty d\varepsilon \frac{\sqrt{\varepsilon}}{\left(\exp \frac{\varepsilon}{T} - 1\right)}. \tag{2.37}$$

Now let us subtract from the total particle density $\frac{N}{V}$ and add to it the quantity $\frac{N_0(T)}{V}$. Then in agreement with Landau–Lifshitz book [2] (problem 1 after the paragraph 62):

$$\frac{N}{V} = \frac{N_0(T)}{V} + \frac{m^{3/2}}{\hbar^3} \frac{\sqrt{2}}{2\pi^2} \int_0^\infty d\varepsilon \sqrt{\varepsilon} \left(\frac{1}{\exp \frac{\varepsilon + |\mu|}{T} - 1} - \frac{1}{\exp \frac{\varepsilon}{T} - 1}\right). \tag{2.38}$$

For temperatures $T \to T_0 + 0$: the absolute value of $|\mu|$ is small and the integral in (2.38) "sits" on the small energies ε. That is why we can expand the exponents under the integral till the linear terms inclusively and get:

$$\begin{aligned}
\frac{N}{V} &= \frac{N_0(T)}{V} - \frac{m^{3/2}}{\hbar^3} \frac{\sqrt{2}}{2\pi^2} \int_0^\infty d\varepsilon \sqrt{\varepsilon} T |\mu| \frac{1}{\varepsilon(\varepsilon + |\mu|)} \\
&= \frac{N_0(T)}{V} - \frac{m^{3/2}}{\hbar^3} \frac{\sqrt{2} T |\mu|}{2\pi^2} \int_0^\infty d\varepsilon \frac{1}{\sqrt{\varepsilon}(\varepsilon + |\mu|)}.
\end{aligned} \tag{2.39}$$

But the integral $\int_0^\infty d\varepsilon \frac{1}{\sqrt{\varepsilon}(\varepsilon + |\mu|)}$ can be easily evaluated and is equal to $\frac{\pi}{\sqrt{|\mu|}}$. Then (2.39) can be written as follows:

$$\frac{N_0(T) - N}{V} = \frac{m^{3/2}}{\hbar^3} \frac{\sqrt{2} T \sqrt{|\mu|}}{2\pi}. \tag{2.40}$$

Taking a square from the l.h.s. and r.h.s. of Eq. (2.40), we derive the expression for $|\mu|$:

$$|\mu| = \frac{2\pi^2\hbar^6}{m^3}\left(\frac{N_0(T) - N}{VT}\right)^2. \qquad (2.41)$$

Now we can use Eq. (2.33) according to which the thermodynamic potential Ω in the ideal Bose gas:

$$\Omega = -PV = -\frac{2}{3}E. \qquad (2.42)$$

But since $N = \frac{\partial\Omega}{\partial\mu}$, then we can write with the same accuracy:

$$\frac{\partial E}{\partial\mu} = -\frac{3}{2}\frac{\partial\Omega}{\partial\mu} = \frac{3}{2}N \approx \frac{3}{2}N_0(T). \qquad (2.43)$$

Hence for temperatures $T \to T_0 + 0$ the energy reads:

$$E \approx E_0(T) + \frac{3}{2}N_0(T)\mu = E_0 - \frac{3}{2}N_0(T)|\mu|$$
$$= E_0 - N_0(T)\frac{3\pi^2\hbar^6}{m^3}\left(\frac{N_0(T) - N}{VT}\right)^2, \qquad (2.44)$$

where in agreement with Eq. (2.28)

$$E_0(T) = E(\mu = 0) = AVT^{5/2}.$$

It is evident, that in a jump in a temperature derivative of the specific heat in the critical point $\Delta\left(\frac{\partial C_V}{\partial T}\right)_V$ contributes only the second temperature derivative from the second term in Eq. (2.44). Performing the calculations and using the fact that for $T = T_0$ the difference $N_0(T) - N = 0$, we obtain:

$$\Delta\left(\frac{\partial C_V}{\partial T}\right)_V = -\frac{6\pi^2\hbar^6}{m^3V^2}\left[N_0\left(\frac{1}{T}\frac{\partial N_0}{\partial T}\right)^2\right]_{T=T_0}. \qquad (2.45)$$

After the substitution of $\frac{\partial N_0}{\partial T} = \frac{3}{2}\frac{N_0}{T}$ in Eq. (2.45) we can represent the jump in the temperature derivative of the specific heat in the following form:

$$\Delta\left(\frac{\partial C_V}{\partial T}\right)_V = -\frac{6\pi^2\hbar^6}{m^3}\frac{N}{T_0}\frac{9}{4}\frac{N^2}{V^2}\frac{1}{T_0^3} = -\frac{27}{2}\frac{N}{T_0}\frac{\pi^2\hbar^6n^2}{m^3}\frac{1}{T_0^3}. \qquad (2.46)$$

After that using Einstein formula (2.13) for the critical temperature $T_0 = \frac{\hbar^2}{m} n^{2/3}$ · 3.31 we finally get the required expression for the jump:

$$\Delta \left(\frac{\partial C_V}{\partial T} \right)_V = - \frac{27\pi^2}{2(3.31)^3} \frac{N}{T_0} \approx -3.66 \frac{N}{T_0}. \tag{2.47}$$

References

1. Y. Kagan, *Lectures on the Solid-state Theoretical Physics* (MEPHI, Moscow, 1981–1982). Part I. Phonons - unpublished
2. L.D. Landau, E.M. Lifshitz, *Statistical Physics, Part I* (Butterworth-Heinemann, Oxford, 1999)
3. S.N. Bose, Zeit. fur Physik **26**, 178 (1924)
4. A. Einstein, Sber. Preuss, Akad. Wiss. Berlin **1**, 3 (1925)
5. K.B. Davis, M.O. Mewes, M.R. Andrews et al., Phys. Rev. Lett. **75**, 3969 (1995)
6. C.C. Bradley, C.A. Sackett, J.J. Tollett, R.G. Hulet, Phys. Rev. Lett. **75**, 1687 (1995)
7. M.H. Anderson, J.R. Ensher, M.R. Mathews et al., Science **269**, 198 (1995)
8. R.P. Feynman, *Statistical Mechanics, A Set of Lectures* (Advanced Books Classics) (Westview Press, Boulder, 1998)
9. R. Kubo, *Statistical Mechanics* (North-Holland Publishing Company, Amsterdam, 1965)
10. V.V. Tolmachev, *The Bose-Gas Theory* (Publishing House of Moscow State University, 1969) (in Russian)
11. P. Nozieres, D. Pines, *The Theory of Quantum Liquids*, in *Superfluid Bose Liquids,* vol. 2 (Addison Wesley Publishing Company, USA, Canada, 1990)
12. L.P. Pitaevskii, S. Stringari, *Bose–Einstein Condensation* (Oxford University Press, 2003)
13. C. Pethick, S. Smith, *Bose–Einstein Condensation in Dilute Gases* (Cambridge University Press, 2002)
14. M.Y. Kagan, *Physics of the Macroscopic Quantum Systems* (Publishing House of the Moscow Energetic Institute (MEI), Moscow, 2014). (in Russian)
15. M.Y. Kagan, R.S. Iksanov, E.A. Popova, *Quantum and Statistical Physics for Chemists* (Publishing House of the Higher School of Economics (HSE), Moscow, 2026). (in press)
16. M.Y. Kagan, *Modern Trends in Superconductivity and Superfluidity*, Lecture Notes in Physics, vol. 874. (Springer, Dordrecht, 2013)
17. L.D. Landau, E.M. Lifshitz, *Quantum Mechanics, Non-Relativistic Theory* (Pergamon Press, New York, 1977)
18. H.J. Lipkin, *Quantum Mechanics* (North-Holland Publishing Company, Amsterdam-New York, New Approaches to selected Topics, 1973)
19. J. Ziman, *Elements of Advanced Quantum Theory* (Cambridge University Press, 1969)
20. J.W. Negele, H. Orland, *Frontiers in Physics, Quantum Many-Particle Physics* (Addison-Wesley Publishing Company, 1988)
21. D.J. Thouless, *The Quantum Mechanics of Many-Body Systems* (Academic, New York and London, 1961)
22. S. Raimes, *Many-Electron Theory* (North-Holland Publishing Company, Amsterdam-London, 1972)
23. D. Ter Haar, *Introduction to the Physics of Many-Body Systems* (Inter-science Publishers Inc., New York, 1958)
24. F.A. Berezin, *The Method of Second Quantization* (Academic, 1966)
25. L. Kadanoff, G. Baym, *Quantum Statistical Mechanics* (CRC Press, 1989)
26. A.L. Fetter, J.D. Walecka, *Quantum Theory of Many Particle Systems* (Dover, New York, 2003)
27. P. Coleman, *Introduction to Many-Body Physics* (Cambridge University Press, 2015)

28. C.P. Enz, *A Course on Many-body Theory Applied to Solid-state Physics*, Lecture Notes in Physics (World Scientific, 1992)
29. N.N. Bogoliubov, J. Phys. USSR **11**, 23 (1947)
30. K. Huang, C.N. Yang, J.M. Luttinger, Phys. Rev. **105**, 776 (1957)
31. K.A. Brueckner, K. Sawada, Phys. Rev. **106**, 1117 (1957)
32. A.A. Abrikosov, L.P. Gor'kov, I.E. Dzyaloshinskii, *Methods of Quantum Field Theory in Statistical Physics* (Prentice Hall, Englewood Cliffs, New Jersey, 1963)
33. V.N. Popov, *Functional Integrals and Collective Excitations* (Cambridge University Press, UK, 1977)
34. E.M. Lifshitz, L.P. Pitaevskii, *Statistical Physics, Part II, Theory of the Condensed State* (Elsevier, 2013)
35. I.M. Khalatnikov, *An Introduction to the Theory of Superfluidity* (Perseus Publishing, Cambridge, 2000)
36. S.J. Putterman, *Superfluid Hydrodynamics*. North Holand Series in Low Temperature Physics (North-Holland, New York, 1974)
37. D.R. Tiley, J. Tiley, *Superfluidity and Superconductivity* (Van Nostrand Reinhold company, New York-Cincinnati-Toronto-London, Melburn, 1974)
38. V.L. Ginzburg, *On superconductivity and Superfluidity, A Scientific Autobiography* (Springer, Heidelberg-Berlin, 2009)

Lecture 3. Interacting (Bogoliubov) Bose Gas

3

Abstract

In this Lecture in the framework of the second quantization technique we diagonalize the Hamiltonian and find the excitation spectrum for the weakly interacting repulsive (Bogoliubov) Bose gas. The spectrum at small wave vectors is linear (has a soundlike character). We apply the obtained results to the search of the possible superfluidity in spin-polarized atomic hydrogen. We analyze the condensate particle number, the correlation functions and the shift of the critical temperature in the interacting Bose gas. We briefly consider two-dimensional case as well.

Let us consider a low-density gas of bosonic particles [1–12] interacting with each other via two-body potential which are described by the secondly quantized Hamiltonian [1,5,6,13–15]:

$$\widehat{H} = \sum_{k} \varepsilon_k^0 a_k^\dagger a_k + \frac{1}{2} \sum_{k_1 k_2 q} \frac{U(q)}{V} a_{k_1}^\dagger a_{k_2}^\dagger a_{k_2+q} a_{k_1-q}, \tag{3.1}$$

where $\varepsilon_k^0 = \frac{\hbar^2 k^2}{2m}$ is the bare kinetic energy of the gas particles, $a_k^\dagger$ and a_k are bosonic creation and annihilation operators of the particles with the wave vector k, and in the second term with potential energy in (3.1) we satisfy the momentum conservation law.

Let us remind the basic commutation relations for the bosonic operators:

$$a_k a_{k_1}^\dagger - a_{k_1}^\dagger a_k = \delta_{kk_1}, \text{ and } a_k a_k^\dagger - a_k^\dagger a_k = 1, \tag{3.2}$$

$$a_{k_1} a_k - a_k a_{k_1} = 0, \tag{3.3}$$

$$a_{k_1}^\dagger a_k^\dagger - a_k^\dagger a_{k_1}^\dagger = 0. \tag{3.4}$$

M. Kagan, *Lecture Notes of Professor Yuri Kagan in Theoretical Solid-State Physics*, Lecture Notes in Physics 1048, https://doi.org/10.1007/978-3-032-14621-2_3

Moreover, for the non-diagonal matrix elements of the bosonic operators which are different from zero we have

$$a_k \left| n_k \right\rangle = \sqrt{n_k} \left| n_k - 1 \right\rangle, \tag{3.5}$$

$$a_k^\dagger \left| n_k \right\rangle = \sqrt{n_k + 1} \left| n_k + 1 \right\rangle. \tag{3.6}$$

Correspondingly, the bilinear operator of the particle number $a_k^\dagger a_k$ is diagonal:

$$a_k^\dagger a_k \left| n_k \right\rangle = n_k \left| n_k \right\rangle. \tag{3.7}$$

Following Bogoliubov [2], we would like to describe the reconstruction of the ground state and excited states in the system.

The energy of the non-interacting gas reads [1, 16]:

$$E = \left\langle \{n_k\} \left| \sum_k \varepsilon_k^0 a_k^\dagger a_k \right| \{n_k\} \right\rangle = \sum_k \varepsilon_k^0 n_k, \tag{3.8}$$

where $n_k = 0, 1, 2, \ldots$ are the occupation numbers.

The total number of particles in the gas is given by

$$N = \left\langle \{n_k\} \left| \sum_k a_k^\dagger a_k \right| \{n_k\} \right\rangle = \sum_k n_k. \tag{3.9}$$

After the temporal averaging:

$$\bar{E}^T = \sum_k \varepsilon_k^0 \bar{n}_k, \tag{3.10}$$

where the average occupation numbers $\bar{n}_k$ are defined by the Bose–Einstein distribution function [16–18]. Note that in the non-interacting Bose gas at $T = 0$ the average energy $\bar{E} = 0$.

At the same time, we know, that for $T = 0$ all the particles in the ideal Bose gas live in the condensate and have zero momentum $k = 0$ and zero energy. Thus, the total number of particles for $T = 0$, equals to the total number of the condensed particles:

$$N = N_0 = \left\langle N_0 | a_0^\dagger a_0 | N_0 \right\rangle. \tag{3.11}$$

Correspondingly, for non-zero matrix elements of the creation and annihilation operators [12–14]:

$$a_0^\dagger \left| N_0 \right\rangle = \sqrt{N_0 + 1} \left| N_0 + 1 \right\rangle, \tag{3.12}$$

$$a_0 \,|N_0\rangle = \sqrt{N_0}\,|N_0 - 1\rangle . \tag{3.13}$$

But since the total number of particles in the condensate is very large $N_0 \sim 10^{24}$, the difference between N_0, $N_0 - 1$ and $N_0 + 1$ is very small, so we can put

$$N_0 - 1 \approx N_0 + 1 \approx N_0$$

in Eqs. (3.5), (3.6). Moreover, since $a_0^\dagger a_0 = N_0$ and $a_0 a_0^\dagger = N_0 + 1$, then in the commutator

$$a_0 a_0^\dagger - a_0^\dagger a_0 = (N_0 + 1) - N_0 = 1 \tag{3.14}$$

with the macroscopic accuracy $\sim 1/N_0$ we can neglect 1 in the r.h.s. and write:

$$\left[a_0 a_0^\dagger\right] = \left[a_0^\dagger a_0\right] = 0. \tag{3.15}$$

Thus, the operators $a_0^\dagger$ and a_0 become commuting [2]. Then in analogy with the electric field in the limit of large occupation numbers of photons (when Maxwell equations [19] can be utilized), we can proceed to the classical limit and replace these operators by the c-numbers, where $|c|^2$ is the statistical weight of the state with the given occupation number.

As a result, we can write

$$a_0^\dagger a_0 = N_0, \text{ and } a_0^\dagger = a_0 = \sqrt{N_0}. \tag{3.16}$$

3.1 Temperatures $T = 0$

Let us consider the gas case:

$$\frac{a}{r_0} \ll 1, \tag{3.17}$$

where a is the size of the particle or the scattering amplitude, r_0 is the distance between the particles. Then the gas approximation is descried by the small parameter (called the gas parameter) [1–6, 11, 20]:

$$a n^{1/3} \ll 1, \text{ or } a^3 n \ll 1. \tag{3.18}$$

The washing out of the particles from the condensate takes place due to the interaction and is governed by the small gas parameter $a^3 n \ll 1$.

Let us consider the interaction (the scattering) to be rare and thus the washing out to be small. It means that:

$$\sum_{k \neq 0} a_k^\dagger a_k \ll N_0. \tag{3.19}$$

Let us solve the problem utilizing the perturbation theory.

(A) The largest term in the potential energy in (3.1) corresponds to the case $k_1 = k_2 = q = 0$ and equals to:

$$\frac{1}{2}\frac{U(q=0)}{V}a_0^\dagger a_0^\dagger a_0 a_0 = \frac{1}{2}\frac{U(q=0)}{V}N_0^2. \tag{3.20}$$

This term describes the particle interaction in the condensate.

Note that all the collisions between particles occur with a small transferred momentum (since the interparticle distances are large).

For the Fourier components of the potential:

$$U(q) = \int U(r)\exp(i\boldsymbol{q}\boldsymbol{r})d^3\boldsymbol{r}. \tag{3.21}$$

In the diluted gas,

$$q \sim \frac{1}{r_0}, \text{ and } qa \sim \frac{a}{r_0} \ll 1. \tag{3.22}$$

Thus, we can replace $U(q)$ on $U(0)$. Note, that we cannot do it for the Coulomb interaction, because $U(r) \sim \frac{1}{r}$, and $\int U(r)d^3\boldsymbol{r} \to \infty$.

Hence:

$$U(q) \to U(0). \tag{3.23}$$

Further on in a more exact theory we should replace the zeroth Fourier component of the potential on the effective particle size (on the scattering length a) according to [1–6,11,20]:

$$U(0) \to \frac{4\pi\hbar^2}{m}a. \tag{3.24}$$

Correspondingly the mean potential energy in the rarified gas (for small gas parameter $\frac{a}{r_0} \sim an^{1/3} \ll 1$) is replaced by

$$nU(0) \to \frac{4\pi\hbar^2}{m}an \sim \frac{\hbar^2 q_{\text{typ.}}^2}{2m}an^{1/3} \ll \frac{\hbar^2 q_{\text{typ.}}^2}{2m} \tag{3.25}$$

and proved to be much smaller than the characteristic kinetic energy

$$\frac{\hbar^2 q_{\text{typ.}}^2}{2m} \sim \frac{\hbar^2}{mr_0^2} \sim \frac{\hbar^2 n^{2/3}}{m}. \tag{3.26}$$

(B) The following terms in the potential energy

1. $k_1 = k_2 = 0$:

$$a_0^\dagger a_0^\dagger a_q a_{-q} = N_0 a_q a_{-q}. \tag{3.27}$$

2. $k_1 - q = 0$, $k_2 + q = 0$, which is equivalent to $k_1 = q = -k_2 = k$. In this case

$$a_0 a_0 a_k^\dagger a_{-k}^\dagger = N_0 a_k^\dagger a_{-k}^\dagger. \tag{3.28}$$

3. $k_1 = 0$, $k_1 - q = 0$, $q = 0$:

$$N_0 a_{k_2}^\dagger a_{k_2}. \tag{3.29}$$

4. $k_1 = 0$, $k_2 + q = 0$, which is equivalent to $k_2 = k_1 - q$. Thus we have again:

$$N_0 a_{k_2}^\dagger a_{k_2}. \tag{3.30}$$

5. $k_2 = 0$, $k_1 - q = 0$, which is equivalent to $k_1 = k_2 + q$. Hence

$$N_0 a_{k_1}^\dagger a_{k_1}. \tag{3.31}$$

6. $k_2 = 0$, $k_2 + q = 0$, which is equivalent to $q = 0$. Then again:

$$N_0 a_{k_1}^\dagger a_{k_1}. \tag{3.32}$$

Gathering the similar terms, we get in the main approximation for the Hamiltonian (3.1):

$$\widehat{H} = \sum_k \varepsilon_k^0 a_k^\dagger a_k + \frac{1}{2} \frac{U(0)N_0}{V} \sum_{k \neq 0} \left(a_k^\dagger a_{-k}^\dagger + a_k a_{-k} + 4 a_k^\dagger a_k \right) + \frac{1}{2} \frac{U(0)N_0^2}{V}, \tag{3.33}$$

where the total number of particles:

$$N = N_0 + \sum_{k \neq 0} a_k^\dagger a_k, \text{ or } N_0 = N - \sum_{k \neq 0} a_k^\dagger a_k. \tag{3.34}$$

Now in the gas limit with the required accuracy we can replace N_0 by N in the second term in the Hamiltonian (3.33), and use the following approximation for the third term:

$$N_0^2 = \left(N - \sum_{k \neq 0} a_k^\dagger a_k \right)^2 \approx N^2 - 2N \sum_{k \neq 0} a_k^\dagger a_k. \tag{3.35}$$

As a result, the Hamiltonian (3.33) looks as follows:

$$\widehat{H} = \frac{1}{2} \frac{U(0)N^2}{V} + \sum_{k \neq 0} \left(\varepsilon_k^0 + \frac{U(0)N}{V} \right) a_k^\dagger a_k + \frac{1}{2} \frac{U(0)N}{V} \sum_{k \neq 0} \left(a_k^\dagger a_{-k}^\dagger + a_k a_{-k} \right). \tag{3.36}$$

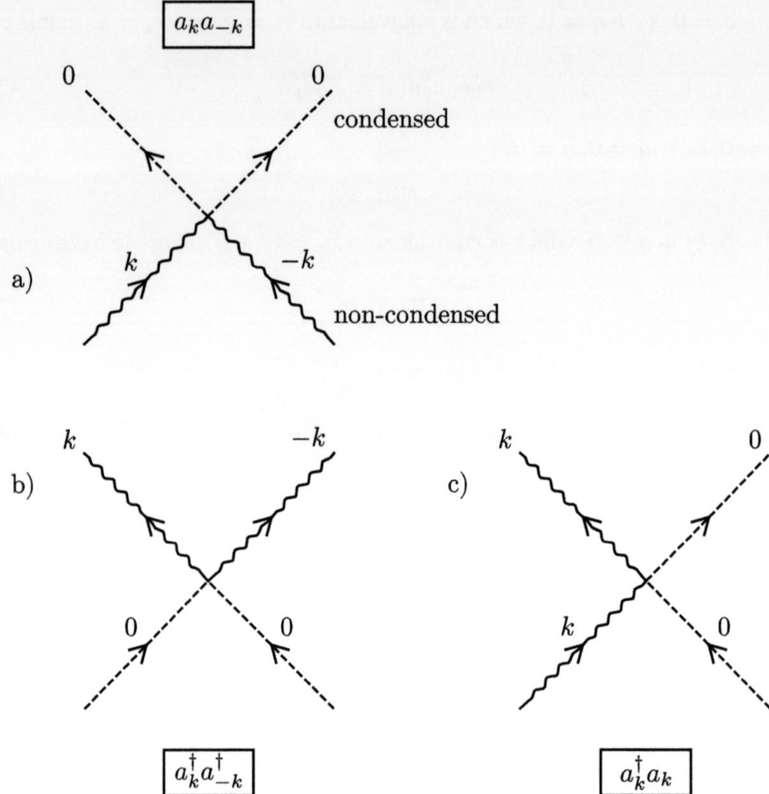

Fig. 3.1 The simplest processes bilinear with respect to creation and annihilation operators, which are proportional to $a_k a_{-k}$ (Fig. a) $a_k^\dagger a_{-k}^\dagger$ (Fig. b) or $a_k^\dagger a_k$ (Fig. c). On these figures the doted lines with symbol 0 correspond to the zero wave vectors and are related to the condensed particles, while the wavy lines represent the non-condensed particles with the wave numbers k and $-k$ [1]

The Hamiltonian (3.36) does not conserve the number of the non-condensed particles and is not a diagonal one. Let us analyze what are the processes which do not conserve the number of the non-condensed particles. The simplest processes are illustrated in Fig. 3.1.

It is important to note that in these processes we effectively fix the number of particles in the condensate, and the condensate starts to play the role of the wall [1]. So non-condensed particles are dancing with the wall now. Something happens to them, but nothing happens to the wall (the wall does not conserve the memory about the collisions) [1].

We can say also that the condensate (especially third term proportional to $a_k^\dagger a_k$ in Fig. c) plays the role of the external field. Thus, the emergence of the Bose condensate in the system changes drastically the particle properties.

We would like to find the <u>new</u> particles, which diagonalize the Hamiltonian (3.36). These novel particles would not interact with each other and thus will have an infinite lifetime.

Following Bogoliubov [2], the creation operator for the new particle $b_k^\dagger$ with the wave vector k in the formalism of the second quantization can be written as a linear combination of the creation operator $a_k^\dagger$ and annihilation operator a_{-k} of the initial ("old") particles with the wave vectors k and $-k$ and the real coefficients u, v:

$$b_k^\dagger = u_k a_k^\dagger - v_k a_{-k}. \tag{3.37}$$

Let us stress, that in accordance with the principle of space homogeneity, the response on the creation of the new particle with the wave vector k in (3.37) can be connected with the Fourier harmonics with the same wave vector.

Moreover, while writing the Bogoliubov transformation [2] in (3.37), we utilized, that the annihilation of the particle with wave vector k is equivalent to the creation of the particle with wave vector $-k$.

Besides that, according to Bogoliubov, the coefficients u_k and v_k should be not only real, but also even functions of k:

$$u_k = u_{-k},\, v_k = v_{-k}. \tag{3.38}$$

In fact, they depend only on the energy.

Correspondingly for the annihilation operator of the novel particle b_k with the wave vector k we get

$$b_k = u_k a_k - v_k a_{-k}^\dagger. \tag{3.39}$$

Further on we use the fact that in agreement with the fundamentals of the quantum statistics, from Bose particles we can obtain only Bose particles again [2,5]. That is why the novel operators b_k and $b_k^\dagger$ should satisfy the same commutation relations, as a_k and $a_k^\dagger$:

$$b_k b_k^\dagger - b_k^\dagger b_k = 1. \tag{3.40}$$

If we substitute the Bogoliubov $u - v$ transformation (3.37) and (3.39) in the commutator (3.40) we get

$$\left(u_k a_k - v_k a_{-k}^\dagger \right)\left(u_k a_k^\dagger - v_k a_{-k} \right) - \left(u_k a_k^\dagger - v_k a_{-k} \right)\left(u_k a_k - v_k a_{-k}^\dagger \right) = \tag{3.41}$$

$$= u_k^2 \left(a_k a_k^\dagger - a_k^\dagger a_k \right) - v_k^2 \left(a_{-k} a_{-k}^\dagger - a_{-k}^\dagger a_{-k} \right) = u_k^2 - v_k^2.$$

Hence, we obtain an important unitarity condition for the Bogoliubov $u - v$ coefficients (for the Bogoliubov transformation matrix) in Bose gas:

$$u_k^2 - v_k^2 = 1. \tag{3.42}$$

After that, to diagonalize the Hamiltonian (3.36) in terms of the novel quasiparticles besides (3.37) and (3.39) we have to derive also the Bogoliubov transformation for the novel creation and annihilation operators with the wave vector $-k$.

Using the fact that Bogoliubov $u - v$ coefficients are even functions of k, we can write [2]:

$$b^{\dagger}_{-k} = u_k a^{\dagger}_{-k} - v_k a_k, \tag{3.43}$$

$$b_{-k} = u_k a_{-k} - v_k a^{\dagger}_k. \tag{3.44}$$

If we multiply now the l.h.s. and the r.h.s. of the Bogoliubov transformation for b_k (3.39) on u_k, and at the same time the l.h.s. and the r.h.s. of the transformation for $b^{\dagger}_{-k}$ on v_k, and sum them up, then we get

$$u_k b_k + v_k b^{\dagger}_{-k} = u_k \left(u_k a_k - v_k a^{\dagger}_{-k} \right) + v_k \left(u_k a^{\dagger}_{-k} - v_k a_k \right) = \left(u_k^2 - v_k^2 \right) a_k = a_k. \tag{3.45}$$

Thus, we derived the inverse Bogoliubov transformation [2], which relates the old annihilation operator a_k with the new ones.

Analogously for the creation operator $a^{\dagger}_k$:

$$u_k b^{\dagger}_k + v_k b_{-k} = u_k \left(u_k a^{\dagger}_k - v_k a_{-k} \right) + v_k \left(u_k a_{-k} - v_k a^{\dagger}_k \right) = \left(u_k^2 - v_k^2 \right) a^{\dagger}_k = a^{\dagger}_k. \tag{3.46}$$

Similar formulas for the inverse Bogoliubov transformation we can derive also for the old operators with the wave vector $-k$:

$$u_k b_{-k} + v_k b^{\dagger}_k = a_{-k}, \tag{3.47}$$

$$u_k b^{\dagger}_{-k} + v_k b_k = a^{\dagger}_{-k}. \tag{3.48}$$

After that it is convenient to introduce the following notations in the Hamiltonian (3.36):

$$\xi_k = \varepsilon_k^0 + \frac{U(0)N}{V} = \varepsilon_k^0 + 2\beta, \tag{3.49}$$

where:

$$\beta = \frac{1}{2}\frac{U(0)N}{V}. \tag{3.50}$$

Let us also define the condensate part of the energy as

$$E_0 = \frac{1}{2}\frac{U(0)N^2}{V}. \tag{3.51}$$

As a result, the Hamiltonian (3.36) reads:

$$\widehat{H} = E_0 + \sum_k \left(\varepsilon_k^0 + 2\beta \right) a^{\dagger}_k a_k + \beta \sum_{k \neq 0} \left(a^{\dagger}_k a^{\dagger}_{-k} + a_k a_{-k} \right)$$

$$= E_0 + \sum_{k \neq 0} \xi_k a^{\dagger}_k a_k + \beta \sum_{k \neq 0} \left(a^{\dagger}_k a^{\dagger}_{-k} + a_k a_{-k} \right). \tag{3.52}$$

Now let us apply the inverse Bogoliubov transformation (3.45)–(3.48) to the Hamiltonian (3.52):

$$\widehat{H} = E_0 + \sum_{k \neq 0} \xi_k \left(u_k b_k^\dagger + v_k b_{-k}\right) \left(u_k b_k + v_k b_{-k}^\dagger\right)$$

$$+\beta \sum_{k \neq 0} \left\{ \left(u_k b_k^\dagger + v_k b_{-k}\right) \left(u_k b_{-k}^\dagger + v_k b_k\right) + \left(u_k b_k + v_k b_{-k}^\dagger\right) \left(u_k b_{-k} + v_k b_k^\dagger\right) \right\}. \tag{3.53}$$

After that let us unite the similar terms under the sum symbol:

$$\widehat{H} = E_0 + \sum_{k \neq 0} \left\{ \left(\xi_k u_k^2 + 2\beta u_k v_k\right) b_k^\dagger b_k + \left(\xi_k v_k^2 + 2\beta u_k v_k\right) b_k b_k^\dagger \right.$$

$$\left. + \left[\xi_k u_k v_k + \beta \left(u_k^2 + v_k^2\right)\right] b_k b_{-k} + \left[\xi_k u_k v_k + \beta \left(u_k^2 + v_k^2\right)\right] b_k^\dagger b_{-k}^\dagger \right\}. \tag{3.54}$$

Let us demand that the coefficients near the non-diagonal bilinear combinations of the novel operators b and $b^\dagger$ should be zero. Then:

$$\xi_k u_k v_k + \beta \left(u_k^2 + v_k^2\right) = 0. \tag{3.55}$$

Since for repulsive sign of the interaction $\beta > 0$ and $\xi_k > 0$, then from (3.55) it follows, that the bilinear combination of the coefficients:

$$u_k v_k < 0. \tag{3.56}$$

Hence, using the commutation relations (3.40) for the new operators, let us substitute $b_k b_k^\dagger = 1 + b_k^\dagger b_k$ in the second term of (3.54) and get the diagonal form of the Hamiltonian:

$$\widehat{H} = E_0 + \sum_{k \neq 0} \left(\xi_k v_k^2 + 2\beta u_k v_k\right) + \sum_{k \neq 0} \left(\xi_k \left(u_k^2 + v_k^2\right) + 4\beta u_k v_k\right) b_k^\dagger b_k$$

$$= E_0 + \sum_{k \neq 0} \left(\xi_k v_k^2 + 2\beta u_k v_k\right) + \sum_{k \neq 0} \tilde{E}_k b_k^\dagger b_k, \tag{3.57}$$

where the energy of the novel quasiparticles

$$\tilde{E}_k = \xi_k \left(u_k^2 + v_k^2\right) + 4\beta u_k v_k. \tag{3.58}$$

Let us find now the explicit expressions for the Bogoliubov coefficients from the requirement on the diagonal form of the Hamiltonian (3.55).

Since it follows from (3.55), that:

$$\xi_k u_k v_k = -\beta \left(u_k^2 + v_k^2\right), \tag{3.59}$$

let us square the l.h.s. and r.h.s. of this relation. As a result, we get

$$\xi_k^2 u_k^2 v_k^2 = \beta^2 \left(u_k^2 + v_k^2\right)^2. \tag{3.60}$$

Substituting in (3.60) the relation between the Bogoliubov coefficients which follows from the unitarity condition [2] (3.42):

$$u_k^2 = 1 + v_k^2, \tag{3.61}$$

let us represent (3.60) in the form of the equation on only one Bogoliubov coefficient:

$$\xi_k^2 \left(1 + v_k^2\right) v_k^2 = \beta^2 \left(1 + 2v_k^2\right)^2. \tag{3.62}$$

Then, it is not difficult to proceed from Eq. (3.62) to the biquadratic equation:

$$v_k^4 \left(\xi_k^2 - 4\beta^2\right) + v_k^2 \left(\xi_k^2 - 4\beta^2\right) - \beta^2 = 0, \ \text{or} \ v_k^4 + v_k^2 - \frac{\beta^2}{\xi_k^2 - 4\beta^2} = 0. \tag{3.63}$$

The solution of this equation is elementary:

$$v_k^2 = -\frac{1}{2} \pm \sqrt{\frac{1}{4} + \frac{\beta^2}{\xi_k^2 - 4\beta^2}} = -\frac{1}{2} \pm \frac{1}{2}\frac{\xi_k}{\gamma_k}, \tag{3.64}$$

where:

$$\gamma_k^2 = \xi_k^2 - 4\beta^2. \tag{3.65}$$

Note that since $\xi_k = \varepsilon_k^0 + 2\beta$ and $\varepsilon_k^0 > 0$ as well as $\beta > 0$, then

$$\xi_k^2 - 4\beta^2 = \left(\varepsilon_k^0\right)^2 + 4\beta\varepsilon_k^0 > 0, \tag{3.66}$$

and we can select $\gamma_k > 0$.

Correspondingly, from the conditions (3.61) and (3.64), the second Bogoliubov coefficient is equal to:

$$u_k^2 = \frac{1}{2} \pm \frac{1}{2}\frac{\xi_k}{\gamma_k}. \tag{3.67}$$

If the interaction is absent, then $\beta = 0$ and $\frac{\xi_k}{\gamma_k} = 1$. In this case $u_k = 1$, $v_k = 0$ and $b_k^\dagger = a_k^\dagger$. That is why we can choose the sign «plus» in (3.64) and (3.67), and finally get [2]:

$$u_k^2 = \frac{1}{2}\left(1 + \frac{\xi_k}{\gamma_k}\right), \tag{3.68}$$

$$v_k^2 = -\frac{1}{2}\left(1 - \frac{\xi_k}{\gamma_k}\right). \tag{3.69}$$

It is important, that a combination

$$u_k^2 v_k^2 = -\frac{1}{4}\left(1 - \frac{\xi_k^2}{\gamma_k^2}\right) = \frac{\beta^2}{\gamma_k^2}, \text{ and } u_k v_k = -\frac{\beta}{\gamma_k} < 0 \text{ for } \gamma_k > 0. \tag{3.70}$$

Thus, something new appears in the problem due to the term $u_k v_k$, which mixes the Bogoliubov coefficients. This term is different from zero only for $\beta \neq 0$, that is only for non-zero interaction between the particles.

Let us return back now to the energy of novel quasiparticles in (3.58). Replacing in the new energy expressions (3.68)–(3.69) for the Bogoliubov coefficients and their mixing squared, we obtain

$$\tilde{E}_k = \xi_k \frac{\xi_k}{\gamma_k} - 4\beta \frac{\beta}{\gamma_k} = \frac{\xi_k^2 - 4\beta^2}{\gamma_k} = \frac{\gamma_k^2}{\gamma_k} = \gamma_k. \tag{3.71}$$

As a result, the Hamiltonian (3.57) can be represented as a sum of the ground state energy and the term with the sum of the new (reconstructed) energies and occupation numbers of novel quasiparticles:

$$E = \langle \widehat{H} \rangle = \tilde{E}_0 + \left\langle \sum_{k \neq 0} \gamma_k b_k^\dagger b_k \right\rangle = \tilde{E}_0 + \sum_{k \neq 0} \gamma_k \langle \tilde{n}_k \rangle, \tag{3.72}$$

where the average occupation numbers

$$\langle \tilde{n}_k \rangle = \left\langle b_k^\dagger b_k \right\rangle = \frac{1}{\exp\left(\frac{\gamma_k}{k_B T}\right) - 1} \tag{3.73}$$

are defined by the Bose–Einstein distribution with the chemical potential equals to zero $\mu = 0$ (number of particles is not fixed), and $\langle \tilde{n}_k \rangle \to 0$ for $T \to 0$. At the same time the reconstructed energy

$$\gamma_k = \sqrt{\xi_k^2 - 4\beta^2} = \sqrt{\varepsilon_k^0 \left(\varepsilon_k^0 + 4\beta\right)} = \sqrt{\frac{\hbar^2 k^2}{2m}\left(\frac{\hbar^2 k^2}{2m} + 2\frac{U(0)N}{V}\right)}. \tag{3.74}$$

It is important to note, that on small wave vectors the reconstructed spectrum of Bogoliubov quasiparticles instead of quadratic becomes a linear one (soundlike or phonon):

$$\gamma_k = \hbar k c_s, \tag{3.75}$$

where the sound velocity squared:

$$c_s^2 = \frac{U(0)N}{mV}. \tag{3.76}$$

Thus, we started from the spectrum of free particles and after switching on the repulsive interaction we get the sound, Moreover, the sound velocity

$$c_s \sim \sqrt{U(0)}. \tag{3.77}$$

The theorists often say that this result for the sound velocity which depends only on the zeroth Fourier component of the potential, corresponds to the homogeneous distribution of the Bose gas or to the jelly model [1,7].

Note, that the sound velocity squared in (3.76) can be written in the canonical form for the thermodynamics, namely as a derivative of pressure with respect to density at fixed entropy:

$$c_s^2 = \left(\frac{\partial P}{\partial \rho}\right)_S, \tag{3.78}$$

where pressure itself is defined by the derivative of the thermodynamic energy with respect to volume at fixed number of particles:

$$P = -\left(\frac{\partial E}{\partial V}\right)_N. \tag{3.79}$$

Substituting the main contribution to the ground state energy $\tilde{E}_0 \approx \frac{1}{2}\frac{U(0)N^2}{V}$ and differentiating with respect to volume, we get the pressure:

$$P \approx \frac{1}{2}\frac{U(0)N^2}{V^2} = \frac{1}{2}U(0)n^2 = \frac{1}{2m^2}U(0)\rho^2, \tag{3.80}$$

where $n = \frac{N}{V}$ is the total particle density, and $\rho = mn$ is the mass density.

Differentiating the pressure with respect to ρ in (3.80), we get

$$c_s^2 = \frac{1}{m^2}U(0)\rho = \frac{1}{m}U(0)n = \frac{U(0)N}{mV}, \tag{3.81}$$

as in (3.76).

For large wave vectors $\hbar k \gg mc_s$ the spectrum becomes quadratic again (as for the case of free particles, see Fig. 3.2):

$$\gamma_k = \frac{\hbar^2 k^2}{2m}. \tag{3.82}$$

It is interesting to note, that the main term in the ground state energy $\frac{1}{2}\frac{U(0)N^2}{V}$ (see Eq. (3.51)) equals to the product of the number of pairs $\frac{N^2}{2}$ and average interaction between particles $\frac{U(0)}{V} = \frac{1}{V}\int d^3r\, U(r)$.

The most important is that the average interaction corresponds to repulsion. For the attractive sign of the zeroth Fourier harmonics of the potential $U(0) < 0$, we have the condensation in the real space (the collapse of the Bose gas).

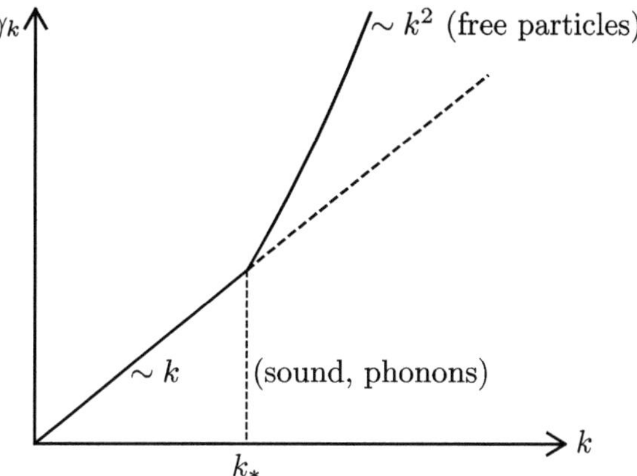

Fig. 3.2 The spectrum of the Bogoliubov quasiparticles has a linear soundlike character on small wave vectors and quadratic character (similar to the spectrum of free particles) on large wave vectors $\hbar k \gg mc_s$ [1]

<u>Comment.</u>

As we already discussed, $\frac{U(0)}{V} = \frac{1}{V} \lim_{q \to 0} \int d^3 r \, U(r) \exp(i\boldsymbol{q}\boldsymbol{r})$.

At the same time according to the Quantum mechanics [14], corrections to the energy in the first two orders of the perturbation theory with respect to interaction are equal to:

$$\Delta E_k = V_{kk} + \sum_{k_1} \frac{V_{kk_1} V_{k_1 k}}{E_k - E_{k_1}}. \tag{3.83}$$

The similar expansion describes the scattering amplitude [1,14]:

$$f_{k \to k'} = -\frac{m}{4\pi\hbar^2} \left\{ V_{kk'} + \sum_{k_1} \frac{V_{kk_1} V_{k_1 k'}}{E_k - E_{k_1}} \right\}, \tag{3.84}$$

where for the scattering on each other of two particles with equal masses, the reduced mass $m^* = \frac{m}{2}$ and wave vectors $|\boldsymbol{k}| = |\boldsymbol{k}'|$ define the elastic scattering problem.

For very small wave vectors $\boldsymbol{k}$ the difference in the expansions of the energy and the scattering amplitude disappears because the momentum transferred in the scattering process $|\boldsymbol{q}| = |\boldsymbol{k} - \boldsymbol{k}'|$ is small. Moreover, for the condensate particles $\boldsymbol{k} \to 0$, and hence $\boldsymbol{k}' = \boldsymbol{k} - \boldsymbol{q} \to \boldsymbol{k} \to 0$. Besides that, for particles with the wave vector k_1 knocked out from the condensate by the interaction, the energy

$$E_{k_1} \gg E_k = E_0. \tag{3.85}$$

As a result, in a more exact theory (with an account of the second order terms with respect to potential in Born approximation) we can replace the Fourier component of the interaction $U(0)$ by the scattering amplitude f as follows [5,6,14,20]:

$$U(0) = -\frac{4\pi\hbar^2}{m} f. \tag{3.86}$$

Further on, for slow particles in vacuum we can replace the scattering amplitude f by the s-wave scattering length a:

$$f = \frac{\exp(2i\delta) - 1}{2ik} \approx \frac{2i\delta}{2ik} = \frac{\delta}{k}, \tag{3.87}$$

where the phase of the s-wave scattering

$$\delta \to -ka \text{ for } k \to 0. \tag{3.88}$$

As a result:

$$f = -a, U(0) = \frac{4\pi\hbar^2}{m}a. \tag{3.89}$$

Note that it is precisely the scattering length (and not the zeroth Fourier component of the potential) which enters in all the observables including expression for the sound velocity squared [5,6,20]:

$$c_s^2 = \frac{4\pi\hbar^2}{m^2}a\frac{N}{V}. \tag{3.90}$$

We discussed already, that in Born approximation [21] the wave function can feel all the potential, and as a result experiences a slight scattering. Moreover, if the potential is alternating in sign, the correct result can be related only to the expression which contains the scattering length a.

Returning back to the problem of spin-polarized hydrogen [8,22–24]], which we began to consider in Lecture 1, let us recollect, that the interaction potential between two spin-polarized hydrogen atoms contains an inaccessible region of strong hardcore repulsion for small r and shallow van der Waals potential well [16,25] with the depth of 6 K on large r (see Fig. 3.3).

Nevertheless, the scattering length for this potential $a = +0.7\,\text{Å} > 0$ corresponds to repulsion [1], and thus, the molecule is not formed in the triplet state.

If we substitute the zeroth Fourier component of the potential $U(0)$ in our calculation for atomic hydrogen, we get obviously incorrect result [1].

Let us consider now the ground state energy. In terms of the scattering length this energy is positive:

$$\tilde{E}_0 \approx \frac{aN^2}{V}\frac{2\pi\hbar^2}{m} > 0. \tag{3.91}$$

Note that at the same time the interaction potential in Fig. 3.3 is negative on large distances $r \to \infty$.

The expression (3.91) for the energy looks like correct and adequate. The dependence of the ground state energy $\tilde{E}_0$ from the volume per one particle $\Omega = \frac{V}{N} = \frac{1}{n}$ described by Eq. (3.91) is shown in Fig. 3.4.

The curve in Fig. 3.4 does not have any minimum, and for $\Omega \to \infty$ (for density $n \to 0$) is described by the monotonically decreasing and positively defined function:

$$\frac{\tilde{E}_0}{V} \approx \frac{2\pi\hbar^2 a}{m}\frac{N^2}{V^2} = \frac{2\pi\hbar^2 a}{m}\frac{1}{\Omega^2} \propto \frac{1}{\Omega^2} \to +0 \text{ for } \Omega \to \infty(n \to 0). \tag{3.92}$$

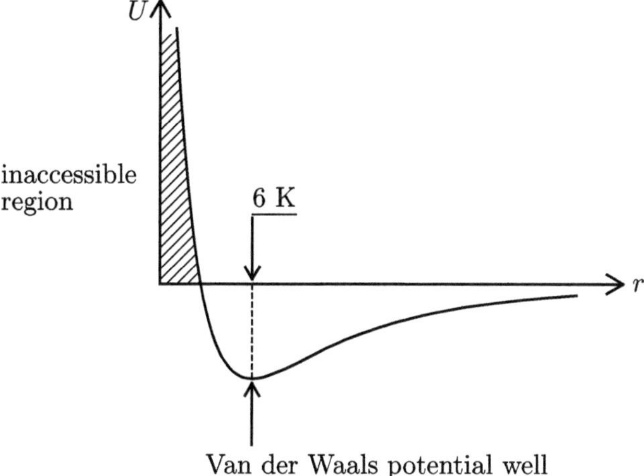

Fig. 3.3 Interaction potential for two spin-polarized hydrogen atoms. We show inaccessible region connected with the hardcore potential at small distances and shallow van der Waals potential well [1,25] at large distances [1]

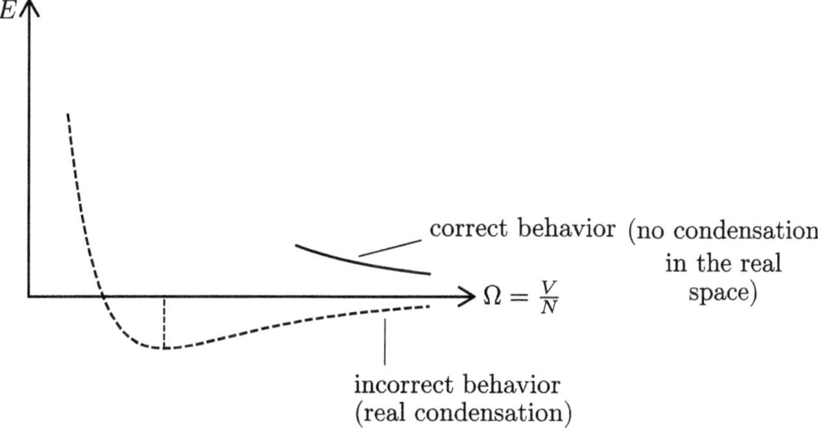

Fig. 3.4 Monotonically decreasing dependence of the ground state energy of the Bogoliubov gas [2] on the volume per one particle $\Omega = \frac{V}{N} = \frac{1}{n}$ for $\Omega \to \infty (n \to 0)$ corresponds to the absence of the condensation in real space at zero temperature (correct behavior) [1]. If the energy dependence would be described by the dotted line, then the real condensation takes place in the system (incorrect behavior) [1]

This type of the energy dependence tells us, that for temperatures $T \to 0$ we do not have molecules or large droplets in the system, or in other words that there is no condensation in the real space [1]. (If the energy dependence would be described by the dotted line in Fig. 3.4, then the real condensation takes place in the system).

Using expression (3.91), let us calculate the chemical potential of the Bogoliubov gas (effectively a mean energy per one particle) [1,2,20]:

$$\mu = \frac{\partial E}{\partial N} = \frac{aN}{V}\frac{4\pi\hbar^2}{m} > 0. \tag{3.93}$$

At first sight this result seems to be very strange. Indeed, we know, that for the ideal Bose gas at high temperatures $\mu < 0$. At the same time if the Bose condensate emerges in the ideal gas, then $\mu = 0$.

On contrary, we have $\mu > 0$, since the gas is interacting. Expression (3.93) describes <u>self-consistent</u> action of all the particles on the given particle (effectively we get the new origin of counting for the chemical potential in the mean field or in the Hartree–Fock approximation [26,27]).

Thus, for temperature $T = 0$ the chemical potential of the ideal gas $\mu_{\text{nonint.}} = 0$, and at the same time $\mu_{\text{int.}} > 0$ for the interacting gas. We can say that $\mu_{\text{int.}}$ is the chemical potential of the real particles and not of the excitations [1].

3.2   The Number of Particles in the Bose Condensate

Thus, in agreement with Eq. (3.34) the total number of particles:

$$N = N_0 + \sum_{k\neq 0}\left\langle a_k^\dagger a_k\right\rangle.$$

Then the number of particles in the condensate:

$$N_0 = N - \sum_{k\neq 0}\left\langle a_k^\dagger a_k\right\rangle = N - \sum_{k\neq 0}\left\langle\left(u_k b_k^\dagger + v_k b_{-k}\right)\left(u_k b_k + v_k b_{-k}^\dagger\right)\right\rangle$$

$$= N - \sum_{k\neq 0}\left\langle\left[u_k^2 b_k^\dagger b_k + v_k^2 b_{-k} b_{-k}^\dagger + u_k v_k\left(b_k^\dagger b_{-k}^\dagger + b_{-k} b_k\right)\right]\right\rangle. \tag{3.94}$$

The third term in the square brackets in Eq. (3.94) is non-diagonal and goes to zero under averaging. In the second term we can use the commutation relation $b_{-k}b_{-k}^\dagger = 1 + b_{-k}^\dagger b_{-k}$. Then:

$$N_0 = N - \sum_{k\neq 0}\left\langle\left[u_k^2 b_k^\dagger b_k + v_k^2 + v_k^2 b_{-k}^\dagger b_{-k}\right]\right\rangle. \tag{3.95}$$

Further on we can use the fact, that the Bogoliubov coefficients [2] are even functions, and rewrite the third term under the sum symbol in Eq. (3.95) in the form $v_k^2 b_k^\dagger b_k$. As a result:

$$N_0 = N - \sum_{k \neq 0} \left\langle \left[\left(u_k^2 + v_k^2 \right) b_k^\dagger b_k + v_k^2 \right] \right\rangle$$

$$= N - \sum_{k \neq 0} v_k^2 - \sum_{k \neq 0} \left(u_k^2 + v_k^2 \right) \langle \tilde{n}_k \rangle . \tag{3.96}$$

For temperature $T = 0$ the average occupation numbers of the novel quasiparticles read: $\langle \tilde{n}_k \rangle = \dfrac{1}{\exp\left(\frac{\gamma_k}{k_B T}\right) - 1} = 0$, and the number of particles in the condensate:

$$N_0 = N - \sum_{k \neq 0} v_k^2 . \tag{3.97}$$

Let us calculate the number of the non-condensed particles at $T = 0$, using the definition of v_k^2 in (3.69) and proceeding in (3.97) from the summation to the integration:

$$\sum_{k \neq 0} v_k^2 = \frac{1}{2} V \int \frac{d^3 k}{(2\pi)^3} \left[-1 + \frac{\xi_k}{\gamma_k} \right] = \frac{1}{2} V \frac{4\pi}{(2\pi)^3} \int dk k^2 \left[\frac{\xi_k}{\gamma_k} - 1 \right] = \frac{V}{4\pi^2} I,$$

$$\tag{3.98}$$

where the integral

$$I = \int dk k^2 \left[\frac{\xi_k}{\gamma_k} - 1 \right] = \int dk k^2 \left[\frac{\varepsilon_k^0 + 2\beta}{\sqrt{\varepsilon_k^0 \left(\varepsilon_k^0 + 4\beta \right)}} - 1 \right] . \tag{3.99}$$

Let us denote the bare energy in (3.99) as

$$\varepsilon_k^0 = \frac{\hbar^2 k^2}{2m} = \alpha^2 k^2, \tag{3.100}$$

where the coefficient α squared

$$\alpha^2 = \frac{\hbar^2}{2m} . \tag{3.101}$$

At the same time the coefficient β in (3.99) after the replacement of the zeroth Fourier component of the potential by the scattering length yields:

$$\beta = \frac{1}{2} \frac{N}{V} \frac{4\pi \hbar^2 a}{m} = \frac{1}{2} \frac{4\pi \hbar^2 a n}{m} . \tag{3.102}$$

Let us evaluate the integral (3.99) with respect to k on the interval from 0 till large limiting values $k_{\max}$, for which the excitation spectrum $\gamma_k = \sqrt{\varepsilon_k^0 \left(\varepsilon_k^0 + 4\beta \right)}$ is already quadratic and not linear, and $\varepsilon_{\max} = \alpha^2 k_{\max}^2 \gg \beta$. Then:

$$I = \int_0^{k_{\max}} dk\, k^2 \frac{\varepsilon_k^0 + 2\beta}{\sqrt{\varepsilon_k^0 \left(\varepsilon_k^0 + 4\beta\right)}} - \int_0^{k_{\max}} dk\, k^2 = \int_0^{k_{\max}} dk\, k^2 \frac{\alpha^2 k^2 + 2\beta}{\alpha k \sqrt{\alpha^2 k^2 + 4\beta}} - \frac{1}{3} k_{\max}^3$$

$$= \int_0^{k_{\max}} dk\, k \frac{\alpha^2 k^2 + 2\beta}{\alpha \sqrt{\alpha^2 k^2 + 4\beta}} - \frac{1}{3} k_{\max}^3 = \frac{1}{2\alpha} \int_0^{k_{\max}^2} dk^2 \frac{\alpha^2 k^2 + 2\beta}{\sqrt{\alpha^2 k^2 + 4\beta}} - \frac{1}{3} k_{\max}^3. \qquad (3.103)$$

We will see, that in the integral (3.99) the terms, divergent on the upper limit as $k_{\max}^3$ and $k_{\max}$, will cancel each other totally. As a result, only the terms which do not depend on $k_{\max}$ or convergent for $k_{\max} \to \infty$, are surviving in (3.99).

Multiplying the variable of the integration k^2 on α^2, we get

$$I = \frac{1}{2\alpha^3} \int_0^{\varepsilon_{\max}} d\varepsilon \frac{\varepsilon + 2\beta}{\sqrt{\varepsilon + 4\beta}} - \frac{1}{3} k_{\max}^3, \qquad (3.104)$$

where $\varepsilon_{\max} = \alpha^2 k_{\max}^2$.

Representing $\varepsilon + 2\beta$ as $(\varepsilon + 4\beta) - 2\beta$ in the nominator of the expression under the integral, we can rewrite (3.104) as follows:

$$I = \frac{1}{2\alpha^3} \int_0^{\varepsilon_{\max}} d\varepsilon \sqrt{\varepsilon + 4\beta} - \frac{\beta}{\alpha^3} \int_0^{\varepsilon_{\max}} d\varepsilon \frac{1}{\sqrt{\varepsilon + 4\beta}} - \frac{1}{3} k_{\max}^3. \qquad (3.105)$$

Both integrals in (3.105) are calculated elementary, and as a result:

$$I = \frac{1}{3\alpha^3} (\varepsilon + 4\beta)^{3/2} \Big|_0^{\varepsilon_{\max}} - \frac{2\beta}{\alpha^3} \sqrt{\varepsilon + 4\beta} \Big|_0^{\varepsilon_{\max}} - \frac{1}{3} k_{\max}^3$$

$$= \frac{1}{3\alpha^3} \left[(\varepsilon_{\max} + 4\beta)^{3/2} - (4\beta)^{3/2} \right] - \frac{2\beta}{\alpha^3} \left(\sqrt{\varepsilon_{\max} + 4\beta} - \sqrt{4\beta} \right) - \frac{1}{3} k_{\max}^3. \qquad (3.106)$$

Now utilizing the fact, that $\varepsilon_{\max} \gg 4\beta$, we can expand $(\varepsilon_{\max} + 4\beta)^{3/2}$ and $\sqrt{\varepsilon_{\max} + 4\beta}$ till the linear terms inclusively with respect to the small parameter $\frac{4\beta}{\varepsilon_{\max}}$. As a result:

$$I \approx \frac{1}{3\alpha^3} \left\{ \left(\varepsilon_{\max}^{3/2} + \frac{3}{2} \frac{4\beta}{\varepsilon_{\max}} \varepsilon_{\max}^{3/2} \right) - (4\beta)^{3/2} \right\} - \frac{2\beta}{\alpha^3} \left\{ \left(\varepsilon_{\max}^{1/2} + \frac{1}{2} \frac{4\beta}{\varepsilon_{\max}} \varepsilon_{\max}^{1/2} \right) \right.$$

$$\left. - (4\beta)^{1/2} \right\} - \frac{1}{3} k_{\max}^3. \qquad (3.107)$$

But $\frac{\varepsilon_{\max}^{3/2}}{3\alpha^3} = \frac{1}{3} k_{\max}^3$ and $\frac{1}{3\alpha^3} \frac{3}{2} \frac{4\beta}{\varepsilon_{\max}} \varepsilon_{\max}^{3/2} = \frac{2\beta}{\alpha^3} \varepsilon_{\max}^{1/2}$. Hence in (3.107) the terms divergent as $k_{\max}^3$ and $k_{\max}$ at $k_{\max} \to \infty$ indeed cancel each other. After the cancellation the integral (3.107) is simplified and becomes equal to:

$$I \approx -\frac{(4\beta)^{3/2}}{3\alpha^3} + \frac{2\beta}{\alpha^3} (4\beta)^{1/2} - \frac{\beta}{\alpha^3} \frac{4\beta}{(\varepsilon_{\max})^{1/2}}. \qquad (3.108)$$

The third term in (3.108) is convergent (going to zero as $\frac{1}{k_{\max}}$ for $k_{\max} \to \infty$). Then finally:

$$I \approx \frac{4\beta^{3/2}}{3\alpha^3} = \frac{4}{3}\left(\frac{\beta}{\alpha^2}\right)^{3/2}. \tag{3.109}$$

Substituting $\alpha^2 = \frac{\hbar^2}{2m}$ and $\beta = \frac{1}{2}\frac{4\pi\hbar^2 an}{m}$, we can rewrite (3.109) in the form:

$$I \approx \frac{4}{3}(4\pi an)^{3/2} = \frac{32}{3}(\pi an)^{3/2}. \tag{3.110}$$

Now we represent the expression (3.97) for the number of the non-condensed particles as

$$N_0 = N - \frac{V}{4\pi^2}I = N - \frac{8V}{3\pi^2}(\pi an)^{3/2} = N - \frac{8V}{3\pi^{1/2}}(an)^{3/2} = N - \frac{8Vn}{3\pi^{1/2}}a^{3/2}n^{1/2}. \tag{3.111}$$

Further on we can use the relation $Vn = N$ and write (3.111) in the canonical form [1–6]:

$$N_0 = N - \frac{8N}{3\pi^{1/2}}a^{3/2}n^{1/2} = N\left(1 - \frac{8}{3}\sqrt{\frac{na^3}{\pi}}\right). \tag{3.112}$$

Equation (3.112) describes the washing of the particles out of the condensate at $T = 0$ in a diluted Bose gas with repulsion between the particles.

It is important that the gas parameter $na^3 \ll 1$ enters in Eq. (3.112).

Namely the square root from the gas parameter $(na^3)^{1/2}$ defines the number of particles knocked out from the Bose condensate by the interaction. Of course, if the gas parameter itself is small, then the square root of it is substantially larger quantity. Hence, the row for the expansion of the number of the non-condensed particles $(N - N_0)$ with respect to the degrees of $(na^3)^{1/2}$ for $T = 0$ can have rather limited radius of the convergence.

3.3 Finite Temperatures $T \neq 0$

For finite temperatures $T \neq 0$ the number of particles in the condensate according to (3.96) reads:

$$N_0(T) = N_0(T = 0) - \sum_{k \neq 0}\left(u_k^2 + v_k^2\right)\langle\tilde{n}_k\rangle$$

$$= N_0(T = 0) - \frac{V}{2\pi^2}\int dk k^2 \frac{\xi_k}{\gamma_k}\frac{1}{\exp\left(\frac{\gamma_k}{k_B T}\right) - 1}. \tag{3.113}$$

For low temperatures $k_B T \ll mc^2$ the spectrum is practically linear $\gamma_k \approx \hbar c k$, $\xi_k \approx 2\beta$, and we can expand the limits of the integration in (3.113) on the interval from 0 till ∞:

$$N_0(T) = N_0(T = 0) - \frac{V}{2\pi^2 \hbar c} \int_0^\infty dk\, k\, \frac{1}{\exp\left(\frac{\hbar c k}{k_B T}\right) - 1}. \qquad (3.114)$$

Introducing the dimensionless variable $x = \frac{\hbar c k}{k_B T}$, we can write (3.114) in the form:

$$N_0(T) = N_0(T = 0) - \frac{V (k_B T)^2}{2\pi^2 (\hbar c)^3} \int_0^\infty dx\, x\, \frac{1}{\exp(x) - 1} = N_0(T = 0) - \eta T^2. \qquad (3.115)$$

Let us remind, that for the ideal Bose gas we have different temperature dependence of the number of the condensed particles, namely:

$$N_0(T) = N \left[1 - \left(\frac{T}{T_0}\right)^{3/2} \right].$$

3.4   Where the λ-Point is Situated?

The critical temperature $k_B T_0 = 3.31 \frac{\hbar^2 n^{2/3}}{m} \sim \frac{\hbar^2 n^{2/3}}{m}$. The transition from the linear to quadratic dispersion law for the Bogoliubov excitations in the rarified gas occurs for the temperatures much less than a critical one:

$$k_B T \sim mc^2 = \frac{4\pi\hbar^2}{m} an = \frac{4\pi\hbar^2 n^{2/3}}{m} \left(an^{1/3}\right) \ll \frac{\hbar^2 n^{2/3}}{m}, \text{ and } k_B T \ll k_B T_0. \qquad (3.116)$$

That is why for the temperatures close to the critical one the spectrum of Bogoliubov excitations practically coincides with the spectrum of the free particles in the ideal Bose gas. Hence for the evaluation of the critical temperature in the diluted Bose gas with repulsion we can use the Einstein formula with an account of the small interaction corrections. Exact and profound calculations of Kashurnikov, Prokof'ev and Svistunov [12] show that in the weakly non-ideal Bose gas with repulsive interaction:

$$k_B T_{0\,\text{int.}} = k_B T_0 \left(1 + 1.3 an^{1/3}\right). \qquad (3.117)$$

3.5   Two-Dimensional Case

In the two-dimensional diluted gas, the number of the condensed particles at low temperatures in terms of the dimensionless variable $x = \frac{\hbar c k}{k_B T}$:

$$N_0(T) = N_0(T = 0) - \frac{\Omega (k_B T)}{2\pi (\hbar c)^2} \int_0^\infty dx\, \frac{1}{\exp(x) - 1}. \qquad (3.118)$$

We can see that for small values of x (on the lower limit) the integral diverges logarithmically as

$$N_0(T = 0) - N_0(T) = \frac{\Omega\,(k_B T)}{2\pi(\hbar c)^2} \int \frac{dk}{k} \sim \frac{\Omega\,(k_B T)}{2\pi(\hbar c)^2} \ln L, \qquad (3.119)$$

where $\Omega \sim L^2$ is the square of the two-dimensional system, L is its size, $k_{\min} = \frac{\pi}{L}$ is the minimally possible wave vector in the system.

Thus, for finite temperatures the Bose condensate is washed out in the 2D system because of the logarithmic divergence on the small wave vectors.

However, in real 2D systems for low temperatures the quasicondensate and finite rigidity (superfluid density) still exist. Meanwhile the system for low temperatures can be described by the Free energy of the vortex molecules (vortex–antivortex pairs with opposite circulations).

When we increase the temperature, the dissociation of the vortex molecules occurs and free vortex filaments appear in the system. For the critical temperature T_C the Berezinskii–Kosterlitz–Thouless (BKT) phase transition [28,29] from the superfluid to the normal state takes place, and the bulk superfluid density ρ_S experiences a jump while decreasing from the finite value to zero.

As a result, the critical temperature can be determined from the condition of the thermodynamic benefit of the creation of the vortex filament:

$$\Delta F = \Delta E - T\Delta S = 0, \text{ and}$$

$$T_C = \frac{\pi d}{2} \frac{\hbar^2 \rho_S}{m^2}, \qquad (3.120)$$

where d is the thickness of the film. (see, e.g., Lifshitz, Pitaevskii [5], paragraph 35, and unpublished Lectures of Yuri Kagan in Leiden in 1992).

In Lecture 4 we will detailly consider the rotating superfluid helium [5,30–34] and the energy of the vortex filament in it [5,32,34]. The full description of the BKT theory we will provide in Lecture 5.

3.6   Correlation Functions for the Bose-Condensed Case

According to the definition, the correlation function of the two particles with the coordinates r_1, r_2 in the mixed state is expressed via the density matrix and can be written as a product:

$$\rho\,(r_1, r_2) = \int dx\, \Psi^*\,(r_1, x)\, \Psi\,(r_2, x). \qquad (3.121)$$

In the homogeneous and isotropic space, the correlation function depends only on the difference $(r_1 - r_2)$:

$$\rho\,(r_1, r_2) \to \rho\,(r_1 - r_2). \qquad (3.122)$$

After the Fourier transformation, we get for the correlation function:

$$\rho(q) = \int \rho\left(r_1 - r_2\right) \exp\left(iq\left(r_1 - r_2\right)\right) d\left(r_1 - r_2\right). \tag{3.123}$$

Let us demonstrate, that the Fourier component of the two-particle correlation function determines the average number of particles with the momentum q:

$$\rho(q) = \left\langle a_q^\dagger a_q \right\rangle = n_q. \tag{3.124}$$

The amplitude of the wave function in the momentum space reads:

$$c_q(x) = \int \Psi(r, x) \exp(iqr) dr. \tag{3.125}$$

Analogously:

$$c_q^\dagger(x) = \int \Psi^*(r, x) \exp(-iqr) dr. \tag{3.126}$$

Correspondingly the absolute value of the wave function squared in the momentum space equals to:

$$\left| c_q(x) \right|^2 = \left| c_q^\dagger(x) c_q(x) \right| = \left| \int \Psi(r, x) \exp(iqr) dr \right|^2. \tag{3.127}$$

Hence the mean value of the probability density in the momentum space is defined as follows:

$$c_q^2 = \frac{1}{V} \int dx \left| c_q(x) \right|^2. \tag{3.128}$$

But this quantity is just n_q. Indeed, after simple transformations, we obtain:

$$c_q^2 = \frac{1}{V} \int dx \left| c_q(x) \right|^2$$

$$= \frac{1}{V} \int dx \left| \int \Psi(r, x) \exp(iqr) dr \right|^2$$

$$= \frac{1}{V} \int dx \int \Psi\left(r_1, x\right) \exp\left(iqr_1\right) dr_1 \int \Psi^*\left(r_2, x\right) \exp\left(-iqr_2\right) dr_2$$

$$= \frac{1}{V} \int dx \iint dr_1 dr_2 \Psi\left(r_1, x\right) \Psi^*\left(r_2, x\right) \exp\left(iq\left(r_1 - r_2\right)\right). \tag{3.129}$$

Now we can use the definition of the correlation function in Eq. (3.121) and get from the last line of Eq. (3.129):

$$\frac{1}{V} \iint dr_1 dr_2 \rho\left(r_1, r_2\right) \exp\left(iq\left(r_1 - r_2\right)\right). \tag{3.130}$$

After that, making the substitution for the integration variable:

$$dr_1 dr_2 \rightarrow d(r_1 - r_2) dr_2, \tag{3.131}$$

we can rewrite (3.130) for the homogeneous and isotropic space in the form:

$$\frac{1}{V} \iint d(r_1 - r_2) dr_2 \rho(r_1 - r_2) \exp(iq(r_1 - r_2)) = \frac{1}{V} \int dr_2 \rho(q) = \rho(q). \tag{3.132}$$

Thus, we showed that:

$$c_q^2 = \frac{1}{V} \int dx \left| c_q(x) \right|^2 = \rho(q) = n_q. \tag{3.133}$$

In the presence of the Bose condensate:

$$\rho(q) = c_q^2 = n_0 \delta_{q0} + n_{q \neq 0} = (2\pi)^3 \delta(q) n_0 + n_{q \neq 0}, \tag{3.134}$$

where n_0 is the density of the condensate particles.

Then the correlation function in the real space:

$$\rho(R = r_1 - r_2) = \int \frac{d^3 q}{(2\pi)^3} \rho(q) \exp(-iqR) = n_0 + \int \frac{d^3 q}{(2\pi)^3} n_{q \neq 0} \exp(-iqR). \tag{3.135}$$

At $R \rightarrow \infty$ the integral in (3.135) goes to zero because of the oscillations (we have seen already that at $T = 0$ the quantity

$$n_{q \neq 0} = \left\langle a_{q \neq 0}^\dagger a_{q \neq 0} \right\rangle = v_q^2 = (n - n_0)$$

determines the number density of the particles kicked out of the condensate by the interaction, and thus the difference of the density of the condensate particles n_0 from the total density n).

Hence, the correlation function for the two infinitely separated points in space r_1 and r_2:

$$\rho(R) \rightarrow n_0 \text{ for } R \rightarrow \infty. \tag{3.136}$$

This behavior of the correlation function corresponds to the presence of the long-range order in the system, and serves, as a main characteristic of the phase transition into the superfluid state (in which the individual degrees of freedom are totally suppressed and there is no viscosity).

We can say that the wave function of the Bose-condensed particles should be symmetrized. Hence, when one particle moves to infinity, the other particle feels this movement since it is necessary to symmetrize the wave function.

Let us emphasize that the condition (3.136) corresponds to the jelly model with the homogeneous distribution of the Bose condensate in space.

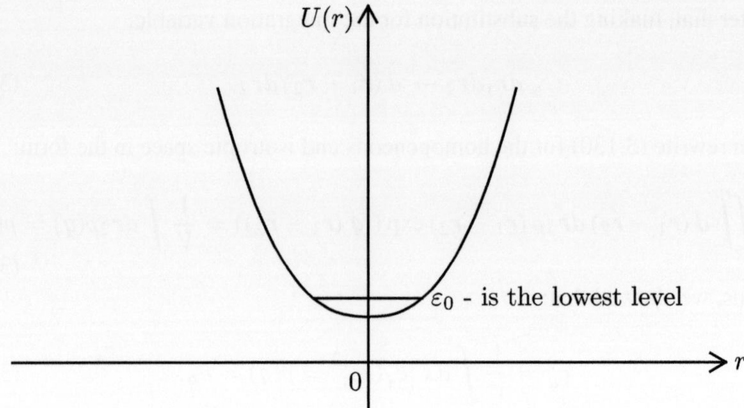

Fig. 3.5 Bose condensation of particles on the lowest level with the energy ε_0 in the potential well [1]

There is a legitimate question: what happens with the correlation function if we have a potential well.

When there is a potential well in the system (see Fig. 3.5), then all the particles (if they do not interact with each other) occupy the lowest level in the well (the Bose condensate now has an energy $\varepsilon_0 \neq 0$ [35–38]).

If there is an additional interaction between the particles, then the lowest level goes up or down in energy. And what happens then with the correlation function?

Let us consider first the free space (in the absence of the well) and write the secondly quantized operators for the annihilation $\widehat{\Psi}(r)$ and creation $\Psi^{\dagger}(r)$ of the particles in the point r in the form of a row with respect to the annihilation and creation operators $\hat{a}_k$ and $\hat{a}_k^{\dagger}$, and wave functions of the plane waves $\varphi_k(r)$, $\varphi_k^*(r)$ [1, 36–38]:

$$\widehat{\Psi}(r) = \sum_k \hat{a}_k \varphi_k(r), \ \widehat{\Psi}^{\dagger}(r) = \sum_k \hat{a}_k^{\dagger} \varphi_k^*(r). \qquad (3.137)$$

If the particles in the coordinate space are located in the points r_i, then the operator of the local density in the point r equals to the sum of delta functions or equivalently to the integral:

$$n(r) = \sum_i \delta(r - r_i) \to V \int d^3 r_1 \delta(r - r_1). \qquad (3.138)$$

The secondly quantized density operator in the point r can be expressed via the local density operator as

$$\hat{n}(r) = V \int d^3 r_1 \widehat{\Psi}^{\dagger}(r_1) \delta(r - r_1) \widehat{\Psi}(r_1) = V \widehat{\Psi}^{\dagger}(r) \widehat{\Psi}(r). \qquad (3.139)$$

Correspondingly the local density in the point r averaged over the states:

$$\langle n(r)\rangle = V\left\langle\left|\widehat{\Psi}^{\dagger}(r)\widehat{\Psi}(r)\right|\right\rangle = V\sum_{k} n_k \varphi_k^*(r)\varphi_k(r), \qquad (3.140)$$

where $n_k = \left\langle \hat{a}_k^{\dagger}\hat{a}_k\right\rangle$.

For the plane waves:

$$\varphi_k(r) = \frac{1}{\sqrt{V}}\exp(ikr) \text{ and } \varphi_k^*(r) = \frac{1}{\sqrt{V}}\exp(-ikr). \qquad (3.141)$$

As a result, $\varphi_k^*(r)\varphi_k(r) = \frac{1}{V}$, and thus the density averaged over the states does not depend on the coordinate r:

$$\langle n(r)\rangle = \sum_{k} n_k. \qquad (3.142)$$

Meanwhile, the two-particle correlation function, which is now in pure and not in the mixed state (for the description in terms of the wave functions and not by the density matrix) reads:

$$\rho(r_1, r_2) = V\left\langle\widehat{\Psi}^{\dagger}(r_1)\widehat{\Psi}(r_2)\right\rangle$$
$$= V\sum_{k} n_k \varphi_k^*(r_1)\varphi_k(r_2) = V n_0 \varphi_{k=0}^*(r_1)\varphi_{k=0}(r_2)$$
$$+V\sum_{k\neq 0} n_k \varphi_k^*(r_1)\varphi_k(r_2). \qquad (3.143)$$

But for the plane waves $\varphi_{k=0}^*(r_1)\varphi_{k=0}(r_2) = \frac{1}{V}\exp(i0) = \frac{1}{V}$. At the same time $\varphi_k^*(r_1)\varphi_k(r_2) = \frac{1}{V}\exp(ik(r_2 - r_1))$. Hence the correlation function in the absence of the potential well:

$$\rho(r_1, r_2) = n_0 + \sum_{k\neq 0} n_k \exp(ik(r_2 - r_1)). \qquad (3.144)$$

Let us introduce a potential well in the system now [1,36–38]. Then, for the secondly quantized operators of the annihilation and creation of the particles at the point r, we can expand them in row with respect to the orthonormalized eigenfunctions $\varphi_m(r)$ (corresponding to the solution of the one-particle Schrodinger equation for the potential well):

$$\widehat{\Psi}(r) = \sum_{m} \hat{a}_m \varphi_m(r), \quad \widehat{\Psi}^{\dagger}(r) = \sum_{m} \hat{a}_m^{\dagger}\varphi_m^*(r). \qquad (3.145)$$

Then the two-particle correlation function in pure (and not in the mixed) state:

$$\rho\left(\boldsymbol{r}_1, \boldsymbol{r}_2\right) = n_0 \varphi_0^*\left(\boldsymbol{r}_1\right) \varphi_0\left(\boldsymbol{r}_2\right) + \sum_{m \neq 0} n_m \varphi_m^*\left(\boldsymbol{r}_1\right) \varphi_m\left(\boldsymbol{r}_2\right), \qquad (3.146)$$

where $\varphi_0(\boldsymbol{r})$ is now the wave function of the lowest level with the energy ε_0 in the potential well in Fig. 3.5.

Note, that the first term $n_0 \varphi_0^*\left(\boldsymbol{r}_1\right) \varphi_0\left(\boldsymbol{r}_2\right)$ in (3.146) corresponds to the pure state, for which there is no loss of the information.

Indeed, since in this term the wave function $\varphi_0\left(\boldsymbol{r}_1\right)$ is a solution of the Schrodinger equation, then knowing $\varphi_0\left(\boldsymbol{r}_1\right)$ we can define $\varphi_0\left(\boldsymbol{r}_2\right)$ in the single valued fashion. It means, that we have again a <u>long-distance correlation</u> on the level of the solution of the Schrodinger equation. Thus, everything which is related to the condensate, can be described by the wave function.

Let us stress, that for the diluted gas $a_0^{\dagger} \approx a_0 = \sqrt{n_0}$, and $a_0^{\dagger} a_0 = n_0$. Thus, for the normalization of the wave function of the ground state on the number of particles we should replace $\varphi_0(\boldsymbol{r}) \rightarrow \sqrt{n_0}\varphi_0(\boldsymbol{r})$.

References

1. Y. Kagan, *Lectures on the Solid-state Theoretical Physics* (MEPHI, Moscow, 1981–1982). Part I. Phonons - unpublished
2. N.N. Bogoliubov, J. Phys. USSR **11**, 23 (1947)
3. K. Huang, C.N. Yang, J.M. Luttinger, Phys. Rev. **105**, 776 (1957)
4. K.A. Brueckner, K. Sawada, Phys. Rev. **106**, 1117 (1957)
5. E.M. Lifshitz, L.P. Pitaevskii, *Statistical Physics, Part II: Theory of the Condensed State* Pergamon, New York, 1980)
6. A.A. Abrikosov, L.P. Gor'kov, I.E. Dzyaloshinskii, *Methods of Quantum Field Theory in Statistical Physics* (Prentice Hall, Englewood Cliffs, New Jersey, 1963)
7. V.N. Popov, *Functional Integrals and Collective Excitations* (Cambridge University Press, UK, 1977)
8. A. Griffin, D.W. Snoke, S. Stringari (eds.), *Bose–Einstein Condensation* (Cambridge University Press, UK, 1995)
9. P. Nozieres, D. Pines, *The Theory of Quantum Liquids*, vol. 2 (Addison Wesley Publishing Company, USA, Canada, Superfluid Bose liquids, 1990)
10. V.V. Tolmachev, *The Bose gas Theory* (Publishing House of Moscow State University, 1969) (in Russian)
11. B. Svistunov, E. Babaev, N. Prokof'ev, *Superfluid States of Matter* (CRS Press, Taylor and Francis group, London, New York, 2015)
12. V.A. Kashurnikov, N.V. Prokof'ev, B.V. Svistunov, Phys. Rev. Lett. **87**, 120402 (2001)
13. F.A. Berezin, *The Method of Second Quantization* (Academic, 1966)
14. L.D. Landau, E.M. Lifshitz, *Quantum Mechanics: Non-Reliativistic Theory* (Pergamon Press, New York, 1977)
15. R.P. Feynman, *Statistical Mechanics: A Set of Lectures* (Advanced Books Classics). (Westview Press, Boulder, 1998)
16. L.D. Landau, E.M. Lifshitz, *Statistical Physics, Part I* (Butterworth-Heinemann, Oxford, 1999)
17. S.N. Bose, Zeit. fur Physik **26**, 178 (1924)
18. A. Einstein, Sber. Preuss, Akad. Wiss. Berlin **1**, 3 (1925)

19. J.C. Maxwell, Phil. Trans. Roy. Soc. London **155**, 459 (1865)
20. S.T. Beliaev, Sov. Phys. JETP **7**, 289 (1958)
21. M. Born, Zeit. fur Physik **38**, 803 (1926)
22. T.J. Greytak, D. Kleppner, in *New Trends in Atomic Physics*, ed. by G. Grynberg, R. Stora, vol. II (North-Holland, New York, 1984), p. 1127
23. I.F. Silvera, J.T.M. Walraven, Phys. Rev. Lett. **44**, 164 (1980)
24. Yu. Kagan, I.A. Vartanyanz, G.V. Shlyapnikov, Sov. Phys. JETP **54**, 590 (1981)
25. J.D. Van der Waals, The Equation of State of Gases and Liquids, in *Nobel Lectures, Physics 1901–1921* (Elsevier, Amsterdam, 1967), p.254
26. D.R. Hartree, Proc. Cambridge Philos. Soc. **24**, 89 (1928)
27. V. Fock, Zeit Phys. **61**, 126 (1930)
28. J.M. Kosterlitz, D.J. Thouless, J. Phys. **6**, 1181 (1973)
29. V.L. Berezinskii, Sov. Phys. JETP **61**, 1144 (1972)
30. R.P. Feynman, Progress Low Temp. Phys. (Chap.2) **1**, 17 (1955)
31. L. Onsager, Nuovo Cimento, **6**(2), 249 (1949)
32. I.M. Khalatnikov, *An Introduction to the Theory of Superfluidity* (Perseus Publishing, Cambridge, 2000)
33. D.R. Tiley, J. Tiley, *Superfluidity and Superconductivity* (Van Nostrand Reinhold company, New York-Cincinnati-Toronto-London, Melburn, 1974)
34. M.Y. Kagan, *Modern Trends in Superconductivity and Superfluidity*, Lecture Notes in Physics, vol. 874. (Springer, Dordrecht, 2013)
35. L.P. Pitaevskii, S. Stringari, *Bose–Einstein Condensation* (Oxford University Press, 2003)
36. Yu. Kagan, B.V. Svistunov, G.V. Shlyapnikov, JETP Lett. **42**, 209 (1985)
37. Yu. Kagan, B.V. Svistunov, G.V. Shlyapnikov, Sov. Phys. JETP **66**, 314 (1987)
38. Y. Kagan, G.V. Shlyapnikov, Phys. Lett. A **130**, 483 (1988)

Lecture 4. Superfluid Helium. Sounds and Vortices

4

Abstract

We start with Landau criterium for superfluidity and consider the excitation spectrum in superfluid He–4 consisting of phonon and roton branches. We introduce the two-fluid description of liquid helium and discuss the classical experiments of Kapitza–Allen–Jones et al., the first and second sound modes in He, as well as the temperature dependence of the normal density and second sound velocity in it. We analyze the formation of the Feynman–Onsager vortex lattices in rotational helium and Bose condensates of quantum gases and determine the first and second critical rotation frequencies.

Let us show, that for the Bose condensate in the Bogoliubov Bose gas with repulsion [2] in contrast to the ideal (noninteracting) Bose gas the Landau criterium for superfluidity is fulfilled [1,3–8].

Let the liquid flows in the capillary with the velocity $\boldsymbol{w}$. Let us sit on the liquid. Then in this reference frame (where the liquid is at a rest) the capillary moves with the velocity $-\boldsymbol{w}$.

According to Landau, the liquid stops if the elementary excitations with momentum parallel to the capillary velocity (and antiparallel to the liquid velocity) start to emerge spontaneously at the wall. Hence in the laboratory frame the capillary will stop (see Fig. 4.1).

If in the reference frame where liquid is at a rest the energy and momentum of the excitations equal to $E_0 = \varepsilon(\boldsymbol{p})$ and $\boldsymbol{P}_0 = \boldsymbol{p}$, then, according to the Galilean transformation, the energy in the laboratory frame reads:

$$E = E_0 + \boldsymbol{w}\boldsymbol{P}_0 + M\frac{w^2}{2} = \varepsilon(\boldsymbol{p}) + \boldsymbol{w}\boldsymbol{p} + M\frac{w^2}{2}, \tag{4.1}$$

where M is the liquid mass.

If the energy difference

$$\Delta\varepsilon = \varepsilon(\boldsymbol{p}) + \boldsymbol{w}\boldsymbol{p} < 0, \tag{4.2}$$

M. Kagan, *Lecture Notes of Professor Yuri Kagan in Theoretical Solid-State Physics*, Lecture Notes in Physics 1048, https://doi.org/10.1007/978-3-032-14621-2_4

57

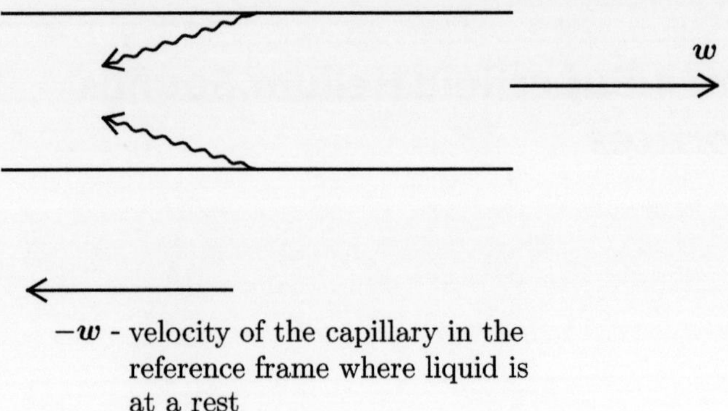

Fig. 4.1 Spontaneous generation of elementary excitations at the wall with the velocities antiparallel to the liquid velocity reduces the kinetic energy of the liquid and the liquid stops [1]

then the kinetic energy of the liquid $M\frac{w^2}{2}$ reduces and the liquid stops. It means that generation of elementary excitations becomes energetically beneficial.

For the momentum $\boldsymbol{p}$ antiparallel to $\boldsymbol{w}$ the scalar product $\boldsymbol{w}\boldsymbol{p} = -wp$. Then for the conservation of the superfluidity:

$$- wp + \varepsilon(\boldsymbol{p}) > 0. \tag{4.3}$$

Correspondingly, we cannot stop the liquid if the velocity:

$$w < \frac{\varepsilon(\boldsymbol{p})}{p}. \tag{4.4}$$

For the Bogoliubov spectrum [2] with the linear part for small wave vectors in Fig. 3.2 for the velocities of the liquid flow smaller than the sound velocity

$$w < c_s, \tag{4.5}$$

it is impossible to excite the system.

4.1   Real Helium

In real dense superfluid He–4 [3–20] the spectrum of the elementary excitations contains not only the phonon part at small wave vectors, but also an additional kink and the roton minimum [3–8, 20–22] (see Fig. 4.2). The critical velocity $v_{\text{cr.}}$ defines by the roton minimum and proves to be smaller than the sound velocity (which is connected with phonons):

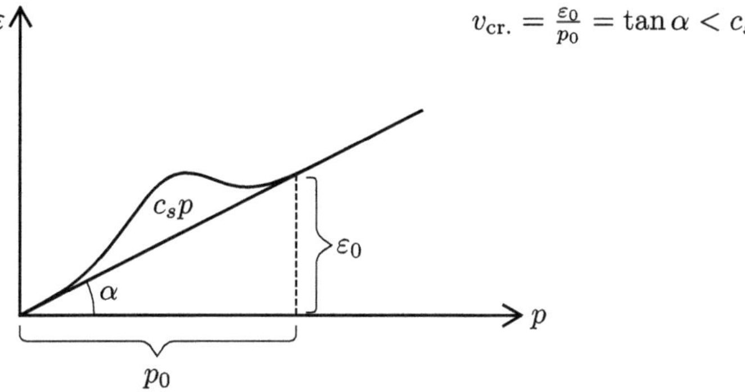

Fig. 4.2 The spectrum of elementary excitations in real He–4. The critical velocity of the destruction of the superfluidity is related to the roton minimum [1]

$$v_{\text{cr.}} = \tan\alpha = \frac{\varepsilon_0}{p_0} < c_s. \tag{4.6}$$

4.2   Non-interacting Bose Gas

The spectrum in the ideal Bose gas [23–25] is quadratic $\varepsilon(p) = \frac{p^2}{2m}$ and thus

$$\frac{\varepsilon(p)}{p} = \frac{p}{2m}. \tag{4.7}$$

As a result, the requirement on the velocity

$$w < \frac{p}{2m} \tag{4.8}$$

is impossible to fulfil, since we can always create an excitation with a very small momentum. Thus, in the ideal Bose gas we never get the superfluidity [1].

4.3   Wave Function of the Condensate. Superfluid Velocity

According to the results of Lecture 3, the doubly quantized operator of particle annihilation at the point r (normalized on the number of particles in the condensate) has a form [1]:

$$\widehat{\Psi}(r) = \frac{1}{\sqrt{V}}\hat{a}_0\varphi_0(r) + \sum_{k\neq0}\ldots$$

If we replace operator $\hat{a}_0$ on $\sqrt{N_0}$, then from the annihilation operator we get the condensate wave function. It is convenient to represent this function as:

$$\Psi_0(r) = \sqrt{n_0}\varphi_0(r). \tag{4.9}$$

The condensate wave function behaves in the thermodynamic way having small fluctuations, and characterizes the system itself.

Thus, we get the description of the system on the level of the wave function.

Let us have now an external field slowly varying in the quasiclassical way with an average distance r_0 between the particles. Suppose the average distance r_0 is small comparing with the size of the system L: $\frac{r_0}{L} \ll 1$.

Then the function $\varphi_0(r)$ in (4.9) instead of the plane wave $\exp(ikr)$ can be written as $\exp(i\chi(r))$. As a result, the condensate wave function reads [6]:

$$\Psi_0(r) = \sqrt{n_0(r)} \exp(i\chi(r)), \tag{4.10}$$

where in the quasiclassical approximation (for slowly varying external field) the condensate density $n_0(r)$ is slowly varying function of the coordinates, while the phase $\chi(r)$ is rapidly varying function.

Let us find now the condensate particles current. It is given by:

$$J_0 = \frac{i\hbar}{2m}\left(\Psi_0\nabla\Psi_0^* - \Psi_0^*\nabla\Psi_0\right) \approx \frac{2i\hbar}{2m}\Psi_0\Psi_0^*(-i\nabla\chi) = \frac{\hbar}{m}\,|\Psi_0|^2\,\nabla\chi = n_0 v_0 \tag{4.11}$$

where:

$$n_0 = |\Psi_0|^2 \tag{4.12}$$

is the condensate density, and

$$v_0 = v_S = \frac{\hbar}{m}\nabla\chi \tag{4.13}$$

is the superfluid velocity [1,6].

Note that in the definition of the superfluid velocity in (4.13) enters the bare (and not the effective) particle mass.

Of course, according to Landau criterium [3–8], we should satisfy the inequality:

$$v_0 = v_S < V_{\text{crit.}} \tag{4.14}$$

to get for the temperatures $T < T_\lambda$ the ground, and not the excited state (where elementary excitations can be produced).

For temperature $T = 0$ we do not have the elementary excitations, and if the velocity of the condensate flow $v_0 < V_{\text{crit.}}$, then they do not appear.

It is interesting, that the correct expression for the superfluid current at $T = 0$ acquires the form:

$$J = nv_S, \tag{4.15}$$

and is determined by the total number of particles n, and not the condensate density n_0. The point is that, though at $T = 0$ the fraction of the particles is washed out from the condensate by the interaction, they still move with the condensate.

Fig. 4.3 The channel with
the friction of the walls [1]

Note also, that the phenomenon of the superfluidity is an analog of the adiabatic perturbation (without transitions to the other excited states) [1].

We can say that adiabatic perturbation is equivalent to the channel with the friction of the walls (see.Fig. 4.3).

4.4 Feynman Considerations

According to the ideas of Feynman [20], if we have Bose particles, then for $T = 0$ their wave function should be symmetrized. Let the function $\Phi(r_1, \ldots, r_N)$ to be the many-particle wave function of the ground state of the Bose system with N particles. Suppose we want to move one particle with the initial coordinate r_1. Then $\Phi \rightarrow \Phi \exp(ikr_1)$. But the necessity to symmetrize the wave function leads us to the replacement:

$$\Phi \rightarrow \Phi \exp(ikr_1 + ikr_2 + \cdots + ikr_N). \tag{4.16}$$

As a result, since the sum of the coordinates is proportional to the center of mass coordinate $(r_1 + \cdots + r_N \sim R_{\text{c.m.}})$ we do not have any individual motions in the system but only collective ones.

4.5 Potential Flow of the Superfluid Liquid

The superfluid velocity is proportional to the gradient of the phase

$$v_S = \frac{\hbar}{m}\nabla\chi = -\nabla\varphi \tag{4.17}$$

where:

$$\varphi = -\frac{\hbar}{m}\chi$$

is the potential of the liquid. That is why the liquid motion is potential and vortexless:

$$\operatorname{curl} v_S = 0. \tag{4.18}$$

According to the Stocks theorem for singly connected area [1]:

$$\iint \operatorname{curl} v_S dS = \oint v_S dl = 0 \tag{4.19}$$

the circulation of the superfluid velocity $\Gamma = \oint v_S dl$ over the closed contour equals to zero.

4.6   Formation of Vortices

Let us consider now not a singly connected area. (see Fig. 4.4). For this doubly connected system:

$$\iint \operatorname{curl} v_S dS = \oint_{C_1} v_S dl - \oint_{C_2} v_S dl = 0, \text{ and } \oint_{C_1} v_S dl = \oint_{C_2} v_S dl \neq 0. \quad (4.20)$$

If both contours C_1 and C_2 in Fig. 4.4 are circular with the radii r_1 and r_2, then Eq. (4.20) acquires a simple form:

$$v_S 2\pi r_1 = v_S 2\pi r_2. \quad (4.21)$$

Hence $v_S \sim \frac{1}{r}$, and the vortexlike motion [21,22] arises in the system:

$$v_S = \frac{\hbar \alpha}{mr}. \quad (4.22)$$

In the polar (cylindrical) coordinates (r, φ) the vector of the superfluid velocity for the vortex solution can be written as follows [1]:

$$v_S = \frac{\hbar}{m} \left(\nabla_r \chi + e_\varphi \frac{1}{r} \frac{\partial \chi}{\partial \varphi} \right), \quad (4.23)$$

where e_φ is the unit vector tangential to circular contours and phase

$$\chi = \varphi \alpha \quad (4.24)$$

because of the cylindrical symmetry.

Fig. 4.4 Doubly connected
system allows the vortex
solutions [1]

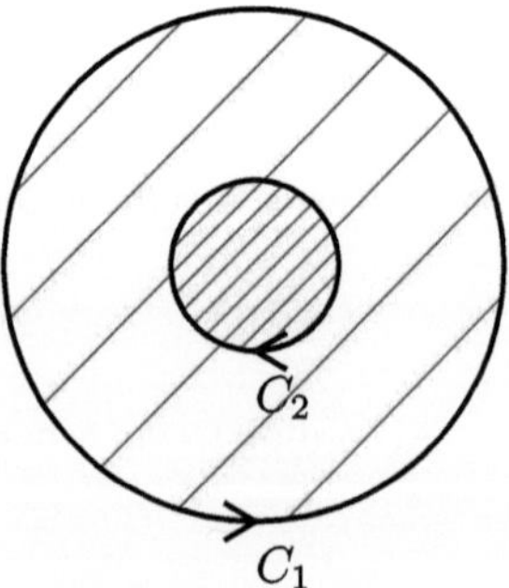

We know however that for the rotation on 2π we go back to the same space point. Correspondingly the wave function gets the phase increase on $2\pi s$, where s is an integer number. Hence, $\alpha = s$ is integer in (4.22) and (4.24).

Thus, the vortex solution for the absolute value of the superfluid velocity reads [6]:

$$v_S = \frac{\hbar s}{mr}.$$

(4.25)

In reality usually $s = 1$.

The velocity circulation:

$$\Gamma = \oint v_S dl = \frac{2\pi \hbar s}{m} = \Gamma_0 s,$$

(4.26)

where:

$$\Gamma_0 = \frac{2\pi \hbar}{m} \sim 10^{-3} \frac{\text{cm}^2}{\text{s}}$$

(4.27)

is the circulation quantum [6].

In He–4 on the atomic contour with the radius $a \sim 3 \cdot 10^{-8}$ cm the superfluid velocity in Eq. (4.25) for $s = 1$ proves to be of the order of sound velocity in helium, that is of the order of the critical velocity corresponding to the destruction of superfluidity:

$$v_{S \text{ on atomic contour}} \sim \frac{\hbar}{ma} \sim \frac{10^{-27}}{7 \cdot 10^{-24} \cdot 3 \cdot 10^{-8}} \sim 10^4 \frac{\text{cm}}{\text{s}} \sim c_S \sim v_{\text{crit.}}.$$

(4.28)

It means that in the vortex core superfluidity is destroyed [6–8].

4.7   Critical Frequency of Rotation

Let us calculate the critical frequency of rotation Ω_C for the cylindrical vessel of the radius R and the height L filled with the superfluid helium (see Fig. 4.5) for which it becomes thermodynamically beneficial to produce the first vortex [1,6] in the middle of the vessel. To do that let us determine the kinetic energy of the liquid and the rotating momentum connected with the vortex.

In cylindrical coordinates the rotating momentum of the vessel for $s = 1$ equals to:

$$M_{\text{vortex}} = L\rho \int r v_s 2\pi r dr = L\rho \int r \frac{\hbar}{mr} 2\pi r dr = L\rho \frac{\hbar}{m} 2\pi \int_a^R r dr \approx L\rho \frac{\hbar}{m} \pi R^2.$$

(4.29)

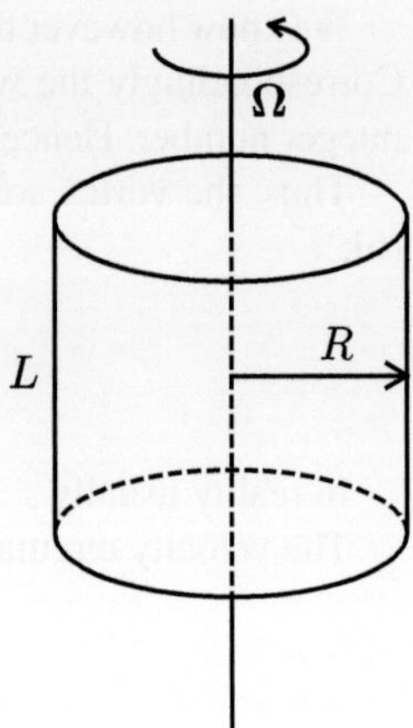

Fig. 4.5 Cylindrical vessel of the radius R and the height L with the superfluid liquid uniformly rotating over the vertical axis with the angular velocity Ω [1,26]

Analogously the kinetic energy of the vessel for $s = 1$:

$$E_{\text{vortex}} = L\rho \int \frac{v_S^2}{2} 2\pi r \, dr = L\rho\pi \int_a^R \frac{\hbar^2}{m^2} \frac{1}{r^2} r \, dr = L\rho \frac{\hbar^2}{m^2}\pi \int_a^R dr \frac{1}{r} = L\rho \frac{\hbar^2}{m^2}\pi \ln \frac{R}{a}. \quad (4.30)$$

Note that the real thermodynamic equilibrium takes place in the rotating frame. The energy variation at $T = 0$ in this frame reads:

$$\Delta E_{\text{rot.}} = E_{\text{vortex}} - M_{\text{vortex}}\Omega. \quad (4.31)$$

Creation of the vortex becomes thermodynamically beneficial for $\Delta E_{\text{rot.}} < 0$. It means that the first vortex appears when we fulfil the condition:

$$\Omega > \Omega_{C1} = \frac{E_{\text{vortex}}}{M_{\text{vortex}}} = \frac{L\rho \frac{\hbar^2}{m^2}\pi \ln \frac{R}{a}}{L\rho \frac{\hbar}{m}\pi R^2} = \frac{\hbar \ln \frac{R}{a}}{m R^2}. \quad (4.32)$$

Note that for $s > 1$ the vortex energy $E_{\text{vortex}} \propto s^2$, while the vortex momentum $M_{\text{vortex}} \propto s$. As a result, the critical frequency increases in s times. Hence the vortices with $s = 1$ are always beneficial [6].

In real helium $R \sim (0.1 \div 1)$ cm and $\ln \frac{R}{a} \sim 15 \div 17$. Correspondingly the critical frequency of rotation $\Omega_{C1} \sim (10^{-3} \div 10^{-1})$ s^{-1}.

For one vortex in the center of the vessel the curl of the superfluid velocity according to the Stocks theorem [1,6,26] is given by:

$$\text{curl } \boldsymbol{v}_S = \Gamma_0 \delta(\boldsymbol{r}) \boldsymbol{e}_z, \quad (4.33)$$

where $\boldsymbol{e}_z$ is the unit vector parallel to the vortex axis.

For the rotation frequencies smaller than the first critical frequency $\Omega < \Omega_{C1}$ the motion of the superfluid liquid for $T = 0$ is vortexless (the superfluid component in contrast to the normal one is not involved in the rotation by the container walls).

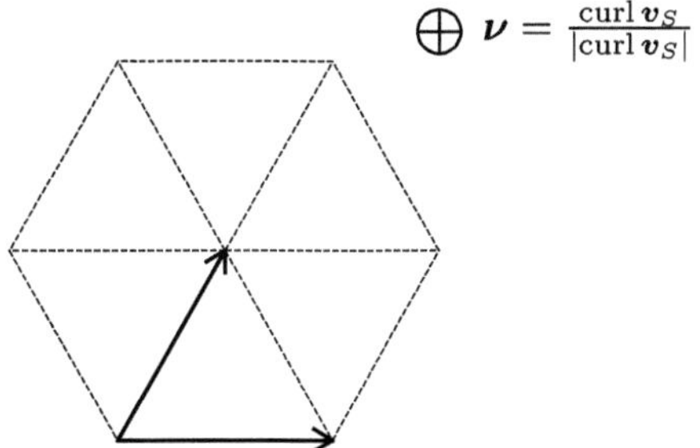

Fig. 4.6 Triangular vortex lattice in superfluid helium for the rotation frequencies of the container with helium $\Omega_{C1} \ll \Omega \ll \Omega_{C2}$ corresponding to the large number of vortices [1,26]

This vortexless flow of the superfluid component in the rotating helium is an analog of the Meissner effect [27] in superconductor.

For $\Omega \gg \Omega_{C1}$ when there are a lot of vortices, they form triangular lattice of Feynman–Onsager [21,22] which is similar to the Abrikosov lattice [28] of magnetic vortices in type-II superconductor (see Fig. 4.6).

The Feynman–Onsager vortex lattice was experimentally photographed in liquid helium [29–33] and visualized in laser experiments with dilute Bose gases of alkali elements Li–7, Na–23, Rb–87 [34–37] (see Fig. 4.7).

For rapid rotation of the container the presence of the large number of vortices imitates the solid-state rotation of the superfluid component [7,26,38–41]. We can say that the solid-state rotation is thermodynamically and energetically beneficial for all types of liquids (both normal and superfluid) [1]. Correspondingly the vortex circulations cancel each other in the bulk of the container and enhance each other on the periphery [42] similar to the circular currents in diamagnetic material [43] (see Fig. 4.8).

If, according to the ideas of Khalatnikov [7], we perform the so-called super-macroscopic averaging on the scales much larger than the mean distance between the vortices and much smaller than the container radius, then in the principal approximation we get for the averaged superfluid velocity an expression for solid-state rotation for $T = 0$ [7,26,38,41]:

$$\langle \boldsymbol{v}_S \rangle \approx [\boldsymbol{\Omega r}].$$

Moreover, for the averaged vorticity $\boldsymbol{\omega} = \langle \mathrm{curl}\,\boldsymbol{v}_S \rangle$ the sum of the delta functions from each of the vortices in the exact expression for the curl of the superfluid velocity is replaced by the averaged expression proportional to the two-dimensional vortex density n_{vortices} [38]:

$$\boldsymbol{\omega} = \langle \mathrm{curl}\,\boldsymbol{v}_S \rangle \approx 2\boldsymbol{\Omega}, \text{ and } \omega = \Gamma_0 n_{\mathrm{vortex}} = 2\Omega. \tag{4.34}$$

Note that just the same averaging is performed for the charge density in the classical macroscopic electrodynamics [43] of the crystalline solid state.

The averaged vorticity satisfies the conservation law for the number of vortices [26,41,45]. Namely:

Fig. 4.7 The lattice of quantum vortices in superfluid He–4 (upper figure) [29–33] and in dilute Bose gases of alkali elements Li–7, Na–23, Rb–87 (lower figure) [1,34–37]

$$\frac{\partial \boldsymbol{\omega}}{\partial t} = \mathrm{curl}\,[\boldsymbol{v}_L \boldsymbol{\omega}]\,, \tag{4.35}$$

where the vortex velocity [26,38,41,45]:

$$\boldsymbol{v}_L = \langle \boldsymbol{v}_S \rangle + \text{ the terms, connected with the rigidity of the vortex lattice.} \tag{4.36}$$

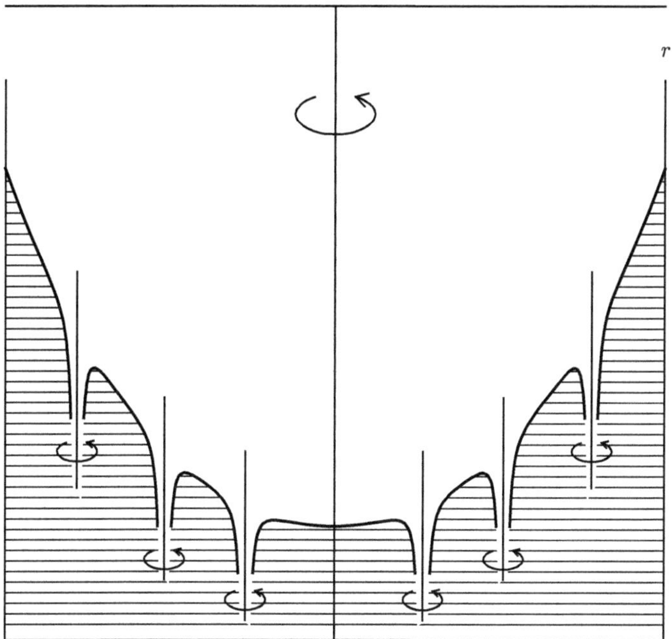

Fig. 4.8 Meniscus with quantum vortices in superfluid rotating helium. The circulations of the vortex velocities cancel each other in the bulk and enhance each other on the periphery of the container [1,20,23]

The second critical frequency of rotation in superfluid helium is connected with the overlapping of the normal vortex cores [6]. As a result of this overlap all the volume of the rotating superfluid experiences a transition to the normal state. The 2D vortex density in this case [6,26]:

$$n_{\text{vortices}} \sim \frac{1}{\pi a^2}, \text{ and } \Omega_{C2} \sim \Gamma_0 \frac{1}{2\pi a^2} \sim \frac{\hbar}{ma^2} \sim 10^{11}\,\text{s}^{-1} \qquad (4.37)$$

The second critical frequency is usually very large and experimentally inaccessible in helium. This frequency is smaller in the rotating Bose condensates of the diluted ultracold gases in the magnetic traps, where the vortex cores have larger size [26,34–37].

4.8   Spectrum of Elementary Excitations in Superfluid Liquid. Phonon Specific Heat

For temperatures $T \neq 0$ we can represent the wave function of the system normalized on the particle density as follows [1]:

$$\Psi = c_0 \Psi_0 + \sum_{k \neq 0} c_k \Psi_k, \qquad (4.38)$$

where the normalization condition yields:

$$|c_0|^2 + \sum_{k \neq 0} |c_k|^2 = n, \tag{4.39}$$

and $|c_0|^2$ in this condition describes a large contribution from the Bose condensate.

As a result, the Hamiltonian of the system diagonalized in terms of the novel quasiparticles reads:

$$\widehat{H} = E_0 + \sum_{k \neq 0} \tilde{\varepsilon}_k b_k^\dagger b_k, \tag{4.40}$$

where the novel energies of quasiparticles $\tilde{\varepsilon}_k = \gamma_k$ at low temperature correspond to the linear part of the spectrum $\tilde{\varepsilon}_k = \hbar c k$.

The distribution function for low temperatures contains the phonon energies $\hbar c k$ and has a Planck form (with chemical potential $\mu = 0$):

$$E_{\text{exc.}} = V \int \frac{d^3 k}{(2\pi)^3} \hbar c k \frac{1}{\exp\left(\frac{\hbar c k}{T}\right) - 1} \sim \eta T^4. \tag{4.41}$$

Thus, the specific heat of the phonon gas [6–8]:

$$C_V = \frac{4E}{T} \propto T^3. \tag{4.42}$$

4.9   Real Helium. Phonons and Rotons. Roton Specific Heat

In real He–4 the spectrum of elementary excitations besides the linear part corresponding to phonons contains the roton minimum [6,7,20,22,26,42,46–48] (see Fig. 4.9). The roton spectrum close to the minimum has the form [6,7,42,46–48]:

$$\varepsilon(p) = \varepsilon_0 + \frac{(p - p_0)^2}{2M^*}. \tag{4.43}$$

The energy of the roton gas equals to:

$$E_{\text{rot.}} = V \int \frac{d^3 p}{(2\pi \hbar)^3} \varepsilon(p) \frac{1}{\exp\left(\frac{\varepsilon(p)}{T}\right) - 1}. \tag{4.44}$$

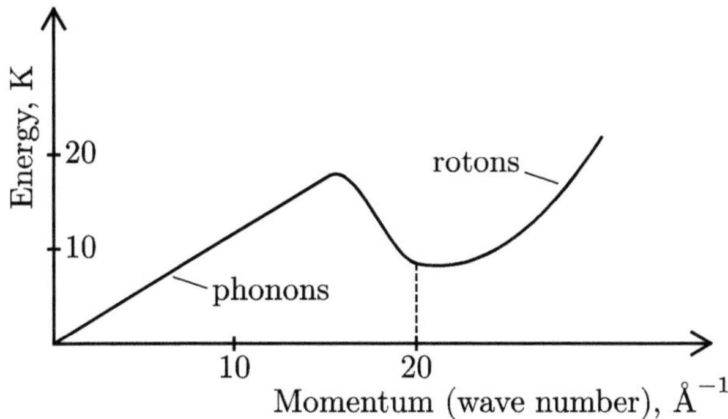

Fig. 4.9 Spectrum of the elementary excitations in the superfluid He–4. The figure illustrates the linear part of the spectrum at small momenta related to phonon and the quadratic roton part close to the minimum on the curve $\varepsilon(p)$, emerging at the momentum values $p = p_0$ [1,6,7,20,22, 26,42,46–48]

In real helium the energy in the roton minimum $\varepsilon_0 \gg T_\lambda$, and thus the distribution function of rotons is the Boltzmann one [6,7,23]:

$$\frac{1}{\exp\left(\frac{\varepsilon(p)}{T}\right) - 1} \approx \exp\left(-\frac{\varepsilon(p)}{T}\right). \tag{4.45}$$

As a result, the roton energy is proportional to: $E_{\text{rot.}} \sim \varepsilon_0 \exp\left(-\frac{\varepsilon_0}{T}\right) (\ldots)$.

Specific heat of the roton gas also depends on the temperature in the exponential fashion [6,7]:

$$C_V \propto \exp\left(-\frac{\varepsilon_0}{T}\right) \tag{4.46}$$

Note that in real helium the roton specific heat prevails over the phonon one for the temperatures $T \geq 0.6$ K.

4.10 Normal Density in Phonon and Roton Temperature Region

Total mass density in helium for temperatures $T \neq 0$, according to the two-fluid description of Landau [3–7], is the sum of superfluid and normal densities:

$$\rho = \rho_S + \rho_n, \tag{4.47}$$

where:

$$\rho_S(T = 0) = \rho, \, \rho_n(T = 0) = 0,$$
$$\rho_S(T \geq T_\lambda) = 0, \, \rho_n(T \geq T_\lambda) = \rho. \tag{4.48}$$

4.11   Classical Experiments of Kapitza, Allen–Jones and Daunt–Mendelssohn [9–15] on the Two-Fluid Model [3–5,16–19] for Helium

In Fig. 4.10 we show the effect of the moving film.

Right Fig. 4.10c describes the effect of the under-barrier tunneling of the superfluid component in the gravitational field. The liquid flow from the test tube rises up along the walls of the vessel in the gravitational field and leaks into the bath.

In Fig. 4.11 we illustrate the thermo-mechanical effect. If in Fig. 4.11a, the temperature of the inner vessel T_2 is higher than the bath temperature $T_1 < T_2$ and the system contains the narrow capillary (where only the frictionless superfluid component penetrates), then from T_1 to T_2 (from cold to hot) there is a flow of superfluid component through the capillary.

In this case there is cooling of the hot part because the superfluid component brings neither the entropy ($S = 0$) nor the heat ($Q = 0$). Since the temperature is the average thermal energy per particle, it decreases due to the increase in the number of particles.

In Fig. 4.11b we show the inverse effect-the effect of helium fountaining when the inner vessel is inverted. The superfluid component fountains overcoming the potential barrier mgh in the gravitational field. At the same time the liquid remaining in the vessel is additionally heated.

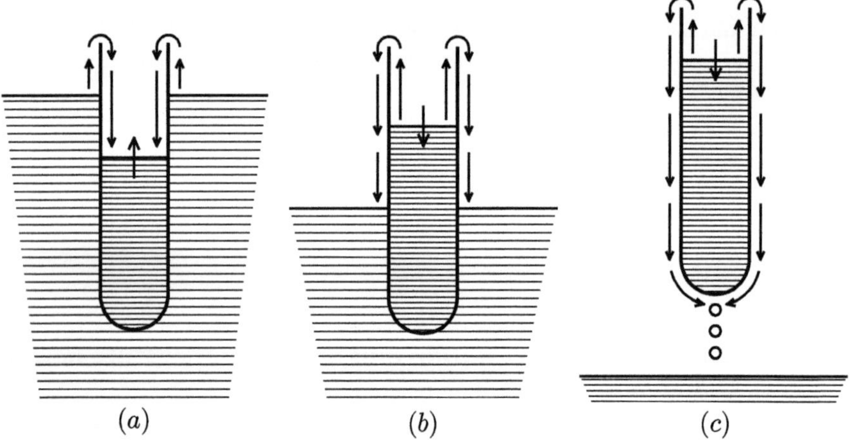

$$(a) \qquad\qquad (b) \qquad\qquad (c)$$

Fig. 4.10 Effect of the moving film. The test tube is placed in a wider vessel filled with liquid helium. At temperatures below the critical temperature, the level of liquid helium in both vessels equalizes (see **a, b**). When the test tube is completely removed from the bath, the liquid from the test tube flows into the bath (**c**). This is how the effect of the under-barrier tunneling of the superfluid component in the gravitational field is manifested itself [1,9–15,42,49]

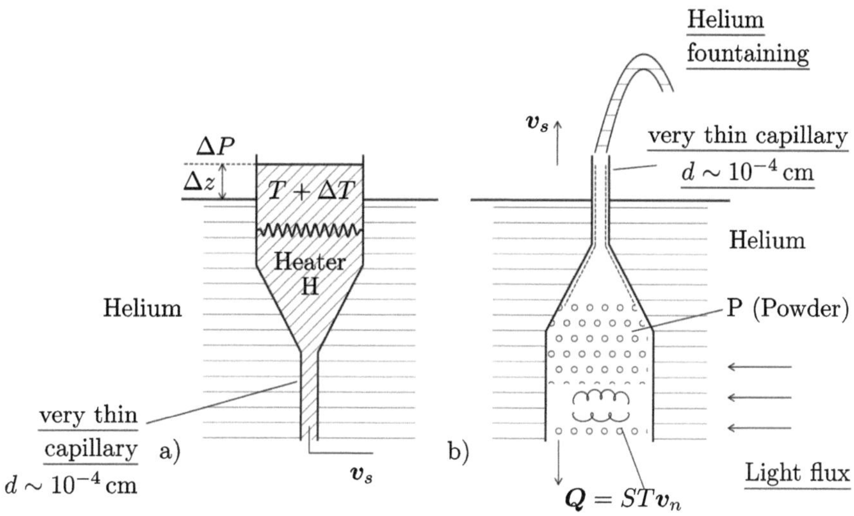

Fig. 4.11 Thermo-mechanical effect. **a**—Level of the liquid in the vessel with the heater H is higher than in the communicating vessel: superfluid liquid with zero entropy flows into the inner vessel through a narrow capillary, cooling helium. **b**—effect of helium fountaining at illumination and heating of powder P located in the vessel with the superfluid helium [1,9–15,42,49]

A fountain of helium up to 20 cm high starts to spurt of the capillary. At the same time the liquid remaining in the vessel is additionally heated.

4.12   Normal Density in Phonon and Roton Temperature Regions

How we can calculate the normal density?

Let us utilize again the Galilean transformation [6] for the total energy and total momentum of the liquid with the total mass M under the transition from one inertial reference frame K_0 to the other inertial frame K:

$$P = P_0 + Mv, \tag{4.49}$$

$$E = E_0 + P_0 v + M\frac{v^2}{2}, \tag{4.50}$$

where the total mass of the liquid $M = $ const (there are no sources and drains in the capillary).

Now let us demand that at the reference frame K_0 the superfluid liquid should be at a rest, and correspondingly the gas of the elementary excitations (together with the capillary) should move relative to the liquid with the velocity v. (see Fig. 4.12).

At the same time let us select the reference frame K, as a system, where vice versa the gas of elementary excitations is at a rest, and liquid is moving with the velocity $(-v)$.

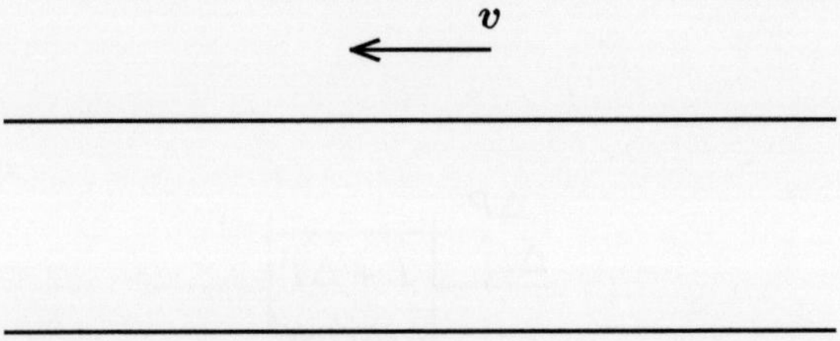

Fig. 4.12 Gas of elementary excitations together with the capillary is moving relative to the superfluid liquid with the velocity v in the reference frame K_0 [1]

Let us stress, that the real thermodynamic equilibrium for the gas of elementary excitations is established in the reference frame K, since the gas is at a rest in it. Then in Eq.'s (4.49–4.50) we should replace $v \to -v$ and get for the total momentum and energy in the frame K:

$$\boldsymbol{P} = \boldsymbol{P}_0 - M\boldsymbol{v}, \tag{4.51}$$

$$E = E_0 - \boldsymbol{P}_0 \boldsymbol{v} + M\frac{v^2}{2}. \tag{4.52}$$

Variation of the total energy in the reference frame K, where the gas of elementary excitations is at a rest, acquires a form:

$$\Delta E = \tilde{\varepsilon}(\boldsymbol{p}) = \varepsilon(\boldsymbol{p}) - \boldsymbol{p}\boldsymbol{v}. \tag{4.53}$$

In fact, Eq. (4.53) describes the energy of one elementary excitation in the frame K. Note that the positive value of this energy $\tilde{\varepsilon}(\boldsymbol{p}) > 0$ guarantees the fulfillment of the Landau criterium on the absolute value of the liquid velocity:

$$v < \frac{\varepsilon(\boldsymbol{p})}{p}. \tag{4.54}$$

At the same time, when we produce one elementary excitation with the momentum $\boldsymbol{p}$ in the reference frame K_0, the total mass of the liquid does not change. Thus, the variation of the total momentum in the reference frames K_0 and K coincides and equals to $\boldsymbol{p}$:

$$\Delta \boldsymbol{P} = \Delta \boldsymbol{P}_0 = \boldsymbol{p}. \tag{4.55}$$

Let us concentrate now on the reference frame K (corresponding to the thermodynamic equilibrium) and determine the total momentum of the gas of elementary excitations in it. According to the ideas of Landau Bose liquid theory [3–8], the gas of elementary excitations is practically an ideal Bose gas and total momentum of the excitations in it in the 3D case equals to:

$$\boldsymbol{P}_{\text{exc.}} = \sum_p \boldsymbol{p} n_B(\tilde{\varepsilon}(\boldsymbol{p})) = V \int \frac{d^3 p}{(2\pi\hbar)^3} \boldsymbol{p} n_B(\tilde{\varepsilon}(\boldsymbol{p}))$$

$$= \int \frac{d^3 p}{(2\pi\hbar)^3} \boldsymbol{p} n_B(\varepsilon(\boldsymbol{p}) - \boldsymbol{p}\boldsymbol{v}) \approx V \int \frac{d^3 p}{(2\pi\hbar)^3} \boldsymbol{p} \left\{ n_B(\varepsilon(\boldsymbol{p})) - \frac{\partial n}{\partial \varepsilon}(\boldsymbol{p}\boldsymbol{v}) \right\}, \tag{4.56}$$

where we expanded the bosonic distribution function of the elementary excitations:

$$n_B(\varepsilon(\boldsymbol{p}) - \boldsymbol{p}\boldsymbol{v}) = \frac{1}{\exp\left(\frac{\varepsilon(\boldsymbol{p}) - \boldsymbol{p}\boldsymbol{v}}{T}\right) - 1} \approx n_B(\varepsilon(\boldsymbol{p})) - \frac{\partial n}{\partial \varepsilon}(\boldsymbol{p}\boldsymbol{v}). \qquad (4.57)$$

The first term in the curly brackets in (4.56) equals to zero due to space isotropy. As a result, the momentum of the excitations gas:

$$\boldsymbol{P}_{\text{exc.}} = -V \int \frac{d^3\boldsymbol{p}}{(2\pi\hbar)^3} \frac{\partial n}{\partial \varepsilon} \boldsymbol{p}(\boldsymbol{p}\boldsymbol{v}), \text{ and } P_{\text{exc.}\,i} = -V \int \frac{d^3\boldsymbol{p}}{(2\pi\hbar)^3} \frac{\partial n}{\partial \varepsilon} p_i \left(p_k v_k\right), \qquad (4.58)$$

where at low temperatures the spectrum is linear $\varepsilon = cp$.

Further on after the averaging over the solid angle:

$$P_{\text{exc.}\,i} = -\frac{V}{3c} \delta_{ik} v_i \int \frac{4\pi\, p^2 dp}{(2\pi\hbar)^3} \frac{\partial n}{\partial p} p^2. \qquad (4.59)$$

Integrating Eq. (4.59) by parts, we get:

$$P_{\text{exc.}\,i} = \frac{V}{3c} v_i \frac{4\pi}{(2\pi\hbar)^3} \int 4p^3 n\, dp. \qquad (4.60)$$

But, since the momentum of the excitations gas equals to:

$$\boldsymbol{P}_{\text{exc.}} = V\rho_n \boldsymbol{v}, \text{ and } \frac{\boldsymbol{P}_{\text{exc.}}}{V} = \rho_n \boldsymbol{v}, \qquad (4.61)$$

where V is the system volume, we finally obtain for the normal density:

$$\rho_n = \frac{2}{3c\pi^2\hbar^3} \int p^3 dp \frac{1}{\exp\left(\frac{cp}{T}\right) - 1}. \qquad (4.62)$$

Hence in the phonon temperature region the normal density:

$$\rho_n \propto T^4. \qquad (4.63)$$

At the same time in roton temperature region:

$$\rho_n \propto \exp\left(-\frac{\varepsilon_0}{T}\right). \qquad (4.64)$$

Fig. 4.13 The ratio of the superfluid density to the total density $\frac{\rho_s}{\rho}$ and normal density to the total density $\frac{\rho_n}{\rho}$ in superfluid helium in phonon and roton temperature regions [1,26,42,46]

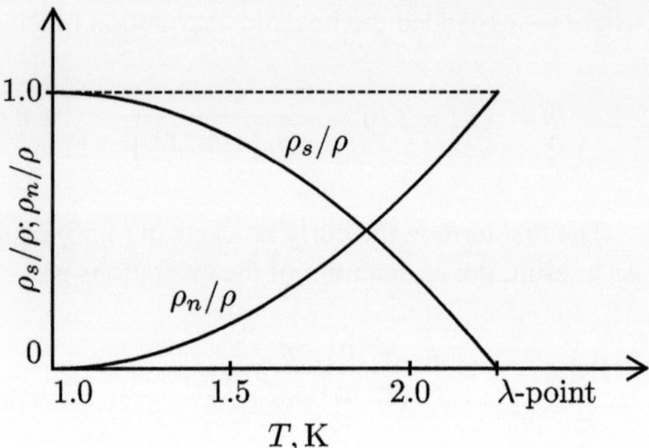

The ratio of the superfluid density to the total density $\frac{\rho_s}{\rho}$ and the normal density to the total density $\frac{\rho_n}{\rho}$ in phonon and roton temperature regions is presented in Fig. 4.13.

References

1. Y. Kagan, *Lectures on the Solid-state Theoretical Physics* (Moscow, MEPHI, 1981–1982). Part I. Phonons - unpublished
2. N.N. Bogoliubov, J. Phys. USSR **11**, 23 (1947)
3. L.D. Landau, J. Phys. (USSR) **5**, 71 (1941)
4. L.D. Landau, J. Phys. (USSR) **11**, 91 (1947)
5. L.D. Landau, Phys. Rev. **75**, 884 (1949)
6. E.M. Lifshitz, L.P. Pitaevskii, *Statistical Physics, Part II, Theory of the Condensed State* (Elsevier, Pergamon, 2013)
7. I.M. Khalatnikov, *An Introduction to the Theory of Superfluidity* (Perseus Publishing, Cambridge, 2000)
8. A.A. Abrikosov, L.P. Gor'kov, I.E. Dzyaloshinskii, *Methods of Quantum Field Theory in Statistical Physics* (Prentice Hall, Englewood Cliffs, New Jersey, 1963)
9. P.L. Kapitza, Nature **141**, 35 (1938)
10. P.L. Kapitza, J. Phys. (USSR) **5**, 59 (1941)
11. P.L. Kapitza, Phys. Rev. **60**, 354 (1941)
12. J.F. Allen, A.D. Misner, Nature **141**, 75 (1938)
13. J.F. Allen, A.D. Misner, Proc. Roy. Soc. **A172**, 467 (1939)
14. J.F. Allen, J. Jones, Nature **141**, 243 (1938)
15. J.G. Daunt, K. Mendelssohn, **143**, 719 (1939)
16. L. Tisza, Nature **141**, 913 (1938)
17. L. Tisza, Compt. Rend. **207**, 1035, 1186 (1938)
18. L. Tisza, J. Phys. Rad. **I**, 164, 350 (1940)
19. L. Tisza, Phys. Rev. **72**, 838 (1947)
20. R.P. Feynman, *Statistical Mechanics, A Set of Lectures, (Advanced Books, Classics)* (Westview Press, Boulder, 1998)
21. R.P. Feynman, Progress Low Temp. Phys. **1**, Chap. 2, 17 (1955)
22. L. Onsager, Nuovo Cimento, **6** (Suppl. 2), 249 (1949)
23. L.D. Landau, E.M. Lifshitz, *Statistical Physics, Part I* (Butterworth-Heinemann, Oxford, 1999)
24. S.N. Bose, Zeit. fur Physik **26**, 178 (1924)
25. A. Einstein, Sber. Preuss, Akad. Wiss. Berlin **1**, 3 (1925)

26. M.Y. Kagan, *Modern Trends in Superconductivity and Superfluidity*, Lecture Notes in Physics, vol. 874. (Springer, Dordrecht, 2013)
27. W. Meissner, R. Ochsenfeld, Naturwissenschaften **21**, 787 (1933)
28. A.A. Abrikosov, Sov. Phys. JETP **5**, 1174 (1957)
29. E. Hall, W.F. Vinen, Proc. Roy. Soc. **A238**, 215 (1956)
30. E. Hall, Adv. Phys. **9**, 89 (1960)
31. R.E. Packard, *XVI International Conference on Low Temperature Physics, Physica B+C*, vol. 109–110 (1982), p. 1474
32. E.J. Yarmchuk, M.J.V. Gordon, R.E. Packard, Phys. Rev. Lett. **43**, 214 (1979)
33. E.J. Yarmchuk, R.E. Packard, J. Low Temp. Phys. **46**, 479 (1982)
34. V. Bretin, S. Stock, Y. Seurin, J. Dalibard, Phys. Rev. Lett. **92**, 050403 (2004)
35. K.W. Madison, F. Chevy, W. Wohlleben, J. Dalibard, Phys. Rev. Lett. **84**, 806 (2000)
36. F. Chevy, K.W. Madison, J. Dalibard, Phys. Rev. Lett. **85**, 2223 (2000)
37. J.R. Abo-Shaeer, C. Raman, J.M. Vogels, W. Ketterle, Science **292**, 476 (2001)
38. I.L. Bekarevich, I.M. Khalatnikov, Sov. Phys. JETP **13**, 643 (1962)
39. E.L. Andronikashvili, Y.G. Mamaladze, Rev. Mod. Phys. **38**, 567 (1966)
40. E.B. Sonin, Rev. Mod. Phys. **59**, 87 (1987)
41. A.F. Andreev, M.Y. Kagan, Sov. Phys. JETP **59**, 318 (1984)
42. D.R. Tiley, J. Tiley, *Superfluidity and Superconductivity* (Van Nostrand Reinhold company, New York-Cincinnati-Toronto-London-Melbourne, 1974)
43. L.D. Landau, E.M. Lifshitz, *Electrodynamics of Continuous Media* (Pergamon Press, New York, 1984)
44. M.Y. Kagan, A.V. Turlapov, Phys. Uspekhi **189**, 225 (2019)
45. I.E. Dzyaloshinskii, G.E. Volovik, Ann. Phys. **125**, 67 (1980)
46. S.J. Putterman, *Superfluid Hydrodynamics* North Holand Series in Low Temperature Physics. (North-Holland, New York, 1974)
47. K.H. Benneman, J. Ketterson (eds.), *The Physics of Liquid and Solid Helium* (Wiley, New York, London, Sydney, Toronto, 1975)
48. P. Nozieres, D. Pines, *The Theory of Quantum Liquids: Superfluid Bose Liquids*, vol. 2 (Addison Wesley Publishing Company, USA, Canada, 1990)
49. P.L. Kapitza, *Experiment, Theory, Practice*, Articles and Addresses, Boston Studies in the Philosophy and History of Science (Springer, Berlin, 1980)

Lecture 5. Superfluidity of the Two-Dimensional Systems. Berezinskii–Kosterlitz–Thouless (BKT) Phase Transition

Abstract

In this Lecture (added to the original course) on the basis of the paragraph 35 of the E. M. Lifshitz, L.P. Pitaevskii monograph *Statistical Physics. Part II. Theory of the Condensed State* [1] and unpublished Lecture of Yu. Kagan in Leiden [2] we discuss the nature of superfluidity, the formation of the vortex–antivortex pairs, the jump in superfluid density and the behavior of the important (phase) correlators in very thin (practically two-dimensional) helium films in the framework of the Berezinskii–Kosterlitz–Thouless (BKT) theory [3,4]. We briefly consider also the Nozieres theory of the BKT-type roughening transition on the surfaces of He-4 quantum crystals.

In Lecture 3, we discussed already, that at finite temperatures in the interacting 2D Bose gas and He-4 monolayers (very thin films of He-4) the two-dimensional density of the non-condensed particles $n_{2\text{D non-cond.}}(T)$ is logarithmically divergent [1–6], and thus for large size of the film $L \to \infty$ all the particles are totally washed out from the Bose condensate:

$$n_{2\text{D non-cond.}}(T) = n_{2\text{D}} - n_{0\,2\text{D}}(T) = \frac{(k_B T)}{2\pi (\hbar c)^2} \int \frac{dk}{k} \sim \frac{(k_B T)}{2\pi (\hbar c)^2} \ln L \sim T \ln L,$$

$$(5.1)$$

where $n_{2\text{D}}$ and $n_{0\,2\text{D}}(T)$ are the two-dimensional total density and density of the condensed particles at finite temperature.

The same divergence of the mean square displacement of atoms $\langle u^2 \rangle \sim T \ln L$, as we will see in the forthcoming lectures, occurs at finite temperatures in the two-dimensional crystal [5], leading to the peculiar character of crystal melting in 2D [7].

Note that the mean square displacement of atoms diverges already at $T = 0$ in the one-dimensional crystal chain [5].

This Lecture is added to the original Lecture course.

© The Author(s), under exclusive license to Springer Nature Switzerland AG 2026   77
M. Kagan, *Lecture Notes of Professor Yuri Kagan in Theoretical Solid-State Physics*,
Lecture Notes in Physics 1048, https://doi.org/10.1007/978-3-032-14621-2_5

We already emphasized in Lecture 3, that the density matrix or the two-particle correlation function in 3D interacting Bose gas behaves as follows [5]:

$$\rho(\boldsymbol{R}) \to n_0 \ \text{for} \ \boldsymbol{R} = \boldsymbol{r}_1 - \boldsymbol{r}_2 \to \infty.$$

This property of the correlation function defines the presence of the long-range order and the rigidity [1,8] of the wave function of the 3D superfluid system.

At the same time in the two-dimensional systems for $\boldsymbol{R} \to \infty$, according to the results of Kane and Kadanov [8] obtained in 1967, the density matrix does not go to the constant limiting value, but is a slowly decreasing one:

$$\rho(\boldsymbol{R}) \to \frac{1}{R^\eta} \ \text{for} \ \boldsymbol{R} \to \infty. \tag{5.2}$$

Let us consider more general case of He-4 thin films of finite width and introduce according to the BKT theory [3,4] the convenient notations:

$$d_S = \rho_S d \tag{5.3}$$

for the two-dimensional superfluid density d_S in the film of thickness d, where $\rho_S = mn_S$ is a standard 3D superfluid (mass) density [1–6,9,10].

For $T = 0$, the condensate wave function has a form [1,5,6,9–11]:

$$\Psi = \sqrt{n_{0\,2D}} \exp(i\,\Phi). \tag{5.4}$$

For $T \neq 0$, the phase Φ is the fluctuating variable. We have to perform the phase averaging with the statistical weight [1]

$$W = \exp\left(-\frac{\Delta F}{T}\right), \tag{5.5}$$

where ΔF is the fluctuation of the Free energy.

We will see soon that in the physics of the 2D systems just <u>the rapid phase fluctuations</u> are washing out the Bose condensate at finite temperatures.

However, the phase of the condensate wave function itself due to its multi-value character cannot enter in the expression for ΔF. Correspondingly, only gradients of phase contribute to the variation of the Free energy. As a result [1–4],

$$\Delta F = d_S \frac{\hbar^2}{2\,m^2} \int (\nabla\Phi)^2 dS, \tag{5.6}$$

where the phase gradient is connected with the superfluid velocity [1]:

$$\frac{\hbar}{m}\nabla\Phi = \boldsymbol{v}_S.$$

Thus, we can represent (5.6) as the kinetic energy of the quasi-2D superfluid [1–4]:

$$\Delta F = \frac{d_S}{2} \int dS v_S^2 = \frac{d \rho_S}{2} \int dS v_S^2. \tag{5.7}$$

The similar expression for the kinetic energy of the vortex filament in the rotating cylindrical vessel with helium (with the substitution of the thickness of the film d on the vessel height L) we have written already in Lecture 4 [5,9,10].

Analogously, only the gradients of displacements contribute to the elastic energy in crystal, since an elastic energy does not change in case of the homogeneous displacements [12]:

$$\Delta F \sim \frac{K}{2} \int dS \, (\nabla_i u_k)^2 , \tag{5.8}$$

where K is an elastic modulus.

The wave function of the condensate for $T \neq 0$ can be defined as follows [1]:

$$\langle \Psi \rangle = \frac{\sum \Psi \exp\left(-\frac{\Delta F}{T}\right)}{\sum \exp\left(-\frac{\Delta F}{T}\right)}. \tag{5.9}$$

But since Bose condensate is absent in 2D system for $T \neq 0$, then $\langle \Psi \rangle = 0$. Let us prove this statement utilizing the well-known formula of Debye–Waller [13,14]. According to this formula, if the operators satisfy the Wick theorem [15] (when the average of the odd number of operators equals to zero and we can split the average of the even number of operators on the pairwise averages), then

$$\langle \exp(y) \rangle = \exp\left(\langle y^2 \rangle /2\right) \tag{5.10}$$

for the averaged values for which $\langle y \rangle = 0$ [1]. Then:

$$\langle \Psi \rangle = \sqrt{n_{0\,2D}} \exp\left(-\frac{\langle (\Delta \Phi)^2 \rangle}{2}\right) = \sqrt{n_{0\,2D}} \exp\left(-\frac{\varphi(0)}{2}\right), \tag{5.11}$$

where $\Delta \Phi = \Phi - \Phi_0$ are the phase fluctuations, and $\Phi_0 = 0$.

Indeed, in the momentum representation, the phase fluctuations:

$$\Delta \Phi = \sum_k \Phi_k \exp(i k r), \tag{5.12}$$

where $\Phi_{-k} = \Phi_k^*$.

Correspondingly, the fluctuation of the Free energy [1]:

$$\Delta F = d_S \frac{\hbar^2 S}{2 m^2} \sum_k k^2 |\Phi_k|^2 = \sum_k \Delta F_k, \tag{5.13}$$

where S is the square of the film surface.

$$\Delta F_k = d_S \frac{\hbar^2 S}{2\,m^2} k^2 \,|\Phi_k|^2 . \tag{5.14}$$

While deriving Eq. (5.13), we took into account that

$$\int d^2 r \,\nabla \left(\sum_k \Phi_k \exp(ikr) \right) \nabla \left(\sum_{k'} \Phi_{k'} \exp\left(ik'r\right) \right)$$

$$= -\sum_{k,k'} \int d^2 r\, \Phi_k \Phi_{k'} \left(kk'\right) \exp\left(i\left(k+k'\right)r\right) \tag{5.15}$$

Performing the integration in Eq. (5.14), we get the required relation:

$$-\sum_{k,k'} \Phi_k \Phi_{k'} \left(kk'\right) \delta_{k,-k'} = \sum_k \Phi_k \Phi_{-k} k^2 = \sum_k \Phi_k \Phi_k^* k^2$$

$$= \sum_k k^2 \,|\Phi_k|^2 . \tag{5.16}$$

Now we can use the general theory of the thermodynamic fluctuations (see, e.g., Landau–Lifshitz [16] paragraph 110) according to which

$$\left\langle x^2 \right\rangle = \int_{-\infty}^{+\infty} x^2 w(x) dx = \frac{1}{\beta}, \tag{5.17}$$

where for the Gauss distributions:

$$w(x) = \sqrt{\frac{\beta}{2\pi}} \exp\left(-\frac{\beta}{2} x^2\right), \ \text{and} \ \int_{-\infty}^{+\infty} w(x) dx = 1. \tag{5.18}$$

Hence, in agreement with [1] (paragraph 28), we obtain

$$\left\langle |\Phi_k|^2 \right\rangle = \frac{m^2 T}{\hbar^2 k^2 d_S}. \tag{5.19}$$

At the same time, for $k \neq k'$,

$$\left\langle \Phi_k \Phi_{k'}^* \right\rangle = 0. \tag{5.20}$$

Now with an account of (5.19) and (5.20) as well as from the condition $\Phi_{-k} = \Phi_k^*$, we can calculate the correlator of the phase fluctuations [1]:

$$\varphi\left(|r_1 - r_2|\right) = \langle \Delta\Phi\left(r_1\right)\Delta\Phi\left(r_2\right)\rangle = \left\langle \sum_k \Phi_k \exp\left(ikr_1\right) \sum_{k'} \Phi_{k'} \exp\left(ik'r_2\right)\right\rangle$$

$$= \left\langle \sum_{k,k'} \Phi_k \Phi_{k'} \exp\left(ikr_1\right)\exp\left(ik'r_2\right)\right\rangle = \left\langle \sum_k |\Phi_k|^2 \exp\left(ik\left(r_1 - r_2\right)\right)\right\rangle$$

$$= \sum_k \frac{m^2 T}{\hbar^2 k^2 d_S} \exp\left(ik\left(r_1 - r_2\right)\right). \tag{5.21}$$

If $r_1 - r_2 = r$, then replacing the summation by the integration, we get

$$\varphi(r) = \frac{m^2 T}{\hbar^2 d_S} \sum_k \frac{1}{k^2} \exp(ikr) = \frac{m^2 T}{\hbar^2 d_S} \int \frac{d^2 k}{(2\pi)^2} \frac{1}{k^2} \exp(ikr). \tag{5.22}$$

For $r = 0$, the correlator of phases is logarithmically divergent on the lower limit of the integration (for $k \to 0$):

$$\varphi(0) = \frac{m^2 T}{\hbar^2 d_S} \int \frac{d^2 k}{(2\pi)^2} \frac{1}{k^2} = \frac{m^2 T}{2\hbar^2 d_S \pi} \int \frac{k\, dk}{k^2} = \frac{m^2 T}{2\hbar^2 d_S \pi} \ln\frac{L}{a}. \tag{5.23}$$

In the general case, the main contribution to the phase correlator $\varphi(r)$ is related to the small values of $kr < 1$ and yields

$$\varphi(r) \approx \frac{m^2 T}{2\hbar^2 d_S \pi} \int \frac{dk}{k} = \frac{m^2 T}{2\hbar^2 d_S \pi} \ln\frac{L}{r} \quad \text{for } \frac{1}{L} < k < \frac{1}{r}. \tag{5.24}$$

We can say that the phase correlator in 2D generates effective Coulomb interaction, which is slowly decaying with distance in a logarithmic fashion [1–4].

Let us show that indeed, $\langle\Psi\rangle = 0$ in Eq. (5.11):

$$\langle\Psi\rangle = \sqrt{n_{0\,2D}}\exp\left(-\frac{\langle(\Delta\Phi)^2\rangle}{2}\right) = \sqrt{n_{0\,2D}}\exp\left(-\frac{\langle\Delta\Phi(r)\Delta\Phi(r)\rangle}{2}\right)$$

$$= \sqrt{n_{0\,2D}}\exp\left(-\frac{\varphi(r-r)}{2}\right) = \sqrt{n_{0\,2D}}\exp\left(-\frac{\varphi(0)}{2}\right)$$

$$= \sqrt{n_{0\,2D}}\lim_{r\to 0}\exp\left(-\frac{m^2 T}{4\hbar^2 d_S \pi}\ln\frac{L}{r}\right)$$

$$= \sqrt{n_{0\,2D}}\lim_{r\to 0}\left(\frac{L}{r}\right)^{-\eta/2} = \sqrt{n_{0\,2D}}\lim_{r\to 0}\left(\frac{r}{L}\right)^{\eta/2} = 0 \text{ for } T \neq 0, \tag{5.25}$$

where [1–4]:

$$\eta = \frac{m^2 T}{2\hbar^2 d_S \pi} > 0. \tag{5.26}$$

Thus, we have shown that $\langle \Psi \rangle = 0$ in 2D for $T \neq 0$.

Let us return back now to the density matrix.

By the definition, in pure (not in the mixed) state, according to Eq. (3.146) in Lecture 3 [5], the two-particle correlation function or the density matrix equals to

$$\rho\left(|\boldsymbol{r}_1 - \boldsymbol{r}_2|\right) = \left\langle \Psi^*\left(\boldsymbol{r}_1\right) \Psi\left(\boldsymbol{r}_2\right)\right\rangle = n_{0\,2D} \left\langle \exp\left(i\Delta\Phi\left(\boldsymbol{r}_1\right) - i\Delta\Phi\left(\boldsymbol{r}_2\right)\right)\right\rangle. \tag{5.27}$$

Now let us apply again the Debye–Waller formula [13, 14]. Then, we obtain

$$\rho\left(|\boldsymbol{r}_1 - \boldsymbol{r}_2|\right) = n_{0\,2D} \exp\left(-\frac{\left\langle\left(\Delta\Phi\left(\boldsymbol{r}_1\right) - \Delta\Phi\left(\boldsymbol{r}_2\right)\right)^2\right\rangle}{2}\right)$$

$$= n_{0\,2D} \exp\left\{\left\langle\Delta\Phi\left(\boldsymbol{r}_1\right)\Delta\Phi\left(\boldsymbol{r}_2\right) - \left(\Delta\Phi\left(\boldsymbol{r}_1\right)\right)^2\right\rangle\right\}, \tag{5.28}$$

where we used the fact that

$$\frac{\left\langle\left(\Delta\Phi\left(\boldsymbol{r}_1\right) - \Delta\Phi\left(\boldsymbol{r}_2\right)\right)^2\right\rangle}{2} = \frac{\left\langle\left(\Delta\Phi\left(\boldsymbol{r}_1\right)\right)^2 + \left(\Delta\Phi\left(\boldsymbol{r}_2\right)\right)^2 - 2\Delta\Phi\left(\boldsymbol{r}_1\right)\Delta\Phi\left(\boldsymbol{r}_2\right)\right\rangle}{2}$$

$$= \frac{\left\langle 2\left(\Delta\Phi\left(\boldsymbol{r}_1\right)\right)^2 - 2\Delta\Phi\left(\boldsymbol{r}_1\right)\Delta\Phi\left(\boldsymbol{r}_2\right)\right\rangle}{2} = \left\langle\left(\Delta\Phi\left(\boldsymbol{r}_1\right)\right)^2 - \Delta\Phi\left(\boldsymbol{r}_1\right)\Delta\Phi\left(\boldsymbol{r}_2\right)\right\rangle. \tag{5.29}$$

As a result, the density matrix reads [1]

$$\rho\left(\boldsymbol{R} = \boldsymbol{r}_1 - \boldsymbol{r}_2\right) = n_{0\,2D} \exp(\varphi(R) - \varphi(0)), \tag{5.30}$$

where the correlators difference:

$$\varphi(R) - \varphi(0) = \frac{m^2 T}{\hbar^2 d_S} \int \frac{d^2 k}{(2\pi)^2} \frac{1}{k^2}(\exp(i\boldsymbol{k}\boldsymbol{R}) - 1). \tag{5.31}$$

Let us analyze thoroughly the integral in Eq. (5.31). This integral is convergent for $kR < 1$. The main contribution to the integral yields the interval $\frac{1}{R} < k < k_{\max}$, where for Bose gas with repulsion $k_{\max}$ at low temperatures is determined by the phonon part of the spectrum: $\hbar c k_{\max} \sim T$ and $k_{\max} \sim \frac{T}{\hbar c}$.

For $kR > 1$, we can neglect the rapidly varying exponent in (5.31) and get

$$\varphi(R) - \varphi(0) \approx -\frac{m^2 T}{2\pi\hbar^2 d_S} \int_{1/R}^{k_{\max}} \frac{dk}{k} = -\frac{m^2 T}{2\pi\hbar^2 d_S} \ln\left(k_{\max} R\right) = \frac{m^2 T}{2\pi\hbar^2 d_S} \ln\frac{1}{k_{\max} R}. \tag{5.32}$$

Correspondingly, the density matrix in 2D Bose gas with repulsion and in dense 2D Bose liquid (in thin films of He-4) in the phonon temperature region, as we proclaimed in Eq. (5.2), is decaying in the power-law fashion [8]:

$$\rho\left(\boldsymbol{R} = \boldsymbol{r}_1 - \boldsymbol{r}_2\right) = n_{0\,2D} \exp\left(\frac{m^2 T}{2\pi\,\hbar^2 d_S}\ln\frac{1}{k_{\max} R}\right)$$

$$= n_{0\,2D}\left(\frac{1}{k_{\max} R}\right)^{\eta}\ \text{for}\ k_{\max} R > 1. \tag{5.33}$$

This behavior of the density matrix at large distances is very peculiar. Usually, the density matrix is exponentially decreasing at large distances.

The main statement of the BKT theory [3,4] is the following: in spite of the fact that the condensate is absent (the condensate wave function goes to zero: $\langle\Psi\rangle = 0$ for $T \neq 0$), the two-dimensional Bose liquid with the density matrix, described by the slowly decaying function, is a superfluid one.

In other words, the superfluid density is non-zero $d_S = \rho_S d \neq 0$ [3,4] in spite of the fact, that the number of the condensate particles goes to zero $n_{0\,2D} = 0$ for $T \neq 0$. Thus, the superfluid density d_S (in fact ρ_S in the film of finite thickness) and the number of the condensate particles (per unit square) are strongly different from each other in quasi-2D situation.

The phenomenon of the superfluidity manifests itself for example, in the potential flow of the 2D liquid with zero circulation $\Gamma = \oint \boldsymbol{v}_S d\boldsymbol{l} = 0$ at low temperatures and slow liquid motions [1,9,10,17].

The existence of the potential flow at slow motion is equivalent to the requirement of the phase unambiguity. This requirement is not violated by the long wavelength phase fluctuations (connected with phonons).

However, the phase unambiguity, (as we know from Lecture 4) can be violated by the generation of vortices inside the closed contour [5,9,10,17–21], going along the boundary of the system (the periphery of the cylindrical vessel or the boundary of the thin film). Note that in the 3D sample creation of the vortex filaments requires the macroscopic energy [1]:

$$E_{\text{vortex}} \propto L,$$

where L is the height of the container with helium.

At the same time, in quasi-2D film the container height L is replaced by the film thickness d, and for the films of the atomic thickness or monolayers $d \sim a$. As a result, vortices in 2D can be thermally generated. Such pointlike vortex filaments serve as an additional branch of elementary excitations in 2D besides the phonon branch (the sound waves) [3,4].

Note that the energy of the 2D vortex filament is proportional to the area of the film. Thus, the filaments can emerge freely only at high temperatures $T > T_C$. The phase transition temperature T_C can be found (as we discussed already in the end of Lecture 3) from the condition that the creation of the vortex filament is energetically beneficial when we vary the Free energy [1–4]:

$$\Delta F = \Delta E - T_C \Delta S = 0, \text{ and } T_C = \frac{\Delta E}{\Delta S}. \tag{5.34}$$

In Lecture 4, we calculated already the energy of the 3D vortex filament [1,5,9, 17,22]. At $T = 0$, it proved to be [1,5]

$$E_{\text{vortex}} = L\rho \frac{\hbar^2}{m^2} \pi \ln \frac{R}{a} = L\rho \frac{\hbar^2}{m^2} \frac{\pi}{2} \ln \frac{R^2}{a^2},$$

where R is the radius of the container with helium, and the density

$$\rho = \rho_S(T = 0).$$

Analogously for $T \neq 0$ in 2D the energy of the vortex reads [1–4]

$$E_{\text{vortex}} = d\rho_S \frac{\hbar^2}{m^2} \frac{\pi}{2} \ln \frac{r_0^2}{a^2} = d\rho_S \frac{\hbar^2}{m^2} \frac{\pi}{2} \ln \frac{\pi r_0^2}{\pi a^2} = d\rho_S \frac{\hbar^2}{m^2} \frac{\pi}{2} \ln \frac{\sigma}{\sigma_0}, \tag{5.35}$$

where $\sigma = \pi r_0^2$ is the area of the helium film and $\sigma_0 = \pi a^2$ is the area of the vortex core.

Thus, the energy variation in Eq. (5.35) $\Delta E = E_{\text{vortex}}$. Similarly, the entropy variation is determined by the logarithm of the number of states, and equals to [3,4]

$$\Delta S = S_{\text{vortex}} = \ln \frac{\sigma}{\sigma_0}. \tag{5.36}$$

As a result, the critical temperature of the transition to a disordered state, connected with the creation of vortices (or, more exactly, with the dissociation of vortex molecules) according to the BKT theory and Eq. (3.120) of Lecture 3 [1–5] is given by

$$T_C = \frac{\hbar^2}{m^2} \frac{\pi}{2} d\rho_s.$$

Note, that the configurational logarithm of the film area also enters in the expression for the entropy of the ordinary excitations (for example, phonons) [1]. However, it is absent in the energy of the ordinary excitation. That is why the creation of the ordinary excitation is always thermodynamically beneficial and some number of ordinary excitations is present in the liquid till zero temperatures $T = 0$. At the same time, the number of vortex excitations for $T < T_C$ rigorously equals to zero [3,4].

Note also, that since an existence of the macroscopic angular momentum of the film $\boldsymbol{M}$ is thermodynamically non-beneficial, then the number of vortices with opposite circulations should be the same in the film $N_{v+} = N_{v-}$ [1–4]. This means, that all the vortices are bound in molecules for $T < T_C$ [3,4]. Creation of vortices for $T > T_C$ is in fact the dissociation of the vortex molecules [3,4].

Since under the emergence of the dissociated vortex filaments the superfluidity (the potential flow) disappears, then the critical temperature $T = T_C$ is the temperature of the destruction of superfluidity, that is the temperature of the phase transition.

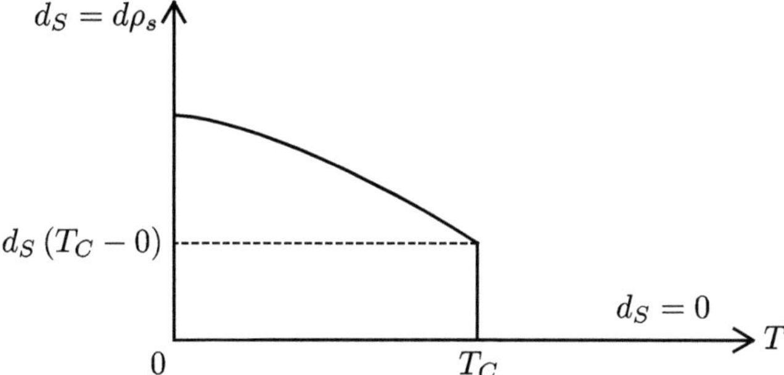

Fig. 5.1 Temperature dependence of the superfluid density $d_S = d\rho_s$ in thin quasi-2D film of He-4. We show in the Figure the jump of the superfluid density at the critical temperature $T = T_C$ [2]

However, this 2D phase transition is rather peculiar. Rigorously speaking, this phase transition is not the transition of the first or second order [1–4].

The superfluid density $d_S = d\rho_s$ in the BKT theory experiences a jump [3,4] at critical temperature. Moreover, as it is shown in Fig. 5.1, the superfluid density goes to zero (and stays zero) at higher temperatures:

$$d_S\,(T_C + 0) \equiv 0. \tag{5.37}$$

At the same time, for the temperatures slightly lower than the critical one, the superfluid density is proportional to T_C and according to Eq. (3.120) reads [1–4]

$$d_S\,(T_C - 0) = \frac{T_C}{\pi}\frac{2\,m^2}{\hbar^2}. \tag{5.38}$$

Correspondingly, since the density matrix at large distances ($k_{\max} R > 1$) according to (5.33) behaves in a power-law fashion:

$$\rho(R) = n_{0\,2D}\left(\frac{1}{k_{\max}R}\right)^{\eta}, \text{ where } \eta = \frac{m^2 T}{2\hbar^2 d_S \pi},$$

then at $T = T_C$ the power η of the correlator decay goes to the universal value [3,4]:

$$\eta\,(T_C) = \frac{m^2 T_C}{2\hbar^2 d_S \pi} = \frac{m^2}{2\hbar^2 d_S \pi}\frac{\hbar^2}{m^2}\frac{\pi}{2}d\rho_S = \frac{1}{4}. \tag{5.39}$$

Thus, the density matrix itself yields [3,4]

$$\rho\,(T_C, R) = n_{0\,2D}\left(\frac{1}{k_{\max}R}\right)^{1/4}, \tag{5.40}$$

where $k_{\max} \sim \frac{T_C}{\hbar c}$ is defined by phonons.

For $T > T_C$, as we discussed already, the superfluid density goes to zero $d_S(T) = 0$ and the density matrix is exponentially decreasing.

We will show soon, that when we are approaching to the critical point T_C from higher temperatures, then the density of the dissociated vortex filaments tends to zero in the exponential fashion.

The vortex contribution in the thermodynamic functions for $T \to T_C$ is also exponentially small, and thus, in contrast to the first- and second-order phase transitions all the derivatives of the thermodynamic functions are continuous at the critical point. [1–4].

Summarizing, we can say again that for low temperatures $T < T_C$ all the vortices are bound in molecules, and only phonons (sound waves) are present in the system. At the same time, for high-temperatures $T > T_C$, there are both phonons and dissociated unpaired vortices in the system. Moreover, the destruction of the superfluidity is connected with the vortices.

Let us emphasize, that the BKT phase transition describes not only the superfluidity in 2D films of He-4 but also the melting of the two-dimensional crystal related to the dislocations motion for high temperatures $T > T_C$. In fact, for $T > T_C$ the 2D crystal (sometimes in the process of two sequential phase transitions) turns into the liquid phase [7].

We can mention in this context also the Nozieres theory [23] of the BKT-type for the roughening transition from atomically smooth to atomically rough surface of He-4 quantum crystals.

5.1 Number of Vortex Filaments for $T > T_C$

Let us determine the number of vortex filaments from the condition that the chemical potential equals zero $\mu = 0$ for $T \to T_C + 0$. To do that, we should understand that when we add one vortex filament to the system, the gain in the Free energy $\Delta F = \Delta E - T\Delta S$ is just the chemical potential. Then, to proceed to the case of a large number of vortex filaments $N_{\text{vort.}}$, we have to replace as usual the total area of the film $S = \pi r_0^2$ by the area per one vortex filament [1]:

$$S_{\text{vortex}} = \frac{\pi r_0^2}{N_{\text{vort.}}}. \tag{5.41}$$

Then utilizing the chemical potential $\mu_{\text{vort.}}$, we get for the "gas" of the vortex filaments containing $N_{\text{vort.}}$:

$$N_{\text{vort.}} \cdot \mu_{\text{vort.}} = \Delta F = N_{\text{vort.}} \cdot E_{\text{vort.}} - N_{\text{vort.}} \cdot T \ln \frac{\pi r_0^2}{N_{\text{vort.}} \pi a^2} + \varepsilon_0 \cdot N_{\text{vort.}}, \tag{5.42}$$

where the chemical constant $\varepsilon_0 \sim T_C$ for the temperatures of the order of the critical one.

Substituting the energy of the single vortex filament with the replacement of $\ln \frac{\pi r_0^2}{\pi a^2}$ in (5.35) by $\ln \frac{\pi r_0^2}{N_{\text{vort.}}\pi a^2}$ in Eq. (5.42), we get [1]

$$N_{\text{vort.}} \cdot \mu_{\text{vort.}} = N_{\text{vort.}} \left[d\rho_S \frac{\hbar^2}{m^2} \frac{\pi}{2} - T \right] \ln \frac{\pi r_0^2}{N_{\text{vort.}}\pi a^2} + \varepsilon_0 N_{\text{vort.}} \qquad (5.43)$$

Now let us introduce a convenient notation:

$$\ln \frac{\pi r_0^2}{N_{\text{vort.}}\pi a^2} = \ln \frac{N_0}{N_{\text{vort.}}}, \qquad (5.44)$$

where

$$N_0 = \frac{S}{S_0} = \frac{\pi r_0^2}{\pi a^2}, \qquad (5.45)$$

and $S_0 = \pi a^2$ is the area of the vortex core.

Then we can write the chemical potential as follows [1]:

$$\mu_{\text{vort.}} = \left[d_S \frac{\hbar^2}{m^2} \frac{\pi}{2} - T \right] \ln \frac{N_0}{N_{\text{vort.}}} + \varepsilon_0. \qquad (5.46)$$

But for $T = T_C$ according to Eq. (3.120)

$$d_S (T_C) \frac{\hbar^2}{m^2} \frac{\pi}{2} - T_C = 0,$$

and the coefficient in front of the logarithm in Eq. (5.46) equals to zero.

At the same time, for $T > T_C$ the coefficient in front of the logarithm is negative. It is impossible to establish from the general considerations the law according to which this coefficient goes to zero since we do not have here the second-order phase transition (when this coefficient goes to zero in linear fashion proportionally to the reduced temperature $\frac{T-T_C}{T_C}$).

Let us suppose, that it goes to zero in a power-law fashion as in the theory of the fluctuational phase transitions [24–27]. Then we get

$$\left[d_S(T) \frac{\hbar^2}{m^2} \frac{\pi}{2} - T \right] \sim (T - T_C)^v , \qquad (5.47)$$

where v is some critical index [24–27].

In this case, the chemical potential goes to zero if the following condition is satisfied:

$$\mu_{\text{vort.}}(T) = 0 = b (T - T_C)^v \ln \frac{N_0}{N_{\text{vort.}}} + \varepsilon_0. \qquad (5.48)$$

As a result, the number of vortices for $T \to T_C + 0$:

$$N_{\text{vort.}} \sim N_0 \exp \left(-\frac{\varepsilon_0}{b (T - T_C)^v} \right) \to 0 \qquad (5.49)$$

tends to zero in the exponential fashion [1].

5.2   Mean Distance Between the Vortices for $T > T_C$

For the temperatures $T > T_C$, the mean distance between the vortex filaments defines the coherence length [1–4]:

$$\xi_0(T) \sim \frac{1}{N_{\text{vort.}}^{1/2}} \sim \frac{1}{N_0^{1/2}} \exp\left(\frac{\varepsilon_0}{2b\,(T - T_C)^v}\right). \qquad (5.50)$$

We can see, that the coherence length is exponentially divergent for $T \to T_C + 0$.

Thus, the correlation function behaves in the exponential fashion on large distances for $T > T_C$ due to the contribution from the vortex excitations [1–4]:

$$\rho\,(T > T_C, R) \sim \exp\left(-\frac{R}{\xi_0(T)}\right). \qquad (5.51)$$

Let us remind, that for $T < T_C$ the correlation function behaves in a power-law fashion due to the phonon contribution [3,4,28]:

$$\rho\,(T < T_C, R) = n_{0\,2D}\left(\frac{1}{k_{\max} R}\right)^{\eta}.$$

In the critical point for temperatures $T = T_C = \frac{\hbar^2}{m^2}\frac{\pi}{2}d\rho_S$. [1–5,28–32] the correlation function acquires a form [3,4,28,29]:

$$\rho\,(T = T_C, R) = n_{0\,2D}\left(\frac{1}{k_{\max} R}\right)^{1/4}.$$

References

1. E.M. Lifshitz, L.P. Pitaevskii, *Statistical Physics, Part II, Theory of the Condensed State* (Pergamon, Elsevier, 2013)
2. Y. Kagan, Lecture in Leiden University on the Grounds of the BKT-Theory (1992)
3. J.M. Kosterlitz, D.J. Thouless, J. Phys. **6**, 1181 (1973)
4. V.L. Berezinskii, Sov. Phys. JETP **61**, 1144 (1972)
5. Y. Kagan, *Lectures on the Solid-State Theoretical Physics* (Moscow, MEPHI, 1981–1982). Part I. Phonons - unpublished
6. P. Nozieres, D. Pines, *The Theory of Quantum Liquids: Superfluid Bose Liquids*, vol. 2 (Addison Wesley Publishing Company, USA, Canada, 1990)
7. V.N. Ryzhov, E.E. Tareeva, Y.D. Fomin, E.N. Tsiok, Berezinskii-Kosterlitz-Thouless transition and two-dimensional melting. Phys. Uspekhi **60**, 857 (2017)
8. J.W. Kane, L.P. Kadanoff, Phys. Rev. **155**, 80 (1967)
9. I.M. Khalatnikov, *An Introduction to the Theory of Superfluidity* (Perseus Publishing, Cambridge, 2000)
10. A.A. Abrikosov, L.P. Gor'kov, I.E. Dzyaloshinskii, *Methods of Quantum Field Theory in Statistical Physics* (Prentice Hall, Englewood Cliffs, New Jersey, 1963)

11. V.L. Ginzburg, *On Superconductivity and Superfluidity: A Scientific Autobiography* (Springer, Heidelberg-Berlin, 2009)
12. L.D. Landau, L.P. Pitaevskii, A.M. Kosevich, E.M. Lifshitz, *Theory of Elasticity, Butterworth-Heinemann* (1986)
13. P. Debye, Ann. Phys. **348**, 49 (1912). ((in German))
14. I. Waller, Zeit. fur Physik A **17**, 398 (1923)
15. G.C. Wick, Phys. Rev. **80**, 268 (1950)
16. L.D. Landau, E.M. Lifshitz, *Statistical Physics, Part I* (Butterworth-Heinemann, Oxford, 1999)
17. M.Y. Kagan, *Modern Trends in Superconductivity and Superfluidity*, Lecture Notes in Physics, vol. 874 (Springer, Dordrecht, 2013)
18. R.P. Feynman, Progress Low Temp. Phys. **1**, Chap. 2, 17 (1955)
19. L. Onsager, Nuovo Cimento **6** (Suppl. 2), 249 (1949)
20. D.R. Tiley, J. Tiley, *Superfluidity and Superconductivity* (Van Nostrand Reinhold Company, New York-Cincinnati-Toronto-London-Melbourne, 1974)
21. I.L. Bekarevich, I.M. Khalatnikov, Sov. Phys. JETP **13**, 643 (1962)
22. E.L. Andronikashvili, Y.G. Mamaladze, Rev. Mod. Phys. **38**, 567 (1966)
23. P. Nozieres, Shape and growth of crystals, in *Solids Far from Equilibrium*, ed. by C. Codreche (Cambridge University Press, UK, 1991), p. 1
24. M. Fisher, *The Nature of Critical Points*, Lectures in Theoretical Physics (University of Colorado Press, Boulder, 1965)
25. L. Kadanoff, *Statistical Physics: Statistics*, Dynamics and Renormalization (World Scientific, 2000)
26. K.G. Wilson, Rev. Mod. Phys. **55**, 583 (1983)
27. A.Z. Patashinskii, V.L. Pokrovskii, *Fluctuation Theory of Phase Transitions* (Pergamon Press, 1979)
28. J.V. Jose (ed.), *40 years of Berezinskii-Kosterliz-Thouless Theory* (World Scientific, Singapore, 2013)
29. A.M. Tsvelik, *Quantum Field Theory in Condensed Matter Physics* (Cambridge University Press, 1995)
30. B. Svistunov, E. Babaev, N. Prokof'ev, *Superfluid States of Matter* (CRC Press, Taylor and Francis Group, 2015)
31. C. Itzykson, J.-M. Drouffe, *Statistical Field Theory*, From Brownian Motion to Renormalization and Lattice Gauge Theory, vol. 1 (Cambridge University Press, 1989)
32. E. Fradkin, *Field Theories of Condensed Matter Physics* (Cambridge University Press, 2013)

Lecture 6. Phonons in the Solid-State Physics

Abstract

In this Lecture we provide the introduction and present the basic ideas concerning the physics of phonons. We discuss the main types and symmetries of the crystalline lattices and the structure of the elementary cells. We introduce the notion of the inverse lattice and explain the adiabaticity principle in the solid-state physics. We write down the Hamiltonian for the atomic displacements and the force matrix in crystal. We derive the equations of motion for the atomic displacements (which describe acoustic and optical phonon modes) and reveal the symmetry properties of the phonon spectra.

6.1 Introduction. Basic Ideas

In solid-state physics an important role plays the parameter of adiabaticity, which is governed by a small ratio of the electronic mass to the nuclear mass [1–4]:

$$\frac{m_{\text{el.}}}{M_{\text{nucl.}}} \ll 1. \tag{6.1}$$

Due to the small value of this parameter (more exactly of the square root from it $\sqrt{\frac{m_{\text{el.}}}{M_{\text{nucl.}}}}$) in the solid-state physics electronic degrees of freedom follow the nuclear degrees of freedom. As a result, the potential energy depends only on nuclear coordinates $U(R_i)$.

In Fig. 6.1 we show electronic and nuclear levels in the solid state. The typical distance between electronic levels is $1\,\text{eV}$ and the distance between nuclear levels is usually 100 times smaller, and equals $0.01\,\text{eV}$.

From the first sight the situation with the adiabatic adjustment in metals seems to be different.

In metal we have free electronic liquid (electron gas) and electrons close to the Fermi surface can be excited by nuclear motions since they need very small energy to proceed to a free state (see Fig. 6.2).

M. Kagan, *Lecture Notes of Professor Yuri Kagan in Theoretical Solid-State Physics*, Lecture Notes in Physics 1048, https://doi.org/10.1007/978-3-032-14621-2_6

Fig. 6.1 Electronic and nuclear energy levels in the solid state. The typical distance between electronic levels is 1 eV and the distance between nuclear levels is usually 100 times smaller, and equals 0.01 eV [1]

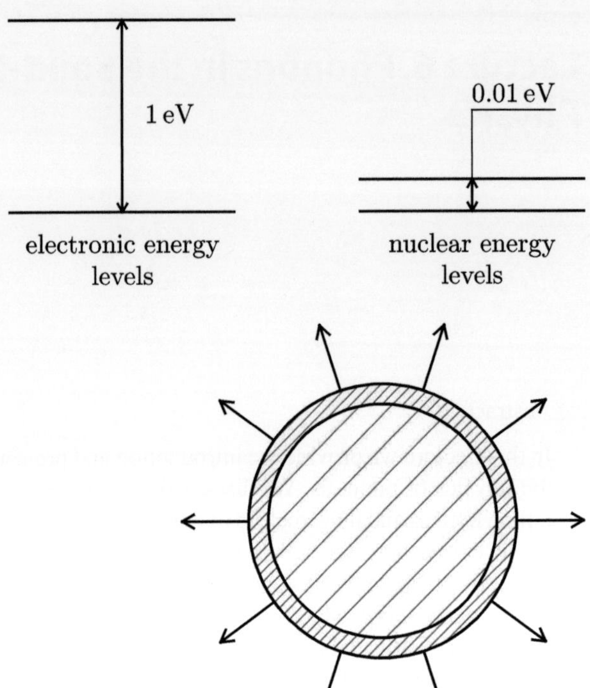

Fig. 6.2 Excitation of the electronic degrees of freedom close to the Fermi surface [1]

However, for the main part of electrons the adiabatic requirement is conserved. Nuclear motions are insufficient for the transfer of electrons to the free state. That is why the main part of electrons adjust themselves to the nuclear motions and hence are not excited. In this way, the adiabatic tracking appears in the system [1,2,5,6].

For electrons close to the Fermi surface this is not the case and hence the condition of adiabaticity is violated. As a result, for the phonon spectrum we get [1–3]:

$$\omega \sim \omega_{\text{ph.}} + \alpha \left(\frac{\omega_{\text{ph.}}}{E_F} \right)^2, \tag{6.2}$$

where the second additional term in Eq. (6.2) represents the renormalization of the phonon spectrum due to electrons living close to the Fermi surface.

Comment 1: All electrons participate in screening effect and in the fulfillment of the electroneutrality condition

Comment 2: We cannot exclude in principle an exotic situation with very slow electron motions when the distance between electronic levels becomes of the order of the energy of the free electron motion. The situation of this type can describe for example the phenomenon of the Electron Polaron Effect in mixed valence heavy fermion compounds (see the papers by Yu. Kagan and N.V. Prokof'ev [3,4]).

Thus, in the adiabatic case $U = U(\boldsymbol{R}_1, \ldots, \boldsymbol{R}_n)$, where $\boldsymbol{R}_1, \ldots, \boldsymbol{R}_n$ are the nuclear coordinates.

The Hamiltonian for the displacements of the atoms in crystal has the form [1,2,5–9]:

$$H = \sum_n \frac{m_n \dot{u}_n^2}{2} + U\left(\boldsymbol{R}_1, \ldots, \boldsymbol{R}_n\right),$$ (6.3)

where $\dot{\boldsymbol{u}}_n$ is the displacement velocity of n-th nucleus of the crystalline lattice, m_n is the mass of n-th nucleus.

Note that for temperatures much smaller than the melting temperature the mean square displacement in classical crystal is much less than the mean square interparticle distance [1,2,9] that is much less than interatomic distance squared:

$$\frac{\overline{u^2}}{a^2} \ll 1.$$ (6.4)

Let us expand the potential energy $U\left(\boldsymbol{R}_1, \ldots, \boldsymbol{R}_n\right)$ close to the equilibrium positions of nuclei $\boldsymbol{R}_1 = \boldsymbol{R}_1^0, \ldots, \boldsymbol{R}_n = \boldsymbol{R}_n^0$. Then:

$$U \approx U\left(\boldsymbol{R}_1^0, \ldots, \boldsymbol{R}_n^0\right) + \sum_n \left(\frac{\partial U}{\partial R_n^i}\right)_0 u_n^i + \frac{1}{2} \sum_{n,n'} A_{n,n'}^{ik} u_n^i u_{n'}^k$$

$$+ \frac{1}{6} \sum_{n,n',n''} B_{n,n',n''}^{i,k,l} u_n^i u_{n'}^k u_{n''}^l,$$ (6.5)

where 0 is an equilibrium position, n is the number of the nucleus, i is Cartesian coordinate, u_n^i is the displacement of n-th nucleus along ith Cartesian coordinate.

In Eq. (6.5) enters the <u>force matrix</u>, which equals [1,2,5–10]:

$$A_{n,n'}^{ik} = \left(\frac{\partial^2 U}{\partial R_n^i \partial R_{n'}^k}\right)_0$$ (6.6)

in the quadratic term, with respect to displacements u_n^i, and the matrix, describing anharmonic terms (connected with phonon interactions) [1,2]:

$$B_{n,n',n''}^{i,k,l} = \left(\frac{\partial^3 U}{\partial R_n^i \partial R_{n'}^k \partial R_{n''}^l}\right)_0.$$ (6.7)

Close to the equilibrium positions:

$$\left(\frac{\partial U}{\partial R_n^i}\right)_0 = 0.$$ (6.8)

If we neglect the anharmonic interactions, then the Hamiltonian in Eq. (6.3) expanded near the equilibrium positions in harmonic approximation reads [1,2,5–11]:

$$H = \sum_n \frac{m_n \dot{u}_n^2}{2} + U\left(R_1^0, \ldots, R_n^0\right) + \frac{1}{2} \sum_{n,n',n''} A_{n,n'}^{ik} u_n^i u_{n'}^k. \tag{6.9}$$

The equations of motion corresponding to the Hamiltonian (6.9) represent the equations for n harmonic oscillators [12]:

$$m_n \ddot{u}_n^i = - \sum_{n'} A_{n,n'}^{ik} u_{n'}^k. \tag{6.10}$$

If the displacement $u_n^i \sim \exp(-i\omega t)$, then the equations of motion acquire the form:

$$m_n u_n^i \omega^2 = \sum_{n'} A_{n,n'}^{ik} u_{n'}^k. \tag{6.11}$$

Further on it is convenient to introduce the renormalized displacements [1,2]:

$$\sqrt{m_n} u_n^i = v_n^i, \tag{6.12}$$

and the renormalized force matrix:

$$\bar{A}_{n,n'}^{ik} = \frac{A_{n,n'}^{ik}}{\sqrt{m_n m_{n'}}}. \tag{6.13}$$

Then we can rewrite the equations of motion as follows:

$$v_n^i \omega^2 = \sum_{n'} \bar{A}_{n,n'}^{ik} v_{n'}^k. \tag{6.14}$$

Note that in Eq. (6.14) the nucleus mass is hidden in the renormalized force matrix $\bar{A}_{n,n'}^{ik}$.

The equations for eigenfrequencies yield [1,2,7]:

$$\left\| \bar{A}_{n,n'}^{ik} - \delta_{nn'} \delta^{ik} \omega^2 \right\| = 0. \tag{6.15}$$

Let us stress that in Eq. (6.15) enters ω^2-the frequency squared. This is typical for bosonic branches of the spectrum [1,2,9].

The force matrix $A_{n,n'}^{ik}$ in Eq. (6.6) is symmetric and real. The same is valid for the square root from the masses $\sqrt{m_n m_{n'}}$. That is why, the renormalized matrix $\bar{A}_{n,n'}^{ik}$ is Hermitian while the squares of the eigenfrequencies are positive $\omega^2 > 0$.

We expand the potential energy close to the minimum, and hence the quadratic form $\sum_{n,n'} A_{n,n'}^{ik} u_n^i u_{n'}^k = \sum_{n,n'} \bar{A}_{n,n'}^{ik} v_n^i v_{n'}^k$ in the Hamiltonian (6.5) is positively determined one [1,2,9].

Thus, all the minors of the renormalized force matrix $\bar{A}_{n,n'}^{ik}$ are positive. If we diagonalize the matrix $\bar{A}_{n,n'}^{ik}$, then it will be equal to

$$
\begin{pmatrix}
\omega_1^2 & 0 & 0 \\
0 & \omega_2^2 & 0 \\
0 & 0 & \omega_3^2
\end{pmatrix},
\tag{6.16}
$$

where

$$
\omega_1^2 > 0,\, \omega_1^2 \omega_2^2 > 0,
\tag{6.17}
$$

and correspondingly $\omega_2^2 > 0$. Analogously $\omega_n^2 > 0$ for all values of n.

It means that all the eigenfrequencies are positive.

Comment: All the above-mentioned considerations are related not only to the crystal, but also to any solid system (for example, to the amorphous one) in which the particles are oscillating close to equilibrium positions [1,2].

If the periodic structure emerges in the system, then it becomes easier for us to solve the equations on the eigenfrequencies.

For the periodic crystal structure any vector of translation, which transforms the crystalline system to the physically equivalent state, can be expanded with three vectors of basic translations as follows [1,2,13]:

$$
\boldsymbol{r}_m = m_1 \boldsymbol{a}_1 + m_2 \boldsymbol{a}_2 + m_3 \boldsymbol{a}_3,
\tag{6.18}
$$

where m_1, m_2, and m_3 are the integer numbers, while the vectors $\boldsymbol{a}_1$, $\boldsymbol{a}_2$, and $\boldsymbol{a}_3$ form the basis of the lattice (see Fig. 6.3).

Let us emphasize that the elementary cell of the crystal can contain one or several atoms.

Note, that utilizing all the possible translations, we can get the Brave lattice [1,2,6] from the elementary cell.

In general case an elementary cell can be described by 6 parameters—3 lengths and 3 angles [14–21].

There are 7 types of symmetry and 14 options for the type of the crystal lattice. For the triclinic symmetry all the 3 lengths and 3 angles are different

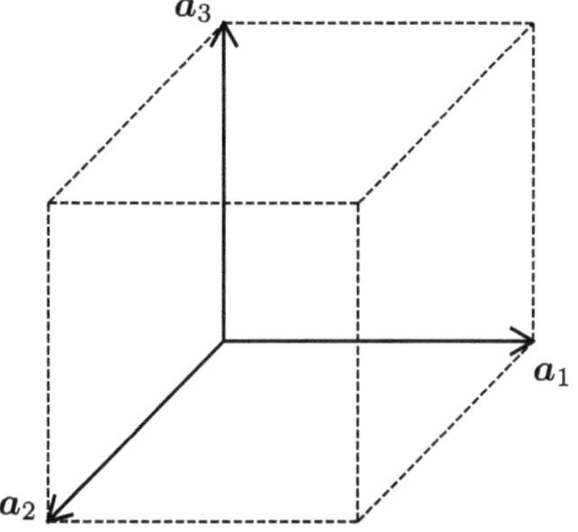

Fig. 6.3 Vectors $\boldsymbol{a}_1$, $\boldsymbol{a}_2$, and $\boldsymbol{a}_3$ form the basis of the crystalline lattice [1]

($a_1 \neq a_2 \neq a_3$, and $\alpha = \gamma = 90°$, $\beta \neq 90°$). Thus, it is described by 6 parameters. For monoclinic symmetry the number of parameters equals 4. ($a_1 \neq a_2 \neq a_3$)

The crystal lattice of the solid system can have different space symmetry: cubic, hexagonal, and so on.

Let us stress that in general case an elementary cell of the three-dimensional crystal is described by 3 vectors of the elementary translations $\boldsymbol{a}_1, \boldsymbol{a}_2, \boldsymbol{a}_3$, which form the basis of the crystalline lattice (see Fig. 6.3), or equivalently as we have said already by 6 parameters—3 lengths (a_1, a_2, a_3) and 3 angles (α, β, γ) with the Cartesian axes.

For the rhombic symmetry the number of parameters equals 3 ($a_1 \neq a_2 \neq a_3$) and ($\alpha = \beta = \gamma = 90°$).

Four other types of symmetry correspond to the uniaxial crystals. Tetragonal symmetry is described by 2 parameters $a_1 = a_2 \neq a_3$ and $\alpha = \beta = \gamma = 90°$ and corresponds to the rotation axis of the 4th order.

Rhombohedral symmetry is related also with the rotation axis of the 4-th order and is described by 2 parameters $a_1 = a_2 = a_3$ and $\alpha = \beta = \gamma$.

The hexagonal symmetry (or the symmetry of the six-faceted prism) corresponds to the rotation axis of the 6-th order and is described by 2 parameters $a_1 = a_2 \neq a_3$ and $\alpha = \beta = 90°$, $\gamma = 120°$.

The simple cubic symmetry is described by 1 parameter $a_1 = a_2 = a_3$ and $\alpha = \beta = \gamma = 90°$.

In total there are 14 types of the crystal lattice as we already stated. Among them there are 3 types of the cubic lattice: simple cubic, body-centered cubic (bcc lattice see Fig. 6.4), and face-centered cubic (fcc).

Besides that, there is 1 type of the hexagonal lattice–simple hexagonal, 1 type of rhombohedral lattice-simple rhombohedral, 1 type of triclinic lattice, and 2 types of monoclinic lattice (simple and base-centered).

Finally, there are 4 types of rhombic lattice (simple, base-centered, face-centered, body-centered) and 2 types of tetragonal lattice–simple and body-centered.

Fig. 6.4 Body-centered cubic lattice (bcc lattice) for α-phase of Fe [1]

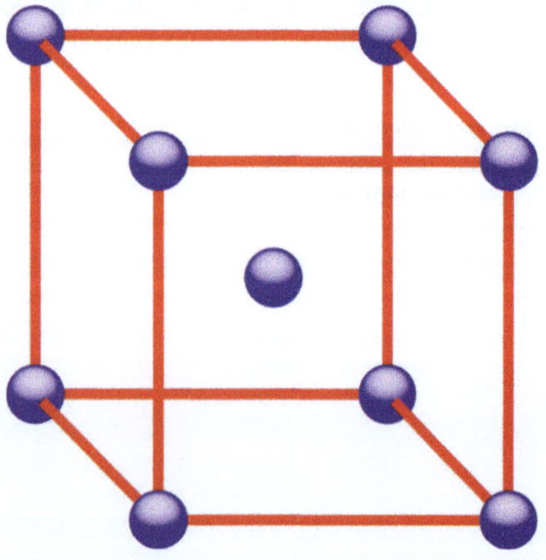

Full Symmetry of the System: Microscopic symmetry, which includes the symmetry of the individual atom in the elementary cell, is described by the crystalline classes [14,20,21].

6.2   Translational Symmetry

$$\Psi\left(r + r_m\right) = \Psi(r). \tag{6.19}$$

Let us expand $\Psi(r)$ in the Fourier series:

$$\Psi(r) = \sum_q \Psi_q \exp(iqr). \tag{6.20}$$

Then:

$$\Psi\left(r + r_m\right) = \sum_q \Psi_q \exp(iqr) \exp\left(iqr_m\right) = \Psi(r). \tag{6.21}$$

It follows from Eq. (6.21), that

$$\exp\left(iqr_m\right) = 1, \text{ or } qr_m = 2\pi p, \tag{6.22}$$

where p is an integer number.

Inverse Lattice: The notion of the inverse lattice was introduced by Gibbs [1, 2,6–8,20–24]. The group of translational symmetries for the inverse lattice $\{\tilde{G}\}$ is determined also on the three-dimensional basis of the inverse lattice vectors b_1, b_2, and b_3. An arbitrary translation vector in the reciprocal space can be represented as follows [1,2,6–8,20–24]:

$$b_S = s_1 b_1 + s_2 b_2 + s_3 b_3, \tag{6.23}$$

where s_1, s_2, s_3 are the integer numbers.

It is convenient also to introduce the vectors:

$$G = 2\pi b. \tag{6.24}$$

In terms of these vectors of the inverse lattice the wave function in crystal is expanded in series as

$$\Psi(r) = \sum_G \Psi_G \exp(iGr). \tag{6.25}$$

There is a simple orthogonality relation between the basis vectors of the direct and inverse lattice:

$$b_i a_k = \delta_{ik}. \tag{6.26}$$

This relation means, that we can represent the basis vectors of the inverse lattice in the following way:

$$b_1 = c\,[a_2a_3]\,,\ b_2 = c\,[a_3a_1]\,,\ b_3 = c\,[a_1a_2]\,. \tag{6.27}$$

Then from the orthogonality condition (6.26) we can determine the coefficient c:

$$(b_1a_1) = c\,[a_2a_3]\,a_1 = c\Omega_0 = 1, \tag{6.28}$$

where $\Omega_0 = \left[\,a_2\ a_3\,\right] a_1$ is the volume of the elementary cell (in the direct space). Hence:

$$c = \frac{1}{\Omega_0}. \tag{6.29}$$

It is possible to demonstrate that the volume of the elementary cell in the reciprocal space reads

$$[b_2b_3]\,b_1 = \frac{1}{\Omega_0}. \tag{6.30}$$

Indeed, by revealing the double vector product, we get

$$[b_2b_3] = \frac{1}{\Omega_0}\,[b_2\,[a_1a_2]] = \frac{1}{\Omega_0}\,\{(b_2a_2)\,a_1 - (b_2a_1)\,a_2\} = \frac{1}{\Omega_0}\,(b_2a_2)\,a_1 = \frac{1}{\Omega_0}a_1. \tag{6.31}$$

As a result, as we assumed:

$$[b_2b_3]\,b_1 = \frac{1}{\Omega_0}\,(a_1b_1) = \frac{1}{\Omega_0}.$$

Let us return back now to the equation on the eigenfrequencies (6.14) and introduce an additional index j for the atom number in the elementary cell [1,2]. Then it is convenient to represent the index n in Eq.'s (6.14) and (6.16) as

$$n = \boldsymbol{n},\,j, \tag{6.32}$$

where vector $\boldsymbol{n}$ corresponds to the center of gravity of the elementary cell.

Then Eq. (6.14) on the eigenfrequencies can be written as follows:

$$v^i_{\boldsymbol{n},j}\omega^2 = \sum_{\boldsymbol{n}',j'} \bar{A}^{ik}_{\boldsymbol{n}j,\boldsymbol{n}'j'} v^k_{\boldsymbol{n}',j'}. \tag{6.33}$$

It is important to stress, however, that the matrix $\bar{A}\left(\boldsymbol{n} - \boldsymbol{n}'\right)$ depends only on the relative positions of the atoms.

Further on let us expand the renormalized displacement $v^i_{n,j}$ in the Fourier series:

$$v^i_{n,j} = \sum_f v^i_j \exp\left(i\,f\,r_n\right). \tag{6.34}$$

Then Eq. (6.33) on eigenfrequencies for one Fourier component (6.33) acquires a form:

$$v^i_j \exp\left(i\,f\,r_n\right) \omega^2 = \sum_{n',j'} \bar{A}^{ik}_{nj,n'j'} v^k_{j'} \exp\left(i\,f\,r_{n'}\right). \tag{6.35}$$

By transferring the multiplier $\exp\left(i\,f\,r_n\right)$ in the right-hand side, we get

$$v^i_j \omega^2 = \sum_{n',j'} \bar{A}^{ik}_{nj,n'j'} v^k_{j'} \exp\left(i\,f\,r_{n'} - i\,f\,r_n\right). \tag{6.36}$$

Let us introduce a matrix convenient for the future calculations [1,2]:

$$c^{ik}_{jj'}(f) = \sum_{n'} \bar{A}^{ik}_{nj,n'j'} \exp\left(-i\,f\left(n - n'\right)\right), \tag{6.37}$$

where $n - n' = r_n - r_{n'}$.

Then in terms of this matrix and the Fourier components of the renormalized displacements we can represent Eq. (6.36) as

$$v^i_j \omega^2 = \sum_{j'} c^{ik}_{jj'}(f) v^k_{j'}. \tag{6.38}$$

Note, that the system of Eqs. (6.37)–(6.38) contains $3jN$ equations for the three-dimensional crystal with N elementary cells and j atoms on one cell.

Note also, that the matrix

$$c^{ik}_{jj'}(f) = \sum_{n'} \bar{A}^{ik}_{j,j'}\left(n - n'\right) \exp\left(-i\,f\left(n - n'\right)\right) = \sum_m \bar{A}^{ik}_{j,j'}(m) \exp(-i\,f\,m), \tag{6.39}$$

does not depend on $m = n - n'$. In derivation of Eq. (6.39) we utilized the fact, that since the matrix $\bar{A}\left(n - n'\right)$ is symmetric and depends only on the difference $n - n'$, then the summation on n' is equivalent to the summation on n.

Thus, the eigenvalues of the frequencies squared depend only on the wave vector. The spectral equation has the form [1,2,7]:

$$\left| c^{ik}_{jj'}(f) - \omega^2 \delta_{jj'} \delta^{ik} \right| = 0, \tag{6.40}$$

The solution of Eq. (6.40) yields the eigenfrequencies squared $\omega^2_\alpha(f)$, where $\alpha = 1 \ldots 3q$ denotes the number of the spectrum branches for each value of the wave vector f, and q-the number of atoms in the elementary cell.

Let us show, that the matrix $c_{jj'}^{ik}(f)$ is Hermitian one. To do that let us consider Hermitian conjugate (that is transposed and complex conjugate) matrix $\left(c_{j'j}^{ki}\right)^*$. According to Eq. (6.37), this matrix equals to:

$$\left(c_{j'j}^{ki}\right)^* = \sum_{n'} \left(\bar{A}_{n'j',nj}^{k,i}\right)^* \exp\left(-if\left(n-n'\right)\right). \tag{6.41}$$

But the matrix $\bar{A}_{nj,n'j'}^{i,k} \sim \left(\dfrac{\partial^2 U}{\partial R_n^i \partial R_{n'}^k}\right)_{0^-}$ is symmetric and real. That is why:

$$\left(\bar{A}_{n'j',nj}^{k,i}\right)^* = \bar{A}_{nj,n'j'}^{i,k} = \bar{A}_{j,j'}^{ik}\left(n-n'\right) \tag{6.42}$$

does not change under the substitution $i, n, j \to k, n', j'$ and sequential complex conjugation. Hence:

$$\left(c_{j'j}^{ki}\right)^* = \sum_{n'} \bar{A}_{j,j'}^{ik}\left(n-n'\right)\exp\left(-if\left(n-n'\right)\right) = \sum_{m} \bar{A}_{j,j'}^{ik}(m)\exp(-ifm) = c_{jj'}^{ik}. \tag{6.43}$$

Thus, we demonstrated, that $c_{jj'}^{ik}(f)$ is Hermitian one. The real frequencies squared correspond to it. Moreover, it does not change under the shift of the wave vector on the vector of the inverse lattice:

$$c_{jj'}^{ik}(f+G) = c_{jj'}^{ik}(f). \tag{6.44}$$

Thus, the frequencies squared also do not change under the shift of the wave vector of the inverse lattice:

$$\omega_\alpha^2(f+G) = \omega_\alpha^2(f). \tag{6.45}$$

We can say that the frequencies are the periodic functions in the reciprocal space of the inverse lattice. In other words, physically different values of the wave vectors are lying in the first elementary cell of the inverse lattice.

6.3   Two Important Remarks

1. Selection of the boundary conditions on the boundary of large cube does not influence the solution with the accuracy $\sim \frac{1}{N}$.
2. As it was shown by Dyson, the influence of the boundary conditions on the density of states expires on the distance of the order of the wavelength λ from the border (note that for long wavelengths this distance can be significant) [1,2].

Let us consider very large crystal. For this crystal from the dimensional (quasi-classical) quantization along each of the Cartesian axes [1,2,25,26]:

$$f_x L_x = 2\pi p_x, \text{ and } f_x = \frac{2\pi p_x}{L_x},$$ (6.46)

where p_x–is an integer number. Then for the three-dimensional crystal:

$$\Delta p_x = \frac{L_x}{2\pi}\Delta f_x, \text{ and } V\frac{d^3 f}{(2\pi)^3}$$ (6.47)

is the density of states.

As a result, the integral from the density of states:

$$V\int \frac{d^3 f}{(2\pi)^3} = V\frac{(2\pi)^3}{(2\pi)^3}\frac{1}{V_{\text{cell}}} = \frac{V}{V_{\text{cell}}} = N_{\text{cell}},$$ (6.48)

where $V_{\text{cell}} = \Omega_0$ equals the number of the elementary cells.

Moreover, the Cartesian components of the wave vector in the first elementary cell of the inverse lattice lie in the interval:

$$-\frac{G_x}{2} < f_x < \frac{G_x}{2}, -\frac{G_y}{2} < f_y < \frac{G_y}{2}, -\frac{G_z}{2} < f_z < \frac{G_z}{2}.$$ (6.49)

Note that $3qN$ is a total number of the degrees of freedom.

One more important property of the phonon spectrum is an invariance of the frequencies with respect to the <u>time reversal</u> $t \to -t$ [1,2] (which in fact is equivalent to the replacement of the wave vector $f \to -f$). Indeed, Eq. (6.38) on eigenfrequencies under the complex conjugation and the simultaneous replacement of the wave vector $f \to -f$ can be written as

$$v_j^{i*}(-f)\omega^2(-f) = \sum_{j'} c_{jj'}^{ik}(f)v_{j'}^{k*}(-f),$$ (6.50)

where we utilized the facts that the eigenfrequencies squared are real and the matrix $c_{jj'}^{ik}(f)$ is Hermitian one. Hence:

$$c^*(-f) = c(f).$$ (6.51)

Thus, the right-hand side of Eq. (6.50) for $v_j^{i*}(-f)$ has the same form, as the right-hand side of Eq. (6.38) for $v_j^{i}(f)$. From that we get the <u>parity of the eigenfrequencies</u> with respect to the replacement of the wave vector $f \to -f$:

$$\omega^2(-f) = \omega^2(f).$$ (6.52)

The reason for the parity of the spectrum is connected with the fact, that under the replacement $t \to -t$ or equivalent replacement $f \to -f$, the energy $E \to E$ since the energy of the wave, running from left to right equals the energy of the wave, running from right to left [1,2].

Let us return now to the initial equations of motion for the displacements $u_n^i = u_{nj}^i$:

$$m_{nj} u_{nj}^i \omega^2 = \sum_{n'j'} A_{nj,n'j'}^{i,k} u_{n'j'}^k,$$ (6.53)

where the set of the masses is the same for each elementary cell n, and thus:

$$m_{nj} = m_j.$$ (6.54)

Let us proceed again to the reciprocal space for the Fourier components of the displacements:

$$u_{n,j}^i = \sum_f u_j^i(f) \exp(i f r_n).$$ (6.55)

Then we can represent the equation of motion (6.53) in the reciprocal space for one Fourier component of the displacements $u_j^i(f)$ in the form, analogous to Eq. (6.36) for $v_j^i(f)$ [1,2]:

$$m_j u_j^i(f)\omega^2 = \sum_{n'j'} A_{j,j'}^{i,k}(n - n') u_{j'}^k(f) \exp(-i f (n - n')).$$ (6.56)

Let us consider now an important limit, when the wave vector $f \to 0$. In this limit the exponent in the right-hand side of Eq. (6.56) $\exp(-i f (n - n')) \to 1$, and:

$$m_j u_j^i(f \to 0)\omega^2 = \sum_{n'j'} A_{j,j'}^{i,k}(n - n') u_{j'}^k(f \to 0).$$ (6.57)

If we sum up left and right-hand sides of this equation by j, we get

$$\sum_j m_j u_j^i(f \to 0)\omega^2 = \sum_j \sum_{n'j'} A_{j,j'}^{i,k}(n - n') u_{j'}^k(f \to 0).$$ (6.58)

But since the force matrix is symmetric and depends only on the difference $(n - n')$, we can replace the summation by n' on the summation by n in the right-hand side of Eq. (6.58) (reversing at the same time summation by the indices j and j').

Further on for the wave vectors $f \to 0$ the displacement:

$$u_j^i = \sum_n u_{jn}^i \exp(-i f r_n) = \sum_n u_{jn}^i.$$ (6.59)

Analogously for $f \rightarrow 0$:

$$u^k_{j'} = \sum_{n'} u^k_{j'n'}.$$ (6.60)

As a result, Eq. (6.58) can be written as follows [1,2]:

$$\sum_j m_j u^i_j \omega^2 = \sum_j \sum_n m_j u^i_{jn} \omega^2 = \sum_{j'} \sum_{nj} \sum_{n'} A^{i,k}_{j,j'} \left(n - n' \right) u^k_{j'n'}.$$ (6.61)

Proceeding now to the more compact notations $n = \{n, j\}$ and $n' = \{n', j'\}$, we can represent Eq. (6.61) as

$$\sum_j m_j u^i_j \omega^2 = -\sum_n m_n \ddot{u}^i_n = \sum_n \sum_{n'} A^{ik}_{nn'} u^k_{n'},$$ (6.62)

At the same time, an equation of motion for center of gravity of the crystal yields

$$\sum_n m_n \ddot{u}^i_n = M_{\text{cryst.}} a^i_{\text{c.g.}},$$ (6.63)

where $M_{\text{cryst.}}$ is the mass of the whole crystal, $a^i_{\text{c.g.}}$ is the Cartesian projection of the acceleration for the center of gravity of the crystal.

In the absence of the external field an acceleration of the center of gravity equals zero. This means that

$$\sum_j m_j u^i_j \omega^2 = -\sum_n m_n \ddot{u}^i_n = 0.$$ (6.64)

Hence, we derive an important condition for the force matrix [1,2]:

$$\sum_n A^{ik}_{nn'} = 0.$$ (6.65)

In fact, this requirement means that the force matrix should be invariant under the translation of the whole crystal. In other words, we can say, that if all the displacements $u^i_n = $ const, then, firstly, all the accelerations $\ddot{u}^i_n = 0$ in Eq. (6.62). Moreover, if all the displacements $u^k_{n'} = $ const are equal in the right-hand side of this equation under the displacement of the whole crystal, then:

$$\sum_{n'} A^{ik}_{nn'} = 0.$$ (6.66)

But since the force matrix is symmetric, then the summation by n and by n' is equivalent, and thus, as a result, we restore the condition (6.65).

Let us emphasize that the requirement (6.64) means also that

$$\omega^2(f) \sum_j m_j u_j^i = 0 \text{ for } f \to 0. \tag{6.67}$$

The condition (6.67) is fulfilled in two cases [1,2], when:

$$(1) \ \omega^2(f) = 0 \text{ for } f \to 0, \tag{6.68}$$

$$(2) \ \sum_j m_j u_j^i = 0 \text{ for } f \to 0. \tag{6.69}$$

In the first case, when $\omega^2 = 0$ it follows from Eq. (6.58) that in the general case:

$$\sum_{n'} A_{nn'}^{ik} u_{j'}^k = 0. \tag{6.70}$$

This relation with an account of the condition (6.66) on the force matrix is fulfilled, when the Cartesian components of the displacements $u_{j'}^k = $ const are equal for all the atoms of the elementary cell. Thus, three displacements (along each of the Cartesian axes) are linearly independent for three-dimensional crystal.

Hence, $\omega^2(f = 0) = 0$ for three acoustic branches (from the total number of phonon branches which equals $3q$). Note that in general case the number of acoustic branches equals the dimension of space.

According to (6.68), the frequency of acoustic phonons goes to zero when the wave vector goes to zero, which means that the crystal, displaced as a whole, does not have an excitation energy. In other words, in acoustic oscillations all the elementary cell goes back and forth (oscillates as a whole) [1,2,5].

At the same time, in the second case $\sum_j m_j u_j^i = 0$, which means that the center of gravity of the elementary cell does not move at $f \to 0$. In this case only relative oscillations of atoms in the elementary cell emerge:

$$\omega^2(f = 0) \neq 0. \tag{6.71}$$

This condition corresponds to the optical phonon branches [1,2,5]. The total number of optical branches in the three-dimensional crystal equals to:

$$(3q - 3). \tag{6.72}$$

If the space dimension is D, then the number of optical branches yields:

$$q(D - 1). \tag{6.73}$$

It is very important to stress, that we need to have more than one atom in the elementary cell for the appearance of the optical branches [1,2,5].

Note that optical branches interact well with light.

For acoustic branches the frequency squared acquires a form for small wave vectors:

$$\omega_\alpha^2(\boldsymbol{f}) = 0 + \gamma^{ik}\left(\frac{f_i f_k}{f^2}\right) f^2 = \gamma^{ik}\left(\frac{f_i f_k}{f^2}\right) f^2, \tag{6.74}$$

where the quadratic form of the spectrum $\gamma^{ik}\left(\frac{f_i f_k}{f^2}\right) f^2$ in Eq. (6.74) satisfies the symmetry condition $\boldsymbol{f} \to -\boldsymbol{f}$.

As a result, for the acoustic spectrum a linear relation between the frequency and the wave vector takes place [1,2,5–9]:

$$\omega_\alpha(\boldsymbol{f}) = c_\alpha\left(\frac{\boldsymbol{f}}{f}\right) f, \tag{6.75}$$

where the sound velocity c_α depends in general case on the angle between the direction of the wave propagation and the crystal axes.

For the spectrum of the optical branches, we get at small wave vectors [1,2,5]:

$$\omega_\alpha^2(\boldsymbol{f}) = \omega_\alpha^2(0) + \tilde{\gamma}^{ik} f_i f_k. \tag{6.76}$$

The quadratic form in Eq. (6.76) for the optical branches can be diagonalized if the degeneracy is absent in the system.

The frequencies of the optical spectrum read

$$\omega_\alpha(\boldsymbol{f}) = \omega_\alpha(0) + \xi^{ik} f_i f_k. \tag{6.77}$$

Note that the phase velocity of the optical phonons goes to infinity [1,2,5]:

$$v_{\text{phase}} = \frac{\omega}{f} \to \infty \text{ for } f \to 0. \tag{6.78}$$

This fact allows optical phonons to interact with light [1,2,5].

At the same time, their group velocity

$$v_{\text{gr.}} \to 0 \tag{6.79}$$

goes to zero for $f \to 0$.

References

1. Y. Kagan, *Lectures on the Solid-State Theoretical Physics* (Moscow, MEPHI, 1969), 178 p. (in Russian)
2. Y. Kagan, *Lectures on the Solid-State Theoretical Physics* (Moscow, MEPHI, 1981–1982). Part I. Phonons - unpublished
3. Y. Kagan, N.V. Prokof'ev, Electron polaron effect and quantum diffusion of heavy particle in metals. Sov. Phys. JETP **63**, 1276 (1986)

4. Y. Kagan, N.V. Prokof'ev, Quantum tunneling in metals and the heavy fermion problems. Sov. Phys. JETP **66**, 211 (1987)
5. J.M. Ziman, *Electrons and Phonons: The Theory of Transport Phenomena in Solids* (Clarendon Press, Oxford, 1960)
6. N.W. Ashcroft, N.D. Mermin, *Solid State Physics* (Holt, Rinehart and Winston, 1976)
7. O. Madelung, *Introduction to Solid-State Theory*, Springer Series in Solid-State Sciences (1995)
8. C. Kittel, *Quantum Theory of Solids* (Wiley, New York-London, 1963)
9. D. Pines, *Elementary Excitations in Solids* (W.A. Benjamin Inc., New York-Amsterdam, 1963)
10. S.V. Vonsovskii, M.I. Kaznelson, *Quantum Solid-State Physics*, Springer Series in Solid-State Sciences (Berlin-Heidelberg- New York, 1989)
11. A. Maradudin, E. Montroll, G. Weiss, I. Ipatova, *Theory of Lattice Dynamics in the Harmonic Approximation*, Solid State Physics, vol. Suppl. 3, 2nd edn. (Academic Press, New York, 1971)
12. L.D. Landau, E.M. Lifshitz, *Mechanics, Course of Theoretical Physics*, vol. 1 (Butterworth-Heinemann, 1976)
13. W.A. Harrison, *Solid State Theory* (McGraw-Hill Book Company, New-York-London-Toronto, 1970)
14. L.D. Landau, E.M. Lifshitz, *Quantum Mechanics: Non-Relativistic Theory* (Pergamon Press, New York, 1977)
15. S. Bhagavantam, T. Venkatarayudu, *Theory of Groups and its Application to Physical Problems* (Andhra University, Waltair, India, 1951)
16. H. Jones, *The Theory of Brillouin Zones and electronic States in Crystals* (North-Holland Publishing Company, Amsterdam, 1962)
17. W. Borchardt-Ott, Crystallography (Springer, Berlin, 1993)
18. W. Mossa, *Crystal Structure Determination* (Springer, Berlin, 1999)
19. F. Hoffmann, *Introduction to Crystallography* (Springer, Berlin, 2020)
20. H.S.M. Coexter, W.O.J. Moser, *Generators and Relations for Discrete Groups* (Springer, New-York, 1980)
21. B.K. Veinstein, *Modern Crystallography, Fundamentals of Crystal Symmetry and Methods of Structural Crystallography*, vol. 1 (Springer, Berlin, 1984)
22. W.A. Harrison, *Electronic Structure and the Properties of Solids* (W.H. Freeman and Company, San Francisco, 1980)
23. A. Fetter, J. Walecka, *Theoretical Mechanics of Particles and Continua* (Dover Books on Physics, 2003)
24. G.D. Mahan, *Many-Particle Physics* (Springer, New York, 1981)
25. A.A. Abrikosov, *Fundamentals of the Theory of Metals* (Elsevier Science Publishers, North-Holland, Amsterdam, 1988)
26. I.M. Lifshitz, M.Y. Azbel', M.I. Kaganov, *Electron Theory of Metals* (Consultants Bureau, New York, 1973)

Lecture 7. Density of the Phonon States. Debye Approximation

7

Abstract

In this Lecture, we introduce the notion of the Brillouin zones of the inverse lattice and consider the behavior of the group velocity on the zone's boundary. We discuss the density of the phonon states in general case and in the Debye approximation. We analyze Van Hove singularities in the phonon spectrum for crystals of different spatial dimensions. We consider one-dimensional chain with two atoms in the elementary cell and illustrate an emergence of acoustic and optical branches of the phonon spectrum in it.

Let us start with the short summary of the previous Lecture [1,2]:

1. At $f \to 0$ for three acoustic branches the phonon spectrum is linear: $\omega_\alpha(f) \sim f$.
2. If we have more than one atom in the elementary cell, then (3q-3) optical branches appear, for which the phonon spectrum [1–3]:

$$\omega_\alpha(f \to 0) \to \omega_\alpha(0) \neq 0 \tag{7.1}$$

3. As a result of the translational symmetry, the phonon spectrum is described by the periodic function [1–4]:

$$\omega_\alpha(f + K) = \omega_\alpha(f), \tag{7.2}$$

where K is the vector of the inverse lattice.
4. The phonon frequency is an even function of the wave vector:

$$\omega_\alpha(f) = \omega_\alpha(-f). \tag{7.3}$$

5. For the set of the wave vectors, which can be obtained from each other by the symmetry transformation of a crystal $f_i = \hat{L} f_0$ (the Wigner umbrella [1,2]), we have the same spectrum

© The Author(s), under exclusive license to Springer Nature Switzerland AG 2026

107

M. Kagan, *Lecture Notes of Professor Yuri Kagan in Theoretical Solid-State Physics*, Lecture Notes in Physics 1048, https://doi.org/10.1007/978-3-032-14621-2_7

$$\omega_\alpha \left(\boldsymbol{f}_i = \hat{L} \boldsymbol{f}_0 \right) = \omega_\alpha \left(\boldsymbol{f}_0 \right). \tag{7.4}$$

Thus, not the whole volume of the phase space of one inverse elementary cell is irreducible. For the cube, the irreducible volume equals to $\frac{1}{48}$ of the total volume of one cell of the inverse lattice [1,2].

7.1   Inverse Lattice

In Fig. 7.1, we show the first and second Brillouin zones [1–5] for the square lattice. Cartesian projections of the wave vectors of the first Brillouin zone lie in the interval:

$$-\frac{K_x}{2} < f_{1x} < \frac{K_x}{2}, \quad -\frac{K_y}{2} < f_{1y} < \frac{K_y}{2}. \tag{7.5}$$

(In this Lecture and further on we will use the notation K for the vector of the inverse lattice).

Since first and second Brillouin zones are equivalent [1–5], the frequencies of the phonon spectrum coincide in them (the phonon spectrum is given by the periodic

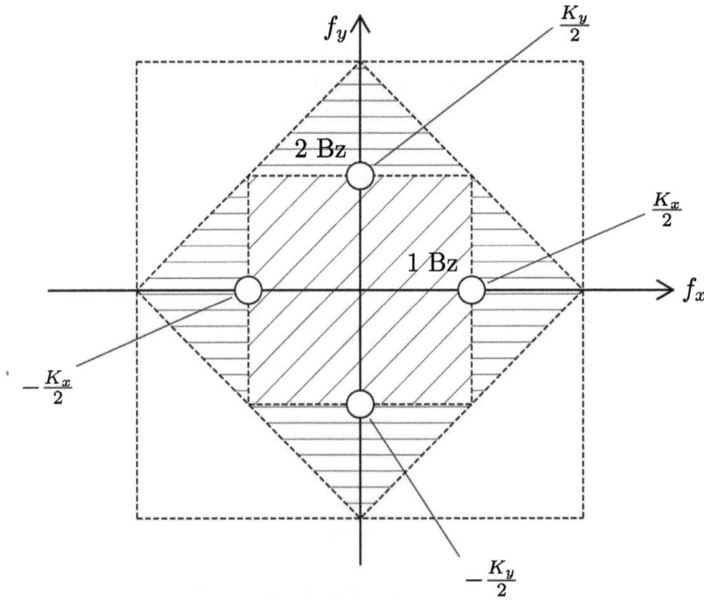

(1, 1) - first Brillouin zone for
square lattice

Fig. 7.1 The first and the second Brillouin zones (Bz) for the square lattice are shown in the figure. Cartesian projections of the wave vectors of the first Brillouin zone lie in the interval given by Eq. 7.5 [1]

function of the wave vector):

$$\omega_\alpha\left(\boldsymbol{f}_1\right) = \omega_\alpha\left(\boldsymbol{f}_2\right). \tag{7.6}$$

From one hand, on the left and right boundaries of the first and second Brillouin zones at $f_x = \frac{K_x}{2}$ and $f_x = -\frac{K_x}{2}$ the group velocities $v_{\mathrm{gr}.x} = \frac{d\omega_\alpha}{df_x}$ should be different in sign:

$$v_{\mathrm{gr}.x\,\mathrm{left}} = -v_{\mathrm{gr}.x\,\mathrm{right}}, \tag{7.7}$$

since according to Eq. (7.3), the frequencies are even functions of the wave vector, while the wave vector itself changes sign. From the other hand, however, the wave vectors f_x on the left and right boundaries differ on the inverse lattice vector K_x. Hence, the group velocities should coincide according to the periodicity condition of the phonon spectrum (7.2):

$$\left(v_{\mathrm{gr}.x\,\mathrm{left}} = \frac{d\omega_\alpha(\boldsymbol{f})}{df_x}\right) = \left(v_{\mathrm{gr}.x\,\mathrm{right}} = \frac{d\omega_\alpha\left(\boldsymbol{f}+\boldsymbol{K}_x\right)}{d\left(f_x+K_x\right)} = \frac{d\omega_\alpha(\boldsymbol{f})}{df_x}\right). \tag{7.8}$$

Thus, we get very important relation, according to which the normal component of the group velocity goes to zero on the boundary of the Brillouin zone [1,2]:

$$v_{\mathrm{gr}.\perp} = \left(\frac{\partial\omega}{\partial\boldsymbol{f}}\right)_{\perp\,\mathrm{Bril.zone}} = 0. \tag{7.9}$$

We can say that the signal does not leave the Brillouin zone [1,2].

At the same time, the tangential component of the group velocity does not go to zero on the boundary of the Brillouin zone [1,2].

7.2   Density of States for the Phonon Spectrum

In Lecture 2, we discussed already the density of states $g(\varepsilon)$ for the ideal Bose gas. The spectral density of states (depending on the frequency) for one branch of the phonon spectrum in the three-dimensional crystal reads

$$g_\alpha(\omega) = \frac{\Omega_0}{(2\pi)^3} \int d^3 f\, \delta\left(\omega - \omega_\alpha(\boldsymbol{f})\right), \tag{7.10}$$

where Ω_0 is the volume of the elementary cell in the real space.

The spectral density of states is normalized on unity [1,2]:

$$\int g_\alpha(\omega)d\omega = \frac{\Omega_0}{(2\pi)^3} \int d^3 f = \frac{\Omega_0}{(2\pi)^3} \frac{(2\pi)^3}{\Omega_0} = 1. \tag{7.11}$$

In the left-hand side of Eq. (7.10) under the integral sign enters the quantity $g_\alpha(\omega)d\omega$. This quantity corresponds to the density of the phonon frequencies in the interval: $\omega \div \omega + d\omega$.

At the same time, the total density of the phonon frequencies (the density of states of all the branches of the phonon spectrum) equals to

$$\sum_\alpha g_\alpha(\omega) \cdot \frac{1}{3q}.$$

(7.12)

7.3 Analytical Properties of the Spectral Density of States $g_\alpha(\omega)$

At small frequencies, only acoustic branches contribute to the total density in Eq. (7.11), since for optical branches the frequency $\omega_\alpha \sim \omega_\alpha(0)$ is not small.

For acoustic branches, the phonon spectrum is linear: $\omega_\alpha = c_\alpha f$. Let us introduce dimensionless variable $x = c_\alpha f$. Then the measure of the integration in the spherical coordinates for the spectral density $g_\alpha(\omega)$ in Eq. (7.10):

$$d^3 f = f^2 df d\Omega = x^2 dx \frac{d\Omega}{c_\alpha^3(f)},$$

(7.13)

where $d\Omega$ is the element of the solid angle, and the sound velocity $c_\alpha\left(\frac{f}{f}\right)$ is the function of the angles. As a result, for the three-dimensional crystal:

$$g_\alpha(\omega) = \frac{\Omega_0}{(2\pi)^3} \int \frac{d\Omega}{c_\alpha^3(f)} \int x^2 dx \delta(\omega - x) = \frac{\Omega_0}{(2\pi)^3} \frac{4\pi}{\bar{c}_\alpha^3} \omega^2 = B\omega^2,$$

(7.14)

where the constant:

$$B = \frac{\Omega_0}{(2\pi)^3} \frac{4\pi}{\bar{c}_\alpha^3},$$

(7.15)

and $\bar{c}_\alpha$ is the sound velocity averaged on the angles.

In the general case, for the D-dimensional space [1,2]:

$$g_\alpha(\omega) \sim \omega^{D-1}.$$

(7.16)

For the two-dimensional space:

$$g_\alpha(\omega) \sim \omega.$$

(7.17)

For the one-dimensional space:

$$g_\alpha(\omega) = \text{const}.$$

(7.18)

7.4   Debye Approximation

In Debye approximation [7], we take into account only acoustic branches (as in the crystal with one atom on the elementary cell), and moreover, we assume that for all the wave vectors: $\omega \sim f$. Then in the 3D crystal, in Debye approximation, the density of states for all the frequencies:

$$g(\omega) = B\omega^2. \tag{7.19}$$

In this approximation, we introduce only the Debye frequency ω_D as a limiting (maximally possible) frequency of the phonon spectrum. Then, from the normalization condition on unity, we get for the spectral density

$$B \int_0^{\omega_D} \omega^2 d\omega = \frac{B}{3}\omega_D^3 = 1. \tag{7.20}$$

Hence the normalization constant:

$$B = \frac{3}{\omega_D^3}, \tag{7.21}$$

and the spectral density equals to

$$g(\omega) = \frac{3\omega^2}{\omega_D^3}. \tag{7.22}$$

Comparing Eq.s (7.15) and (7.21), we see that, for the normalization constant:

$$B = \frac{3}{\omega_D^3} = \frac{\Omega_0}{(2\pi)^3}\frac{4\pi}{\bar{c}_\alpha^3}. \tag{7.23}$$

From Eq. (7.23), we can determine the Debye frequency:

$$\omega_D^3 = \frac{3}{B} = \frac{3(2\pi)^3\bar{c}_\alpha^3}{4\pi\Omega_0} = \frac{6\pi^2\bar{c}_\alpha^3}{\Omega_0}, \text{ or } \omega_D = \frac{\sqrt[3]{6\pi^2}\bar{c}_\alpha}{\Omega_0^{1/3}} \approx \frac{3.9\bar{c}_\alpha}{\Omega_0^{1/3}}. \tag{7.24}$$

With an accuracy of the order of one, we can substitute Ω_0 in Eq. (7.24) by the volume of the spherical elementary cell:

$$\Omega_0 = \frac{4\pi a^3}{3}, \text{ and } \Omega_0^{1/3} = \sqrt[3]{\frac{4\pi}{3}}a \approx 1.6a, \tag{7.25}$$

where a is the size of the elementary cell.

Then the Debye frequency reads

$$\omega_D \approx 2.4\frac{\bar{c}_\alpha}{a}. \tag{7.26}$$

Note that the typical values of the Debye frequencies are of the order of

$$\omega_D \sim \frac{10^5}{10^{-8}} \sim 10^{13} \text{ Hz.} \qquad (7.27)$$

At the same time, the Debye temperature:

$$\theta_D = \frac{\hbar \omega_D}{k_B} \sim 10^2 \text{ K.} \qquad (7.28)$$

<u>Comment 1.</u> The Debye model works well at low temperatures $T \ll \theta_D$. For $T \geq \theta_D$, the Debye model can be used as an approximation. But, since in all the models, the density of states is normalized on unity ($\int g(\omega)d\omega = 1$), the Debye model can serve as a reasonable approximation even in the region of large frequencies $\frac{\hbar \omega_{\text{typ.}}}{k_B} \sim T \geq \theta_D$ (the normalization condition saves us). Usually, in crystals the maximal frequency (the upper limit of the integration in the normalization condition for the density of states $\int_0^{\omega_{\max}} g(\omega)d\omega = 1$) by order of magnitude equals to $\frac{\hbar \omega_{\max}}{k_B} \sim \theta_D \sim 10^2$ K. [1,2]

<u>Comment 2.</u> In principle, we can satisfy the normalization condition even in the model, where the maximal frequency $\omega_{\max} \to \infty$. However, this situation is impossible, since we have a finite coupling constant $c_{jj'}^{ik}(\boldsymbol{f})$ in the model, and thus the maximal frequency $\omega_{\max}$ in the system of Eq. (6.40) is also finite [1,2] as we can see from

$$\left| c_{jj'}^{ik}(\boldsymbol{f}) - \omega^2 \delta_{jj'} \delta^{ik} \right| = 0.$$

<u>Intermediate conclusion.</u> In all the realistic models (for example, in the Debye model), we have its own value of the maximal frequency $\omega_{\max}$, which enters in the normalization condition for the density of states as an upper limit of the integration. Generally speaking, this value can differ from the real value for the maximal frequency of the phonon spectrum in the real crystal.

7.5   Real Density of States

Real density of states in the three-dimensional crystal with several atoms on the elementary cell (and optical branches of the spectrum) is presented in Fig. 7.2.

Close to the maximal frequency ω_0, the phonon spectrum for the optical branches can be expanded in series as follows [1–3]:

$$\omega_\alpha(\boldsymbol{f}) = \omega_0 - \gamma^{ik}(f_i - f_{0i})(f_k - f_{0k}), \qquad (7.29)$$

where the maximal frequency $\omega_0 = \omega_\alpha(\boldsymbol{f}_0)$ is obtained for the values of the wave vector $\boldsymbol{f} = \boldsymbol{f}_0$. Note that it is the frequency ω (and not the density of states $g(\omega)$), which has maximum as a function of the wave vector $\boldsymbol{f}$ at $\boldsymbol{f} = \boldsymbol{f}_0$.

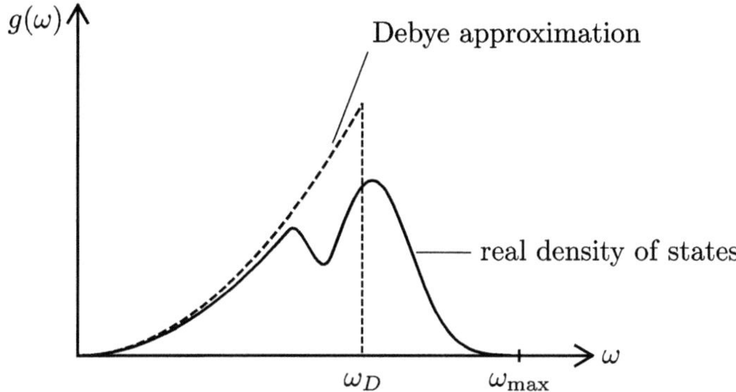

Fig. 7.2 Real density of states in the three-dimensional crystal with several atoms on the elementary cell (the solid line). The dashed line in the Figure corresponds to the density of states in the Debye approximation [1]

The spectrum in the main axes acquires a form:

$$\omega_\alpha(f) = \omega_0 - \sum_{i=1}^{3} \gamma^i \left(\tilde{f}_i \right)^2, \tag{7.30}$$

where $\tilde{f}_i = f_i - f_{0i}$.

Since we can place the origin in the point f_0, the measure of the integration does not change in novel coordinates $\tilde{f}_i$:

$$d^3 f \to d\tilde{f}_1 d\tilde{f}_2 d\tilde{f}_3. \tag{7.31}$$

Let us introduce the novel variables:

$$\sqrt{\gamma_i}\,\tilde{f}_i = x_i. \tag{7.32}$$

Then the density of states in terms of novel variable reads

$$\begin{aligned}
g_\alpha(\omega) &= \frac{\Omega_0}{(2\pi)^3} \frac{1}{\sqrt{\gamma_1\gamma_2\gamma_3}} \int dx_1 dx_2 dx_3 \delta\left(\omega - \omega_0 + x_1^2 + x_2^2 + x_3^2 \right) \\
&= \frac{\Omega_0}{(2\pi)^3} \frac{1}{\sqrt{\gamma_1\gamma_2\gamma_3}} 4\pi \int dx\, x^2 \delta\left(\omega - \omega_0 + x^2 \right),
\end{aligned} \tag{7.33}$$

where $x^2 = x_1^2 + x_2^2 + x_3^2$.

Further on, let us introduce the variable: $u = x^2$. Then:

$$2x dx = du, \text{ and } dx = \frac{du}{2x} = \frac{du}{2\sqrt{u}}. \tag{7.34}$$

As a result, the density of states [1,2]

$$g_\alpha(\omega) = \frac{\Omega_0}{(2\pi)^3} \frac{1}{\sqrt{\gamma_1\gamma_2\gamma_3}} 2\pi \int du\sqrt{u}\,\delta\left(\omega - \omega_0 + u\right)$$

$$= \frac{\Omega_0}{(2\pi)^3} \frac{1}{\sqrt{\gamma_1\gamma_2\gamma_3}} 2\pi\sqrt{\omega_0 - \omega}\,\theta\left(\omega_0 - \omega\right) \sim \sqrt{\omega_0 - \omega}\,\theta\left(\omega_0 - \omega\right) \qquad (7.35)$$

is described by the analytical function, which is finite and equals to zero at the frequency: $\omega = \omega_0$. At the same time, all the derivatives of this function are divergent (go to infinity) at $\omega = \omega_0$ (see Fig. 7.3a).

Note that, in the two-dimensional case, the density of states [1,2] for $\omega \to \omega_0$:

$$g_\alpha(\omega) = \text{const}\,\theta\left(\omega_0 - \omega\right). \qquad (7.36)$$

For the frequency $\omega \to \omega_0 - 0$, the density of states $g_\alpha(\omega) = \text{const}$, and at the same time $g_\alpha(\omega) = 0$ for $\omega > \omega_0$ (see Fig. 7.3b).

Finally, in the one-dimensional case, the density of states has an integrable singularity [1,2] for $\omega = \omega_0$ (see Fig. 7.3c)

$$g_\alpha(\omega) \sim \frac{1}{\sqrt{\omega_0 - \omega}}\theta\left(\omega_0 - \omega\right). \qquad (7.37)$$

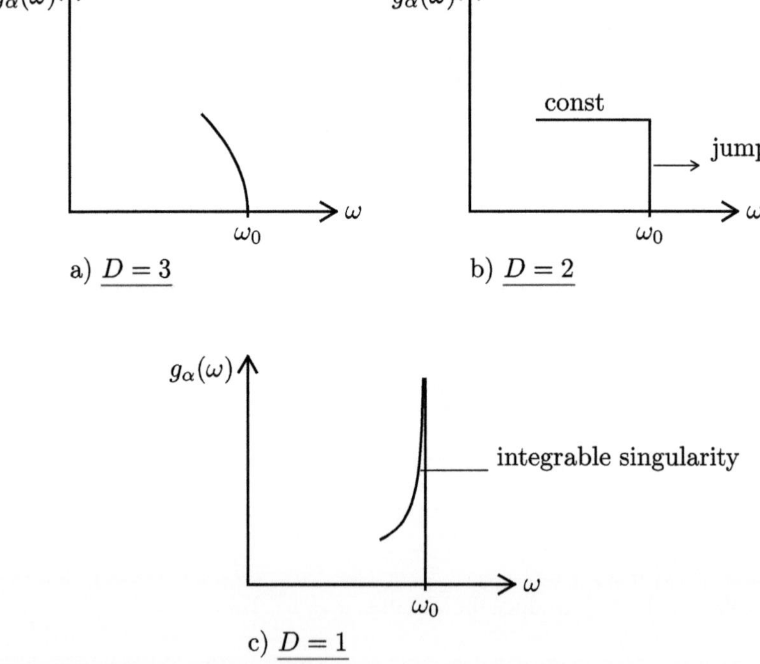

Fig. 7.3 Density of states at the frequency $\omega \to \omega_0$ for **a** three-dimensional crystal **b** for two-dimensional crystal **c** for one-dimensional crystal [1]

In the general case, the density of states of the D-dimensional crystal for $\omega \to \omega_0$:

$$g_\alpha(\omega) \sim (\omega_0 - \omega)^{\frac{D-2}{2}} \, \theta\,(\omega_0 - \omega)\,. \tag{7.38}$$

7.6   Van Hove Singularities

There is a theorem in mathematical analysis, according to which, if the analytical function is a periodic one in the three-dimensional space, then, besides minimums and maximums, it has the saddle points of the two kinds as well, namely: (max, max, min) and (max, min, min) [1,2]. The saddle points correspond to the Van Hove singularities [1,2,8,9]. In the three-dimensional crystal, for the expansion in the main axes of the type of (7.30) for the phonon spectrum close to Van Hove singularities, two coefficients γ^i are positive, while the third one is negative, or vice versa.

The saddle point (min–max) also exists in the two-dimensional case for the periodic function [1,2,8–11]. It resembles the mountain pass (see Fig. 7.4). Close to the saddle point the phonon spectrum acquires a form:

$$w_\alpha(f) = \omega_0 - |\gamma_1|\,\tilde{f}_1^2 + |\gamma_2|\,\tilde{f}_2^2\,. \tag{7.39}$$

Then the density of states for the 2D crystal yields

$$g_\alpha(\omega) = \frac{S_0}{(2\pi)^2}\,\frac{1}{\sqrt{|\gamma_{1||}\gamma_2|}}\int dx_1 dx_2 \delta\left(\omega - \omega_0 + x_1^2 - x_2^2\right), \tag{7.40}$$

where S_0 is the area of the elementary cell in 2D.

Further, on let us introduce the variable $u = x_2^2$, and correspondingly $dx_2 = \frac{du}{2\sqrt{u}}$.

As a result:

$$g_\alpha(\omega) = \frac{S_0}{(2\pi)^2}\,\frac{1}{\sqrt{|\gamma_{1||}\gamma_2|}}\,\frac{1}{2}\int dx_1\,\frac{1}{\sqrt{\omega - \omega_0 + x_1^2}}\,\theta\left(\omega - \omega_0 + x_1^2\right)$$

$$= \frac{S_0}{(2\pi)^2}\,\frac{1}{\sqrt{|\gamma_{1|}|\gamma_2\,|}}\,\frac{1}{2}I\,, \tag{7.41}$$

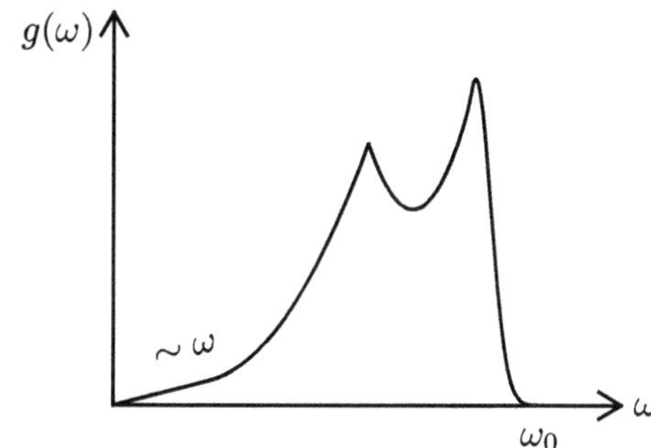

Fig. 7.4 The saddle point of the type of (min–max) in the two-dimensional case resembles the mountain pass. Van Hove singularity in the 2D crystal has a logarithmic character [1]

where for $\omega > \omega_0$ the integral I reads

$$I = \ln \frac{x_{1\,\text{max}} + \sqrt{\omega - \omega_0 + x_{1\,\text{max}}^2}}{\sqrt{\omega - \omega_0}}. \tag{7.42}$$

Note that in the expression (7.42) $x_{1\text{max}}$ is an upper limit of the integration. At the same time, the lower limit of the integration equals to zero. Close to the singularity $x_{1\text{max}}^2 \gg (\omega - \omega_0)$, and the integral:

$$I \approx \ln \frac{2x_{1\,\text{max}}}{\sqrt{\omega - \omega_0}}. \tag{7.43}$$

If $\omega < \omega_0$, then the integral in Eq. (7.41) can be calculated in the limits from $\sqrt{\omega_0 - \omega}$ till $x_{1\text{max}}$, and for $x_{1\text{max}}^2 \gg (\omega_0 - \omega)$ equals to

$$I \approx \ln \frac{2x_{1\,\text{max}}}{\sqrt{\omega_0 - \omega}}. \tag{7.44}$$

Finally, the density of states, close to the saddle-point singularity for $\omega \to \omega_0$, is given by (see Fig. 7.4)

$$g_\alpha(\omega) \approx \frac{S_0}{(2\pi)^2} \frac{1}{\sqrt{|\gamma_1||\gamma_2|}} \frac{1}{2} \ln \frac{2x_{1\,\text{max}}}{\sqrt{|\omega - \omega_0|}} \propto \ln \frac{1}{|\omega - \omega_0|}. \tag{7.45}$$

We can say that the Van Hove singularity in the two-dimensional systems always has a logarithmic character [1,2].

7.7   Van Hove Singularity in the Three-Dimensional Case

For the three-dimensional crystal the saddle point acquires a form (max-max-min) [1,2]. The phonon spectrum close to the saddle point reads in the main axes:

$$\omega_\alpha(f) = \omega_0 - |\gamma_1|\,\tilde{f}_1^2 - |\gamma_2|\,\tilde{f}_2^2 + |\gamma_3|\,\tilde{f}_3^2. \tag{7.46}$$

The density of states after the introduction of the novel variables $\sqrt{|\gamma_i|}\tilde{f}_i = x_i$ is defined by the integral:

$$g_\alpha(\omega) = \frac{\Omega_0}{(2\pi)^3} \frac{1}{\sqrt{|\gamma_1||\gamma_2||\gamma_3|}} \int dx_1 dx_2 dx_3 \delta\left(\omega - \omega_0 + x_1^2 + x_2^2 - x_3^2\right). \tag{7.47}$$

Let us calculate the integral I in Eq. (7.47) in cylindrical coordinates:

$$x_1^2 + x_2^2 = r^2, \quad x_3 = z. \tag{7.48}$$

Then the measure of the integration:

$$dx_1 dx_2 dx_3 = 2\pi r dr dz, \tag{7.49}$$

and the density of states equals to

$$
\begin{aligned}
g_\alpha(\omega) &= \frac{\Omega_0}{(2\pi)^3} \frac{1}{\sqrt{|\gamma_1||\gamma_2||\gamma_3|}} \int_0^\infty 2\pi r dr \int_{-\infty}^{+\infty} dz \delta\left(\omega - \omega_0 + r^2 - z^2\right) \\
&= \frac{\Omega_0}{(2\pi)^3} \frac{\pi}{\sqrt{|\gamma_1||\gamma_2||\gamma_3|}} \int_0^\infty dr^2 \int_0^{+\infty} dz^2 \frac{1}{\sqrt{z^2}} \delta\left(\omega - \omega_0 + r^2 - z^2\right) \\
&= \frac{\Omega_0}{(2\pi)^3} \frac{\pi}{\sqrt{|\gamma_1||\gamma_2||\gamma_3|}} \int_0^\infty dr^2 \frac{1}{\sqrt{\omega - \omega_0 + r^2}} \theta\left(\omega - \omega_0 + r^2\right).
\end{aligned} \tag{7.50}
$$

In fact, at the expense of the θ function we have to satisfy the condition:

$$r^2 > \omega_0 - \omega. \tag{7.51}$$

Note that:

(1) for $\omega_0 - \omega < 0$: the variable r^2 runs over all the values from 0 till ∞. The density of states in this case yields

$$
\begin{aligned}
g_\alpha(\omega) &= \frac{\Omega_0}{(2\pi)^3} \frac{\pi}{\sqrt{|\gamma_1||\gamma_2||\gamma_3|}} \int_0^\infty dr^2 \frac{1}{\sqrt{\omega - \omega_0 + r^2}} \\
&= \frac{\Omega_0}{(2\pi)^3} \frac{2\pi}{\sqrt{|\gamma_1||\gamma_2||\gamma_3|}} \sqrt{\omega - \omega_0 + r^2}.
\end{aligned} \tag{7.52}
$$

However, in the real crystal we have maximal (large but final) value of $r_{max} \gg \omega - \omega_0$, and hence [1,2]:

$$
\begin{aligned}
g_\alpha(\omega) &= \frac{\Omega_0}{(2\pi)^3} \frac{2\pi}{\sqrt{|\gamma_1||\gamma_2||\gamma_3|}} \sqrt{\omega - \omega_0 + r^2} \Big|_0^{r_{max}} \\
&\approx \frac{\Omega_0}{(2\pi)^3} \frac{2\pi}{\sqrt{|\gamma_1||\gamma_2||\gamma_3|}} \left(r_{max} - \sqrt{\omega - \omega_0}\right) \\
&= \frac{\Omega_0}{(2\pi)^3} \frac{2\pi}{\sqrt{|\gamma_1||\gamma_2||\gamma_3|}} \left(\text{const} - \sqrt{\omega - \omega_0}\right).
\end{aligned} \tag{7.53}
$$

Thus, for $\omega_0 \to \omega - 0$, the density of states itself is finite, but at the same time, all the derivatives from the density of states are divergent at $\omega = \omega_0$.

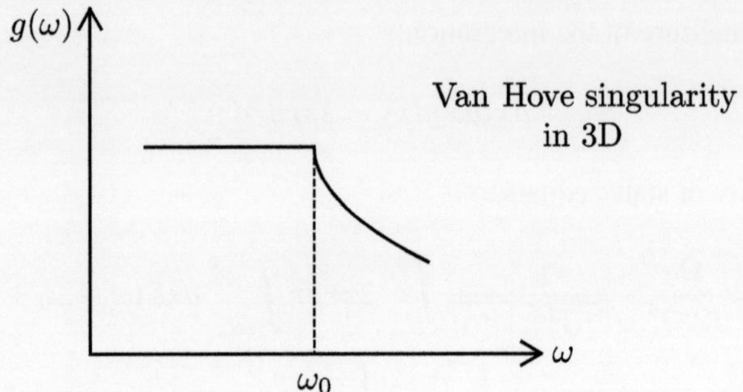

Fig. 7.5 Van Hove singularity in the 3D crystal [1]

(2) For $\omega_0 - \omega > 0$, the variable r^2 spans the interval from $\omega_0 - \omega$ till $r_{\max}$. The density of states in this case [1,2]:

$$g_\alpha(\omega) = \frac{\Omega_0}{(2\pi)^3} \frac{\pi}{\sqrt{|\gamma_1||\gamma_2||\gamma_3|}} \int_{\omega_0-\omega}^{r_{\max}} dr^2 \frac{1}{\sqrt{\omega - \omega_0 + r^2}}$$

$$\frac{\Omega_0}{(2\pi)^3} \frac{\pi}{\sqrt{|\gamma_1||\gamma_2||\gamma_3|}} \sqrt{\omega - \omega_0 + r^2}\,\Big|_{\omega_0-\omega}^{r_{\max}} = \frac{\Omega_0}{(2\pi)^3} \frac{\pi}{\sqrt{|\gamma_1||\gamma_2||\gamma_3|}} r_{\max}$$

$$= \frac{\Omega_0}{(2\pi)^3} \frac{2\pi}{\sqrt{|\gamma_1||\gamma_2||\gamma_3|}}\, \text{const}. \tag{7.54}$$

As a result, Van Hove singularity in the 3D crystal acquires a form illustrated in Fig. 7.5.

7.8   Useful Expression for the Density of States

It is convenient to write the measure of the integration $d^3 f$ for the density of states in the three-dimensional system via the element of the surface dS_ω, which corresponds to the constant frequency ω in the momentum space (see Fig. 7.6), and the absolute

Fig. 7.6 The surface element dS_ω, which corresponds to the constant frequency ω in the momentum space [1]

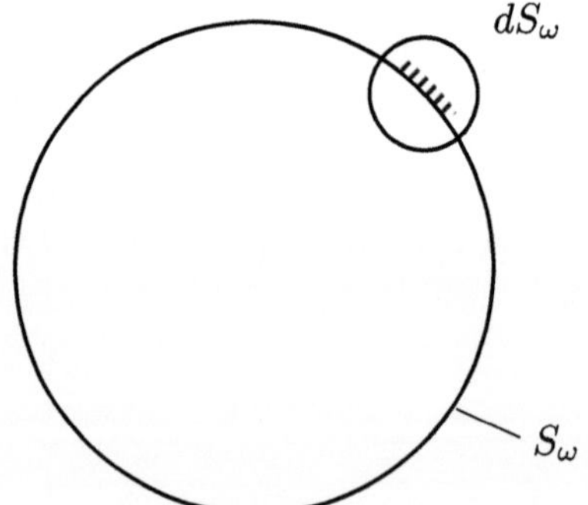

value of the component of the group velocity, normal to the surface [1,2]:

$$|v_\perp|_{\text{gr.}} = \frac{\partial \omega}{\partial f_\perp} :$$

$$d^3 f = dS_\omega \frac{df_\perp}{d\omega} d\omega = \frac{dS_\omega}{|v_\perp|_{\text{gr.}}} d\omega. \tag{7.55}$$

As a result, the density of states reads [1,2]

$$g_\alpha(\omega) = \frac{\Omega_0}{(2\pi)^3} \int d^3 f \delta\left(\omega - \omega_\alpha(f)\right) = \frac{\Omega_0}{(2\pi)^3} \int \frac{dS_\omega}{|v_\perp|_{\text{gr.}}} d\omega \delta\left(\omega - \omega_\alpha(f)\right)$$

$$= \frac{\Omega_0}{(2\pi)^3} \frac{dS_\omega}{|v_\perp|_{\text{gr.}}}. \tag{7.56}$$

It is important to note that the normal component of the group velocity in the extremal point:

$$|v_\perp|_{\text{gr.}} = \frac{\partial \omega}{\partial f_\perp} = 0. \tag{7.57}$$

Correspondingly, the density of states in the general case can be represented as a sum of regular and singular contributions:

$$g(\omega) = g_{\text{reg.}}(\omega) + g_{\text{sing.}}(\omega). \tag{7.58}$$

Moreover, an expansion of the eigenfrequencies $\omega_\alpha(f)$ in series on the wave vector f is valid for the singular contribution, yielding

$$\omega_\alpha(f) = \omega_0 - \gamma^{ik} f_i f_k$$

close to the singular points.

7.9 One-Dimensional Chain with Two Atoms in the Elementary Cell

Let us consider the one-dimensional chain with two atoms in the elementary cell (see [1–3] and Fig. 7.7). Let the size of the cell equals to a. Correspondingly, the distance between the atoms 1 and 2 in the cell equals to $\frac{a}{2}$. Let us consider only onsite and the nearest neighbors interaction. Then the spectral Eq. (6.40)

$$\left| c_{jj'}^{ik}(f) - \omega^2 \delta_{jj'} \delta^{ik} \right| = 0$$

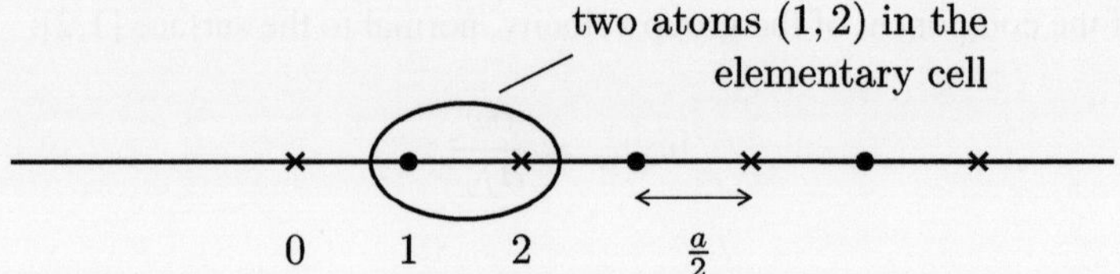

Fig. 7.7 One-dimensional chain with two atoms $(1, 2)$ in the elementary cell. The size of the cell is a, the distance between the atoms 1 and 2 in the cell equals to $\frac{a}{2}$ [1]

is simplified and acquires the form [1,2]:

$$\begin{vmatrix} c_{11}^{xx}(f) - \omega^2 & c_{12}^{xx}(f) \\ c_{21}^{xx}(f) & c_{22}^{xx}(f) - \omega^2 \end{vmatrix} = 0. \tag{7.59}$$

Explicitly writing an expression for the determinant in (7.59), we get the biquadratic equation [1,2]:

$$\omega^4 - \omega^2 (c_{11} + c_{22}) + c_{11}c_{22} - c_{12}c_{21} = 0. \tag{7.60}$$

Let us return back now to the initial force matrix $A_{jj'}^{xx} \left(\boldsymbol{m} = \boldsymbol{n} - \boldsymbol{n}' \right)$ and utilize the fact that according to Eq. (6.66), for the homogeneous displacement of a crystal as a whole, the sum:

$$\sum_{\boldsymbol{m} j'} A_{jj'}^{ik}(\boldsymbol{m}) = 0.$$

Let us put $j = 1$. Then, for the one-dimensional chain in Fig. 7.7 with the interaction on the same and neighboring sites, only the components of the force matrix $A_{11}^{xx}(0)$, $A_{12}^{xx}(0)$ and $A_{12}^{xx}(-a)$ differ from zero, where a is the size of the elementary cell. As a result, the sum in Eq. (6.66) reduces to

$$A_{11}^{xx}(0) + A_{12}^{xx}(0) + A_{12}^{xx}(-a) = 0. \tag{7.61}$$

Let us assume that diagonal interaction constants on the sites 1 and 2 of the same cell equal to each other:

$$A_{11}^{xx}(0) = A_{22}^{xx}(0) = 2\gamma. \tag{7.62}$$

Non-diagonal interaction constants in the neighboring cells should be also equal

$$A_{12}^{xx}(0) = A_{12}^{xx}(-a).$$

As a result, from the condition (7.61), we obtain

$$A_{12}^{xx}(0) = A_{12}^{xx}(-a) = -\gamma. \tag{7.63}$$

Recollecting the relation (6.37) $c_{jj'}^{ik}(f) = \sum_m \bar{A}_{nj,n'j'}^{ik} \exp(i f m)$ (where $m = n - n'$) between the force constants $c_{jj'}^{ik}(f)$ and the renormalized force matrix:

$$\bar{A}_{nj,n'j'}^{ik} = \frac{A_{nj,n'j'}^{i,k}}{\sqrt{m_j m_{j'}}},$$

we get from Eq. (7.62) for the diagonal coefficients [1,2]:

$$c_{11}^{xx}(f) = \bar{A}_{1,1}^{xx}(0) = \frac{A_{11}^{xx}(0)}{m_1} = \frac{2\gamma}{m_1}, \, c_{22}^{xx}(f) = \frac{2\gamma}{m_2}. \tag{7.64}$$

Analogously, for the non-diagonal coefficients it follows from Eq. (7.63):

$$c_{12}^{xx}(f) = \frac{A_{12}^{xx}(0) + A_{12}^{xx}(-a)\exp(-ifa)}{\sqrt{m_1 m_2}} = -\frac{\gamma}{\sqrt{m_1 m_2}}\left(1 + e^{-ifa}\right),$$

$$c_{21}^{xx}(f) = \frac{A_{21}^{xx}(0) + A_{21}^{xx}(a)\exp(ifa)}{\sqrt{m_1 m_2}} = -\frac{\gamma}{\sqrt{m_1 m_2}}\left(1 + e^{ifa}\right). \tag{7.65}$$

As a result, the combinations of the coefficients which enter in the biquadratic Eq. (7.60), equal to [1,2]

$$c_{11}^{xx}(f) + c_{22}^{xx}(f) = 2\gamma\left(\frac{1}{m_1} + \frac{1}{m_2}\right), \tag{7.66}$$

$$c_{11}^{xx}(f)c_{22}^{xx}(f) = \frac{4\gamma^2}{m_1 m_2}, \tag{7.67}$$

$$c_{12}^{xx}(f)c_{21}^{xx}(f) = \frac{\gamma^2}{m_1 m_2}\left(1 + e^{-ifa}\right)\left(1 + e^{ifa}\right) = \frac{\gamma^2}{m_1 m_2}\left(2 + e^{-ifa} + e^{ifa}\right)$$

$$= \frac{2\gamma^2}{m_1 m_2}(1 + \cos fa) = \frac{\gamma^2}{m_1 m_2}\cos^2\left(\frac{fa}{2}\right). \tag{7.68}$$

The roots of the biquadratic equation acquire the form [1–3]:

$$\omega^2 = \gamma\left(\frac{1}{m_1} + \frac{1}{m_2}\right) \pm \sqrt{\gamma^2\left(\frac{1}{m_1} + \frac{1}{m_2}\right)^2 - \frac{4\gamma^2}{m_1 m_2} + \frac{4\gamma^2}{m_1 m_2}\cos^2\left(\frac{fa}{2}\right)}$$

$$= \gamma\left(\frac{1}{m_1} + \frac{1}{m_2}\right) \pm \sqrt{\gamma^2\left(\frac{1}{m_1} - \frac{1}{m_2}\right)^2 + \frac{4\gamma^2}{m_1 m_2}\cos^2\left(\frac{fa}{2}\right)}. \tag{7.69}$$

One of the roots, which enters in Eq. (7.69) with the minus sign (ω_-^2) corresponds to the acoustic phonons [1–3]:

$$\omega_-^2 = \omega_{ac.}^2 = 0 \text{ for } f = 0. \tag{7.70}$$

The second root, entering with the sign plus, is related to the optical phonons [1–3]:

$$\omega_+^2 = \omega_{\text{opt.}}^2 = 2\gamma \left(\frac{1}{m_1} + \frac{1}{m_2} \right) \quad \text{for } f = 0. \tag{7.71}$$

For equal masses $m_1 = m_2 = m$, we have effectively one atom in the elementary cell with the size $\tilde{a} = \frac{a}{2}$, and hence only one acoustic branch of the phonon spectrum [1–3]:

$$\omega^2 = \frac{2\gamma}{m} - \sqrt{\frac{4\gamma^2}{m^2} \cos^2(f\tilde{a})} = \frac{2\gamma}{m}(1 - \cos(f\tilde{a})) = \frac{4\gamma}{m} \sin^2 \left(\frac{f\tilde{a}}{2} \right). \tag{7.72}$$

Thus, the eigenfrequency reads [1–3]

$$\omega = 2\sqrt{\frac{\gamma}{m}} \left| \sin \left(\frac{f\tilde{a}}{2} \right) \right|. \tag{7.73}$$

The frequency of the acoustic branch depends linearly from the wave vector for $f \to 0$

$$\omega = \sqrt{\frac{\gamma}{m}} \tilde{a} f = c_s f, \tag{7.74}$$

where $c_s = \sqrt{\frac{\gamma \tilde{a}^2}{m}}$ is the sound velocity.

Note that the force constant γ in the expression for the spectrum and the sound velocity $c_s \sim \frac{d^2 U}{dx^2}$, hence the sound velocity squared has a correct dimensionality $[c_s^2] = \left[\frac{\gamma \tilde{a}^2}{m} \right] = \left[\frac{U}{m} \right] = [v^2]$.

Thus, the sound velocity is proportional to

$$c_s \sim \frac{1}{\sqrt{m}},$$

and at the same time:

$$c_s \sim \sqrt{\frac{d^2 U}{dx^2}}.$$

It is interesting to note that even in case of only the onsite interaction and the interaction between the nearest neighbors, we get an excellent sound (or large wavelength phonons) [1,2].

It occurs that we do not need the long-range interaction for the existence of the sound mode [1,2]. It is enough to have the coupling with media which can be realized even by the onsite interactions and the interactions between the neighbors on the interatomic distance.

The spectrum of the acoustic phonons in Eq. (7.73) for one-dimensional chain is presented in Fig. 7.8.

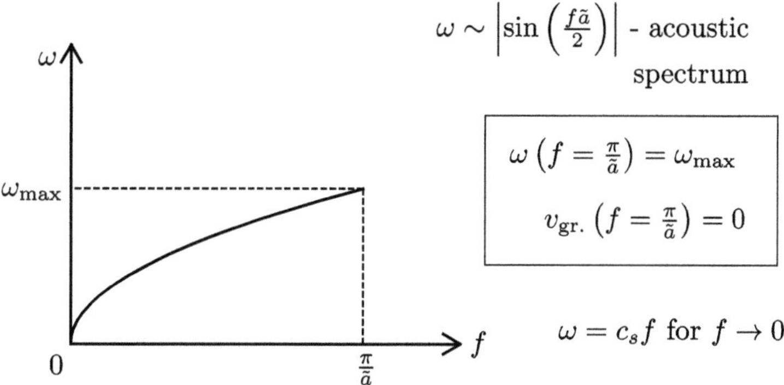

Fig. 7.8 Spectrum of the acoustic phonon for the one-dimensional chain with the size of the elementary cell equals to $\tilde{a} = \frac{a}{2}$ [1]

Let us emphasize that the group velocity for the acoustic phonons goes to zero on the boundary of the Brillouin zone for the wave vector values $f = \frac{\pi}{\tilde{a}}$:

$$v_{\mathrm{gr.}} = \frac{d\omega}{df} = \sqrt{\frac{\gamma}{m}} \tilde{a} \cos\left(\frac{f\tilde{a}}{2} \right) = 0 \text{ for } f = \frac{\pi}{\tilde{a}}. \tag{7.75}$$

The integration measure of the one-dimensional problem:

$$df = \frac{\partial f}{\partial \omega} d\omega = g(\omega) d\omega, \tag{7.76}$$

where the density of states:

$$g(\omega) = \frac{\partial f}{\partial \omega} = \frac{1}{v_{\mathrm{gr.}}} = \frac{1}{\sqrt{\frac{\gamma}{m}} \tilde{a} \cos\left(\frac{f\tilde{a}}{2} \right)}. \tag{7.77}$$

But, the cosine function in Eq. (7.75) for the group velocity equals to

$$\cos\left(\frac{f\tilde{a}}{2} \right) = \sqrt{1 - \sin^2\left(\frac{f\tilde{a}}{2} \right)} = \sqrt{1 - \frac{m\omega^2}{4\gamma}} = \sqrt{\frac{m}{4\gamma}} \sqrt{\frac{4\gamma}{m} - \omega^2}. \tag{7.78}$$

Hence the density of states [1,2]:

$$g(\omega) = \frac{1}{\sqrt{\frac{\gamma}{m}} \tilde{a} \sqrt{\frac{m}{4\gamma}} \sqrt{\frac{4\gamma}{m} - \omega^2}} = \frac{2}{\tilde{a} \sqrt{\frac{4\gamma}{m} - \omega^2}} = \frac{2}{\tilde{a} \sqrt{\omega_{\max}^2 - \omega^2}}, \tag{7.79}$$

where $\omega_{\max}^2 = \frac{4\gamma}{m}$ is the maximal frequency of the phonon spectrum squared.

For the one-dimensional chain, the maximal frequency is realized for $f = \frac{\pi}{\tilde{a}}$, that is just on the boundary of the Brillouin zone.

The density of states for $\omega \to \omega_{\text{max}}$ is given by

$$g(\omega) = \frac{2}{\tilde{a}\sqrt{\omega_{\text{max}} - \omega}} \frac{1}{\sqrt{\omega_{\text{max}} + \omega}} \approx \frac{2}{\tilde{a}\sqrt{\omega_{\text{max}} - \omega}} \frac{1}{\sqrt{2\omega_{\text{max}}}} = \frac{\text{const}}{\sqrt{\omega_{\text{max}} - \omega}}, \tag{7.80}$$

as it has to be close to the extremal (maximum) point in the one-dimensional case according to Eq. (7.37).

Let us return back now to the general form of the phonon spectrum in Eq. (7.69) for the one-dimensional chain with two atoms having different masses in the elementary cell. In this case, we get one acoustic and one optical branch of the spectrum [1–3].

For small wave vectors at $f \to 0$, the eigenfrequencies of acoustic and optical phonons equal to [1–3]

$$\omega_{\text{ac.}}^2 (f = 0) = 0, \tag{7.81}$$

$$\omega_{\text{opt.}}^2 (f = 0) = 2\gamma \left(\frac{1}{m_1} + \frac{1}{m_2} \right). \tag{7.82}$$

At the same time, on the boundary of the Brillouin zone for $f = \frac{\pi}{a}$ and $m_1 < m_2$ (that is now for the cell with the size $a = 2\tilde{a}$) the eigenfrequencies read [1–3]:

$$\omega_{\text{opt.}}^2 \left(f = \frac{\pi}{a} \right) = \frac{2\gamma}{m_1}, \tag{7.83}$$

$$\omega_{\text{ac.}}^2 \left(f = \frac{\pi}{a} \right) = \frac{2\gamma}{m_2} < \omega_{\text{opt.}}^2 \left(f = \frac{\pi}{a} \right). \tag{7.84}$$

The phonon spectrum for the one-dimensional chain in case of the different masses is illustrated in Fig. 7.9. We show the forbidden gap (the forbidden range of frequencies) in this figure, which appears in the 1D crystal for the different masses of atoms in the cell. The excitations with these frequencies cannot emerge in the system. They are reflecting, that is they are exponentially decreasing deep in the system [1,2].

The width of the forbidden gap in Fig. 7.9 corresponds to the minimal difference of the eigenfrequencies [1–3]. For the spectrum in Fig. 7.9 the width of the gap equals to

$$\Delta = \omega_{\text{opt.}} \left(f = \frac{\pi}{a} \right) - \omega_{\text{ac.}} \left(f = \frac{\pi}{a} \right) = \sqrt{\frac{2\gamma}{m_1}} - \sqrt{\frac{2\gamma}{m_2}}. \tag{7.85}$$

Note that, for equal masses $m_1 = m_2$, the eigenfrequencies coincide on the boundary of the Brillouin zone:

$$\omega_{\text{ac.}}^2 \left(f = \frac{\pi}{a} \right) = \omega_{\text{opt.}}^2 \left(f = \frac{\pi}{a} \right) \text{ for } m_1 = m_2, \tag{7.86}$$

and hence the width of the forbidden gap $\Delta = 0$.

In the general case, the branches are intersecting on the zone boundary or in other points of a very high symmetry (the branches intersection not on the zone boundary is highly unlikely) [12–14].

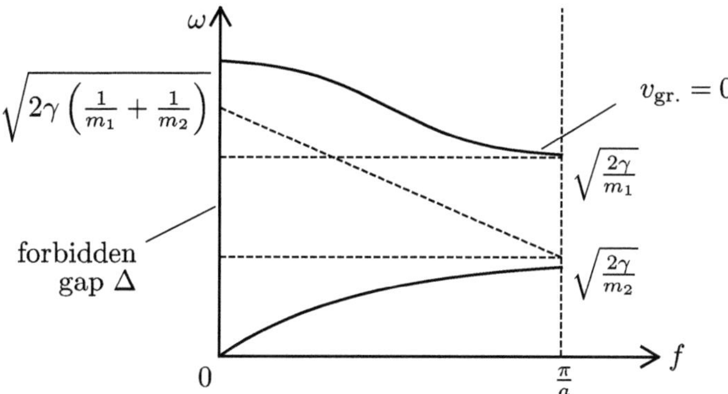

Fig. 7.9 Acoustic and optical branch of the phonon spectrum for the one-dimensional chains with two atoms having the different masses ($m_1 < m_2$) in the elementary cell. The phonon spectrum for two different masses acquires a forbidden gap (forbidden band of frequencies) $\Delta = \omega_{\mathrm{opt.}}\left(f = \frac{\pi}{a}\right) - \omega_{\mathrm{ac.}}\left(f = \frac{\pi}{a}\right) = \sqrt{\frac{2\gamma}{m_1}} - \sqrt{\frac{2\gamma}{m_2}}$. We show by the dashed line the spectrum for equal masses when the gap does not emerge [1]

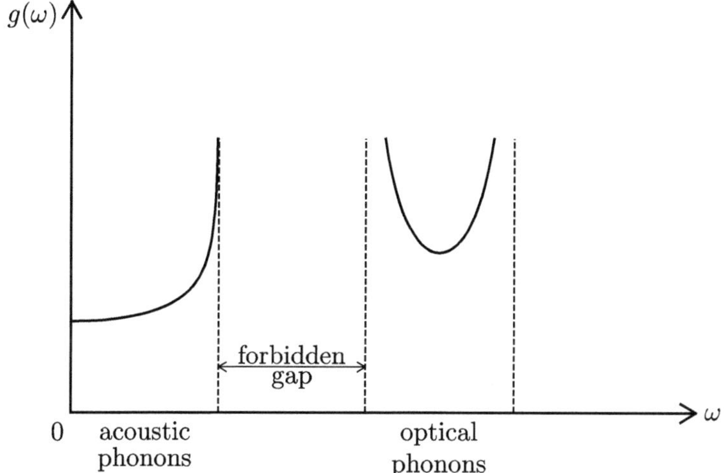

Fig. 7.10 Typical density of states for acoustic and optical branches of the phonon spectrum for the one-dimensional chain with two atoms (having different masses) on the elementary cell. In this case, the forbidden frequency band appears where $g(\omega) = 0$ [1]

The typical density of states for the one-dimensional chain with two atoms (having different masses) on the elementary cell has the form presented in Fig. 7.10.

Let us stress that in the three-dimensional case the acoustic branch in one direction can lie higher than the optical branch in the other direction. Then, the density of states does not have the gap. In general case, the width of the gap is proportional to the mass difference. If $m_1 = m_2$ (as it occurs in cadmium, tin, magnesium, etc.), then the gap does not arise [1,2].

Let us find now the ratio of the atomic displacements in acoustic and optical modes. For the renormalized displacements in the one-dimensional chain with the two atoms in the cell (and with the interaction on the same and neighboring sites), an equation of motion acquires a form [1,2]:

$$v_1^x \omega^2 = c_{11}^{xx} v_1^x + c_{12}^{xx} v_2^x.$$

(7.87)

Then, with an account of the relations (7.64)–(7.65) at $f \to 0$, we get for the ratio of the renormalized atomic displacements:

$$\frac{v_1^x}{v_2^x} = \frac{c_{12}^{xx}(f)}{\omega^2 - c_{11}^{xx}(f)} = -\frac{\frac{2\gamma}{\sqrt{m_1 m_2}}}{\omega^2 - \frac{2\gamma}{m_1}}.$$

(7.88)

For the acoustic mode, the frequency squared $\omega^2 \to 0$ for $f \to 0$, and thus:

$$\left(\frac{v_1^x}{v_2^x}\right)_{ac.} = \frac{\frac{2\gamma}{\sqrt{m_1 m_2}}}{\frac{2\gamma}{m_1}} = \sqrt{\frac{m_1}{m_2}}.$$

(7.89)

At the same time, for the atomic displacements in the optical mode, the frequency squared $\omega_{opt.}^2 \to 2\gamma \left(\frac{1}{m_1} + \frac{1}{m_2} \right)$ for $f \to 0$, and hence:

$$\left(\frac{v_1^x}{v_2^x}\right)_{opt.} = -\frac{\frac{2\gamma}{\sqrt{m_1 m_2}}}{2\gamma \left(\frac{1}{m_1} + \frac{1}{m_2} \right) - \frac{2\gamma}{m_1}} = -\frac{\frac{2\gamma}{\sqrt{m_1 m_2}}}{\frac{2\gamma}{m_2}} = -\sqrt{\frac{m_2}{m_1}}.$$

(7.90)

The initial displacements are related to the renormalized ones as follows:

$$v_1^x = \sqrt{m_1} u_1^x, \, v_2^x = \sqrt{m_2} u_2^x.$$

(7.91)

Then for the ratio of the initial atomic displacements in the acoustic wave, we get [1,2]:

$$\left(\frac{u_1^x}{u_2^x}\right)_{ac.} = 1, \text{ and } u_1^x = u_2^x.$$

(7.92)

It means that the elementary cell shifts as a whole in the acoustic mode for $f \to 0$. The ratio of the initial displacements in the optical mode yields [1,2]

$$\left(\frac{u_1^x}{u_2^x}\right)_{opt.} = -\frac{m_2}{m_1}, \text{ and } m_1 u_1^x + m_2 u_2^x = 0.$$

(7.93)

Thus, in the optical mode, the center of gravity of the cell does not shift for $f \to 0$ [1–3].

References

1. Y. Kagan, *Lectures on the Solid-State Theoretical Physics* (Moscow, MEPHI, 1969), 178 p. (in Russian)
2. Y. Kagan, *Lectures on the Solid-State Theoretical Physics* (Moscow, MEPHI, 1981–1982). Part I. Phonons - unpublished
3. J.M. Ziman, *Electrons and Phonons, The Theory of Transport Phenomena in Solids* (Clarendon Press, Oxford, 1960)
4. O. Madelung, *Introduction to Solid-State Theory*, Springer Series in Solid-State Sciences (1995)
5. C. Kittel, *Quantum Theory of Solids* (Wiley, New York-London, 1963)
6. N.W. Ashcroft, N.D. Mermin, *Solid State Physics* (Holt, Rinehart and Winston, 1976)
7. P. Debye, Ann. Phys. **39**, 789 (1912). ((in German))
8. L. Van Hove, Phys. Rev. **89**(6), 1189 (1953)
9. J. Ziman, *Principles of the Theory of Solids* (Cambridge University Press, 1972)
10. E.G. Brovman, Y. Kagan, Phonons in non-transition metals, in Lattice dynamics, ed. by A. Maradudin, G.K. Horton (Amsterdam, North-Holland, 1974)
11. E.G. Brovman, Yu. Kagan, *Neutron Inelastic Scattering*, vol. 1 (IAEA, Vienna, 1968), p.3
12. S. Bhagavantam, T. Venkatarayudu, *Theory of Groups and its Application to Physical Problems* (Andhra University, Waltair, India, 1951)
13. H.S.M. Coexter, W.O.J. Moser, *Generators and Relations for Discrete Groups* (Springer, New-York, 1980)
14. B.K. Veinstein, *Fundamentals of Crystal Symmetry and Methods of Structural Crystallography and Modern Crystallography*, vol. 1 (Springer, Berlin, 1984)

Lecture 8. Phonon Spectrum in Crystals with Pairwise Interaction: Quantization of the Lattice Vibrations

Abstract

In this Lecture we consider the phonon spectrum in crystals with the pairwise interactions. We analyze the long wavelength part of the phonon spectrum and show the emergence of the longitudinal and transverse sound modes in this case. We discuss an interplay of the sound modes with static compressibility and shear modulus of the crystal and show the formation of sound in metals. In the end of the Lecture, we derive the phonon Hamiltonian in the second quantization technique thus quantizing the phonon spectrum.

Let us consider the phonon spectrum in crystal with the pairwise interaction between atoms [1]. Potential energy for the pairwise interaction acquires a form:

$$U = \frac{1}{2} \sum_{n \neq n'} \varphi \left(\boldsymbol{R}_n - \boldsymbol{R}_{n'} \right). \tag{8.1}$$

Note that in principle the forces of not the pairwise nature can emerge in the real crystal (the interaction between the atoms 1 and 2 exerts an influence on the interaction between the atoms 1 and 3) [1,2].

Let us denote $\boldsymbol{R}_n - \boldsymbol{R}_{n'} = \boldsymbol{R}_{nn'}$ and expand the component of the pairwise potential $\varphi \left(\boldsymbol{R}_{nn'} \right)$ in the Fourier series:

$$\varphi \left(\boldsymbol{R}_{nn'} \right) = \frac{1}{V} \sum_{q} \varphi(\boldsymbol{q}) \exp \left(i \boldsymbol{q} \, \boldsymbol{R}_{nn'} \right). \tag{8.2}$$

M. Kagan, *Lecture Notes of Professor Yuri Kagan in Theoretical Solid-State Physics*, Lecture Notes in Physics 1048, https://doi.org/10.1007/978-3-032-14621-2_8

For one atom in the elementary cell only one term $\varphi\left(\boldsymbol{R}_{nn'}\right)$ for $n \neq n'$ (and $\varphi\left(\boldsymbol{R}_{n'n}\right)$ as well) contributes to the force matrix of the crystal in case of the pairwise interaction between the atoms:

$$
\begin{aligned}
A_{nn'}^{ik} &= \frac{\partial^2 U}{\partial R_n^i \partial R_{n'}^k} = \frac{1}{V} \sum_{\boldsymbol{q}} \varphi(\boldsymbol{q})(-i\boldsymbol{q})^k (i\boldsymbol{q})^i \exp\left(i\boldsymbol{q}\,\boldsymbol{R}_{nn'}\right) \\
&= \frac{1}{V} \sum_{\boldsymbol{q}} \varphi(\boldsymbol{q}) q^k q^i \exp\left(i\boldsymbol{q}\,\boldsymbol{R}_{nn'}\right).
\end{aligned}
\tag{8.3}
$$

In Lecture 7 we showed, that due to translational invariance for the displacement of the crystal as a whole $\sum_{n'} A_{nn'}^{ik} = 0$.

Then the diagonal term in the force matrix A_{nn}^{ik} for $n = n'$ equals to [1]:

$$
A_{nn}^{ik} = - \sum_{n'(\neq n)} A_{nn'}^{ik} = - \sum_{n'(\neq n)} \frac{1}{V} \sum_{q} \varphi(\boldsymbol{q}) q^k q^i \exp\left(i\boldsymbol{q}\,\boldsymbol{R}_{nn'}\right).
\tag{8.4}
$$

If the atom mass equals to M, then the force coefficient $c^{ik}(\boldsymbol{f})$, which depends on the wave vector, is defined as follows [1]:

$$
\begin{aligned}
c^{ik}(\boldsymbol{f}) &= \frac{1}{M} \sum_{n'} A_{nn'}^{ik} \exp\left(-i\boldsymbol{f}\,\boldsymbol{R}_{nn'}\right) \\
&= \frac{1}{M} \sum_{n'(\neq n)} \frac{1}{V} \sum_{q} \varphi(\boldsymbol{q}) q^k q^i \exp\left(i\boldsymbol{q}\,\boldsymbol{R}_{nn'}\right) \exp\left(-i\boldsymbol{f}\,\boldsymbol{R}_{nn'}\right) + \frac{1}{M} A_{nn}^{ik} \exp(-i\boldsymbol{f}\,\boldsymbol{0}) \\
&= \frac{1}{M} \sum_{n'(\neq n)} \frac{1}{V} \sum_{q} \varphi(\boldsymbol{q}) q^k q^i \exp\left(i\boldsymbol{q}\,\boldsymbol{R}_{nn'}\right) \exp\left(-i\boldsymbol{f}\,\boldsymbol{R}_{nn'}\right) \\
&\quad + \exp(-i\boldsymbol{f}\,\boldsymbol{0}) \left(- \sum_{n'(\neq n)} \frac{1}{VM} \sum_{q} \varphi(\boldsymbol{q}) q^k q^i \exp\left(i\boldsymbol{q}\,\boldsymbol{R}_{nn'}\right) \right).
\end{aligned}
\tag{8.5}
$$

But the terms with $n = n'$ equal each other in the first and second sum of Eq. (8.5). Hence, we can replace the summation by $n'(\neq n)$ on the summation by n' (adding the term with $n = n'$ to each of the two terms in Eq. (8.5)). As a result:

$$
\begin{aligned}
c^{ik}(\boldsymbol{f}) &= \sum_{n'} \frac{1}{VM} \sum_{q} \varphi(\boldsymbol{q}) q^k q^i \exp\left(i(\boldsymbol{q} - \boldsymbol{f})\boldsymbol{R}_{nn'}\right) \\
&\quad - \sum_{n'} \frac{1}{VM} \sum_{q} \varphi(\boldsymbol{q}) q^k q^i \exp\left(i\boldsymbol{q}\,\boldsymbol{R}_{nn'}\right).
\end{aligned}
\tag{8.6}
$$

Further on let us consider the term:

$$\sum_{n'} \exp\left(i(q-f)R_{nn'}\right) = \sum_{m} \exp\left(i(q-f)R_{m}\right), \qquad (8.7)$$

where $R_{nn'} = R_m$, entering in the first of the two sums in Eq. (8.6)

In the solid-state physics this term is called the lattice sum.

Let $\kappa = q - f$. Then the lattice sum equals to [1]:

$$\sum_{m} \exp\left(i\kappa R_m\right). \qquad (8.8)$$

It is easy to observe, that the lattice sum does not change at the interatomic shift (at the shift on the lattice constant) [1]:

$$R_m \rightarrow R_m + a. \qquad (8.9)$$

It is just the renumeration of nodes in the initial expression (8.7), which goes on, and, in particular, the diagonal node with $n' = n$ is replaced by the node with $n' = n + 1$. However, this is not important under the summation over all n'. Then, the lattice sum for this shift:

$$\sum_{m} \exp\left(i\kappa R_m\right) = \exp(i\kappa a) \sum_{m} \exp\left(i\kappa R_m\right), \ \text{ and } \exp(i\kappa a) = 1. \qquad (8.10)$$

Hence:

$$\kappa = (0, K), \qquad (8.11)$$

where K is the vector of the inverse lattice, and $\exp(iKa) = 1$.

As a result, the lattice sum yields [1]:

$$\sum_{m} \exp\left(i\kappa R_m\right) = \sum_{K} N\delta_{\kappa K}, \qquad (8.12)$$

where N is the number of nodes (or the number of the elementary cells in the crystal with one atom per cell). Note, that for $\kappa = (0, K)$:

$$\kappa R_m = 2\pi n, \qquad (8.13)$$

where n is an integer number.

In the initial notations the lattice sum can be written as [1]:

$$\sum_{m} \exp\left(i(q-f)R_m\right) = \sum_{K} \delta_{q-f, K} N. \qquad (8.14)$$

Substituting an expression for the lattice sum in (8.14) into the relation (8.6), we get:

$$c^{ik}(f) = \frac{N}{VM} \sum_{q,K} \varphi(q) q^k q^i \delta_{q-f,K} - \frac{N}{VM} \sum_{q,K} \varphi(q) q^k q^i \delta_{qK}$$

$$= \frac{N}{VM} \left(\sum_K \varphi(f+K)(f+K)^i (f+K)^k - \sum_K \varphi(K)(K)^i (K)^k \right). \qquad (8.15)$$

At the same time, the system of equations on the eigenfunctions $\left| c^{ik}(f) - \omega^2 \delta^{ik} \right| = 0$ for one atom in the cell corresponds to 3×3 matrix and contains three acoustic branches $\omega_\alpha^2(f)$, which are the functions of the Fourier components of the pairwise interaction $\varphi(q)$.

8.1   Behavior of the Long Wavelength Part of the Spectrum

Equations of motion for the renormalized displacements have the form: $v^i \omega^2 = c^{ik}(f) v^k$. Let us consider the limit of small wave vectors $f \to 0$ (or equivalently of large wavelengths $\lambda \to \infty$).

At $f \equiv 0$ the force coefficients $c^{ik}(0) = 0$ due to the translational symmetry in the problem. (The shift of the crystal as a whole goes on without the force impact).

Terms linear in f equal to zero in Eq. (8.15):

$$\sum_K \varphi(K) \left(f^i K^k + f^k K^i \right) = 0, \qquad (8.16)$$

since under the summation symbol stands an odd function of K (K^i is odd, $\varphi(K)$ is even), and correspondingly for each term in the sum with the K value of the wave vector we have a term with the $-K$ value. As a result, these terms cancel each other identically.

Note, that the parity of the function $\varphi(K)$ is related to the real values of the potential φ, depending only on the distance in real space.

Let us stress, that Eq. (8.15) contains one more type of the linear terms proportional to [1]:

$$\sum_K K^i K^k \frac{\partial \varphi}{\partial K^\gamma} f^\gamma = 0. \qquad (8.17)$$

These terms equal to zero also, since under the sum over all the values of the K vector stands the product of the even function $K^i K^k$ and odd function $\frac{\partial \varphi}{\partial K^\gamma}$.

8.2   Terms Quadratic in f

Let us collect now the terms quadratic in f in Eq. (8.15):

$$
c^{ik}(\boldsymbol{f}) = \frac{1}{MV_0} \left\{ f^i f^k \varphi(0) + f^i f^k \sum_{\boldsymbol{K} \neq 0} \varphi(\boldsymbol{K}) + \sum_{\boldsymbol{K}} \left(f^i K^k + f^k K^i \right) \frac{\partial \varphi(\boldsymbol{K})}{\partial K^\gamma} f^\gamma \right.
$$
$$
\left. + \frac{1}{2} \sum_{\boldsymbol{K}} K^i K^k \frac{\partial^2 \varphi(\boldsymbol{K})}{\partial K^\gamma \partial K^\delta} f^\gamma f^\delta \right\},
$$

$$(8.18)$$

where $V_0 = \frac{V}{N}$ is the volume of the elementary cell in the real space (Ω_0 in our early notations).

Let us emphasize, that only one term proportional to $\varphi(0)$ enters in Eq. (8.18).

All the other terms are proportional to $\varphi(\boldsymbol{K})$ (or, equivalently to the first and second derivatives of $\varphi(\boldsymbol{K})$).

But the minimal value of the inverse lattice vector $K_{\min} = \frac{2\pi}{a}$ is large enough.

We know at the same time, that the Fourier components do not like large exponents (they are rapidly damping in this case). Hence, it would be natural to drop out all the terms proportional to $\varphi(\boldsymbol{K})$. The force coefficient in this case:

$$
c^{ik}(\boldsymbol{f}) = \frac{1}{MV_0} f^i f^k \varphi(0), \tag{8.19}
$$

and equation of motion acquires a simple form [1]:

$$
v^i \omega^2 = \frac{\varphi(0)}{MV_0} f^i f^k v^k = \frac{\varphi(0)}{MV_0} f^i (\boldsymbol{f}\boldsymbol{v}). \tag{8.20}
$$

Thus, the displacement vector is parallel to the wave vector ($\boldsymbol{v}\|\boldsymbol{f}$), and we have only longitudinal oscillations [1].

Multiplying left and right-hand sides of Eq. (8.20) on f^i, we obtain the linear spectrum of the longitudinal sound [1]:

$$
\omega^2 = \frac{\varphi(0)}{MV_0} f^2. \tag{8.21}
$$

It is important, that the frequencies of the longitudinal sound squared in Eq. (8.21) are proportional to the mean potential $\omega^2 \sim \varphi(0)$, and we have here an analogy with weakly non-ideal Bose gas [1].

Note, however, that in the Bose gas we have only longitudinal sound [7–10]. At the same time, in the solid state there is also a transverse sound. [1,3–6,11]. Its emergence is a consequence of the translational symmetry and is connected only with the presence of non-zero vector of the inverse lattice $\boldsymbol{K} \neq 0$ [1].

We can show, that the term proportional to $f^i K^k \frac{\partial \varphi(\boldsymbol{K})}{\partial K^\gamma} f^\gamma$ under the sum symbol in Eq. (8.18) does not yield the transverse sound. Indeed, in this term enters the derivative $\frac{\partial \varphi(\boldsymbol{K})}{\partial K^\gamma}$. But the Fourier component of the potential depends only on the absolute value of the inverse lattice vector $\varphi = \varphi(|\boldsymbol{K}|)$. Hence, the derivative:

$$\frac{\partial \varphi(K)}{\partial K^\gamma} = \frac{\partial \varphi}{\partial K} \frac{\partial K}{\partial K^\gamma} = \varphi'(K) \frac{K^\gamma}{K}. \tag{8.22}$$

As a result, the term interesting for us reads:

$$f^i K^k \frac{\partial \varphi(K)}{\partial K^\gamma} f^\gamma = \varphi'(K) \frac{K^\gamma f^\gamma}{K} f^i K^k = \varphi'(K) \frac{(\boldsymbol{K} \boldsymbol{f})}{K} f^i K^k. \tag{8.23}$$

We can remove also all the summation signs in Eq. (8.18), considering only the first term in the sums over K, corresponding to the minimal value of the inverse lattice vector $K_{\min} = \frac{2\pi}{a}$. The contribution of this term to the force coefficient is given by:

$$c^{ik}(\boldsymbol{f}) = \frac{\varphi'(K)}{M V_0} \frac{(\boldsymbol{K} \boldsymbol{f})}{K} f^i K^k, \tag{8.24}$$

where $K = K_{\min}$, and thus, the equation of motion [1]:

$$v^i \omega^2 = \frac{\varphi'(K)}{M V_0} \frac{(\boldsymbol{K} \boldsymbol{f})}{K} f^i K^k v^k = \frac{\varphi'(K)}{M V_0} \frac{(\boldsymbol{K} \boldsymbol{f})}{K} (\boldsymbol{K} \boldsymbol{v}) f^i \tag{8.25}$$

corresponds again to the longitudinal sound with $\boldsymbol{v} \| \boldsymbol{f}$.

At the same time, the second from the two terms in Eq. (8.18), containing the first derivative of the pairwise potential $\frac{\partial \varphi(\boldsymbol{K})}{\partial K^\gamma}$, has the form:

$$f^k K^i \frac{\partial \varphi(K)}{\partial K^\gamma} f^\gamma = \varphi'(K) \frac{(\boldsymbol{K} \boldsymbol{f})}{K} f^k K^i, \tag{8.26}$$

and its contribution in the force coefficient can be written as [1]:

$$c^{ik}(\boldsymbol{f}) = \frac{\varphi'(K)}{M V_0} \frac{(\boldsymbol{K} \boldsymbol{f})}{K} f^k K^i. \tag{8.27}$$

Correspondingly, the contribution of this term to the equation of motion [1]:

$$v^i \omega^2 = \frac{\varphi'(K)}{M V_0} \frac{(\boldsymbol{K} \boldsymbol{f})}{K} f^k K^i v^k = \frac{\varphi'(K)}{M V_0} \frac{(\boldsymbol{K} \boldsymbol{f})}{K} (\boldsymbol{f} \boldsymbol{v}) K^i. \tag{8.28}$$

We can see, that for this term $\boldsymbol{v} \| \boldsymbol{K}$, and the displacement is directed along $\boldsymbol{K}$, and not along $\boldsymbol{f}$ [1].

Moreover, it yields non-zero contribution to the renormalized displacement v^i in Eq. (8.28) under the conditions:

$$(\boldsymbol{K} \boldsymbol{f}) \neq 0, (\boldsymbol{f} \boldsymbol{v}) \neq 0. \tag{8.29}$$

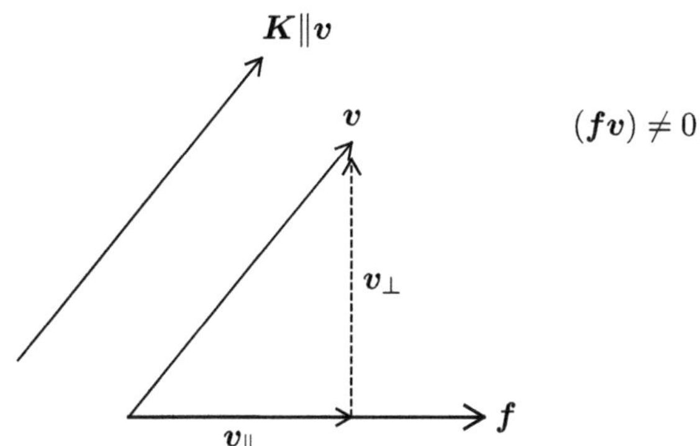

Fig. 8.1 Configuration of the three vectors: the displacement vector v, vector of the inverse lattice K and wave vector f in the equations of motion, which contribute both to the longitudinal and the transverse sound under the account of the non-zero Fourier components of the pairwise potential $\varphi(K\neq 0)$ and its derivatives [1]

The second condition in Eq. (8.29) $(fv)\neq 0$ in fact is fulfilled automatically for $(Kf)\neq 0$ and $v\|K$.

Configuration of the three vectors v, K and f is presented in Fig. 8.1. We can see, that in general case the displacement vector v is directed under the finite angle to the wave vector f, and has projections both parallel and perpendicular to f.

Thus, equation of motion (8.28) contributes both to the longitudinal and transverse sound [1].

Multiplying left and right-hand side of Eq. (8.28) on f^i, for $(fv)\neq 0$ we get the following contribution to the spectrum [1]:

$$\omega^2 \sim \frac{\varphi'(K)}{MV_0}\frac{(Kf)^2}{K}. \tag{8.30}$$

Analogous picture occurs for the term with the second derivative of the pairwise interaction proportional to $K^i K^k \frac{\partial^2\varphi(K)}{\partial K^\gamma \partial K^\delta} f^\gamma f^\delta$.

In particular, for $\varphi = \varphi(|K|)$ the second derivative contains the contribution:

$$\frac{\partial^2\varphi(K)}{\partial K^\gamma \partial K^\delta} \sim \frac{\partial^2\varphi}{\partial K^2}\frac{K^\gamma}{K}\frac{K^\delta}{K} = \varphi''\frac{K^\gamma}{K}\frac{K^\delta}{K}. \tag{8.31}$$

Then, for this term we obtain:

$$K^i K^k \frac{\partial^2\varphi(K)}{\partial K^\gamma \partial K^\delta} f^\gamma f^\delta \sim \varphi''\frac{K^\gamma}{K}\frac{K^\delta}{K} f^\gamma f^\delta K^i K^k$$

$$= \varphi''\frac{(Kf)}{K}\frac{(Kf)}{K} K^i K^k = \varphi''\frac{(Kf)^2}{K^2} K^i K^k. \tag{8.32}$$

The contribution of this term to the force coefficient reads [1]:

$$c^{ik}(f) = \frac{\varphi''(K)}{MV_0}\frac{(Kf)^2}{K^2} K^i K^k. \tag{8.33}$$

Thus, for the contribution of this term to the equation of motion we get [1]:

$$v^i \omega^2 = \frac{\varphi''(K)}{M V_0} \frac{(\boldsymbol{K} \boldsymbol{f})^2}{K^2} K^i K^k v^k = \frac{\varphi''(K)}{M V_0} \frac{(\boldsymbol{K} \boldsymbol{f})^2}{K^2} (\boldsymbol{K} \boldsymbol{v}) K^i, \tag{8.34}$$

and correspondingly the displacement $\boldsymbol{v} \| \boldsymbol{K}$ is directed again along the vector $\boldsymbol{K}$ as in Eq. (8.28), and not along $\boldsymbol{f}$.

The contribution of (8.34) to the renormalized displacement v^i does not equal to zero also under the fulfilment of the conditions (8.29), and this equation contributes again both to the longitudinal and transverse sound [1].

Multiplying left and right-hand side of Eq. (8.34) on K^i, we get the contribution to the spectrum of the term with the second derivative of the pairwise potential [1]:

$$\omega^2 \sim \frac{\varphi''(K)}{M V_0} (\boldsymbol{K} \boldsymbol{f})^2. \tag{8.35}$$

We can see, that both contributions in Eqs. (8.30) and (8.35) to the frequency squared, connected with the first and second derivatives of the pairwise interaction, are proportional to [1]:

$$\omega^2 \sim (\boldsymbol{K} \boldsymbol{f})^2.$$

Let us stress, that for the longitudinal sound both contributions of Eqs. (8.30) and (8.35) in the frequency spectrum represent small corrections to the main contribution in (8.21), related to the zeroth Fourier component of the pairwise potential $\varphi(0)$.

Note also, that the transverse sound in crystal emerges under the account of the non-zero Fourier components of the pairwise interaction $\varphi(\boldsymbol{K} \neq 0)$ and its derivatives [1].

At the same time, the longitudinal sound is connected with the zeroth Fourier component $\varphi(0)$, which is substantially larger than $\varphi(\boldsymbol{K})$. Thus, the velocity of longitudinal sound is usually larger than the velocity of the transverse sound [1,3–6,11].

We can say, that the transverse sound is connected with the shear displacements, [1,3–6,11] and in principle its velocity can be very small.

At the same time, the longitudinal sound is connected with the finite compressibility $(\frac{\partial P}{\partial V} < 0)$. The finite compressibility of the system dictates the final values for the velocity of the longitudinal sound [1].

8.3 Coulomb Interaction

As we know, the crystal is electrically neutral as a whole. The Coulomb potential is long-ranged. Electron liquid in metals guarantees electroneutrality [1,6,12–14].

We consider only ions. The Fourier component of the bare Coulomb interaction (if we neglect screening) for the ion charge $Z = 1$ equals to:

$$\varphi(q) = \frac{4\pi e^2}{q^2}. \tag{8.36}$$

For the bare Coulomb interaction $\varphi(q)$ has a singularity at zero.

The combination entering in the equations of motion for the longitudinal displacements reads in case of Coulomb interaction [1]:

$$f^i f^k \varphi(f \to 0) = \frac{4\pi e^2}{f^2} f^i f^k. \tag{8.37}$$

For $f \to 0$ this expression has a finite limit [1]. Correspondingly, if we consider the Coulomb interaction of ions only, the spectrum of the longitudinal oscillations acquires a finite gap, which equals to the Langmuir ion frequency [1,3,5,6,14,16]:

$$\omega^2 \to \frac{4\pi e^2}{MV_0} = \frac{4\pi e^2 n}{M}, \tag{8.38}$$

where M is an ion mass, and from the electroneutality condition the ion density for $Z = 1$ equals to the electron density: $n = n_i = n_e$.

It seems, that this result contradicts to the presence of the translational symmetry, which leads to the existence of the longitudinal sound.

It occurs, that the result of Eq. (8.38) is not exact. Electrons in metals adjust to the ionic motion and the screening appears in the system [1,3–6,14,16]. As a result, the longitudinal sound restores.

Indeed, with an account of screening the Fourier component of the pairwise interaction has a form [1,3–6,14,16,19]:

$$\varphi(q) = \frac{4\pi e^2}{q^2 \varepsilon(q)}, \tag{8.39}$$

where $\varepsilon(q)$ is the dielectric permittivity of medium depending on the wave vector [1,17,18].

As a consequence of Maxwell equations, the dielectric permittivity for $q \to 0$ reads [1,17,18]:

$$\varepsilon(q) \to 1 + \frac{\kappa^2}{q^2}, \tag{8.40}$$

where κ is the Thomas–Fermi wave vector [20,21].

As a result, the Fourier component of the screened Coulomb interaction has a form, resembling the Fourier component of the Yukawa potential [22] in nuclear physics:

$$\varphi(q) = \frac{4\pi e^2}{q^2 + \kappa^2}. \tag{8.41}$$

For $q \rightarrow 0$ the Fourier component of the screened Coulomb interaction:

$$\varphi(q) \rightarrow \frac{4\pi e^2}{\kappa^2} = \varphi(0),\qquad(8.42)$$

where $\varphi(0)$ is finite.

Thus, an excellent sound exists in metal [1].

We can say, that electrons play a main role in the formation of sound in metal. We will consider sound formation in metals more detailly in Lecture 19.

Note, that if electrons are immobile, it means, that they are much heavier than ions. In this case there are two atoms in the elementary cell—heavy electron and light ion. Hence, we have now more than one atom in the elementary cell. As a result, the optical branches of the spectrum (which are connected with the relative motion of ion with respect to electron) should appear in the system [1].

Comment. Let us demonstrate, that it is possible to get the longitudinal sound in crystal just starting from the potential energy for the pairwise interaction in the static limit (without oscillations) [1].

The crystal energy without oscillations reads:

$$U = \frac{1}{2} \sum_{n \neq n'} \varphi\left(\boldsymbol{R}_{n,n'}\right),$$

where the pairwise potential can be expanded in the Fourier series as:

$$\varphi\left(\boldsymbol{R}_{n,n'}\right) = \frac{1}{V} \sum_{q} \varphi(q) \exp\left(i\boldsymbol{q}\,\boldsymbol{R}_{n,n'}\right).$$

As a result, the potential energy is given by:

$$U = \frac{1}{2} \sum_{n \neq n'} \frac{1}{V} \sum_{q} \varphi(q) \exp\left(i\boldsymbol{q}\,\boldsymbol{R}_{n,n'}\right)$$

$$= \frac{1}{2} \sum_{n,n'} \frac{1}{V} \sum_{q} \varphi(q) \exp\left(i\boldsymbol{q}\,\boldsymbol{R}_{n,n'}\right) - \frac{1}{2} \sum_{n} \frac{1}{V} \sum_{q} \varphi(q),\qquad(8.43)$$

where the second term in the last line of Eq. (8.43) corresponds to $n = n'$ and $\boldsymbol{R}_{n,n'} = 0$.

Further on let us fulfil summation over n' (which is equivalent to the summation over $m = n - n'$) in the first term in Eq. (8.43). While doing that let us utilize expression (8.12) for the lattice sum:

$$\sum_{n'} \exp\left(i\boldsymbol{q}\,\boldsymbol{R}_{n,n'}\right) = \sum_{K} N\delta_{q\boldsymbol{K}}.$$

Then the first term in Eq. (8.45) acquires a form:

$$\frac{1}{2}\sum_n \frac{1}{V}\sum_{q,K} N\delta_{qK}\varphi(q) = \frac{1}{2}\sum_n \frac{N}{V}\sum_K \varphi(K). \qquad (8.44)$$

The residual summation over n in the first term yields the total number of the elementary cells N. Correspondingly Eq. (8.44) can be written as:

$$\frac{1}{2}\frac{N^2}{V}\sum_K \varphi(K). \qquad (8.45)$$

Analogously implementing the summation over n in the second term of Eq. (8.43), we obtain:

$$\frac{1}{2}\frac{N}{V}\sum_q \varphi(q). \qquad (8.46)$$

But the summation by q in the second term is equivalent to the calculation of the integral $\frac{V}{(2\pi)^3}\int d^3q$. As a result, this term equals to:

$$\frac{1}{2}\frac{N}{V}\frac{V}{(2\pi)^3}\int d^3q\,\varphi(q) = \frac{1}{2}\frac{N}{(2\pi)^3}\int d^3q\,\varphi(q), \qquad (8.47)$$

and does not depend on the system volume V.

Further on we can find the pressure [1]:

$$P = -\frac{\partial U}{\partial V} = \frac{1}{2}\frac{N^2}{V^2}\sum_K \varphi(K) = \frac{1}{2}n^2\sum_K \varphi(K). \qquad (8.48)$$

Thus, we get for the sound velocity squared [1]:

$$c^2 = \frac{\partial P}{\partial \rho} = \frac{1}{M}\frac{\partial P}{\partial n} = \frac{n}{M}\sum_K \varphi(K) + \cdots = \frac{1}{MV_0}\sum_K \varphi(K) + \cdots, \qquad (8.49)$$

where $n = \frac{1}{V_0}$.

Note that additional terms in the expression for the sound velocity squared are connected with the variation of the inverse lattice vectors when we vary the system volume $\frac{\partial K}{\partial V} \neq 0$ [1].

Let us stress, that the calculation of the sound velocity in the static limit is equivalent to the calculation of the sound velocity starting from the equation of motion and the expression for the force coefficient [1]:

$$c^{ik}(f) = \frac{1}{MV_0} f^i f^k \sum_K \varphi(K). \qquad (8.50)$$

In this (dynamic) approach we also get for the sound velocity squared:

$$\omega^2 = c^2 f^2, \text{ and } c^2 = \frac{1}{M V_0} \sum_{K} \varphi(K).$$

(8.51)

8.4   Quantization of the Oscillations

Let us quantize the lattice vibrations [1,3–6,23,24]. The Hamiltonian of the system reads:

$$H = T + U,$$

(8.52)

where the crystal kinetic energy [1,3–6]:

$$T = \sum_{n,j} \frac{M_j}{2} \dot{u}_{nj}^2,$$

(8.53)

and potential energy [1,3–6]:

$$U = \frac{1}{2} \sum_{n,j,n',j'} A_{n,j,n',j'}^{ik} u_{nj}^i u_{n'j'}^k.$$

(8.54)

At the same time, equations of motion for the renormalized displacements have a form:

$$v_j^i \omega^2 = c_{jj'}^{ik}(f) v_{j'}^k,$$

where v_j^i is the displacement of the j-th atom of the elementary cell in the i-th direction (here and further on to simplify the reading of the formulas we are not making bold the indices n and n').

As we discussed already, for all the values of the wave vector f we get $3q$ branches of the phonon spectrum $\omega^2(f, \alpha)$ and $3q$ eigenvectors of polarization $e_j^i(f, \alpha)$, where index α is related to the number of the given phonon branch.

The polarization vectors correspond to the displacements v_j^i and are orthonormalized. It means, that for the given value of the wave vector f the following conditions are satisfied [1]:

$$\sum_{j \text{ (by atoms)}, \, i \text{ (by directions)}} e_j^i(f, \alpha) e_j^{i*}(f, \alpha') = \delta_{\alpha\alpha'},$$

(8.55)

$$\sum_{\alpha \text{ (by branches)}} e_j^i(f, \alpha) e_{j'}^{k*}(f, \alpha) = \delta^{ik} \delta_{jj'}.$$

(8.56)

Let us emphasize again, that index i connected with the direction of displacements runs over 3 values, while index j related to the number of atoms in the cell runs over q values. Finally index α corresponding to the number of the spectrum branch runs over $3q$ values.

Let us introduce the eigenfunctions of the vibrations of the atom, occupying the lattice site $\boldsymbol{R}_n$ [1]:

$$\Psi_{f,\alpha}\left(\boldsymbol{R}_n, i, j\right) = \frac{1}{\sqrt{N}} e_j^i(f, \alpha) \exp\left(i f \boldsymbol{R}_n\right). \tag{8.57}$$

Let us demonstrate, that for the eigenfunctions the normalization condition is fulfilled [1]:

$$\sum_{n,j,i} \Psi_{f,\alpha}\left(\boldsymbol{R}_n, i, j\right) \Psi_{f',\alpha'}^*\left(\boldsymbol{R}_n, i, j\right) = \delta_{ff'}\delta_{\alpha\alpha'}. \tag{8.58}$$

Indeed, revealing the left-hand side of the equality (8.58) and utilizing the definition of the lattice sum, we get:

$$\sum_{n,j,i} \frac{1}{N} e_j^i(f, \alpha) e_j^{i*}\left(f', \alpha'\right) \exp\left(i f \boldsymbol{R}_n\right) \exp\left(-i f' \boldsymbol{R}_n\right)$$

$$= \sum_{ji} e_j^i(f, \alpha) e_j^{i*}\left(f', \alpha'\right) \sum_n \frac{1}{N} \exp\left(i\left(f - f'\right) \boldsymbol{R}_n\right)$$

$$= \sum_{ji} e_j^i(f, \alpha) e_j^{i*}\left(f', \alpha'\right) \sum_{K} \frac{N}{N} \delta_{f-f',K} = \sum_{ji} e_j^i(f, \alpha) e_j^{i*}\left(f', \alpha'\right) \sum_{K} \delta_{f-f',K}.$$

$$\tag{8.59}$$

If we work in the one cell of the inverse lattice, then in Eq. (8.59) the summation over $\boldsymbol{K}$ is removed, and only one main term proportional to $\delta_{f-f',0}$ (or $\delta_{ff'}$) survives in the sum. As a result, the left-hand side of Eq. (8.58) yields [1]:

$$\sum_{n,j,i} \Psi_{f,\alpha}\left(\boldsymbol{R}_n, i, j\right) \Psi_{f',\alpha'}^*\left(\boldsymbol{R}_n, i, j\right) = \sum_{ji} e_j^i(f, \alpha) e_j^{i*}\left(f', \alpha'\right) \delta_{ff'}, \tag{8.60}$$

Further on, we can utilize orthogonality condition (8.55) for the polarization vectors and get the desired orthogonality condition for the wave functions in the right-hand side of Eq. (8.58) [1]:

$$\sum_{n,j,i} \Psi_{f,\alpha}\left(\boldsymbol{R}_n, i, j\right) \Psi_{f',\alpha'}^*\left(\boldsymbol{R}_n, i, j\right) = \delta_{ff'}\delta_{\alpha\alpha'}.$$

Let us return back now to the quantization of the system of Eqs. (8.52)–(8.54) for the initial displacements $u\sqrt{M} = v$ (or $u = \frac{v}{\sqrt{M}}$). For the reduction of the Hamiltonian (8.52) to the sum of the Hamiltonians of the harmonic oscillators it is convenient to expand the displacements u^i_{nj} in series on the polarization vectors $e^i_j(\beta)$ and normal coordinates Q_β [1,22,23]:

$$u^i_{nj} = \sum_\beta \frac{1}{\sqrt{NM_j}} \left(e^i_j(\beta) \exp\left(i f R_n\right) Q_\beta(t) + e^{i*}_j(\beta) \exp\left(-i f R_n\right) Q^*_\beta(t) \right),$$

$$(8.61)$$

where $\beta = (f, \alpha)$.

Let us remind, that the force matrix in the expression for the potential energy runs over $3qN$ values, where N is the number of the elementary cells, and $3q$ is the number of the spectrum branches. Note that just the same number of values (N) acquires the wave vector f in the crystal.

If the normal coordinates depend on time as:

$$Q_\beta(t) \sim \exp(-i\omega t),$$

then the velocities of the lattice displacements $\dot{u}^i_{nj}$ read [1]:

$$\dot{u}^i_{nj} = \sum_\beta \frac{1}{\sqrt{NM_j}} \left(e^i_j(\beta) \exp\left(i f R_n\right) Q_\beta \left(-i\omega_\beta\right) + e^{i*}_j(\beta) \exp\left(-i f R_n\right) Q^*_\beta \left(i\omega_\beta\right) \right). \quad (8.62)$$

As a result, expression for the kinetic energy yields [1]:

$$
\begin{aligned}
T &= \sum_{n,j,i} \frac{M_j}{2} \dot{u}^2_{nj} = \sum_{n,j} \frac{M_j}{2} \dot{u}^i_{nj} \dot{u}^i_{nj} \\
&= \sum_{n,j,i} \frac{M_j}{2} \sum_{\beta\beta'} \frac{1}{NM_j} \left[e^i_j(\beta) \exp\left(i f R_n\right) Q_\beta \left(-i\omega_\beta\right) \right. \\
&\quad \left. + e^{i*}_j(\beta) \exp\left(-i f R_n\right) Q^*_\beta \left(i\omega_\beta\right) \right] \times \left[\beta \to \beta' \right] \\
&= \sum_{n,j,i} \frac{1}{2N} \sum_{\beta\beta'} \left\{ -Q_\beta Q_{\beta'} e^i_j(\beta) e^i_j(\beta') \exp\left(i f R_n + i f' R_n\right) \right. \\
&\quad - Q^*_\beta Q^*_{\beta'} e^{i*}_j(\beta) e^{i*}_j(\beta') \exp\left(-i f R_n - i f' R_n\right) \\
&\quad \left. + 2 Q_\beta Q^*_{\beta'} e^i_j(\beta) e^{i*}_j(\beta') \exp\left(i f R_n - i f' R_n\right) \right\} \omega_\beta \omega_{\beta'}. \quad (8.63)
\end{aligned}
$$

Further on we can use the expressions for the eigenfunctions in Eq. (8.57) and represent Eq. (8.63) as [1]:

$$T = \frac{1}{2} \sum_{n,j,i} \sum_{\beta\beta'} \left\{ -Q_\beta Q_{\beta'} \frac{1}{\sqrt{N}} e^i_j(\beta) \exp\left(i f R_n\right) \frac{1}{\sqrt{N}} e^i_j(\beta') \exp\left(i f' R_n\right) \right.$$

$$- Q^*_\beta Q^*_{\beta'} \frac{1}{\sqrt{N}} e^{i*}_j(\beta) \exp\left(-i f R_n\right) \frac{1}{\sqrt{N}} e^{i*}_j(\beta') \exp\left(-i f' R_n\right)$$

$$\left. + 2 Q_\beta Q^*_{\beta'} \frac{1}{\sqrt{N}} e^i_j(\beta) \exp\left(i f R_n\right) \frac{1}{\sqrt{N}} e^{i*}_j(\beta') \exp\left(-i f' R_n\right) \right\} \omega_\beta \omega_{\beta'}$$

$$= \frac{1}{2} \sum_{n,j,i} \sum_{\beta\beta'} \left\{ -Q_\beta Q_{\beta'} \Psi_\beta \Psi_{\beta'} - Q^*_\beta Q^*_{\beta'} \Psi^*_\beta \Psi^*_{\beta'} + 2 Q_\beta Q^*_{\beta'} \Psi_\beta \Psi^*_{\beta'} \right\} \omega_\beta \omega_{\beta'}.$$

$$(8.64)$$

Now we can utilize an important condition, according to which the displacements u^i_{nj} take the real values in Eq. (8.61). It follows from this condition that for the polarization vectors [1]:

$$e^{i*}_j(f, \alpha) = e^i_j(-f, \alpha). \tag{8.65}$$

From this relation and the definition of the eigenfunctions [1]:

$$\Psi_\beta = \frac{1}{\sqrt{N}} e^i_j(\beta) \exp\left(i f R_n\right)$$

we obtain [1]:

$$\Psi^*_\beta = \frac{1}{\sqrt{N}} e^{i*}_j(\beta) \exp\left(-i f R_n\right) = \Psi_{-\beta}, \tag{8.66}$$

where $-\beta = (-f, \alpha)$.

With an account of the condition (8.66) on the eigenfunctions, we can represent expression (8.64) for the kinetic energy in the following convenient form [1]:

$$T = \frac{1}{2} \sum_{n,j,i} \sum_{\beta\beta'} \left\{ -Q_\beta Q_{\beta'} \Psi_\beta \Psi^*_{-\beta'} - Q^*_\beta Q^*_{\beta'} \Psi_{-\beta} \Psi^*_{\beta'} + 2 Q_\beta Q^*_{\beta'} \Psi_\beta \Psi^*_{\beta'} \right\} \omega_\beta \omega_{\beta'}.$$

$$(8.67)$$

Let us utilize now the powerful orthogonality condition (8.8) for the eigenfunctions. Note that, when we calculate the lattice sums in terms of $\beta\beta'$ in the main approximation (in the first inverse lattice cell) the orthogonality condition can be written in a compact form [1]:

$$\sum_{n,j,i} \Psi_\beta(R_n, i, j) \Psi^*_{\beta'}(R_n, i, j) = \delta_{\beta\beta'}. \tag{8.68}$$

As a result, all the summations except one by β can be removed in the expression for the kinetic energy and thus [1]:

$$T = \frac{1}{2} \sum_{\beta} \left\{ -Q_{\beta} Q_{-\beta} - Q_{\beta}^{*} Q_{-\beta}^{*} + 2 Q_{\beta} Q_{\beta}^{*} \right\} \omega_{\beta}^{2}, \tag{8.69}$$

where we used the relation $\omega_{\beta} = \omega_{-\beta}$, connected with the parity of the phonon frequencies as the functions of the wave vector.

Let us quantize now expression (8.54) for the potential energy [1,4]:

$$U = \frac{1}{2} \sum_{n,j,n',j'} A^{ik}_{n,j,n',j'} u^{i}_{nj} u^{k}_{n'j'}$$

by normal coordinates.

To do that we can utilize equations of motion [1,4]:

$$M_{j} \ddot{u}^{i}_{nj} = - \sum_{n'j'} A^{ik}_{n,j,n',j'} u^{k}_{n'j'},$$

and express the force matrix in Eq. (8.54) via the acceleration of the lattice displacements. Then, the potential energy can be reduced to the form which is convenient for the quantization [1,4]:

$$U = -\frac{1}{2} \sum_{n,j,i} M_{j} \ddot{u}^{i}_{nj} u^{i}_{nj}. \tag{8.70}$$

Substituting in Eq. (8.70) expression for the displacements (8.61) and for their accelerations [1,4]:

$$\ddot{u}^{i}_{nj} = - \sum_{\beta} \frac{\omega_{\beta}^{2}}{\sqrt{N M_{j}}} \left(e^{i}_{j}(\beta) \exp\left(i \boldsymbol{f} \boldsymbol{R}_{n}\right) Q_{\beta} + e^{i*}_{j}(\beta) \exp\left(-i \boldsymbol{f} \boldsymbol{R}_{n}\right) Q_{\beta}^{*} \right), \tag{8.71}$$

and taking into account relation (8.66) $\Psi_{\beta}^{*} = \Psi_{-\beta}$ for the eigenfunctions [1], we get for the potential energy:

$$U = \sum_{n,j,i} \frac{M_{j}}{2} \sum_{\beta\beta'} \frac{\omega_{\beta}^{2}}{N M_{j}} \left(e^{i}_{j}(\beta) \exp\left(i \boldsymbol{f} \boldsymbol{R}_{n}\right) Q_{\beta} + e^{i*}_{j}(\beta) \exp\left(-i \boldsymbol{f} \boldsymbol{R}_{n}\right) Q_{\beta}^{*} \right)$$

$$\times \left(\beta \to \beta' \right) = \frac{1}{2} \sum_{n,j,i} \sum_{\beta\beta'} \left\{ Q_{\beta} Q_{\beta'} \Psi_{\beta} \Psi_{\beta'} + Q_{\beta}^{*} Q_{\beta'}^{*} \Psi_{\beta}^{*} \Psi_{\beta'}^{*} + 2 Q_{\beta} Q_{\beta'}^{*} \Psi_{\beta} \Psi_{\beta'}^{*} \right\} \omega_{\beta}^{2}$$

$$= \frac{1}{2} \sum_{n,j,i} \sum_{\beta\beta'} \left\{ Q_{\beta} Q_{\beta'} \Psi_{\beta} \Psi_{-\beta'}^{*} - Q_{\beta}^{*} Q_{\beta'}^{*} \Psi_{-\beta} \Psi_{\beta'}^{*} + 2 Q_{\beta} Q_{\beta'}^{*} \Psi_{\beta} \Psi_{\beta'}^{*} \right\} \omega_{\beta}^{2}. \tag{8.72}$$

Let us use again the orthogonality condition for the eigenfunctions (8.68). Then the potential energy equals to [1,4,22–24]:

$$U = \frac{1}{2} \sum_{\beta} \left\{ Q_\beta Q_{-\beta} + Q_\beta^* Q_{-\beta}^* + 2 Q_\beta Q_\beta^* \right\} \omega_\beta^2. \tag{8.73}$$

Thus, the Hamiltonian of the system [1,4,22,23]:

$$H = T + U = \frac{1}{2} \sum_{\beta} 4 Q_\beta Q_\beta^* \omega_\beta^2 = 2 \sum_{\beta} Q_\beta Q_\beta^* \omega_\beta^2. \tag{8.74}$$

Now, according to the Quantum mechanics prescriptions, we can introduce canonical coordinates q_β and momenta p_β for the generalized oscillators [1,23–25]:

$$q_\beta = Q_\beta + Q_\beta^*, \tag{8.75}$$

$$\dot{q}_\beta = -i\omega_\beta Q_\beta + i\omega_\beta Q_\beta^* = -i\omega_\beta \left(Q_\beta - Q_\beta^* \right) = p_\beta. \tag{8.76}$$

Correspondingly, initial normal coordinates are given by [1,22–24]:

$$Q_\beta = \frac{1}{2} \left(q_\beta - \frac{p_\beta}{i\omega_\beta} \right), \; Q_\beta^* = \frac{1}{2} \left(q_\beta + \frac{p_\beta}{i\omega_\beta} \right). \tag{8.77}$$

As a result, the Hamiltonian of the system in terms of the canonical coordinates and momenta can be represented as a sum of the Hamiltonians of the generalized oscillators [1,22,23]:

$$H = \frac{1}{2} \sum_{\beta} \left(p_\beta^2 + \omega_\beta^2 q_\beta^2 \right), \tag{8.78}$$

where $\beta = (f, \alpha)$ runs over $3qN$ values (we have $3qN$ oscillators in our system).

For the quantization of the Hamiltonian (8.78) we have to introduce the creation and annihilation operators and establish their commutation relations [1,23–25]. To do that we should write down explicitly the non-zero matrix elements (subdiagonal and superdiagonal) for the canonical operators $\hat{q}_\beta$ and $\hat{p}_\beta$ [1,23–25]:

$$\left(q_\beta \right)_{n-1,n} = \left(q_\beta \right)_{n,n-1} = \sqrt{\frac{\hbar}{2\omega_\beta}} \sqrt{n}, \tag{8.79}$$

$$\left(p_\beta \right)_{n-1,n} = -i\omega_\beta \left(q_\beta \right)_{n-1,n}, \tag{8.80}$$

$$\left(p_\beta \right)_{n,n-1} = i\omega_\beta \left(q_\beta \right)_{n,n-1}. \tag{8.81}$$

Then non-zero matrix elements of the initial normal coordinates read [1,23–25]:

$$(Q_\beta)_{n-1,n} = \frac{1}{2}\left(q_\beta - \frac{p_\beta}{i\omega_\beta}\right)_{n-1,n} = (q_\beta)_{n-1,n}, \tag{8.82}$$

$$(Q_\beta)_{n,n-1} = 0. \tag{8.83}$$

Now we can introduce creation operators $\hat{a}_\beta^\dagger$ and annihilation operators $\hat{a}_\beta$ for phonons as [1,22–24]:

$$\hat{Q}_\beta = \sqrt{\frac{\hbar}{2\omega_\beta}}\,\hat{a}_\beta, \text{ and } (\hat{a}_\beta)_{n-1,n} = \sqrt{n}, \tag{8.84}$$

$$\hat{Q}_\beta^\dagger = \sqrt{\frac{\hbar}{2\omega_\beta}}\,\hat{a}_\beta^\dagger, \text{ and } \left(\hat{a}_\beta^\dagger\right)_{n,n-1} = \sqrt{n}. \tag{8.85}$$

It is easy to understand that the creation and annihilation operators which we introduce satisfy bosonic commutation relations [1,23–25]:

$$\langle n|\left[\hat{a}_\beta\hat{a}_\beta^\dagger\right]|n\rangle = \langle n|\hat{a}_\beta\hat{a}_\beta^\dagger - \hat{a}_\beta^\dagger\hat{a}_\beta|n\rangle$$
$$= (\hat{a}_\beta)_{n,n+1}\left(\hat{a}_\beta^\dagger\right)_{n+1,n} - \left(\hat{a}_\beta^\dagger\right)_{n,n-1}(\hat{a}_\beta)_{n-1,n} = \sqrt{n+1}\sqrt{n+1} - \sqrt{n}\sqrt{n}$$
$$= n + 1 - n = 1. \tag{8.86}$$

Then, the Hamiltonian of the system can be written in the canonical form [1,22–24]:

$$H = 2\sum_\beta Q_\beta Q_\beta^* \omega_\beta^2 = \sum_\beta \left(\hat{Q}_\beta\hat{Q}_\beta^\dagger + \hat{Q}_\beta^\dagger\hat{Q}_\beta\right)\omega_\beta^2$$
$$= \tfrac{1}{2}\hbar\sum_\beta \omega_\beta\left(\hat{a}_\beta\hat{a}_\beta^\dagger + \hat{a}_\beta^\dagger\hat{a}_\beta\right). \tag{8.87}$$

But from the commutation relation (8.86) it follows that:

$$\hat{a}_\beta\hat{a}_\beta^\dagger = \hat{a}_\beta^\dagger\hat{a}_\beta + 1. \tag{8.88}$$

Hence the Hamiltonian of the system [1,23–25]:

$$H = \sum_\beta \hbar\omega_\beta\left(\hat{a}_\beta^\dagger\hat{a}_\beta + \frac{1}{2}\right) = \sum_\beta \hbar\omega_\beta\left(\hat{n}_\beta + \frac{1}{2}\right), \tag{8.89}$$

where $\hat{n}_\beta = \hat{a}_\beta^\dagger\hat{a}_\beta$ is the number of particles (number of phonons) operator.

We should emphasize, that the nth oscillator level with the given frequency ω_β in the Hamiltonian (8.89), as well as the n-th line or column number in the matrix elements (8.86), correspond to the state with n phonons:

$$\langle n|\hat{n}_\beta|n\rangle = \langle n|a_\beta^\dagger a_\beta|n\rangle = n. \tag{8.90}$$

At the same time, the operator of the atom displacement at the lattice node $\boldsymbol{R}_n$ in terms of phonon creation and annihilation operators has a form [1]:

$$\hat{u}_{nj}^i = \sum_\beta \sqrt{\frac{\hbar}{2M_j N \omega_\beta}} \left[e_j^i(\beta) \exp\left(i\boldsymbol{f}\boldsymbol{R}_n\right) \hat{a}_\beta + e_j^{i*}(\beta) \exp\left(-i\boldsymbol{f}\boldsymbol{R}_n\right) \hat{a}_\beta^\dagger \right]. \tag{8.91}$$

We can see, that the displacement operators in Eq. (8.91) contain the polarization vectors $e_j^i(\beta)$ and phonon frequencies ω_β, depending not only on the wave vectors $\boldsymbol{f}$, but also on the branch number of the phonon spectrum α.

At the same time, in case of photons (for the quantization of the field of the electromagnetic wave [26,27]) only vectors of the field polarization emerge. Meanwhile the photon frequencies depend only on the wave vector $\boldsymbol{f}$ (and not on the spectrum branch number α).

References

1. Y. Kagan, *Lectures on the Solid-State Theoretical Physics* (Moscow, MEPHI, 1981–1982). Part I. Phonons - unpublished
2. E.G. Brovman, Yu. Kagan, Phonons in non-transition metals. Sov. Physics Uspekhi **17**, 25 (1974)
3. C. Kittel, *Quantum Theory of Solids* (Wiley, New York-London, 1963)
4. O. Madelung, *Introduction to Solid-State Theory*, Springer Series in Solid-State Sciences (1995)
5. N.W. Ashcroft, N.D. Mermin, *Solid State Physics* (Holt, Rinehart and Winston, 1976)
6. J.M. Ziman, *Electrons and Phonons: The Theory of Transport Phenomena in Solids* (Clarendon Press, Oxford, 1960)
7. N.N. Bogoliubov, J. Phys. USSR **11**, 23 (1947)
8. A.A. Abrikosov, L.P. Gor'kov, I.E. Dzyaloshinskii, *Methods of Quantum Field Theory in Statistical Physics* (Prentice Hall, Englewood Cliffs, New Jersey, 1963)
9. P. Nozieres, D. Pines, *The Theory of Quantum Liquids: Superfluid Bose Liquids*, vol. 2 (Addison Wesley Publishing Company, USA, Canada, 1990)
10. V.V. Tolmachev, *The Bose Gas Theory* (Publishing House of Moscow State University, 1969) (in Russian)
11. L.D. Landau, A.M. Kosevich, E.M. Lifshitz, L.P. Pitaevskii, Theory of elasticity, in *Course of Theoretical Physics*, vol. 7 (Elsevier, 1986)
12. A.A. Abrikosov, *Fundamentals of the Theory of Metals* (Elsevier Science Publishers, North-Holland, Amsterdam, 1988)
13. I.M. Lifshitz, M.Y. Azbel', M.I. Kaganov, *Electron Theory of Metals* (Consultants Bureau, New York, 1973)
14. D. Pines, *Elementary Excitations in Solids* (W.A. Benjamin Inc., New York-Amsterdam, 1963)
15. L. Tonks, I. Langmuir, Phys. Rev. **33**, 195 (1929)
16. E.M. Lifshitz, L.P. Pitaevskii, *Physical Kinetics* (Butterworth-Heinemann, 2012)

17. L.D. Landau, E.M. Lifshitz, *Electrodynamics of Continuous Media* (Pergamon Press, New York, 1984)
18. D. Pines, P. Nozieres, *The Theory of Quantum Liquids*, vol. 1 (Normal Fermi Liquids, W.A. Benjamin Inc. New York-Amsterdam, 1966)
19. S.N. Burmistrov, *Physical Kinetics: Classical and Quantum Problems and Solutions* (Springer, Berlin, 2022)
20. L.H. Thomas, Math. Proc. Cambridge Philos. Soc. **23**, 542 (1927)
21. E. Fermi, Rend. Accad. Naz. Lincei **6**, 602 (1927)
22. H. Yukawa, Proc. Phys.-Mat. Soc. Japan **17**, 48 (1935)
23. R.P. Feynman, *Statistical Mechanics, A Set of Lectures (Advanced Books, Classics)* (Westview Press, Boulder, 1998)
24. L.D. Landau, E.M. Lifshitz, *Statistical Physics, Part I* (Butterworth-Heinemann, Oxford, 1999)
25. L.D. Landau, E.M. Lifshitz, *Quantum Mechanics: Non-Relativistic Theory* (Pergamon Press, New York, 1977)
26. P.A.M. Dirac, Proc. Royal Soc. London **A114**, 243 (1927)
27. R.P. Feynman, *QED: The Strange Theory of Light and Matter* (Princeton University Press 1985)

Lecture 9. Phonon Distribution Function: Specific Heat of the Crystal

9

Abstract

We introduce the Planck distribution function for phonons and calculate the phonon contribution to the thermodynamic quantities and derivatives (including Gruneisen constant and the coefficient of thermal expansion). We find low-temperature (quantum) and high-temperature (classical) behavior of the specific heat and mean squared atom displacement in the Debye model. We illustrate important infrared divergencies of the mean squared displacement in crystals of reduced dimensionality.

Summarizing the results of the previous Lecture, we can write the phonon Hamiltonian as [1–10]

$$\widehat{H} = \sum_{\beta} \hbar\omega_{\beta} \left(\hat{a}_{\beta}^{\dagger}\hat{a}_{\beta} + \frac{1}{2} \right), \tag{9.1}$$

where $\beta = (\boldsymbol{f}, \alpha)$, and $-\beta = (-\boldsymbol{f}, \alpha)$.

At the same time, the energy of the system reads [1–10]

$$E = \langle \{n_{\beta}\} | \widehat{H} | \{n_{\beta}\} \rangle = \sum_{\beta} \hbar\omega_{\beta} \left(n_{\beta} + \frac{1}{2} \right), \tag{9.2}$$

where n_{β} are the integer numbers.

Correspondingly, averaged over temperature (thermodynamically equilibrium) energy is given by

$$\bar{E} = \sum_{\beta} \hbar\omega_{\beta} \left(\bar{n}_{\beta} + \frac{1}{2} \right), \tag{9.3}$$

where $\bar{n}_{\beta}$ are the averaged filling numbers (which are not integer now).

The averaged filling numbers are described by the Planck distribution function [1–10]:

$$\bar{n}_\beta = \frac{1}{\exp\left(\frac{\hbar\omega_\beta}{T}\right) - 1}. \tag{9.4}$$

Note that zero value of the chemical potential ($\mu = 0$) in Eq. (9.4) is connected with the fact that the number of phonons is not fixed [1,4].

Proceeding in Eq. (9.3) from the summation over β to the integral, we can write the average energy in the three-dimensional case as

$$\bar{E} = \frac{V}{(2\pi)^3} \sum_\alpha \int d^3 f \, \hbar\omega_\alpha(f) \left(\frac{1}{\exp\left(\frac{\hbar\omega_\alpha(f)}{T}\right) - 1} + \frac{1}{2} \right). \tag{9.5}$$

Recollecting now the definition of the average density of states $g(\omega)$ of the phonon system [1]:

$$g(\omega) = \frac{1}{3q} \sum_\alpha g_\alpha(\omega), \tag{9.6}$$

where q is the number of atoms in the cell, $g_\alpha(\omega)$ is the density of states for the branch of the phonon spectrum with index α, we can write the measure of integration in Eq. (9.5) in the following form:

$$\frac{V}{(2\pi)^3} \int d^3 f = 3Nq \int d\omega g(\omega), \tag{9.7}$$

where N is the number of the elementary cells.

As a result, the average energy equals to [1]

$$\bar{E} = 3Nq \int_0^{\omega_{\max}} d\omega g(\omega) \hbar\omega \left(\frac{1}{\exp\left(\frac{\hbar\omega}{T}\right) - 1} + \frac{1}{2} \right)$$

$$= E_{\text{oscil.}}(T) + E_{\text{zero oscil.}}, \tag{9.8}$$

where $\omega_{\max}$ is the maximal frequency of the phonon spectrum, and $g(\omega) = 0$ for $\omega > \omega_{\max}$ (here and further on we put $k_B = 1$).

The temperature-dependent part of the oscillation energy in Eq. (9.8) reads [1,9]

$$E_{\text{oscil.}}(T) = 3Nq \int_0^{\omega_{\max}} d\omega g(\omega) \frac{\hbar\omega}{\exp\left(\frac{\hbar\omega}{T}\right) - 1}, \tag{9.9}$$

meanwhile, the energy of zero oscillations, which is temperature-independent, yields [1,9]

$$E_{\text{zero oscil.}} = 3Nq \int_0^{\omega_{\max}} d\omega g(\omega) \frac{\hbar\omega}{2}. \tag{9.10}$$

If we consider low temperatures $T \ll \hbar\omega_{max}$, then the typical frequencies $\hbar\omega_{typ.} \sim T \ll \hbar\omega_{max}$, and the Debye model [11] works pretty well (in this case only low frequencies with the linear spectrum are excited). As we already discussed, in the Debye approximation, the average density of states in the 3D case: $g(\omega) = \frac{3\omega^2}{\omega_D^3}$, and number of atoms in the cell $q = 1$ (in this approximation we consider only acoustic branches as in the crystal with one atom in the elementary cell). At the same time, the Debye temperature (for $k_B = 1$) [11]:

$$\theta_D = \hbar\omega_D \sim 100 \text{ K}.$$

As a result, introducing dimensionless variable $x = \frac{\hbar\omega}{T}$, for $T \ll \theta_D$, we get for the temperature-dependent part of the oscillation energy [1,9]:

$$E_{oscil.}(T) = 3N \int_0^\infty d\omega \frac{3\omega^2}{\omega_D^3} \frac{\hbar\omega}{\exp\left(\frac{\hbar\omega}{T}\right) - 1}$$

$$= 3N \frac{3}{(\hbar\omega_D)^3} T^4 \int_0^\infty \frac{dx\, x^3}{\exp(x) - 1} \propto T^4. \tag{9.11}$$

For the arbitrary dimension, the density of states in the Debye approximation $g(\omega) \sim \omega^{D-1}$ [11], and temperature-dependent part of the oscillation energy:

$$E_{oscil.}^{(D)}(T) \sim \int d\omega\omega g(\omega) \propto T^{D+1}. \tag{9.12}$$

Correspondingly, at low temperatures, the phonon contribution to the specific heat in the crystal of arbitrary dimensions reads [1–10]

$$C_V = \left(\frac{\partial E}{\partial T}\right)_V \propto T^D, \tag{9.13}$$

where we utilized the fact that, for the constant crystal volume, we have constant frequencies and the inverse lattice vectors do not change as well.

In the opposite high-temperature limit, for $\hbar\omega \leq \hbar\omega_{max} \ll T$, we can replace the Planck distribution function by its classical limit while calculating the temperature-dependent part of the energy [1,4]:

$$\frac{1}{\exp\left(\frac{\hbar\omega}{T}\right) - 1} \to \frac{T}{\hbar\omega}. \tag{9.14}$$

Then, using the normalization condition for the density of states:

$$\int g(\omega)d\omega = 1,$$

we get for the energy from Eq. (9.9) [1,9]:

$$E_{\text{oscil.}}(T) = 3Nq\frac{T}{\hbar\omega}\hbar\omega = 3NqT \propto T \tag{9.15}$$

for all dimensionalities.

Hence, the phonon specific heat at high temperatures equals to the number of the freedom degrees [1,4,9]:

$$C_V = 3Nq. \tag{9.16}$$

For the D-dimensional crystal $3 \rightarrow D$ in Eq. (9.16), and specific heat yields

$$C_V = DNq. \tag{9.17}$$

Let us analyze the next terms in the expansion of the Planck distribution function, in order to understand how the temperature-dependent part of the energy approaches to $3NqT$, and correspondingly how the specific heat goes to $3Nq$ at high temperatures [1].

To do that, it is convenient to utilize the general Formula (9.8) for the total energy without separating it on two contributions (from the temperature-dependent and zero oscillations parts). Thus, it is more convenient to expand in series on small parameter $\frac{\hbar\omega}{T}$ not the Planck distribution function itself, but the sum $\left(\frac{1}{\exp\left(\frac{\hbar\omega}{T}\right)-1} + \frac{1}{2} \right)$, which enters in the formula for the total energy [1]:

$$\left(\frac{1}{\exp\left(\frac{\hbar\omega}{T}\right) - 1} + \frac{1}{2} \right) \rightarrow \frac{T}{\hbar\omega} \left(\frac{1}{1 + \frac{1}{2}\frac{\hbar\omega}{T} + \frac{1}{6}\left(\frac{\hbar\omega}{T}\right)^2} \right) + \frac{1}{2}$$

$$= \frac{T}{\hbar\omega} \left\{ 1 - \frac{1}{2}\frac{\hbar\omega}{T} - \frac{1}{6}\left(\frac{\hbar\omega}{T}\right)^2 + \left(\frac{1}{2}\frac{\hbar\omega}{T} + \frac{1}{6}\left(\frac{\hbar\omega}{T}\right)^2 \right)^2 \right\} + \frac{1}{2}$$

$$\approx \frac{T}{\hbar\omega} - \frac{1}{2} + \frac{1}{2} - \frac{1}{6}\left(\frac{\hbar\omega}{T}\right)^2 + \frac{1}{4}\left(\frac{\hbar\omega}{T}\right)^2 = \frac{T}{\hbar\omega}\left(1 + \frac{1}{12}\frac{\hbar^2\omega^2}{T^2} \right). \tag{9.18}$$

As a result, the total energy is given by [1]

$$\bar{E} = 3NqT \int d\omega g(\omega) \left(1 + \frac{1}{12}\frac{\hbar^2\omega^2}{T^2} \right). \tag{9.19}$$

Hence the phonon specific heat at high temperatures reads [1]

$$C_V = 3Nq \left(1 - \frac{1}{12}\frac{\hbar^2 \langle\omega^2\rangle}{T^2} \right), \tag{9.20}$$

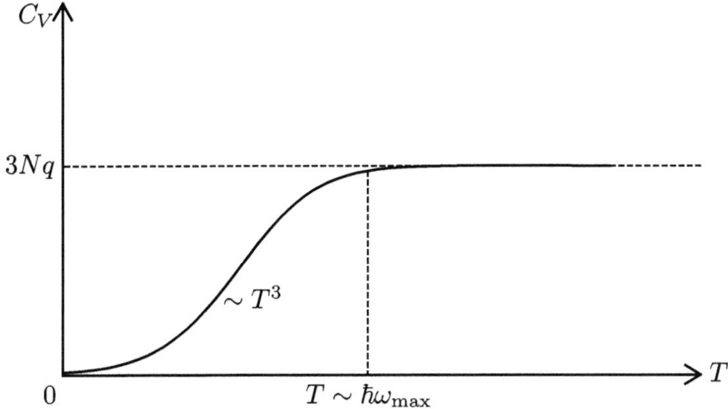

Fig. 9.1 Temperature dependence of the phonon specific heat in the three-dimensional crystal. At low temperatures specific heat $C_V \propto T^3$. It goes on the saturation value (which equals to the number of freedom degrees) at higher temperatures [1]

where the average frequency squared equals to [1]

$$\langle \omega^2 \rangle = \int d\omega g(\omega)\omega^2. \tag{9.21}$$

Let us stress that, if we know the force constants of the lattice, we can determine exactly the quantity $\langle \omega^2 \rangle$ [1].

It is interesting to note that the correction to the classical limit $C_V = 3Nq$ for the specific heat is negative and tends to zero for $\hbar \to 0$ [1]. Moreover, this correction is proportional to $\hbar^2$ as usual [1].

Let us emphasize that the phonon specific heat reaches the classical limit at the temperatures $T \sim \hbar\omega_{\max}$ (and not at the higher temperatures $T \gg \hbar\omega_{\max}$) due to the small numerical coefficient $\frac{1}{12}$ [1] in front of the correction term $-\frac{1}{12}\frac{\hbar^2\langle\omega^2\rangle}{T^2}$ in Eq. (9.20).

The temperature dependence of the phonon specific heat in the three-dimensional crystal is presented in Fig. 9.1.

9.1   Statistical Sum and Phonon Contribution to the Free Energy

To finalize the thermodynamic description of the phonon subsystem, we should determine also the phonon contribution to the Free energy F and the coefficient of the thermal expansion proportional to $\left(\frac{\partial V}{\partial T}\right)$ [1].

According to the general prescriptions of the Statistical physics, the Free energy, corresponding to the crystalline lattice oscillations, is given by [1,2,4]

$$F_{\text{oscil.}} = -T \ln Z, \qquad (9.22)$$

where Z is the statistical sum.

If the statistical sum is the product of the partial statistical sums:

$$Z = \prod_{\beta} Z_{\beta},$$

then the Free energy is the sum of the logarithms of the partial contributions [1,2,4] (multiplied by temperature):

$$F_{\text{oscil.}} = -T \ln Z = -T \sum_{\beta} \ln Z_{\beta}. \qquad (9.23)$$

The partial statistical sum Z_{β}, corresponding to the harmonic oscillator with the frequency ω_{β} and the spectrum $E_{\beta} = \hbar \omega_{\beta} \left(n_{\beta} + \frac{1}{2} \right)$, equals to the sum of the geometrical progression [1,2,4]:

$$
\begin{aligned}
Z_{\beta} &= \exp\left(-\frac{\hbar\omega_{\beta}}{2T}\right) + \exp\left(-\frac{3\hbar\omega_{\beta}}{2T}\right) + \exp\left(-\frac{5\hbar\omega_{\beta}}{2T}\right) + \cdots \\
&= \exp\left(-\frac{\hbar\omega_{\beta}}{2T}\right) \sum_{n_{\beta}=0}^{\infty} \exp\left(-\frac{n_{\beta}\hbar\omega_{\beta}}{T}\right) = \frac{\exp\left(-\frac{\hbar\omega_{\beta}}{2T}\right)}{1 - \exp\left(-\frac{\hbar\omega_{\beta}}{T}\right)}.
\end{aligned} \qquad (9.24)
$$

Then, the Free energy yields [1]

$$
\begin{aligned}
F_{\text{oscil.}} &= -T \sum_{\beta} \ln Z_{\beta} = -T \sum_{\beta} \ln\left(\frac{\exp\left(-\frac{\hbar\omega_{\beta}}{2T}\right)}{1 - \exp\left(-\frac{\hbar\omega_{\beta}}{T}\right)} \right) \\
&= \sum_{\beta} \frac{\hbar\omega_{\beta}}{2} + T \sum_{\beta} \ln\left(1 - \exp\left(-\frac{\hbar\omega_{\beta}}{T}\right) \right) \\
&= F_{\text{zero oscil.}}(V) + F_{\text{oscil.}}(V,T),
\end{aligned} \qquad (9.25)
$$

where the Free energy of zero oscillations:

$$F_{\text{zero oscil.}}(V) = E_{\text{zero oscil.}} = \sum_{\beta} \frac{\hbar\omega_{\beta}}{2}$$

coincides with the energy of zero oscillations.

At the same time:

$$F_{\text{oscil.}}(V, T) = T \sum_{\beta} \ln\left(1 - \exp\left(-\frac{\hbar\omega_\beta}{T}\right)\right) \tag{9.26}$$

is the temperature-dependent part of the Free energy [1].

Let us emphasize that both the temperature-independent energy of zero oscillations and the temperature-dependent part of the Free energy, depend also on the system volume V [1].

The Free energy F is connected with the thermodynamic energy $\bar{E}$ by the well-known relation [4]:

$$F = \bar{E} - TS. \tag{9.27}$$

The differential of the thermodynamic energy reads [4]

$$d\bar{E} = TdS - PdV. \tag{9.28}$$

Hence, the differential of the Free energy is given by [4]

$$dF = -SdT - PdV \tag{9.29}$$

Correspondingly, the pressure of the system equals to [1,4]

$$P = -\left(\frac{\partial F}{\partial V}\right) = -\left(\frac{\partial F_0}{\partial V}\right) - \left(\frac{\partial F_{\text{zero oscil.}}}{\partial V}\right) - \left(\frac{\partial F_{\text{oscil.}}}{\partial V}\right)_T$$

$$= -\left(\frac{\partial F_0}{\partial V}\right) - \sum_{\beta} \frac{\hbar}{2} \frac{\partial \omega_\beta}{\partial V} - T \sum_{\beta} \frac{\exp\left(-\frac{\hbar\omega_\beta}{T}\right)}{1 - \exp\left(-\frac{\hbar\omega_\beta}{T}\right)} \frac{\hbar}{T} \frac{\partial \omega_\beta}{\partial V}. \tag{9.30}$$

After simple transformations we get for the pressure [1]:

$$P = -\left(\frac{\partial F_0}{\partial V}\right) - \sum_{\beta} \frac{\hbar}{2} \frac{\partial \omega_\beta}{\partial V} - \sum_{\beta} \frac{\hbar}{\exp\left(\frac{\hbar\omega_\beta}{T}\right) - 1} \frac{\partial \omega_\beta}{\partial V}$$

$$= P_0 + P_{\text{zero oscil.}} + \frac{1}{V} \sum_{\beta} \hbar\omega_\beta \gamma_\beta \bar{n}_\beta, \tag{9.31}$$

where the first contribution to the pressure in Eq. (9.31):

$$P_0 = -\left(\frac{\partial F_0}{\partial V}\right) \tag{9.32}$$

is related to the fact that the crystal has the equilibrium density [1]. At the same time, the zero oscillations pressure in (9.31) is written as [1]

$$P_{\text{zero oscil.}} = -\sum_{\beta} \frac{\hbar}{2} \frac{\partial \omega_{\beta}}{\partial V} = \frac{1}{V} \sum_{\beta} \frac{\hbar \omega_{\beta}}{2} \gamma_{\beta}. \tag{9.33}$$

Note that in Eqs. (9.31) and (9.33), we introduced the partial Gruneisen constant [1,6,12,13]:

$$\gamma_{\beta} = -\frac{\partial \omega_{\beta}}{\partial V} \frac{V}{\omega_{\beta}} = -\frac{\partial \ln \omega_{\beta}}{\partial \ln V}. \tag{9.34}$$

If we introduce also the Gruneisen constant [1,6,12,13] $\gamma = \bar{\gamma}_{\beta}$ averaged over all the spectrum branches, then the pressure in Eq. (9.31) can be represented as follows [1]:

$$P = P_0 + \gamma \frac{E_{\text{zero oscil.}}}{V} + \gamma \frac{E_{\text{oscil.}}(T)}{V}, \tag{9.35}$$

where the temperature-dependent part of the oscillations energy $E_{\text{oscil.}}(T)$ in the last term of Eq. (9.35) is determined by the Formula (9.9).

It is interesting to note that the last temperature-dependent term in the expression for pressure in (9.35) is analogous to the pressure of the classical ideal gas of particles [1]. In the three-dimensional case, the pressure of the ideal gas [1,4]:

$$P = N k_B T = \frac{\left(\frac{3}{2} k_B T\right)\left(\frac{2}{3} N\right)}{V} \sim \frac{2E}{3V}. \tag{9.36}$$

Let us stress that the typical value of the Gruneisen constant in crystal $\gamma \approx 2$ [1,6,12,13]. We can say that the phonons behave themselves as the phonon gas [1]. Let us write more explicitly the derivative $\left(\frac{\partial V}{\partial T}\right)_P$, which determines the thermal expansion of the crystal. According to the general rules of the Statistical physics [4] for the derivatives of the thermodynamic quantities, we obtain [1]

$$\left(\frac{\partial V}{\partial T}\right)_P = \frac{\partial(V, P)}{\partial(T, P)} = \frac{\partial(V, P)}{\partial(V, T)} \frac{\partial(V, T)}{\partial(T, P)} = \left(\frac{\partial P}{\partial T}\right)_V \left(\frac{\partial V}{\partial P}\right)_T. \tag{9.37}$$

But we can see that the derivative [1]:

$$\left(\frac{\partial P}{\partial T}\right)_V = \frac{\gamma}{V} \left(\frac{\partial E_{\text{oscil.}}(T)}{\partial T}\right)_V = \frac{\gamma}{V} C_V(T). \tag{9.38}$$

At the same time, in the main approximation, the derivative:

$$\left(\frac{\partial P}{\partial V}\right)_T \approx \frac{\partial P_0}{\partial V} \tag{9.39}$$

is related to the equilibrium compressibility of the crystal [1].

Hence, we get for the thermal expansion [1]

$$\left(\frac{\partial V}{\partial T}\right)_P \approx \frac{\gamma}{V} C_V(T) \left(\frac{\partial P_0}{\partial V}\right)^{-1} \propto C_V(T). \tag{9.40}$$

We can say that the temperature dependence of the thermal expansion is completely determined by the phonon gas [1].

9.2 Mean Square of Atom Displacement

In the three-dimensional crystal with j-atoms in the elementary cell, it is important to calculate also the mean square of atom displacement:

$$\left\langle u_{nj}^2 \right\rangle = \left\langle \left(u_{nj}^x\right)^2 + \left(u_{nj}^y\right)^2 + \left(u_{nj}^z\right)^2 \right\rangle, \tag{9.41}$$

where the displacement operator, as it was shown in the previous Lectures, equals to [1]

$$\hat{u}_{nj}^i = \sum_\beta \sqrt{\frac{\hbar}{2M_j\omega_\beta N}} \left[e_j^i \exp\left(if\boldsymbol{R}_n\right) \hat{a}_\beta + e_j^{i*} \exp\left(-if\boldsymbol{R}_n\right) \hat{a}_\beta^\dagger\right]. \tag{9.42}$$

Thus, for the square of the x-component of the displacement $\left(u_{nj}^x\right)^2$, we have to calculate the average of [1]

$$\left\langle \left(u_{nj}^x\right)^2 \right\rangle = \left\langle \{n\} \left| \sum_{\beta\beta_1} \frac{\hbar}{2M_j N} \frac{1}{\sqrt{\omega_\beta \omega_{\beta_1}}} \left[e_j^x \exp\left(if\boldsymbol{R}_n\right) \hat{a}_\beta \right.\right.\right.$$
$$\left.\left.\left. + \ e_j^{x*} \exp\left(-if\boldsymbol{R}_n\right) \hat{a}_\beta^\dagger\right] \times [\beta \rightarrow \beta_1] \right| \{n\} \right\rangle. \tag{9.43}$$

Let us utilize the relations for the different bilinear combinations of the creation and the annihilation operators. The diagonal matrix elements for first two bilinear combinations [2,3]:

$$a_\beta a_{\beta_1} = 0,$$

$$a_\beta^\dagger a_{\beta_1}^\dagger = 0 \tag{9.44}$$

always (even for $\beta = \beta_1$) equal to zero

At the same time, the diagonal matrix elements for two other bilinear combinations are non-zero for $\beta = \beta_1$, namely [2,3]:

$$a_\beta a_{\beta_1}^\dagger \rightarrow \delta_{\beta\beta_1} \left(n_\beta + 1\right), \tag{9.45}$$

$$a_\beta^\dagger a_{\beta_1} \to \delta_{\beta\beta_1} n_\beta. \tag{9.46}$$

As a result, using the orthogonality condition for the eigenfunctions $\Psi_\beta = \frac{1}{\sqrt{N}} e_j^x(\beta) \exp\left(i f R_n\right)$, considered in Lecture 8, we get much more compact expression [1,9]:

$$\left\langle \left(u_{nj}^x\right)^2 \right\rangle = \sum_\beta \frac{\hbar \left|e_j^x\right|^2}{2 M_j N \omega_\beta}\left(n_\beta + 1 + n_\beta\right) = \frac{\hbar}{2 M_j N} \sum_\beta \frac{\left|e_j^x\right|^2}{\omega_\beta}\left(2 n_\beta + 1\right). \tag{9.47}$$

Performing now the temperature averaging, we receive the final result for the mean square of the displacement along the x-axis [1,9]:

$$\overline{\left\langle \left(u_{n_J}^x\right)^2 \right\rangle} = \frac{\hbar}{2 M_j N} \sum_\beta \frac{\left|e_j^x\right|^2}{\omega_\beta} \overline{\left(2 n_\beta + 1\right)} = \frac{\hbar}{2 M_j N} \sum_\beta \frac{\left|e_j^x\right|^2}{\omega_\beta}\left(2 \bar{n}_\beta + 1\right), \tag{9.48}$$

where $\bar{n}_\beta$ is the Planck distribution function for the branch of the phonon spectrum with the frequency $\omega_\alpha(f)$.

Analogous expressions we get for the mean square of the displacements $\overline{\left\langle \left(u_{n_J}^y\right)^2 \right\rangle}$ and $\overline{\left\langle \left(u_{n_J}^z\right)^2 \right\rangle}$ along y and z axes, respectively, with the replacement of $\left|e_j^x\right|^2$ by $\left|e_j^y\right|^2$ and $\left|e_j^z\right|^2$.

It is important to note that, for temperature $T \to 0$, mean square of the displacement differs from zero [1,9]:

$$\overline{\left\langle \left(u_{n_J}^x\right)^2 \right\rangle} = \frac{\hbar}{2 M_j N} \sum_\beta \frac{\left|e_j^x\right|^2}{\omega_\beta}, \tag{9.49}$$

and corresponds to zero oscillations. Zero oscillations have a rigorously defined meaning. They are the consequence of the uncertainty principle [1]. We cannot exactly determine the coordinate and the momentum of the atom.

For high temperatures $T \sim \hbar\omega_{\max}$, we reach the classical limit for the Planck distribution function [4,14]:

$$\bar{n}_\beta + \frac{1}{2} \to \frac{T}{\hbar\omega}. \tag{9.50}$$

As a result, the mean square of the displacement is linearly proportional to the temperature [1,9]:

$$\overline{\left\langle \left(u_{n_J}^x\right)^2 \right\rangle} = \frac{T}{M_j N} \sum_\beta \frac{\left|e_j^x\right|^2}{\omega_\beta^2} \propto T. \tag{9.51}$$

Let us consider first the simplest case of one atom per elementary cell ($j = 1$). Then, in the three-dimensional crystal, we have three acoustic branches of the atom oscillations ($\alpha = 1, 2, 3$).

Let us utilize the orthonormality condition for the polarization vectors. Then, according to the Pythagorean theorem, we get for their Cartesian projections:

$$\left|e^x\right|^2 + \left|e^y\right|^2 + \left|e^z\right|^2 = 1. \tag{9.52}$$

Correspondingly, the mean square of the displacement reads [1,9]

$$\begin{aligned}
\overline{\langle u_n^2 \rangle} &= \overline{\left\langle \left(u_n^x\right)^2 \right\rangle} + \overline{\left\langle \left(u_n^y\right)^2 \right\rangle} + \overline{\left\langle \left(u_n^z\right)^2 \right\rangle} \\
&= \frac{\hbar}{2MN} \sum_{\alpha, f} \frac{\left|e^x\right|^2 + \left|e^y\right|^2 + \left|e^z\right|^2}{\omega_\alpha(f)} \left(2\bar{n}\left(\omega_\alpha(f)\right) + 1\right) \\
&= \frac{\hbar}{2MN} \sum_{\alpha, f} \frac{1}{\omega_\alpha(f)} \left(2\bar{n}\left(\omega_\alpha(f)\right) + 1\right).
\end{aligned} \tag{9.53}$$

Let us proceed from the summation over the wave vectors to the integral in Eq. (9.53):

$$\sum_f \to \frac{V}{(2\pi)^3} \int d^3 f.$$

Then the mean square of the displacement in (9.53) can be written as [1,9]

$$\begin{aligned}
\overline{\langle u_n^2 \rangle} &= \frac{\hbar}{2MN} \sum_{\alpha=1,2,3} \frac{V}{(2\pi)^3} \int d^3 f \frac{1}{\omega_\alpha(f)} \left(2\bar{n}\left(\omega_\alpha(f)\right) + 1\right) \\
&= \frac{\hbar}{2M} \sum_{\alpha=1,2,3} \frac{V_0}{(2\pi)^3} \int d^3 f \frac{1}{\omega_\alpha(f)} \left(2\bar{n}\left(\omega_\alpha(f)\right) + 1\right),
\end{aligned} \tag{9.54}$$

where $V_0 = \frac{V}{N}$ is the volume of the elementary cell.

We can introduce now the phonon density of states for three acoustic branches of the spectrum:

$$3 \int g(\omega) d\omega = \frac{V_0}{(2\pi)^3} \sum_{\alpha=1,2,3} \int d^3 f. \tag{9.55}$$

As a result, we finally get for the mean square of the displacement [1,9]:

$$\overline{\langle u_n^2 \rangle} = \frac{3\hbar}{2M} \int g(\omega) d\omega \frac{(2\bar{n}(\omega) + 1)}{\omega}. \tag{9.56}$$

Correspondingly, the mean squares of the displacements along 3 Cartesian axes equal to [1,9]

$$\overline{\left\langle (u_n^x)^2 \right\rangle} = \overline{\left\langle (u_n^y)^2 \right\rangle} = \overline{\left\langle (u_n^z)^2 \right\rangle} = \frac{1}{3}\overline{\langle u_n^2 \rangle} = \frac{\hbar}{2M} \int g(\omega)d\omega \frac{(2\bar{n}(\omega) + 1)}{\omega}. \qquad (9.57)$$

At $T = 0$, the mean square of the displacement connected only with the averaging of the inverse frequency with the density of states $g(\omega)$ [1,9]:

$$\overline{\left\langle (u_n^x)^2 \right\rangle} = \frac{\hbar}{2M} \int g(\omega)d\omega \frac{1}{\omega} = \frac{\hbar}{2M} \left\langle \frac{1}{\omega} \right\rangle^g. \qquad (9.58)$$

At high temperatures $T \sim \hbar\omega_{\max}$, the combination

$$\bar{n}(\omega) + \frac{1}{2} = \frac{T}{\hbar\omega},$$

and we reach the classical limit for the mean square of the displacement, which does not contain the Planck constant $\hbar$, and is related only to the averaging of the inverse frequency squared with the density of states [1,9]:

$$\overline{\left\langle (u_n^x)^2 \right\rangle} = \frac{T}{M} \int g(\omega)d\omega \frac{1}{\omega^2} = \frac{T}{M} \left\langle \frac{1}{\omega^2} \right\rangle^g. \qquad (9.59)$$

Substituting the density of states $g(\omega) = \frac{3\omega^2}{\omega_D^3}$ in the Debye approximation [11], we get for the averaged inverse frequency:

$$\left\langle \frac{1}{\omega} \right\rangle^g = \int_0^{\omega_D} g(\omega)d\omega \frac{1}{\omega} = \int_0^{\omega_D} \frac{3\omega^2}{\omega_D^3} d\omega \frac{1}{\omega} = \frac{3}{2\omega_D}. \qquad (9.60)$$

Thus, for $T = 0$, we obtain [1,9]

$$\overline{\left\langle (u_n^x)^2 \right\rangle} = \frac{3\hbar}{4M\omega_D}. \qquad (9.61)$$

Analogously for the averaged inverse frequency squared [1,9]:

$$\left\langle \frac{1}{\omega^2} \right\rangle^g = \int_0^{\omega_D} \frac{3\omega^2}{\omega_D^3} d\omega \frac{1}{\omega^2} = \frac{3}{\omega_D^2}. \qquad (9.62)$$

As a result, for $T \sim \hbar\omega_D$ [1,9]:

$$\overline{\left\langle (u_n^x)^2 \right\rangle} = \frac{3T}{M\omega_D^2}. \qquad (9.63)$$

9.3   Two-Dimensional Crystal at Zero Temperature

For the two-dimensional crystal, the density of states in the Debye approximation yields [1]:

$$g(\omega) = \frac{2\omega}{\omega_D^2} \propto \omega. \tag{9.64}$$

Then, for $T = 0$, we get the convergent result for the mean square of the displacement [1]:

$$\langle \boldsymbol{u}^2(T=0) \rangle = \frac{2\hbar}{M\omega_D^2} \int_0^{\omega_D} d\omega \frac{\omega}{\omega} = \frac{2\hbar}{M\omega_D^2} \int_0^{\omega_D} d\omega = \frac{2\hbar}{M\omega_D}. \tag{9.65}$$

9.4   Two-Dimensional Crystal at Finite Temperature

For the two-dimensional crystal, at finite temperature, we obtain [1]

$$\begin{aligned} \langle \boldsymbol{u}^2 \rangle = \langle \boldsymbol{u}^2(T) \rangle + \langle \boldsymbol{u}^2(T=0) \rangle &= \frac{\hbar}{M} \int_0^{\omega_D} d\omega g(\omega) \frac{2\bar{n}(\omega)}{\omega} \\ + \langle \boldsymbol{u}^2(T=0) \rangle &= \frac{\hbar}{M} \int_0^{\omega_D} d\omega \frac{2\bar{n}(\omega)}{\omega} \frac{2\omega}{\omega_D^2} \\ + \langle \boldsymbol{u}^2(T=0) \rangle & \\ &= \frac{4\hbar}{M\omega_D^2} \int_0^{\omega_D} d\omega \frac{1}{\left(\exp\left(\frac{\hbar\omega}{T} \right) - 1 \right)} + \langle \boldsymbol{u}^2(T=0) \rangle. \end{aligned} \tag{9.66}$$

Let us utilize the fact that the frequency integral from the Planck distribution function [14] in the first term of Eq. (9.66) is logarithmically divergent at low frequencies [1]. We can say that the integral is effectively determined by the frequencies which are less or of the order of the temperature ($\hbar\omega \leq T$). For these frequencies, the high-temperature expansion $\bar{n}(\omega) \approx \frac{T}{\hbar\omega}$ is valid for the Planck distribution function. As a result, the temperature-dependent mean square of the displacement equals to [1]

$$\langle \boldsymbol{u}^2(T) \rangle = \frac{4\hbar}{M\omega_D^2} \int_0^{\sim T/\hbar} d\omega \frac{T}{\hbar\omega} = \frac{4T}{M\omega_D^2} \int_0^{\sim T/\hbar} \frac{d\omega}{\omega}. \tag{9.67}$$

Logarithmic singularity at low frequencies and large wavelengths in (9.67) is often called the infrared catastrophe [15] in the condensed matter physics [1,4]. For the two-dimensional crystal of finite size $L \times L$, we have to cut the logarithmic singularity

Fig. 9.2 Two-dimensional or one-dimensional crystal (one-dimensional atomic chain) of the finite size (of the finite length) L [1]

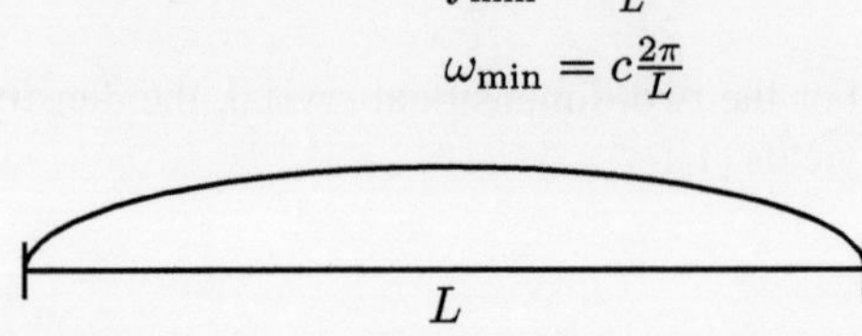

on the minimal possible wave vector $f_{\min} \sim \frac{2\pi}{L}$ (see Fig. 9.2), or, equivalently, on the minimal possible frequency:

$$\omega_{\min} \sim cf_{\min} \sim \frac{2\pi c}{L}, \tag{9.68}$$

where c is the average sound velocity.

Then, we finally get for the temperature-dependent mean square of the displacement [1,4]:

$$\langle u^2(T) \rangle = \frac{4\,T}{M\omega_D^2} \ln \frac{T}{\hbar\omega_{\min}} = \frac{4\,T}{M\omega_D^2} \ln \frac{T\,L}{\hbar 2\pi c} \propto T \ln L. \tag{9.69}$$

Note that the logarithmic singularity in the two-dimensional crystal is equivalent to the singularity in the density of the non-condensed particles, washed out of the condensate in the two-dimensional films of the superfluid He $-$ 4. We consider this type of singularity in Lecture 5 in the framework of the Berezinskii–Kosterlitz–Thouless (BKT) theory [16,17]. In solid-state physics, it is responsible for the anomalous melting of the two-dimensional crystal [18].

9.5   One-Dimensional Crystal at Zero Temperature

For the one-dimensional crystal, the density of states in the Debye approximation [11] is constant [1]:

$$g(\omega) = \frac{1}{\omega_D} = \text{const}. \tag{9.70}$$

Hence, the mean square of the displacement at $T = 0$ reads [1]

$$\begin{aligned}
\langle u^2(T=0) \rangle &= \frac{\hbar}{2M} \int_0^{\omega_D} d\omega g(\omega)\frac{1}{\omega} = \frac{\hbar}{2M\omega_D} \int_0^{\omega_D} \frac{d\omega}{\omega} \\
&= \frac{\hbar}{2M\omega_D} \ln \frac{\omega_D}{\omega_{\min}} = \frac{\hbar}{2M\omega_D} \ln \frac{\omega_D L}{2\pi c}.
\end{aligned} \tag{9.71}$$

9.6   **One-Dimensional Crystal at Finite Temperatures**

At the same time, the temperature-dependent mean square of the displacement [1]:

$$
\begin{aligned}
\langle u^2(T)\rangle &= \frac{\hbar}{2M}\int_0^{\omega_D} d\omega\, g(\omega)\frac{2\bar{n}(\omega)}{\omega}\\
&= \frac{\hbar}{2M\omega_D}\int_0^{\omega_D} d\omega\frac{2\bar{n}(\omega)}{\omega} \approx \frac{\hbar}{M\omega_D}\int_0^{\sim T/\hbar} d\omega\frac{T}{\hbar\omega}\frac{1}{\omega}\\
&= \frac{T}{M\omega_D}\int_0^{\sim T/\hbar} d\omega\frac{1}{\omega^2}.
\end{aligned}
\tag{9.72}
$$

We see that for the 1D chain, the temperature-dependent mean square of the displacement is strongly divergent (proportionally to $\frac{1}{\omega}$) on the lower limit of the integration [1]. We have to cut it again on the minimal frequency for the chain of the finite length L. As a result:

$$
\langle u^2(T)\rangle \approx \frac{T}{M\omega_D}\frac{1}{\omega_{\min}} = \frac{T}{M\omega_D}\frac{L}{2\pi c}.
\tag{9.73}
$$

Thus, the mean square of the displacement of the one-dimensional chain diverges proportionally to the size of the system at finite temperatures [1,4].

We can often read in the textbooks that 1D crystal does not exist [4]. Atoms are not oscillating around the well-defined positions in space.

For the one-dimensional crystal we have useful analogy with the caterpillar [1]. If we paint the piece of the caterpillar, then, after some time, this piece will move on a large distance from the initial position, as it is shown in Fig. 9.3. At the same time, the bonds between the pieces of the caterpillar are not broken—we do not have discontinuities in the system [1].

In other words, the displacements are accumulating on the large length-scales [1]. However, the distances between the nearest neighbor atoms are conserved (the displacements are accumulating between the distant atoms) [1].

Important Remark. The higher is the space dimensionality, the better is the convergence of the mean square of the displacement [1,4,19–24]. For low-dimensional systems, the long wavelength fluctuations damage the picture leading to the infrared divergences [1,4,15–24].

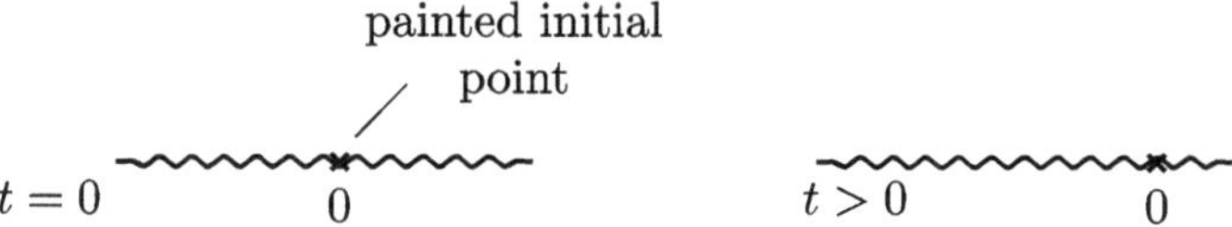

Fig. 9.3 If in the initial moment at $t = 0$ we paint the point 0 on the caterpillar body on the left Figure, then in the next moments at $t > 0$, this point can move on a large distance from its initial position (see the right Figure) [1]

9.7   Displacement Correlators in One-Dimensional and Two-Dimensional Crystals

To understand better the accumulation of the displacements on large distances, it is useful to calculate the correlator for the mean square of the difference between two displacements on the lattice nodes n and n_1 [1]:

$$\overline{\left\langle \left(u_n^x - u_{n_1}^x \right)^2 \right\rangle}, \tag{9.74}$$

where the distance between the nodes is governed by the vector $\boldsymbol{R}_{n,n_1} = \boldsymbol{R}_n - \boldsymbol{R}_{n_1}$. The operator of the difference between the displacements in Eq. (9.74) has a form [1]:

$$u_n^x - u_{n_1}^x = \sum_\beta \sqrt{\frac{\hbar}{2M\omega_\beta N}} e^x \left\{ \exp\left(if\boldsymbol{R}_n\right) \left[\hat{a}_\beta + \hat{a}_{-\beta}^\dagger \right] - \exp\left(if\boldsymbol{R}_{n_1}\right) \left[\hat{a}_\beta + \hat{a}_{-\beta}^\dagger \right] \right\}$$

$$= \sum_\beta \sqrt{\frac{\hbar}{2M\omega_\beta N}} e^x \exp\left(if\boldsymbol{R}_n\right) \left(\hat{a}_\beta + \hat{a}_{-\beta}^\dagger \right) \left(1 - \exp\left(-if\boldsymbol{R}_{n,n_1}\right) \right). \tag{9.75}$$

Further on let us utilize the fact that only bilinear combinations

$$a_\beta a_{-\beta_1}^\dagger \quad \text{and} \quad a_{-\beta}^\dagger a_{\beta_1} \quad \text{for } \beta_1 = -\beta$$

yield non-zero contributions to the diagonal matrix elements of the correlator [1–3] in (9.74) when we take the square from the difference of the displacements in (9.75) and then perform the temperature averaging.

As a result, for $T \neq 0$, the displacement correlator in the two-dimensional crystal is given by [1]

$$\overline{\left\langle \left(u_n^x - u_{n_1}^x \right)^2 \right\rangle} \sim \int \frac{d^2 f}{\omega} (2\bar{n} + 1) \left(1 - \exp\left(if\boldsymbol{R}_{n,n_1}\right) \right) \left(1 - \exp\left(-if\boldsymbol{R}_{n,n_1}\right) \right). \tag{9.76}$$

Transforming the product of second and third multipliers in Eq. (9.76) with the help of simple trigonometric relations:

$$\left(1 - \exp\left(if\boldsymbol{R}_{n,n_1}\right) \right) \left(1 - \exp\left(-if\boldsymbol{R}_{n,n_1}\right) \right) = 2 - 2\cos\left(f\boldsymbol{R}_{n,n_1} \right)$$

$$= 4\sin^2\left(\frac{f\boldsymbol{R}_{n,n_1}}{2} \right), \tag{9.77}$$

we get the final expression for the correlator at $T \neq 0$ [1]:

$$\overline{\left\langle \left(u_n^x - u_{n_1}^x \right)^2 \right\rangle} \sim \int \frac{d^2 f}{\omega} (2\bar{n} + 1) \sin^2\left(\frac{f\boldsymbol{R}_{n,n_1}}{2} \right). \tag{9.78}$$

Then, expanding again in series the Planck distribution function [14] $\bar{n}(\omega) \approx \frac{T}{\hbar\omega}$ at small frequencies $\hbar\omega \leq T$, we can write the displacement correlator as follows [1]:

$$\overline{\left\langle \left(u_n^x - u_{n_1}^x\right)^2\right\rangle} \sim T \int \frac{d^2 f}{\omega^2} \sin^2\left(\frac{fR_{n,n_1}}{2}\right). \tag{9.79}$$

On the small wave vectors $\sin^2\left(\frac{fR_{n,n_1}}{2}\right) \sim f^2$, and thus, for the linear sound spectrum $\omega = cf$, the displacement correlator in Eq. (9.79) is convergent at the lower limit of integration at $f \to 0$ [1].

Let the distance between the nodes to be large and $R_{n,n_1} \to \infty$. At $f = 0$, we still do not have the singularity. However, for the values of $fR_{n,n_1} \sim \pi$ (when we can replace $\sin^2\left(\frac{fR_{n,n_1}}{2}\right)$ by $\frac{1}{2}$), the logarithmic singularity appears again in the expression for the correlator (9.79) [1]:

$$\overline{\left\langle \left(u_n^x - u_{n_1}^x\right)^2\right\rangle} \sim T \int \frac{2\pi\, f\, df}{c^2 f^2} \frac{1}{2} \approx \frac{\pi T}{c^2} \int_{f_{min}}^{\sim T/\hbar c} \frac{df}{f}, \tag{9.80}$$

where $f_{min} \sim \frac{\pi}{R_{n,n_1}}$.

Hence, the displacement correlator at large distances between the nodes [1]:

$$\overline{\left\langle \left(u_n^x - u_{n_1}^x\right)^2\right\rangle} \sim T \ln \frac{TR_{n,n_1}}{\hbar\pi c} = T \ln\left(f_T R_{n,n_1}\right), \tag{9.81}$$

where $f_T = \frac{T}{\hbar\pi c}$.

Performing the analogous calculations in the one-dimensional case, we can represent the displacement correlator for the atomic chain at $T \neq 0$ as [1]:

$$\overline{\left\langle \left(u_n - u_{n_1}\right)^2\right\rangle} \sim T \int \frac{df}{\omega^2} \sin^2\left(\frac{fR_{n,n_1}}{2}\right) \approx \frac{T}{c^2} \int_{f_{min}}^{\sim T/\hbar c} \frac{df}{f^2} \approx \frac{T}{c^2 f_{min}}. \tag{9.82}$$

Substituting in Eq. (9.82), the minimal value of the wave vector $f_{min} \sim \frac{\pi}{R_{n,n_1}}$, we finally get for the correlator in the one-dimensional case [1]:

$$\overline{\left\langle \left(u_n - u_{n_1}\right)^2\right\rangle} \sim T R_{n,n_1}. \tag{9.83}$$

We see that the correlations are accumulated at large distances [1]. In other words, the displacement correlator is proportional to the distance between the nodes $R_{n,n'}$ (see Fig. 9.4). On small distances, the correlator is convergent again acquiring the finite value as in the two-dimensional crystal [1].

Fig. 9.4 Correlator of the difference between two displacements at the lattice nodes n and n_1 is proportional to the distance between the nodes R_{n,n_1} in the one-dimensional crystal (in the 1D atomic chain) [1]

References

1. Y. Kagan, *Lectures on the Solid-State Theoretical Physics*, Moscow, MEPHI, 1981–1982. Part I. Phonons (Unpublished)
2. R.P. Feynman, *Statistical Mechanics: A Set of Lectures*. Advanced Books, Classics (Westview Press, Boulder, 1998)
3. L.D. Landau, E.M. Lifshitz, *Quantum Mechanics: Non-relativistic Theory* (Pergamon Press, New York, 1977)
4. L.D. Landau, E.M. Lifshitz, *Statistical Physics, Part I* (Butterworth-Heinemann, Oxford, 1999)
5. O. Madelung, *Introduction to Solid-State Theory*. Springer Series in Solid-State Sciences (1995)
6. C. Kittel, *Quantum Theory of Solids* (Wiley, New York–London, 1963)
7. J.M. Ziman, *Electrons and Phonons: The Theory of Transport Phenomena in Solids* (Clarendon Press, Oxford, 1960)
8. N.W. Ashcroft, N.D. Mermin, *Solid State Physics* (Rinehart and Winston, Holt, 1976)
9. D. Pines, *Elementary Excitations in Solids* (W.A. Benjamin Inc., New York–Amsterdam, 1963)
10. E.G. Brovman, Yu. Kagan, Phonons in non-transition metals. Sov. Physics Uspekhi **17**, 25 (1974)
11. P. Debye, Ann. Phys. **39**, 789 (1912) (in German)
12. E. Gruneisen, Ann. Phys. **344**, 257 (1912) (in German)
13. L.D. Landau, A.M. Kosevich, E.M. Lifshitz, L.P. Pitaevskii, *Theory of Elasticity, Course of Theoretical Physics*, vol. 7 (Elsevier, 1986)
14. M. Planck, N. Lecture, June 2, 1920, in *Les Prix Nobel*. Nobel Lectures, Physics (Elsevier Publishing Company, Amsterdam, 1967), pp. 1901–1921
15. C. Itzykson, J.-B. Zuber, *Quantum Field Theory* (McGraw-Hill, 1980)
16. J.M. Kosterlitz, D.J. Thouless, J. Phys. **6**, 1181 (1973)
17. V.L. Berezinskii, Sov. Phys. JETP **61**, 1144 (1972)
18. V.N. Ryzhov, E.E. Tareeva, Y.D. Fomin, E.N. Tsiok, Berezinskii–Kosterlitz–Thouless transition and two-dimensional melting. Physics Uspekhi **60**, 857 (2017)
19. M. Fisher, *The Nature of Critical Points*. Lectures in Theoretical Physics (University of Colorado Press, Boulder, 1965)
20. L. Kadanoff, *Statistical Physics: Statistics, Dynamics and Renormalization* (World Scientific, 2000)
21. K.G. Wilson, Rev. Mod. Phys. **55**, 583 (1983)
22. A.Z. Patashinskii, V.L. Pokrovskii, *Fluctuation Theory of Phase Transitions* (Pergamon Press, 1979)
23. J.V. Jose (ed.), *40 years of Berezinskii-Kosterliz-Thouless Theory* (World Scientific, Singapore, 2013)
24. A.M. Tsvelik, *Quantum Field Theory in Condensed Matter Physics* (Cambridge University Press, 1995)

Lecture 10. Mossbauer Effect: Diagonal Processes with Respect to Phonon Occupation Numbers

10

Abstract

In this Lecture we provide the theoretical explanation of the main features of the Mossbauer effect (including the width and the position of the spectral lines). Our analysis is based on the investigation of the diagonal processes with respect to phonon occupation numbers. We get the expression for the Debye–Waller factor and find the temperature dependence for the probability of the phononless processes.

We know, that in nuclei so-called <u>isomeric states</u> [1–3] <u>exist</u> (see Fig. 10.1). These states are <u>long-living</u>. Their life-time is $\tau \sim 10^{-8}$s [1–3]. Correspondingly, the distance between the ground and excited states for them is of the order of $10\,\mathrm{keV}$ [1–3].

The linewidth for this state is <u>extremely narrow</u> and yields [1–4]:

$$\Gamma \sim \frac{\hbar}{\tau} \sim \frac{10^{-27}\mathrm{erg} \cdot \mathrm{s}}{10^{-8}\mathrm{s}} \sim 10^{-19}\ \mathrm{erg} \cdot \mathrm{s},\ \mathrm{or}\ \Gamma \sim \frac{10^{-19}}{1.6 \cdot 10^{-12}} \sim 10^{-7}\mathrm{eV}. \quad (10.1)$$

At the same time, the recoil energy $R = \frac{\hbar^2 k^2}{2M}$ [5] for the photon injection to the nucleus with the energy of the order of $10\,\mathrm{keV}$ is 4 orders higher [1]:

$$R = \frac{\hbar^2 k^2}{2M} = \frac{\hbar^2 k^2 c^2}{2Mc^2} = \frac{E_\gamma^2}{2Mc^2} \sim \frac{10^8}{10^{11}} \sim 10^{-3}\ \mathrm{eV} \gg \Gamma, \quad (10.2)$$

where $E_\gamma = \hbar k c \sim 10^4 \mathrm{eV}$ is the energy of the γ-quantum, $2Mc^2 \sim 10^{11}\mathrm{eV}$ is the doubled rest energy of the nucleus.

For the photon emission with the energy $E \sim 10$ eV, and the transition of the system from the excited to the ground state, the emission and absorption lines [4,5] respectively with the energies $E - R$ (smaller on the recoil energy) and $E + R$ (larger on the recoil energy) will be located on the energy scale in the way shown in Fig. 10.2 [1,6–9]. The distance between the emission and absorption lines is twice the recoil

© The Author(s), under exclusive license to Springer Nature Switzerland AG 2026

167

M. Kagan, *Lecture Notes of Professor Yuri Kagan in Theoretical Solid-State Physics*, Lecture Notes in Physics 1048, https://doi.org/10.1007/978-3-032-14621-2_10

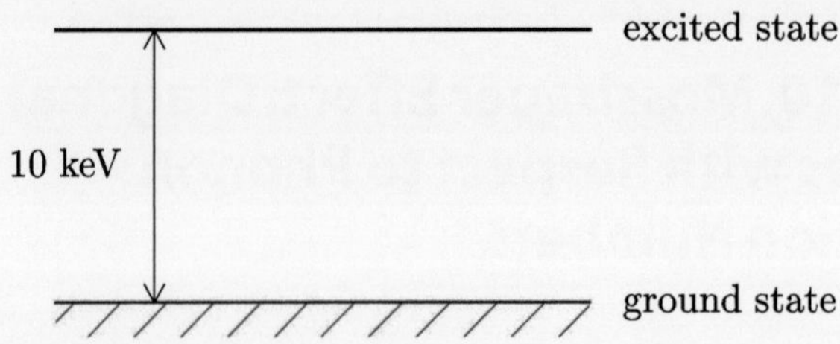

Fig. 10.1 Long-living isomeric states in nuclei [1]

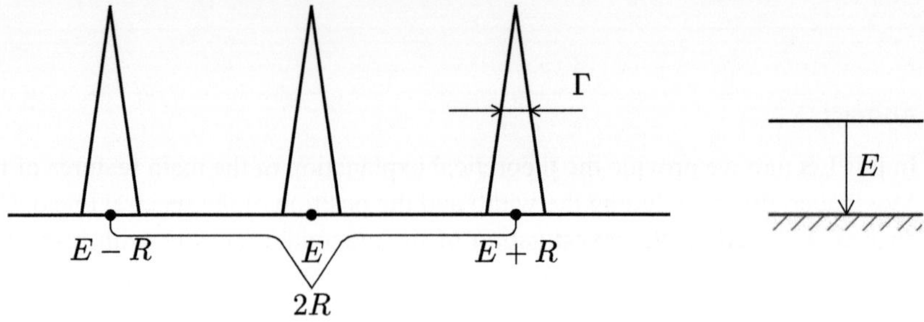

Fig. 10.2 Emission and absorption lines with the energies $E - R$ and $E + R$ for the photon emission with the energy $E \sim 10$ eV and the transition of the system from the excited to the ground state. The distance between the emission and the absorption lines is twice the recoil energy $2R$. We also show in the figure the central line with the energy E and the width Γ, which is not shifted on the recoil energy [1]

energy $2R$ [1,6–9]. We also show on this Figure the central line with the energy E and the width Γ, which is not shifted on the recoil energy [1].

Thus, we do not have <u>the resonance condition</u> between the emitting and absorbing nuclei [1]. In other words, the resonance between two free nuclei is impossible (there is no resonant interaction between the nuclei) [1].

If we heat the gas (and increase the average velocities of the particles), then the <u>Doppler lines broadening</u> emerges [6]. The lines overlap appears (see Fig. 10.3), when the <u>Doppler broadening</u> (proportional to $\frac{v}{c}E_\gamma$) [6] becomes of the order of the recoil energy [1,4,6–9]:

$$\Gamma \sim \frac{v}{c}E_\gamma \sim R. \tag{10.3}$$

It follows from this estimate, that the average velocity [1]:

$$\frac{v}{c} \sim \frac{R}{E_\gamma} \sim \frac{10^{-3}}{10^4} \sim 10^{-7}, \text{ and } v \sim 3 \cdot 10^3 \frac{\text{cm}}{\text{s}} \sim 30\frac{\text{m}}{\text{s}}. \tag{10.4}$$

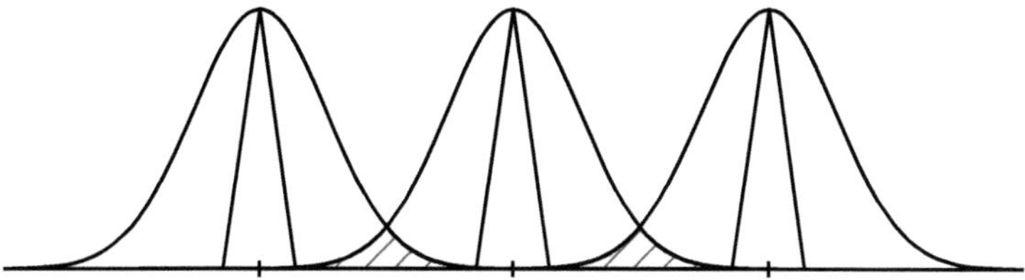

Fig. 10.3 The overlap of the lines broadened due to the Doppler effect [1]

Rudolf Mossbauer got everything vice versa in his experiments [10–12]:

1. Increasing the temperature, we decrease the resonance.
2. Decreasing the temperature, we increase the resonance.
3. There are lines in the crystal which do not shift at all.
4. Let us take two crystals: one–absorber and another one–emitter. If one of the crystals moves slowly, then due to the Doppler frequency shift the resonance is detuned.

The Hamiltonian for the interaction of the electromagnetic wave with the crystal acquires a form [1, 13, 14]:

$$H_{\text{int.}} = \frac{1}{c} \sum_a j_a A(r_a).$$ (10.5)

If the vector potential $A(r_a)$ corresponds to the plane wave falling on the crystal, then [1]:

$$A(r_a) \sim \exp(\pm ikr_a) = \exp(\pm ik(r_a - R_n)) \exp(\pm ikR_n).$$ (10.6)

In the right-hand side of Eq. (10.6) we distinguish the exponential multiplier $\exp(\pm ikR_n)$. It is related to the wave hitting the center of gravity of the nucleus [1]. As a result, the interaction Hamiltonian can be written as [1, 15–18]:

$$H_{\text{int.}} = \exp(\pm ikR_n) \tilde{H}_{\text{int.}}(r_a - R_n).$$ (10.7)

We need to consider the case when the state of the crystal (and phonon occupation numbers) do not change [1, 15–18]. It means in fact, that we have to calculate the following matrix element [1, 15–18]:

$$\langle 1, i | H_{\text{int.}} | 0, i \rangle,$$ (10.8)

where the occupation numbers 1 and 0 correspond to photons (one photon falling on the nucleus), while the occupation numbers $| i >$ are related to the phonon states before and after the wave incidence.

Let us stress, that on the language of the excitations the recoil process corresponds to phonons [1, 15–18]. Hence if there are no phonons at all (or if the phonon occupation numbers do not change in all the states of the crystal), then we do not have the recoil in our system [1, 15–18].

Thus, we have to calculate the diagonal matrix element with respect to the phonon states [1, 15–18]:

$$\langle \{n\}| \exp\left(\pm i k \boldsymbol{R}_n\right) |\{n\}\rangle. \tag{10.9}$$

If the γ-quantum falling on the nucleus transfers the momentum $\boldsymbol{p} = -\hbar\boldsymbol{k}$ to the nucleus, then

$$W(\boldsymbol{p}, \Delta E = 0) \tag{10.10}$$

is the probability of the zero-energy transfer to the nucleus [1, 15–18].

As we discussed already, the line width is of the order of $\Gamma \sim \left(10^{-7} \div 10^{-10}\right)$ eV. At the same time, the typical Debye frequency [1, 19] for the phonon spectrum is much larger: $\hbar\omega_D \sim 10^{-2}$eV $\gg \Gamma$. The density of the phonon states in the three-dimensional crystal at low temperatures is proportional to: $g(\omega) \sim \omega^2$ [1, 20–26]. Thus, the line width Γ in the energy units corresponds to the negligibly small fraction from the total number of phonons (from the bandwidth of the phonon spectrum) [1].

Hence, we have to consider the phononless process [1, 15–18], that is truly elastic process, for which all the quantum numbers (including spins and etc.) are the same.

Comment 1. Inelastic process even in the limit of zero energy- transfer yields absolutely different result. Its probability goes to zero together with the phonon density of states $g(\omega) \sim \omega^2$ [1].

Comment 2. Matrix element

$$\langle \{n\}| \exp\left(i k \left(\boldsymbol{R}_0 + \hat{\boldsymbol{u}}\right)\right) |\{n\}\rangle, \tag{10.11}$$

where $\hat{\boldsymbol{u}}$ is displacement operator, represents truly elastic element, corresponding to the non-excitation of phonons, while $\boldsymbol{R}_0$ is the coordinate of one (not shifted) nucleus [1, 15–18].

Let us check the correctness of the normalization condition. For the sum of the squares of the phonon matrix elements we get [1]:

$$\sum_f |\langle f| \exp\left(i k \boldsymbol{R}_0\right) |i\rangle|^2 = \int \Pi(dR)\Pi\left(dR'\right) \Phi_f^*(\boldsymbol{R}) \exp\left(i k \boldsymbol{R}_0\right) \Phi_i(\boldsymbol{R})$$

$$\cdot \Phi_f\left(\boldsymbol{R}'\right) \exp\left(-i k \boldsymbol{R}_0'\right) \Phi_i^*\left(\boldsymbol{R}'\right), \tag{10.12}$$

where $\Pi(dR)$ is the product of the differentials over all the ions, $\boldsymbol{R}_0$ is the coordinate of one nucleus, $\Phi_i(\boldsymbol{R})$ is the wave function of the initial state depending on the coordinates of all the ions.

Let us utilize now the completeness relation for the wave functions which reads [1]:

$$\sum_f \Phi_f^*(\boldsymbol{R})\Phi_f\left(\boldsymbol{R}'\right) = \Pi\delta\left(\boldsymbol{R} - \boldsymbol{R}'\right). \tag{10.13}$$

Then, expression (10.12) is considerably simplified [1]:

$$\sum_f |\langle f|\exp\left(i\boldsymbol{k}\boldsymbol{R}_0\right)|i\rangle|^2 = \int \Pi(dR)\,|\Phi_i(\boldsymbol{R})|^2 = 1. \tag{10.14}$$

Distinguishing the square of the diagonal matrix element, we can write the normalization condition as follows [1]:

$$|\langle i|\exp\left(i\boldsymbol{k}\boldsymbol{R}_0\right)|i\rangle|^2 = 1 - \sum_{f\neq i}|\langle f|\exp\left(i\boldsymbol{k}\boldsymbol{R}_0\right)|i\rangle|^2. \tag{10.15}$$

Let us stress again, that according to Eq. (10.11) it is precisely the diagonal matrix element $\langle\{n\}|\exp\left(i\boldsymbol{k}\left(\boldsymbol{R}_0 + \hat{\boldsymbol{u}}\right)\right)|\{n\}\rangle$, where $\boldsymbol{R}_0$ is the equlibrium coordinate, which corresponds to the true probability amplitude of phonons non-excitation [1].

Let us write down this amplitude more detailly with respect to the phonon operators, taking the equilibrium part of the exponent out of the averaging sign [1]:

$$\exp\left(i\boldsymbol{k}\boldsymbol{R}_0\right)\langle\{n\}|\exp\left(i\boldsymbol{k}\hat{\boldsymbol{u}}\right)|\{n\}\rangle = \exp\left(i\boldsymbol{k}\boldsymbol{R}_0\right)\prod_\beta\langle n_\beta|\exp\left(i\boldsymbol{k}\hat{\boldsymbol{u}}_\beta\right)|n_\beta\rangle, \tag{10.16}$$

where, as we know, the displacement operators are given by [1]:

$$\hat{\boldsymbol{u}} = \sum_{f,\alpha}\hat{\boldsymbol{u}}_{f,\alpha} = \sum_\beta\hat{\boldsymbol{u}}_\beta,\ \beta = \{f,\alpha\}, \tag{10.17}$$

and for $\beta \neq \beta_1$ the following commutation relation is fulfilled for them:

$$\left[\hat{\boldsymbol{u}}_\beta, \hat{\boldsymbol{u}}_{\beta_1}\right] = 0. \tag{10.18}$$

At the same time, the displacement operator for one atom in the elementary cell according to Eq. (8.90) has the following form for one value of β [1]:

$$\hat{\boldsymbol{u}}_\beta = \sum_\beta\sqrt{\frac{\hbar}{2MN\omega_\beta}}\left[\boldsymbol{e}_\beta\exp\left(i\boldsymbol{f}\boldsymbol{R}_n^0\right)\hat{a}_\beta + \boldsymbol{e}_\beta^*\exp\left(-i\boldsymbol{f}\boldsymbol{R}_n^0\right)\hat{a}_\beta^\dagger\right]. \tag{10.19}$$

Note, that for $\beta \neq \beta_1$ the commutation relations for phonon creation and annihilation operators read [1,20,27,28]:

$$\left[\hat{a}_\beta \hat{a}_{\beta_1}\right] = \left[\hat{a}_\beta^\dagger \hat{a}_{\beta_1}\right] = \left[\hat{a}_\beta^\dagger \hat{a}_{\beta_1}^\dagger\right] = 0. \tag{10.20}$$

However, we know from Quantum mechanics, that for two commuting operators $\hat{A}$ and $\hat{B}$ the following relation holds [28]:

$$\exp(\hat{A} + \hat{B}) = \exp(\hat{A})\exp(\hat{B}), \ \text{if} \ [\hat{A}\hat{B}] = 0. \tag{10.21}$$

As a result, in Eq. (10.16) the relation:

$$\exp(i\boldsymbol{k}\hat{\boldsymbol{u}}) = \prod_\beta \exp\left(i\boldsymbol{k}\hat{\boldsymbol{u}}_\beta\right) \tag{10.22}$$

is fulfilled indeed.

Thus, in Eq. (10.16) we have to calculate only the average [1]:

$$\langle n_\beta | \exp\left(i\boldsymbol{k}\hat{\boldsymbol{u}}_\beta\right) | n_\beta \rangle. \tag{10.23}$$

Let us expand the exponent $\exp\left(i\boldsymbol{k}\hat{\boldsymbol{u}}_\beta\right)$ in series till the quadratic terms inclusively. Then, the average [1]:

$$\langle n_\beta | \exp\left(i\boldsymbol{k}\hat{\boldsymbol{u}}_\beta\right) | n_\beta \rangle = \langle n_\beta | 1 + i\boldsymbol{k}\hat{\boldsymbol{u}}_\beta - \frac{1}{2}\left(\boldsymbol{k}\hat{\boldsymbol{u}}_\beta\right)^2 | n_\beta \rangle + O\left(\frac{1}{N^{3/2}}\right), \tag{10.24}$$

where we utilized the normalization on $\frac{1}{\sqrt{N}}$ for the displacement operator in Eq. (10.19):

$$\hat{\boldsymbol{u}}_\beta \propto \frac{1}{\sqrt{N}}, \ \hat{\boldsymbol{u}}_\beta^2 \propto \frac{1}{N}, \ \hat{\boldsymbol{u}}_\beta^3 \propto \frac{1}{N^{3/2}}. \tag{10.25}$$

Note, that the second term of the expansion in Eq. (10.24) (linear in the phonon creation and annihilation operators) goes to zero, since [1,20–25,28]:

$$\langle n_\beta | \hat{\boldsymbol{u}}_\beta | n_\beta \rangle \sim \langle n_\beta | n_\beta \pm 1 \rangle = 0. \tag{10.26}$$

As a result [1]:

$$\langle n_\beta | \exp\left(i\boldsymbol{k}\hat{\boldsymbol{u}}_\beta\right) | n_\beta \rangle = \langle n_\beta | 1 - \frac{1}{2}\left(\boldsymbol{k}\hat{\boldsymbol{u}}_\beta\right)^2 | n_\beta \rangle + O\left(\frac{1}{N^{3/2}}\right)$$

$$\approx 1 - \frac{1}{2}\langle n_\beta | \left(\boldsymbol{k}\hat{\boldsymbol{u}}_\beta\right)^2 | n_\beta \rangle \approx \exp\left(-\frac{\langle n_\beta | \left(\boldsymbol{k}\hat{\boldsymbol{u}}_\beta\right)^2 | n_\beta \rangle}{2}\right). \tag{10.27}$$

This result for the averaging of the probability amplitude is equivalent to the well-known Debye–Waller formula [29,30] considered in Lecture 5 on BKT theory [31,32].

Let us emphasize also, that in the product (10.16) the zero order terms together with quadratic terms in the expansion of each exponent (with the account of the smallness of the higher order terms) can be collected in the sum according to the formula [1]:

$$\prod_\beta \left(1 + \alpha u_\beta^2\right) = \left(1 + \alpha u_1^2\right)\left(1 + \alpha u_2^2\right)\cdots \approx 1 + \alpha\left(u_1^2 + u_2^2 + \cdots\right) = 1 + \sum_\beta \alpha u_\beta^2.$$

(10.28)

But the total number of terms in the sum (10.28) is of the order of N (in the three-dimensional crystal with one atom in the elementary cell the index β runs over $3N$ values).

Thus, with the account of the normalization for the mean square of the displacement $\hat{u}_\beta^2 \propto \frac{1}{N}$, we get for each nucleus [1]:

$$\sum_\beta u_\beta^2 \propto N \cdot \frac{1}{N} \propto 1.$$

(10.29)

As a result, the smallness appears already for the cubic terms in the sum [1]:

$$\sum_\beta u_\beta^3 \propto N \cdot \frac{1}{N^{3/2}} \propto \frac{1}{N^{1/2}}.$$

(10.30)

Note, that in the absence of the motion smearing on many modes (degrees of freedom), in the situation when each atom in crystal oscillates in its own mode, the normalization condition $\hat{u}_\beta \propto \frac{1}{\sqrt{N}}$ does not hold, and our expansion is not valid [1].

The possibility of the phononless process is determined by the square of the absolute value of the amplitude and equals to [1,15–18]:

$$W_0 = \overline{|\langle\{n\}|\exp(\imath k\hat{u})|\{n\}\rangle|}^2,$$

(10.31)

where the line in the expression (10.31) means the temperature averaging.

With the account of Eqs. (10.16) and (10.27) we can write the expression for the probability in the following form [1]:

$$W_0 = \left|\overline{\prod_\beta \langle n_\beta|\exp\left(\imath k\hat{u}_\beta\right)|n_\beta\rangle}\right|^2 = \left|\overline{\prod_\beta \left(1 - \frac{1}{2}\langle n_\beta|\left(k\hat{u}_\beta\right)^2|n_\beta\rangle\right)}\right|^2.$$

(10.32)

But with an account of the self-conjugation condition for the displacement operator in Eq. (10.19), we can remove the modular brackets in (10.32) and get for the probability [1]:

$$W_0 = \prod_\beta \overline{\left(1 - \frac{1}{2}\langle n_\beta| \left(k\hat{u}_\beta\right)^2 |n_\beta\rangle\right)}^2 \approx \prod_\beta \overline{\left(1 - \langle n_\beta| \left(k\hat{u}_\beta\right)^2 |n_\beta\rangle\right)}, \quad (10.33)$$

where we utilized the expansion $(1 - x)^2 \approx 1 - 2x$, if $x \ll 1$.

Now with the required accuracy we can convert again the multipliers under the product sign in (10.33) into the exponents, and get an elegant formula for the probability [1]:

$$W_0 \approx \prod_\beta \overline{\exp\left(-\langle n_\beta| \left(k\hat{u}_\beta\right)^2 |n_\beta\rangle\right)} = \prod_\beta \exp\left(-\overline{\langle n_\beta| \left(k\hat{u}_\beta\right)^2 |n_\beta\rangle}\right)$$

$$= \exp\left(-\sum_\beta \overline{\langle n_\beta| \left(k\hat{u}_\beta\right)^2 |n_\beta\rangle}\right).$$

Further on we introduce the unit vector [1]:

$$q = \frac{k}{k}. \quad (10.34)$$

Then, the probability reads [1, 15–18]:

$$W_0 \approx \exp\left(-k^2 \sum_\beta \overline{\langle n_\beta| \left(q\hat{u}_\beta\right)^2 |n_\beta\rangle}\right). \quad (10.35)$$

But in the Cartesian coordinates with the account of the conditions on the orthonormality of the polarization vectors (8.55)–(8.56) [1]:

$$\sum_\beta \overline{\langle n_\beta| \left(q\hat{u}_\beta\right)^2 |n_\beta\rangle} = \sum_\beta \overline{\langle n_\beta| \left(q_x^2\hat{u}_{\beta x}^2 + q_y^2\hat{u}_{\beta y}^2 + q_z^2\hat{u}_{\beta z}^2\right) |n_\beta\rangle}$$

$$= q_x^2 \sum_\beta \overline{\langle n_\beta| \left(\hat{u}_{\beta x}^2\right) |n_\beta\rangle} + q_y^2 \sum_\beta \overline{\langle n_\beta| \left(\hat{u}_{\beta y}^2\right) |n_\beta\rangle}$$

$$+ q_z^2 \sum_\beta \overline{\langle n_\beta| \left(\hat{u}_{\beta z}^2\right) |n_\beta\rangle}.$$

Let us utilize now Eq. (9.48) for the mean square of the displacement along each of the Cartesian axes. We get then [1]:

$$\sum_\beta \overline{\langle n_\beta | \left(\hat{u}_{\beta x}^2 \right) | n_\beta \rangle} = \left\langle \overline{\hat{u}_x^2} \right\rangle = \frac{\hbar}{2m} \sum_\beta |e_{\beta x}|^2 \frac{(1 + 2\bar{n}_\beta)}{\omega_\beta}, \tag{10.36}$$

where $\bar{n}_\beta$ is the Planck distribution function [32] for phonons.

Analogously [1]:

$$\sum_\beta \overline{\langle n_\beta | \left(\hat{u}_{\beta y}^2 \right) | n_\beta \rangle} = \left\langle \overline{\hat{u}_y^2} \right\rangle, \quad \sum_\beta \overline{\langle n_\beta | \left(\hat{u}_{\beta z}^2 \right) | n_\beta \rangle} = \left\langle \overline{\hat{u}_z^2} \right\rangle. \tag{10.37}$$

For the cubic crystal [1]:

$$\left\langle \overline{\hat{u}_x^2} \right\rangle = \left\langle \overline{\hat{u}_y^2} \right\rangle = \left\langle \overline{\hat{u}_z^2} \right\rangle = \frac{1}{3} \left\langle \overline{u^2} \right\rangle. \tag{10.38}$$

As a result, the probability equals to [1]:

$$W_0 \approx \exp\left(-\frac{1}{3} k^2 \left\langle \overline{u^2} \right\rangle \left(q_x^2 + q_y^2 + q_z^2 \right) \right) = \exp\left(-\frac{1}{3} k^2 \left\langle \overline{u^2} \right\rangle \right). \tag{10.39}$$

Further on let us use expression (9.56) for the mean square of the displacement, according to which [1, 25]:

$$\frac{1}{3} \left\langle \overline{u^2} \right\rangle = \frac{\hbar}{2M} \int d\omega\, g(\omega) \frac{1}{\omega} \left(2\bar{n}_\omega + 1 \right), \tag{10.40}$$

where $g(\omega)$ is the average density of states of the phonon system. Recollecting, that the recoil energy $R = \frac{\hbar^2 k^2}{2M}$, let us write Eq. (10.39) in the canonical form [1, 15–18]:

$$W_0 = \exp\left(-R \int d\omega\, g(\omega) \frac{1}{\hbar\omega} \left(2\bar{n}_\omega + 1 \right) \right). \tag{10.41}$$

As we already discussed in the beginning of this Lecture, the energy of the γ-quantum is: $E_\gamma = \hbar k c = c p_\gamma$, and hence the recoil energy $R = \frac{p_{\text{nucl.}}^2}{2M}$ due to momentum conservation law $p_{\text{nucl.}} = p_\gamma$ equals to [1, 4–9]: $R = \frac{p_\gamma^2}{2M} = \frac{E_\gamma^2}{2Mc^2}$.

At $T = 0$ the probability of the phononless process [1, 15–18]:

$$W_0 = \exp\left(-R \int d\omega\, g(\omega) \frac{1}{\hbar\omega} \right) = \exp\left(-R \left\langle \frac{1}{\hbar\omega} \right\rangle \right). \tag{10.42}$$

In the Debye model [19], the density of states for the three-dimensional crystal according to Eq. (7.22) has a form: $g(\omega) = \frac{3\omega^2}{\omega_D^3}$ [1, 19–25], and hence (for $k_B = 1$):

$$\left\langle \frac{1}{\hbar\omega} \right\rangle = \frac{3}{2\hbar\omega_D} = \frac{3}{2\theta_D}. \tag{10.43}$$

The corresponding probability is given by [1, 15–18]:

$$W_0 = \exp\left(-\frac{3R}{2\theta_D}\right). \tag{10.44}$$

At high temperatures $T \geq \theta_D$ in the expression (10.41):

$$2\bar{n}_\omega + 1 \approx 2\bar{n}_\omega = \frac{2T}{\hbar\omega} \geq 1. \tag{10.45}$$

Thus, the probability in the Debye model [1, 15–19]:

$$W_0 = \exp\left(-2RT\left\langle \frac{1}{(\hbar\omega)^2} \right\rangle\right) = \exp\left(-\frac{6RT}{\theta_D^2}\right). \tag{10.46}$$

It is important to note, that the probability of the phononless process at high temperatures does not depend on $\frac{1}{N}$ [1]. Hence, Eq. (10.46) describes the macroscopically finite probability [1].

Note also, that we can measure the probability till the small values of the order of $W_0 \sim 10^{-2}$ [1]. If we carefully analyze the energies of the γ-quanta, then we see that the critical ones are the energies of the order of $100\,\mathrm{keV}$. In other words, if $E_\gamma \gg 100\,\mathrm{keV}$, then we do not have the Mossbauer effect [10–12]).

Comment 1. The diagonal matrix element in Quantum mechanics makes the contribution of the same order as all non-diagonal elements. Each non-diagonal element makes the contribution of the order of $\frac{1}{N}$, and only their sum has the order of magnitude as the diagonal element [1, 28].

Let us stress, that if the photon momentum is accepted by all the system (by the crystal as a whole), then the recoil energy goes to zero [1, 15–18]:

$$R = \frac{\hbar^2 k^2}{2M_{\mathrm{cryst.}}} \to 0, \tag{10.47}$$

and the probability is of the order of 1.

At the same time, if the photon momentum is accepted by one nucleus, then the recoil energy $R = \frac{\hbar^2 k^2}{2M_{\mathrm{nucl.}}}$ is finite, and the probability decreases [1, 15–18].

If the crystal is very small, then in the phononless process it moves away as a whole [1]. The probability decreases with the temperature increase [1, 15–18]:

$$W_0 \sim \exp\left(-\frac{6RT}{\theta_D^2}\right).$$

If the temperature increases, then the phonon density increases. Hence, since phonons are the bosonic particles [1,20,27,28], then the probability of the phonon emission increases [1]. Thus, the probability not to emit the phonons strongly decreases [1].

Note, that all our considerations are valid for the amorphous systems as well [1,34]. The only condition is the finite value of the displacement squared [1].

Let us emphasize, that the Mossbauer effect [10–12] is absent in the liquid, since there is diffusion in this case [1,35]. As a result, the mean square of the displacement grows linearly with time [1]:

$$\left\langle \overline{u^2} \right\rangle = 6Dt, \tag{10.48}$$

where D is the diffusion coefficient [35].

Correspondingly, the probability of the phononless process [1,35]:

$$W_0 = \exp\left(-\frac{1}{3}k^2 \left\langle \overline{u^2} \right\rangle\right) \to \exp\left(-2k^2 Dt\right). \tag{10.49}$$

Comment 2. The nucleus emits during all the life-time in the metastable state (and not discretely in the given time-moment) [1].

Thus, since the nucleus emits during all the life-time, it is necessary that the mean square displacement in Eqs. (10.48)–(10.49) on the times of the order of the lifetime of the metastable state $\left\langle \overline{u^2} \right\rangle \propto Dt$, where $t \sim t_{\text{life-time}}$, should be much less than the inverse square of the photon wave vector [1,35]:

$$\left\langle \overline{u^2} \right\rangle \propto Dt_{\text{life-time}} \ll \frac{1}{k^2_{\text{photon}}}. \tag{10.50}$$

Condition (10.50) is fulfilled in very viscous liquids with small diffusion coefficient D [1,34]. The broadened lines with the width:

$$\Gamma \sim \hbar k^2 D \tag{10.51}$$

emerge in this case [1], but nevertheless the Mossbauer effect [10–12] is observed.

Let us stress, that the lines broadening appears due to the fact, that particles are moving in liquid and hence come to the neighborhood of the observation point with different phase shifts $\exp(ik\Delta r)$ [1]. Hence, after the averaging and the utilization of the Debye–Waller formula [29,30] for the probability amplitude, we get [1]:

$$\langle \exp(ik\Delta r) \rangle = \exp\left(-\frac{\langle (k\Delta r)^2 \rangle}{2}\right) = \exp\left(-\frac{k^2 \langle (\Delta r)^2 \rangle}{6}\right). \tag{10.52}$$

As a result, the line broadening emerges [1,35]:

$$(\Delta r)^2 \propto D t_{\text{life-time}}.$$

References

1. Y. Kagan, *Lectures on the Solid-State Theoretical Physics* (Moscow, MEPHI, 1981–1982). Part I. Phonons - unpublished
2. P.W. Walker, Z. Podolyak, Nuclear isomers, *Handbook of Nuclear Physics*, p. 487, published online 05 Sept 2023
3. S. Gorg, B. Maheswari, B. Singh, Y. Sun, A. Goel, A.K. Jain, Atlas of nuclear isomers, 2nd Edition. Atomic Data Nucl. Data Tables **150**, 101546 (2023)
4. V. Krainov, H. Reiss, B. Smirnov, *Radiative Processes in Atomic Physics* (Wiley 1997)
5. C. Cohen-Tannoudji, *Atoms in Electromagnetic Fields* (World Scientific, 2004)
6. C. Doppler, Publisher: Abhandlungen der Königl. Böhm. Gesellschaft der Wissenschaften, V. Folge, Bd. 2, S. 465, Prague, 1842 [Proceedings of the Royal Bohemian Society of Sciences (Part V, vol. 2, p. S 465)], Prague, 1842 (Reissued, 1903)
7. W.D. Phillips, Nobel lecture: laser cooling and trapping of neutral atoms. Rev. Mod. Phys. **70**, 721 (1998)
8. T.W. Hansch, A.L. Schawlow, Opt. Commun. **13**, 68 (1975)
9. V.S. Letokhov, *Laser Control of Atoms and Molecules* (Oxford Academic, 2007)
10. R. Mossbauer, Z. Phys. **151**, 124 (1958); Naturwissenschaften **45**, 538 (1959) (in German)
11. R. Mossbauer, Zeit. Naturforsch **14**, 211 (1959) (in German)
12. R. Mossbauer, *Recoilless Nuclear Resonance Absorption of Gamma Radiation, Nobel Lecture, December 11, 1961, Nobel Lectures, Physics, 1942–1962* (Elsevier Publishing Company, Amsterdam, 1964)
13. P.A.M. Dirac, Proc. Royal Soc. London **A114**, 243 (1927)
14. R.P. Feynman, *QED: The Strange Theory of Light and Matter* (Princeton University Press, 1985)
15. Y.M. Kagan (Author) and I.S. Lyubutin (ed.), Applications of the Mossbauer effect, vol. 2 Spectroscopy and Magnetism, vol. 3 Chemistry (Gordon and Breach Science Publishers, 1985)
16. Y. Kagan, Y. Iosilevskii, The Mossbauer effect for the impurity nucleus in crystal. Sov. Phys. JETP **44**, 284 (1963)
17. Y. Kagan, Use of the anomalous passing effect to obtain stimulated emission of gamma-quanta in crystal. JETP Lett. **20**, 12 (1974)
18. A.A. Maradudin, P.A. Flinn, J.M. Rodcliffe, The Mossbauer effect for one impurity nucleus. Ann. Phys. **26**, 81 (1964)
19. P. Debye, Ann. Phys. **39**, 789 (1912) (in German)
20. L.D. Landau, E.M. Lifshitz, *Statistical Physics* (Part I, Butterworth-Heinemann, Oxford, 1999)
21. O. Madelung, *Introduction to Solid-State Theory*, Springer Series in Solid-State Sciences (1995)
22. C. Kittel, *Quantum Theory of Solids* (John Wiley and Sons, New York-London, 1963)
23. J. M. Ziman, Electrons and phonons, *The Theory of Transport Phenomena in Solids* (Clarendon Press, Oxford, 1960)
24. N.W. Ashcroft, N.D. Mermin, *Solid State Physics* (Holt, Rinehart and Winston, 1976)
25. D. Pines, *Elementary Excitations in Solids* (W.A. Benjamin Inc., New York-Amsterdam, 1963)
26. E.G. Brovman, Y. Kagan, Phonons in non-transition metals. Sov. Physics Uspekhi **17**, 25 (1974)
27. R.P. Feynman, *Statistical Mechanics, A Set of Lectures*, Advanced Books, Classics, (Westview Press, Boulder, 1998)
28. L.D. Landau, E.M. Lifshitz, *Quantum Mechanics* (Pergamon Press, New York, Non-Relativistic Theory, 1977)

29. P. Debye, Ann. Phys. **348**, 49 (1912) (in German)
30. I. Waller, Zeit. fur Physik A **17**, 398 (1923) (in German)
31. J.M. Kosterlitz, D.J. Thouless, J. Phys. **6**, 1181 (1973)
32. V.L. Berezinskii, Sov. Phys. JETP **61**, 1144 (1972)
33. M. Planck, June 2, 1920, in *Les Prix Nobel*. Nobel Lectures, Physics. (Elsevier Publishing Company, Amsterdam, 1967), pp. 1901–1921
34. M.F. Thorpe, L. Tichy, *Properties and Applications of Amorphous Materials* (Springer, Dordrecht, 2001)
35. L.D. Landau, E.M. Lifshitz, Fluid mechanics, in *Course of Theoretical Physics*, vol. 6 (Butterworth-Heinemann, 1987)

Lecture 11. Neutron Scattering on Crystals

11

Abstract

In this Lecture we discuss neutron scattering on crystals. We calculate the lattice sums considering both coherent and incoherent scattering processes. We evaluate the cross sections for elastic and inelastic (one-phonon) processes in mono- and polycrystals. We show inter-connection of the peaks of the doubly differential cross section for the inelastic processes with the phonon spectrum.

11.1 Elastic Scattering of Neutrons on the Crystal

Let us consider the plane-wave (describing neutrons) incident on the crystal

$$\frac{1}{\sqrt{V}} \exp(i\boldsymbol{k}\boldsymbol{r}), \tag{11.1}$$

where V is the system volume. And let us suppose that the incident wave interacts with all the atoms in the crystal with the potential [1,2]:

$$V(\boldsymbol{r}) = \sum_n V(\boldsymbol{r} - \boldsymbol{R}_n). \tag{11.2}$$

The incident wave has an energy $\varepsilon = \frac{\hbar^2 k^2}{2m}$ and the wave vector $\boldsymbol{k}$. The scattered wave then possesses the energy $\varepsilon' = \frac{\hbar^2 k'^2}{2m}$ and the wave vector $\boldsymbol{k}'$ (see Fig. 11.1).

© The Author(s), under exclusive license to Springer Nature Switzerland AG 2026

M. Kagan, *Lecture Notes of Professor Yuri Kagan in Theoretical Solid-State Physics*, Lecture Notes in Physics 1048, https://doi.org/10.1007/978-3-032-14621-2_11

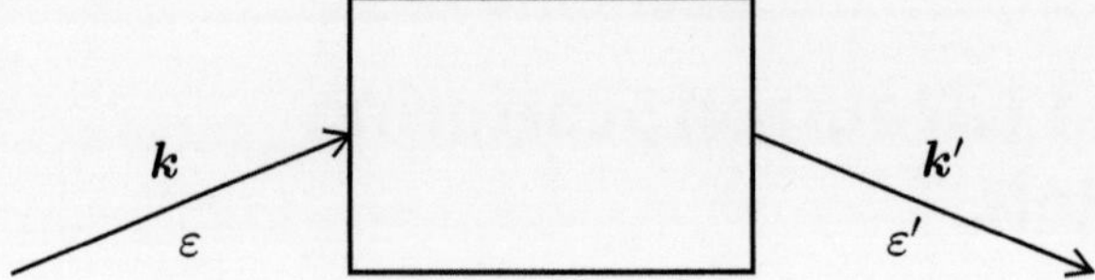

Fig. 11.1 The plane wave incident on the crystal and the scattered wave are shown. The incident wave (describing neutrons) has the energy $\varepsilon = \frac{\hbar^2 k^2}{2m}$ and the wave vector $\boldsymbol{k}$. The scattered wave possesses the energy $\varepsilon' = \frac{\hbar^2 k'^2}{2m}$ and the wave vector $\boldsymbol{k}'$ [1]

Interaction potential in momentum space acquires a form [1,2]:

$$V_{k'k} = \frac{1}{V} \int d^3 r \exp\left(-i \boldsymbol{k}' \boldsymbol{r}\right) \sum_n V\left(\boldsymbol{r} - \boldsymbol{R}_n\right) \exp(i \boldsymbol{k} \boldsymbol{r})$$

$$= \frac{1}{V} \sum_n \int d^3 r \exp\left(i \left(\boldsymbol{k} - \boldsymbol{k}'\right) \left(\boldsymbol{r} - \boldsymbol{R}_n + \boldsymbol{R}_n\right)\right) V\left(\boldsymbol{r} - \boldsymbol{R}_n\right)$$

$$= \sum_n \exp\left(i \left(\boldsymbol{k} - \boldsymbol{k}'\right) \boldsymbol{R}_n\right) \frac{1}{V} \int d^3 r \exp\left(i \left(\boldsymbol{k} - \boldsymbol{k}'\right) \left(\boldsymbol{r} - \boldsymbol{R}_n\right)\right) V\left(\boldsymbol{r} - \boldsymbol{R}_n\right)$$

$$= \sum_n \exp\left(i \left(\boldsymbol{k} - \boldsymbol{k}'\right) \boldsymbol{R}_n\right) V\left(\boldsymbol{k} - \boldsymbol{k}'\right). \tag{11.3}$$

Introducing convenient notation $\boldsymbol{\kappa} = \boldsymbol{k} - \boldsymbol{k}'$, we can get finally:

$$V_{k'k} = \sum_n \exp\left(i \boldsymbol{\kappa} \boldsymbol{R}_n\right) V(\boldsymbol{\kappa}). \tag{11.4}$$

If under the wave action the phonon states in the crystal are changed (from the set of states $\{n\}$ till the set of states $\{n'\}$), then the matrix element of the potential describing this interaction can be written as follows [1]:

$$V_{k'\{n'\},k\{n\}} = V(\boldsymbol{\kappa})\langle\{n'\} \mid \sum_n \exp\left(i \boldsymbol{\kappa} \boldsymbol{R}_n\right) \mid\{n\}\rangle. \tag{11.5}$$

Note, that in Eq. (11.5) the term in angular brackets is responsible for the modifications of the phonon states. Moreover, we have to keep in mind the condition for the validity of Born approximation [1–3], which assumes that the amplitude of the scattered wave should be much less than the amplitude of the incident wave (condition of weak scattering) [1–3]:

$$|\psi_{\text{scat.}}| \ll |\psi_{\text{inc.}}|. \tag{11.6}$$

In fact, it is the condition on the width of the crystal (of our system, in the other words).

Let us introduce the probability differential for the process in crystal when we consider the phonon states transition from the initial set of states $\{n\}$ to all the final sets of states $\{n'\}$ (and to all the final values of the wave vector of the scattered wave). If Born approximation is valid, then we can utilize the Fermi Golden rule [1,2,4–7] of Quantum mechanics, according to which the probability differential is proportional to the sum (to the integral) over all the final states from the square of the interaction matrix element multiplied by the delta function governing the energy conservation law:

$$dW = \frac{2\pi}{\hbar} \sum_{\{n'\}} \left|V_{k'\{n'\},k\{n\}}\right|^2 \delta\left(\varepsilon + E(\{n\}) - \varepsilon' - E\left(\{n'\}\right)\right) \frac{V d^3 k'}{(2\pi)^3}, \qquad (11.7)$$

where statistical weight for neutron scattering in crystal [1,8,12] is given by:

$$d^3 k' = mk' d\varepsilon' d\Omega_{k'}, \qquad (11.8)$$

and m is the neutron mass.

Further on it is convenient to introduce the current density for the number of neutrons:

$$J = \frac{\hbar k}{m} \frac{1}{V}, \qquad (11.9)$$

and define the differential scattering cross section as:

$$d\sigma = \frac{dW}{j}. \qquad (11.10)$$

It is useful also to introduce the doubly differential cross section [1,8,12]

$$\frac{d^2\sigma}{d\varepsilon' d\Omega'}, \qquad (11.11)$$

as follows:

$$d\sigma = \frac{dW}{j} = \frac{d^2\sigma}{d\varepsilon' d\Omega'} d\varepsilon' d\Omega'. \qquad (11.12)$$

It is interesting, that in the purely elastic neutron scattering for $\varepsilon = \varepsilon'$ we get in Eq. (11.5) exactly the same matrix element, as in the Mossbauer effect [1,8,13,16]:

$$\sum_n \exp\left(i\kappa R_n\right) = \sum_n \exp\left(i\kappa R_n^0\right) \exp\left(i\kappa u_n\right). \qquad (11.13)$$

Moreover, in the sum over n' in Eq. (11.7) only diagonal matrix element survives for $n' = n$ which is proportional [1,8,16] to:

$$\langle\{n\}|\exp\left(i\kappa u_n\right)|\{n\}\rangle. \qquad (11.14)$$

Correspondingly, the probability of the diagonal (phononless) process contains Debye–Waller factor [1,13–15], as in the Mossbauer effect [1,8,16,19]:

$$W_{0\,\text{Moss}} = \exp\left(-Z(\kappa)\right), \tag{11.15}$$

where the probability amplitude [1,16,19] reads:

$$\exp\left(-\frac{1}{2}Z(\kappa)\right) = \langle\{n\}|\exp\left(i\kappa u_n\right)|\{n\}\rangle. \tag{11.16}$$

As a result, Eq. (11.7) up to a constant is given by:

$$dW = \cdots |V(\kappa)|^2 \exp\left(-Z(\kappa)\right)\left|\sum_n \exp\left(i\kappa R_n^0\right)\right|^2 \delta\left(\varepsilon - \varepsilon'\right) k d\varepsilon d\Omega, \tag{11.17}$$

where an emergence of the absolute value squared of the lattice-sum [1,16,20,23] $\left|\sum_n \exp\left(i\kappa R_n^0\right)\right|^2$ is connected with the summation over n ($\sum_n \langle\{n\}|...|\{n\}\rangle$ in Eq. (11.5)), and besides that corresponds to the diagonal element $E(\{n\}) = E\left(\{n'\}\right)$

Graphically the elastic process for neutron scattering on the crystal nuclei is shown in Fig. 11.2. Note that in case of the original Mossbauer effect [1,13–15] in the Debye-Waller factor $\exp(-Z(\kappa))$ [1,16–19] for resonance scattering of the γ-quantum on nucleus, instead of the momentum $\kappa = k - k'$ transferred by the incident neutron to the nucleus, enters just the momentum k of the incident photon. The point is that nucleus just absorbs the γ-quantum and that is all [1].

Let us calculate now the doubly differential scattering cross section [1,8–12], utilizing expressions (11.9–11.12). As a result:

$$\frac{d^2\sigma}{d\varepsilon' d\Omega'} = |V(\kappa)|^2 \cdot \frac{2\,m^2 V^2}{(2\pi\hbar)^3} \exp\left(-Z(\kappa)\right)\left|\sum_n \exp\left(i\kappa R_n^0\right)\right|^2 \delta\left(\varepsilon - \varepsilon'\right) \cdot \frac{k}{k'}$$

$$\sim \sigma_0(\kappa) \exp\left(-Z(\kappa)\right)\left|\sum_n \exp\left(i\kappa R_n^0\right)\right|^2 \delta\left(\varepsilon - \varepsilon'\right) \cdot \frac{k}{k'}, \tag{11.18}$$

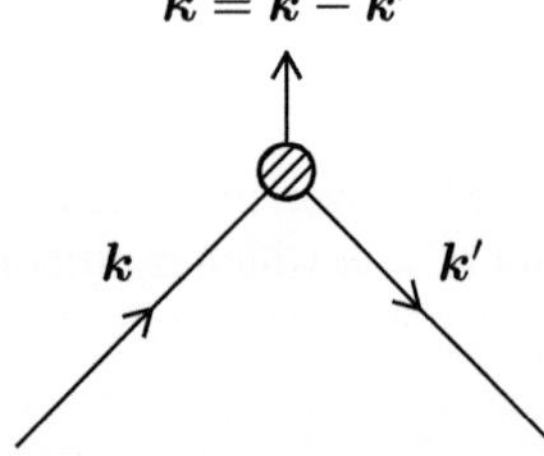

Fig. 11.2 The schematical diagram for the elastic neutron scattering on the nucleus. In the scattering process the incident particle transfers the momentum $\kappa = k - k'$ to the nucleus [1]

where from the delta function with respect to energy in Eq. (11.18) it follows, that $\varepsilon_k = \varepsilon_{k'}$, and, correspondingly, $k^2 = k'^2$, and $k = k'$.

At the same time the bare quantum-mechanical cross section $\sigma_0(\kappa)$, which enters in Eq. (11.18) reads:

$$\sigma_0(\kappa) = 4\pi |f(\kappa)|^2, \qquad (11.19)$$

where $f(\kappa) \approx -\frac{mV}{2\pi\hbar^2} V(\kappa)$ is the amplitude of the weak (Born) scattering [1–7].

Full cross section of the elastic scattering, according to Eqs. (11.12) and (11.18), is given by [1–7]:

$$\sigma_{\text{el.}}^{\text{total}} = \int \frac{d^2\sigma_{\text{el.}}}{d\varepsilon_{k'}d\Omega_{k'}} d\varepsilon_{k'}d\Omega_{k'} \sim \sigma_0(\kappa)\exp\left(-Z(\kappa)\right)$$

$$\times \int d\Omega_{k'} \left| \sum_n \exp\left(i\kappa R_n^0\right) \right|^2 . \qquad (11.20)$$

Let us investigate now the angular dependence of the scattering cross section. To do that let us write the absolute value squared of the lattice sum as follows [1, 20–23]:

$$\left| \sum_n \exp\left(i\kappa R_n^0\right) \right|^2 = \sum_{n,n'} \exp\left(i\kappa \left(R_n^0 - R_{n'}^0\right)\right) = \sum_{n,n'} \exp\left(i\kappa R_{n''}^0\right), \qquad (11.21)$$

where $R_{n''}^0 = R_n^0 - R_{n'}^0$ and index n'' is running again over all the integer values which numerate the lattice nodes.

Then:

$$\sum_{n,n'} \exp\left(i\kappa R_{n''}^0\right) = \left(\sum_n 1 \right) \sum_{n''} \exp\left(i\kappa R_{n''}^0\right) = N \sum_{n''} \exp\left(i\kappa R_{n''}^0\right), \qquad (11.22)$$

where again stands the lattice sum which equals to:

$$\sum_{n''} \exp\left(i\kappa R_{n''}^0\right) = N \sum_K \delta_{\kappa,K}. \qquad (11.23)$$

Thus, finally [1, 20–23]:

$$\left| \sum_n \exp\left(i\kappa R_n^0\right) \right|^2 = N^2 \sum_K \delta_{\kappa,K}. \qquad (11.24)$$

As a result, full cross section of the elastic scattering [1, 8–10]:

$$\sigma_{\text{el.}}^{\text{total}} \sim N^2 \sigma_0(\kappa)\exp\left(-Z(\kappa)\right) \sum_K \delta_{\kappa,K} \propto N^2. \qquad (11.25)$$

Moreover, the scattering emerges only at the values of the transferred momentum [1,8–10]:

$$\boldsymbol{\kappa} = \boldsymbol{k} - \boldsymbol{k'} = \boldsymbol{K}, \tag{11.26}$$

where $\boldsymbol{K}$ is the inverse lattice vector.

Squaring Eq. (11.26), we get in accordance with the cosine theorem:

$$k^2 + k'^2 - 2kk' \cos \theta = K^2. \tag{11.27}$$

For the elastic scattering $k = k'$, and thus:

$$2k^2(1 - \cos \theta) = K^2. \tag{11.28}$$

Hence from the elementary trigonometry we get the famous formula for the Bragg maxima [24]:

$$2k \sin \frac{\theta}{2} = K. \tag{11.29}$$

Let us remind that in the general case we can expand the vector $\boldsymbol{K}$ over the basic vectors of the inverse lattice $\boldsymbol{b}_1, \boldsymbol{b}_2, \boldsymbol{b}_3$ [1,20–23]:

$$\boldsymbol{K} = 2\pi \left(m_1 \boldsymbol{b}_1 + m_2 \boldsymbol{b}_2 + m_3 \boldsymbol{b}_3 \right), \tag{11.30}$$

where m_1, m_2, m_3 are the integer numbers.

Moreover, the Gibbs relations hold between the vectors of the direct and inverse lattices, according to which [1,20–23]:

$$\boldsymbol{b}_1 = \frac{1}{V_0} \left[\boldsymbol{a}_2 \times \boldsymbol{a}_3 \right], \tag{11.31}$$

where $\boldsymbol{a}_2$ and $\boldsymbol{a}_3$ are the basic vectors of the direct lattice, V_0 is the volume of the elementary cell. Thus basic vectors of the inverse lattice with respect to dimensionality are given by [1]:

$$[b] = \left[\frac{1}{a} \right] = \left[\frac{1}{\text{intersite distance}} \right]. \tag{11.32}$$

If d is the distance between the planes of the crystalline lattice and λ is the wavelength of the particle incident on the crystal, then in Eq. (11.29):

$$k = \frac{2\pi}{\lambda}, \ K = \frac{2\pi m}{d}, \tag{11.33}$$

where m is an integer number.

As a result, the condition for Bragg maxima can be written as follows [1,24]:

$$2\frac{2\pi}{\lambda} \sin \frac{\theta}{2} = \frac{2\pi m}{d}, \ \text{or} \ \frac{2}{\lambda} \sin \frac{\theta}{2} = \frac{m}{d}. \tag{11.34}$$

Thus, we get Bragg–Wolf interference maxima well known in optics and crystallography [1,11,12,24–29]:

$$2d \sin \frac{\theta}{2} = m\lambda. \tag{11.35}$$

For clarity we can write the condition (11.26) on the transferred momentum in different form [1] as:

$$\boldsymbol{k'} = \boldsymbol{k} - \boldsymbol{K}. \tag{11.36}$$

Then, after squaring we get:

$$k'^2 = k^2 + K^2 - 2\boldsymbol{k}\boldsymbol{K}, \tag{11.37}$$

or for the elastic scattering:

$$K^2 - 2\boldsymbol{k}\boldsymbol{K} = 0. \tag{11.38}$$

It is convenient to write the condition (11.35) in the equivalent form [1]:

$$\boldsymbol{K} \left(\boldsymbol{k} - \frac{\boldsymbol{K}}{2} \right) = 0. \tag{11.39}$$

Hence, we should have for the scattering:

$$\boldsymbol{K} \perp \left(\boldsymbol{k} - \frac{\boldsymbol{K}}{2} \right) \tag{11.40}$$

where $\boldsymbol{k}$ is the wave vector of the incident wave (see Fig. 11.3a).

In other words, the vector $\boldsymbol{k} - \frac{\boldsymbol{K}}{2}$ should lie on the boundary of the Brillouin zone [1,20–23,30] (see Fig. 11.3b).

We can say, that elastic scattering is impossible if the absolute value of the incident wave vector is less than half a value of the minimal vector of the inverse lattice [1]:

$$|\boldsymbol{k}| < \frac{|\boldsymbol{K}_{\min}|}{2}. \tag{11.41}$$

For the emission with these wave vectors the crystal is absolutely transparent [1].

The neutron scattering on the crystal is schematically illustrated in Fig. 11.4.

Note that elastic scattering is a coherent one (the sum of the amplitudes is proportional to the number of scatterers N, and the probability to the number of scatterers squared N^2).

It means that all the scatterers scatter in-phase [1,2]:

$$A = \exp\left(i\varphi_{\text{tot.}}\right)\left(A_1 + \cdots + A_N\right) \propto N, \tag{11.42}$$

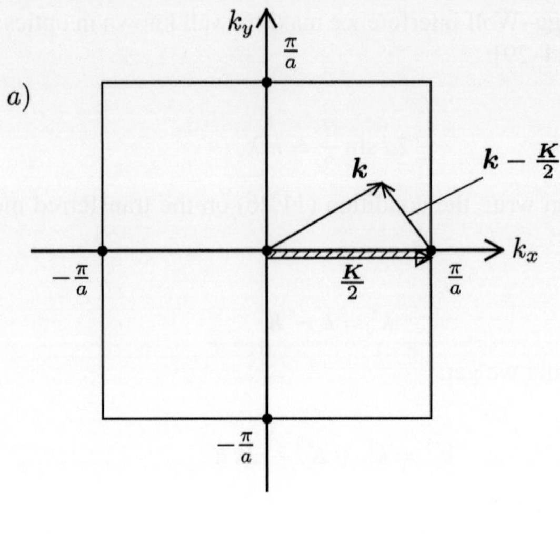

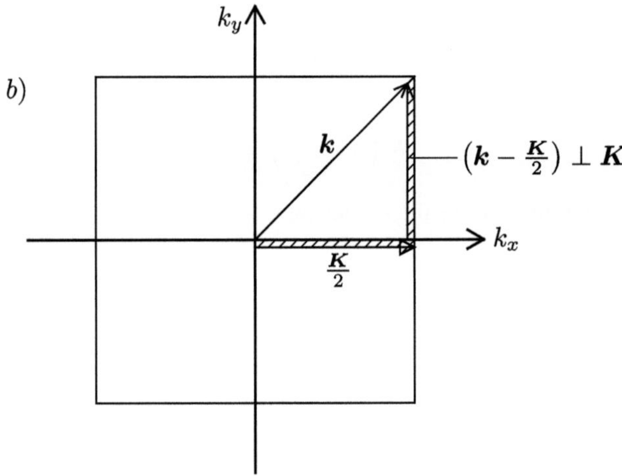

Fig. 11.3 Graphical illustration of the condition for the Bragg maxima: **a** General case for the location of the vector $k - \frac{K}{2}$ in the first Brillouin zone. For clarity of the illustration, we select the square lattice of the two-dimensional crystal **b** the vector $k - \frac{K}{2}$ is located on the boundary of the Brillouin zone to satisfy the condition $K \perp \left(k - \frac{K}{2}\right)$ [1]

$$W = |A|^2 \propto N^2. \tag{11.43}$$

The oscillations of each atom is a linear combination of the normal oscillators. Only for the oscillations in the definite directions all the normal oscillators oscillate in-phase [1,2].

In reality we do not have purely coherent scattering [1,8–10], since, first of all, the crystal is finite. Secondly, we get the result in Born approximation [3], considering

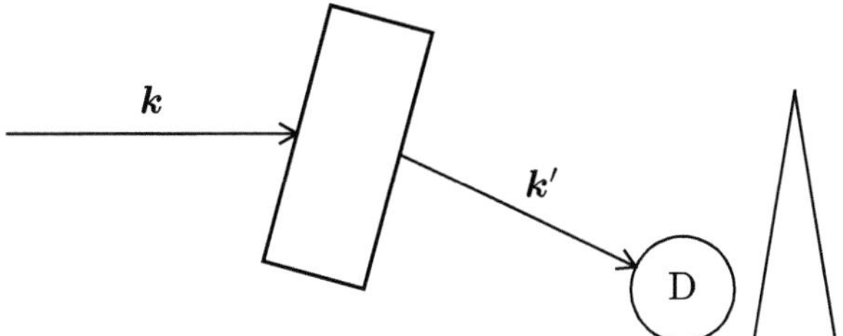

Fig. 11.4 Schematic illustration for the elastic scattering of the monochromatic neutron beam on the crystal. The scattered neutrons are fixed by the detector, Angular dependence of the scattering cross section acquires a sharp peak at a definite scattering angle determined by the condition (11.26) [1]

potential as a perturbation. Here Born approximation means, that $|\psi_{\text{scat.}}| \ll |\psi_{\text{inc.}}|$ [1–3], or in fact, that the number of scattered neutrons is much less than a number of incident neutrons [1,8–10].

As we discussed already, the Born approximation is the condition on the thickness of the crystal (If the crystal is very thick, then the number of scatterers tends to infinity, and in this case the number of scattered neutrons equals by order of magnitude to the number of incident neutrons). Thus, the validity of Born approximation demands the crystal to have finite thickness [1].

The crystal volume $V_{\text{cryst.}} = L_x L_y L_z$ in the three-dimensional case. Thus, from the uncertainty principle the transferred momentum has a finite smearing $\Delta \kappa_i \sim \frac{1}{L_i}$ [1,8]. The incident and scattered wave vectors $\boldsymbol{k}$ and $\boldsymbol{k}'$ are also defined not exactly (are smeared). Suppose the condition (11.36) on the wave vector of the scattered wave as a result of the smearing is replaced by [1]:

$$\boldsymbol{k}' = \boldsymbol{k} - \boldsymbol{K} + \boldsymbol{q}, \tag{11.44}$$

where the absolute value of the "smearing" vector $\boldsymbol{q}$ equals by order of magnitude to the inverse size of the crystal [1]:

$$|\boldsymbol{q}| \sim \frac{1}{L}. \tag{11.45}$$

At the same time, as we discussed already (see Eq. (11.33)) the absolute value of the wave vector of the incident wave equals by order of magnitude to the inverse wavelength: $|\boldsymbol{k}| \sim \frac{1}{\lambda}$, where λ is of the order of several interparticle distances. Hence the following inequality always is satisfied for the macroscopic crystal [1]:

$$\frac{\lambda}{L} \ll 1, \tag{11.46}$$

and correspondingly the smearing is always small [1,8]:

$$\frac{|\boldsymbol{q}|}{|\boldsymbol{k}|} \ll 1. \tag{11.47}$$

Squaring the condition (11.44), with an accuracy of the quadratic terms in q, we obtain [1]:

$$k'^2 \cong (k - K)^2 + 2q(k - K) + O\left(q^2\right).\tag{11.48}$$

Utilizing the fact, that according to Eq. (11.23) for the intensive scattering we have to fulfil the condition: $k'^2 \approx (k - K)^2$, we get with the required accuracy from (11.48) [1]:

$$q(k - K) = 0.\tag{11.49}$$

Thus, the vectors q lie in the plane perpendicular to vectors k and K, and smear the exact conservation law [1]. As a result, the cone with a small scattering angle appears (see Fig. 11.5). Moreover, the possible values of the smearing vector q lie in the cone base [1].

Full cross section of the elastic neutron scattering, even with the account of smearing, according to Eq. (11.20) reads [1,8]:

$$\sigma_{\text{el.}}^{\text{total}} = \int \frac{d^2\sigma_{\text{el.}}}{d\varepsilon_{k'}d\Omega_{k'}} d\varepsilon_{k'} d\Omega_{k'} = \sigma_0(\kappa)\exp\left(-Z(\kappa)\right)\int d\Omega_{k'}\left|\sum_n \exp\left(i\kappa R_n^0\right)\right|^2.$$

Let us analyze again the behavior of the absolute value squared of the lattice sum [1,20–23] $\left|\sum_n \exp\left(i\kappa R_n^0\right)\right|^2$ in the full cross-section. The transferred momentum with the account of smearing yields [1]:

$$\kappa = k - k' = k - (k - K) - q = K - q.\tag{11.50}$$

$$k' = k - K + q$$
$$|k'| \approx |k - K|$$

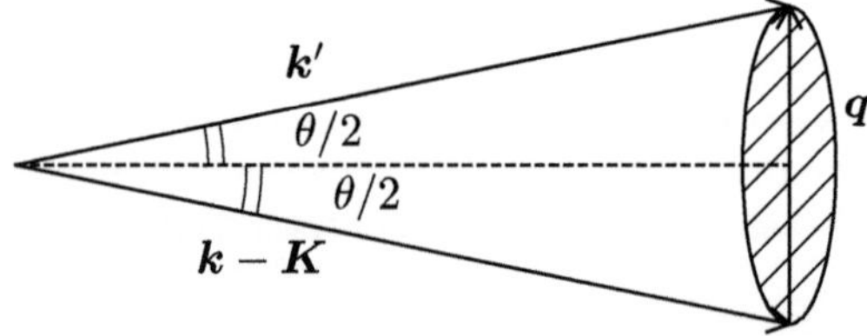

Fig. 11.5 The cone with a small scattering angle, formed by 3 vectors k', $k - K$ and q, where $k' = (k - K) + q$. The absolute values of the two vectors are approximately equal: $|k'| \approx |k - K|$. As a result, we get the isosceles triangle with angle θ at the vertex and base q (which equals to the diameter of the cone base). The angles between the equal sides of the triangle and the triangle height equal to $\frac{\theta}{2}$ [1]

But by the definition of the direct and inverse lattice we get [1]:

$$\boldsymbol{K}\boldsymbol{R}_n^0 = 0. \tag{11.51}$$

Hence, the absolute value squared of the lattice sum reads [1]:

$$\left|\sum_n \exp\left(i\boldsymbol{\kappa}\,\boldsymbol{R}_n^0\right)\right|^2 = \left|\sum_n \exp\left(-i\boldsymbol{q}\,\boldsymbol{R}_n^0\right)\right|^2 = \left(\sum_n 1\right)\sum_{n''} \exp\left(i\boldsymbol{q}\,\boldsymbol{R}_{n''}^0\right)$$
$$= N \sum_{n''} \exp\left(i\boldsymbol{q}\,\boldsymbol{R}_{n''}^0\right). \tag{11.52}$$

Further on we can replace summation by the integral and get [1]:

$$\left|\sum_n \exp\left(i\boldsymbol{\kappa}\,\boldsymbol{R}_n^0\right)\right|^2 = \frac{N}{V_0}\int d\boldsymbol{R}_n^0 \exp\left(i\boldsymbol{q}\,\boldsymbol{R}_n^0\right). \tag{11.53}$$

Note that in the exponent in Eq. (11.53) stands the small smearing which by order of magnitude equals to $|\boldsymbol{q}| \sim \frac{1}{L}$. But, at the same time $\left|\boldsymbol{R}_n^0\right| \sim L$ is large. Hence, expression in the exponent is of the order of 1, and we cannot expand it [1].

Let us choose the axis $z\|(\boldsymbol{k} - \boldsymbol{K})$ in Eq. (11.49) (see Fig. 11.5). Then, from the condition on intensive scattering, the smearing vector $\boldsymbol{q} \perp z$-axis. Hence in exponent (11.53) contribute only two-dimensional vectors $\boldsymbol{R}_n^0 = \boldsymbol{\rho}$, lying in (xy)-plane ($\perp z$-axis)[1]. As a result, in cylindrical frame $(z, \boldsymbol{\rho})$, in the cone-geometry with small scattering angle [1]:

$$\left|\sum_n \exp\left(i\boldsymbol{\kappa}\,\boldsymbol{R}_n^0\right)\right|^2 = \frac{N}{V_0}\int d\boldsymbol{\rho}\,dz \exp(i\boldsymbol{q}\boldsymbol{\rho}) = \frac{NL_z}{V_0}(2\pi)^2\frac{1}{(2\pi)^2}\int d\boldsymbol{\rho}\exp(i\boldsymbol{q}\boldsymbol{\rho}). \tag{11.54}$$

But, by the definition at small values of $|\boldsymbol{q}| \sim \frac{1}{L}$:

$$\frac{1}{(2\pi)^2}\int d\boldsymbol{\rho}\exp(i\boldsymbol{q}\boldsymbol{\rho}) = \delta(\boldsymbol{q}). \tag{11.55}$$

Thus finally [1]:

$$\left|\sum_n \exp\left(i\boldsymbol{\kappa}\,\boldsymbol{R}_n^0\right)\right|^2 = \frac{NL_z}{V_0}(2\pi)^2\delta(\boldsymbol{q}). \tag{11.56}$$

Now we can perform the integration over the solid angle in the expression for the total cross section. The element of the solid angle:

$$d\Omega_{k'} = \sin\theta d\theta d\varphi. \tag{11.57}$$

Then the absolute value of vector $\boldsymbol{q}$ in Fig. 11.5 according to the sine theorem for elastic scattering (at $k = k'$) equals to:

$$q = 2k \sin \frac{\theta}{2}. \tag{11.58}$$

Correspondingly, since in Eq. (11.58) $k = \text{const}$:

$$dq = k \cos \frac{\theta}{2} d\theta, \text{ and } qdq = 2k^2 \sin \frac{\theta}{2} \cos \frac{\theta}{2} d\theta = k^2 \sin \theta d\theta. \tag{11.59}$$

As a result [1]:

$$\sin \theta d\theta = \frac{qdq}{k^2}, \tag{11.60}$$

and the element of the solid angle reads [1]:

$$d\Omega_{k'} = \frac{qdqd\varphi}{k^2}. \tag{11.61}$$

Going to Cartesian coordinates in (xy)-plane where the vector $\boldsymbol{q}$ is lying we get [1]:

$$d\Omega_{k'} = \frac{dq_x dq_y}{k^2} = \frac{d^2 \boldsymbol{q}}{k^2}. \tag{11.62}$$

Note, that this expression can be written also via the element of the cone surface $d\Omega_{k'} = \frac{dS}{R^2}$ [1].

Returning back to Eq. (11.62) and utilizing (11.56), we can represent the integral of the absolute value squared of the lattice sum by the solid angle, with the account of smearing, in the total cross section of the elastic scattering as [1]:

$$\sigma_{\text{el.}}^{\text{total}} \sim \int d\Omega_{k'} \left| \sum_n \exp\left(i\boldsymbol{\kappa} \boldsymbol{R}_n^0\right) \right|^2 = \frac{NL_Z}{V_0} (2\pi)^2 \int \delta(\boldsymbol{q}) \frac{d^2 \boldsymbol{q}}{k^2} = \frac{NL_Z}{V_0} \frac{(2\pi)^2}{k^2}. \tag{11.63}$$

However, since the crystal size in z-direction [1]:

$$L_z \sim V^{1/3} = (V/V_0)^{1/3} V_0^{1/3} \sim V_0^{1/3} N^{1/3}, \tag{11.64}$$

then the total cross section of the elastic scattering with the account of smearing now depends on the scatterers number in the different fashion, namely as follows [1]:

$$\sigma_{\text{el.}}^{\text{total}} \sim N^{4/3} \sigma_0(\boldsymbol{\kappa}) \exp\left(-Z(\boldsymbol{\kappa})\right) \propto N^{4/3}. \tag{11.65}$$

Moreover, now the peak in the angular dependence of the scattering cross section in Fig. 11.4 is not an infinitely high, but acquires a finite width [1] because the crystal has a finite size and conservation laws are fulfilled not exactly now.

It is interesting, that in ideal polycrystal [1] (in the absence of smearing) the transferred momentum $\kappa = K$, but K vector is randomly oriented (in each monocrystal the K values are the same by absolute value but randomly oriented). Let us average now the doubly differential cross section $\frac{d^2\sigma_{\mathrm{el}}}{d\varepsilon_{k'}d\Omega_{k'}}$ over the direction of the unit vector $\frac{K}{K}$. As a result, the expression for the absolute value squared of the lattice sum in the ideal polycrystal again as in Eq. (11.22) is simplified and becomes proportional to the lattice sum in the first power [1]:

$$\left|\sum_n \exp\left(i\kappa R_n^0\right)\right|^2 = N \sum_n \exp\left(i\kappa R_n^0\right).$$

Now we can proceed from the summation to the integration in the expression for the lattice sum just as in the finite-size crystal [1]:

$$\sum_n \exp\left(i\kappa R_n^0\right) = \frac{1}{V_0} \int dR \exp(i\kappa R) = \frac{(2\pi)^3}{V_0} \sum_K \delta(\kappa - K)$$

$$= \sum_K \delta_{\kappa,K}. \tag{11.66}$$

As a result, we get for the polycrystal [1]:

$$\left|\sum_n \exp\left(i\kappa R_n^0\right)\right|^2 = N \sum_K \delta_{\kappa,K} = N\frac{(2\pi)^3}{V_0} \sum_K \delta(\kappa - K). \tag{11.67}$$

Now we have to average the obtained expression over the direction of the unit vector $\frac{K}{K}$, or, in other words, over the solid angle Ω_K.

In fact, for polycrystal we should calculate the mean over the solid angle [1]:

$$\left\langle \left|\sum_n \exp\left(i\kappa R_n^0\right)\right|^2 \right\rangle_{\Omega_K} = \frac{N}{V_0}(2\pi)^3 \left\langle \sum_K \delta(\kappa - K) \right\rangle_{\Omega_K}$$

$$= \frac{N}{V_0}(2\pi)^3 \sum_K \langle \delta(\kappa - K)\rangle_{\Omega_K}. \tag{11.68}$$

Hence for polycrystal (taking into account, that in each monocrystal the K vector is the same by absolute value, but randomly oriented) we get [1]:

$$\langle \delta(\kappa - K)\rangle_{\Omega_K} = \frac{1}{4\pi} \int d\Omega_K \delta(\kappa - K) = \delta(\kappa - K) \int d\Omega_K \delta\left(\frac{\kappa}{\kappa} - \frac{K}{K}\right)$$

$$= B\delta(\kappa - K). \tag{11.69}$$

Then, from the condition of the delta function normalization, we can determine the value of B [1]:

$$\int d^3 K \delta(\kappa - K) = 1 = \iint d\Omega_K K^2 dK \delta(\kappa - K) = 4\pi B \int K^2 dK \delta(\kappa - K)$$

$$= 4\pi B \kappa^2. \tag{11.70}$$

Hence [1]:

$$B = \frac{1}{4\pi \kappa^2}. \tag{11.71}$$

As a result, the total cross section of the neutron elastic scattering on the ideal polycrystal after the averaging over the solid angle reads [1]:

$$\left\langle \sigma_{\text{el.}}^{\text{total}} \right\rangle_{\Omega_K} \sim \sigma_0(\kappa) \exp\left(-Z(\kappa)\right) \left\langle \left| \sum_n \exp\left(i\kappa R_n^0\right) \right|^2 \right\rangle_{\Omega_K} = \frac{N}{V_0} \sigma_0(\kappa)$$

$$\cdot \exp\left(-Z(\kappa)\right) (2\pi)^3 \sum_K \delta(\kappa - K) \frac{1}{4\pi \kappa^2} \propto N \sum_K \delta(\kappa - K) \frac{1}{\kappa^2}. \tag{11.72}$$

Dependence of the averaged cross section in ideal polycrystal for the absolutely elastic scattering from the transferred momentum is shown in Fig. 11.6. We can see that the cross section equals to zero for small transferred momenta $\kappa < K_{\min}$ [1]. Moreover, for the successive wave vectors of the inverse lattice $K > K_{\min}$, we observe the jumps in the dependence of the cross section on the transferred momentum κ due to the presence of the delta function in Eq. (11.72) [1].

At the same time, between the two neighboring values of the inverse lattice wave vector K the dependence of the cross section on κ which is proportional to $\frac{1}{\kappa^2}$ is conserved [1].

Thus in fact the crystal serves as a filter in the scattering process for the cutoff of the extremely slow (ultracold) neutrons for which the incident wave vector $k < \frac{K_{\min}}{2}$. They pass through the crystal without scattering (see Fig. 11.7). At the same time, more rapid (Do «temporal») neutrons scatter on the crystal [1].

11.2 Inelastic Neutron Scattering on the Crystal: One-Phonon Process

Let us consider now one-phonon process for the inelastic neutron scattering on the crystal [1,8,9]. Let the wave vector of the incident neutron equals to k, and the wave vector of the scattered neutron—to k'. The crystal in the final state $|f\rangle$ contains one phonon more or one phonon less than in the initial state $|i\rangle$ [1,8] (see Fig. 11.8).

The probability amplitude of the one-phonon process $A(\kappa)$ can be calculated according to the relation (11.5) [1,8,14]:

$$A(\kappa) = V(\kappa) \sum_n \exp\left(i\kappa R_n^0\right) \langle k', \{n'\} | \exp\left(i\kappa \hat{u}_n\right) | k, \{n\}\rangle, \tag{11.73}$$

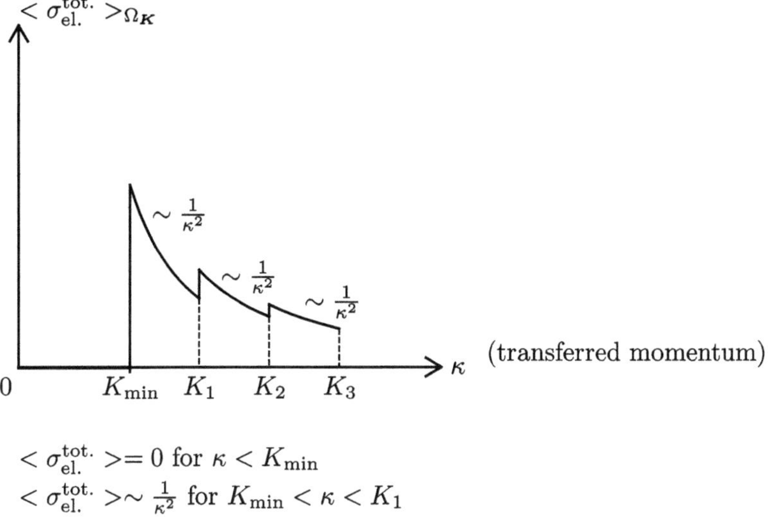

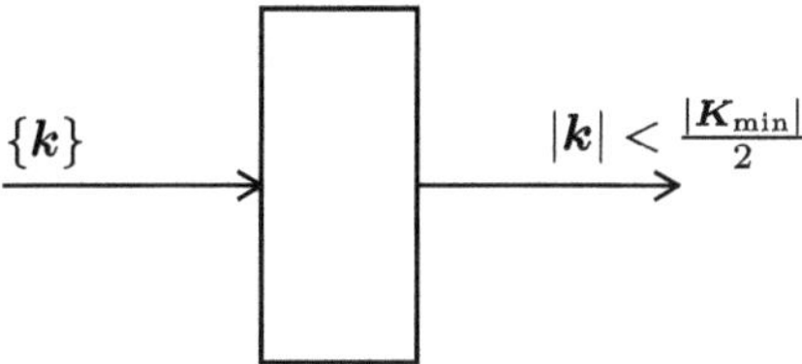

Fig. 11.6 Dependence of the averaged cross section of the absolutely elastic scattering in ideal polycrystal on the magnitude of the transferred momentum. Cross section equals to zero for small transferred momenta $\kappa < K_{\min}$. Moreover, for the successive wave vectors of the inverse lattice $K > K_{\min}$ we observe the jumps in the dependence of the cross section on the transferred momentum κ. At the same time, between the two neighboring values of the inverse lattice wave vector K the dependence of the cross section on κ proportional to $\frac{1}{\kappa^2}$ is conserved [1]

Fig. 11.7 In the scattering process the crystal serves as a filter for the cutoff of the extremely slow (ultracold) neutrons for which the incident wave vector $k < \frac{K_{\min}}{2}$. They pass through the crystal without scattering. At the same time, more rapid («temporal») neutrons scatter on the crystal [1]

where the displacement operator can be represented as a sum of the displacement operators for the normal oscillators [1,14]:

$$\hat{u}_n = \sum_\beta \hat{u}_{n,\beta}, \; \beta = \{\alpha, f\}.$$

At the same time, the initial state of the phonon subsystem in crystal can be described by the set of the occupation numbers for the phonon states $\{n\} = \{n_\beta\}$. If we distinguish from this set a specific occupation number n_β (which in the process of

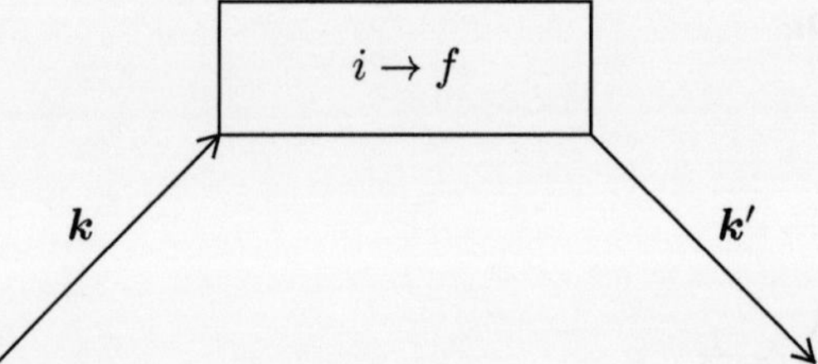

Fig. 11.8 One-phonon process for the inelastic neutron scattering on the crystal. The wave vector of the incident neutron equals to k, and the wave vector of the scattered neutron equals to k'. In the final state $|f\rangle$ the crystal contains one phonon more or one phonon less than in the initial state $|i\rangle$ [1]

neutron scattering changes on1), and denote the set of all the rest occupation numbers as $\{\tilde{n}_\beta\}$, then we can write [1, 8, 16]:

$$\{n\} = \{n_\beta\} = \left(\{\tilde{n}_\beta\}, n_\beta\right). \tag{11.74}$$

Correspondingly, the set of the occupation numbers in the final state in Eq. (11.73) reads:

$$\{n'\} = \{n'_\beta\} = \left(\{\tilde{n}_\beta\}, n_\beta \pm 1\right). \tag{11.75}$$

Then the probability amplitude [1, 8, 14]:

$$V(\kappa) \sum_n \exp\left(i\kappa R_n^0\right) \prod_\beta \langle k', \{\tilde{n}_\beta\}, n_\beta \pm 1| \exp\left(i\kappa \hat{u}_{n,\beta}\right) |k, \{\tilde{n}_\beta\}, n_\beta\rangle. \tag{11.76}$$

Let us expand again the exponent $\exp\left(i\kappa \hat{u}_{n,\beta}\right)$ in Eq. (11.75) in the Taylor series. Then, as in Lecture 10, we get [1, 8]:

$$\prod_\beta \exp\left(i\kappa \hat{u}_{n,\beta}\right) = \prod_\beta \left(1 + i\kappa \hat{u}_{n,\beta} - \frac{1}{2}\left(\kappa \hat{u}_{n,\beta}\right)^2 + O\left(\frac{1}{N^{3/2}}\right)\right). \tag{11.77}$$

Further on, we can utilize the fact that in the product over β for all the oscillators $\tilde{\beta} \neq \beta$ (besides that specific oscillator, for which the occupation number changes on 1 in the scattering process) the transition would be purely diagonal. Hence, we can represent the product as [1]:

$$\prod_\beta \exp\left(i\kappa \hat{u}_{n,\beta}\right) = \left(\prod_{\tilde{\beta}} \exp\left(i\kappa \hat{u}_{n,\tilde{\beta}}\right)\right) \cdot \exp\left(i\kappa \hat{u}_{n,\beta}\right). \tag{11.78}$$

For the diagonal transition in the expansion (11.77) the term linear in the displacement operator yields zero contribution according to the results of Lecture 10 [1]. Hence:

$$\prod_{\widetilde{\beta}} \langle k', \{\tilde{n}\}| \exp\left(i\kappa\hat{u}_{n,\widetilde{\beta}}\right) |k, \{\tilde{n}\}\rangle \approx \prod_{\widetilde{\beta}} \langle\{\tilde{n}\}| \left(1 - \frac{1}{2}\left(\kappa\hat{u}_{n,\widetilde{\beta}}\right)^2\right) |\{\tilde{n}\}\rangle$$

$$\approx \exp\left(-\frac{Z(\kappa)}{2}\right), \tag{11.79}$$

where $Z(\kappa)$ is the Debye—Waller factor [1,17–19]. In fact, from the $3qN$ branches of the phonon spectrum, only one branch participates in the non-diagonal transition, while all the rest branches with the macroscopic accuracy reproduce the result of Eq. (11.79) [1].

At the same time, for the non-diagonal transition (with the variation of the occupation number on 1), both the contributions from the first term (which does not depend on the displacement operator) and from the third term, quadratic in the displacement operator (the last term always changes the number of states on 0 or 2) equal to zero [1]. As a result, it is precisely the second term linear with respect to $\hat{u}_{n,\beta}$ which gives the main contribution in the one-phonon process [1]:

$$\langle n_\beta \pm 1|i\kappa\hat{u}_{n,\beta}|n_\beta\rangle. \tag{11.80}$$

Correspondingly, the total amplitude of the one-phonon process $A(\kappa)$ is given by [1,8]:

$$A(\kappa) = V(\kappa)\exp\left(-\frac{Z(\kappa)}{2}\right)\sum_n \exp\left(i\kappa R_n^0\right)\langle n_\beta \pm 1|i\kappa\hat{u}_{n,\beta}|n_\beta\rangle. \tag{11.81}$$

Further on we can substitute in Eq. (11.80) the representation of the displacement operator in crystal with one atom on the elementary cell via the operators of the phonon creation and annihilation (see Eq. (9.42)):

$$\hat{u}_n^i = \sum_\beta \sqrt{\frac{\hbar}{2M\omega_\beta N}}\left[e_\beta^i \exp(if n)\hat{a}_\beta + e_\beta^{i*}\exp(-if n)\hat{a}_\beta^\dagger\right] = \sum_\beta \hat{u}_{n,\beta}^i.$$

Then the probability amplitude reads [1,8]:

$$A(\kappa) = V(\kappa)\exp\left(-\frac{Z(\kappa)}{2}\right)\sum_n \exp(i\kappa n)i\sqrt{\frac{\hbar}{2M\omega_\beta N}}$$

$$\kappa\left(e_\beta \exp(if n)\langle n_\beta \pm 1|\hat{a}_\beta|n_\beta\rangle + e_\beta^*\exp(-if n)\langle n_\beta \pm 1|\hat{a}_\beta^\dagger|n_\beta\rangle\right), \tag{11.82}$$

where for the sake of brevity $R_n^0 = n$.

But for the operator of the phonon annihilation in Eq. (11.82) [1,2,31,32]:

$$\hat{a}_\beta \left|n_\beta\right\rangle = \sqrt{n_\beta - 1}\left|n_\beta - 1\right\rangle. \tag{11.83}$$

At the same time, for the creation operator [1,2,31,32]:

$$\hat{a}_\beta^\dagger \left|n_\beta\right\rangle = \sqrt{n_\beta + 1}\left|n_\beta + 1\right\rangle. \tag{11.84}$$

Thus, for the transition $n_\beta \to n_\beta - 1$ with the phonon absorption, the following matrix element works [1,2,31,32]:

$$\left\langle n_\beta - 1\right|\hat{a}_\beta\left|n_\beta\right\rangle = \sqrt{n_\beta - 1}. \tag{11.85}$$

At the same time, for the transition with the phonon emission, the matrix element [1,2,31,32]:

$$\left\langle n_\beta + 1\right|\hat{a}_\beta^\dagger\left|n_\beta\right\rangle = \sqrt{n_\beta + 1} \tag{11.86}$$

is important.

Moreover, the probability amplitude for these transitions reads [1,8]:

$$A(\kappa) = i\, V(\kappa)\exp\left(-\frac{Z(\kappa)}{2}\right)\sqrt{\frac{\hbar}{2M\omega_\beta N}}\left\{\begin{array}{l} \sum_n \exp(i(\kappa + f)n)\left(\kappa e_\beta\right)\sqrt{n_\beta - 1} \\ \sum_n \exp(i(\kappa - f)n)\left(\kappa e_\beta^*\right)\sqrt{n_\beta + 1} \end{array}\right. . \tag{11.87}$$

Note that in Eq. (11.87) the upper expression in the curly brackets corresponds to the phonon absorption while the lower one—to the phonon emission [1,8].

Proceeding from the amplitude to the probability of the inelastic scattering, according to (11.7) we get for the probability element of all the one-phonon processes [1,8]:

$$dW = |V(\kappa)|^2 \exp\left(-Z(\kappa)\right)\sum_\beta \frac{\hbar}{2M\omega_\beta N}\left|\kappa e_\beta\right|^2 \frac{V d^3 k'}{(2\pi)^3}$$

$$\times \left\{\begin{array}{l} (n_\beta - 1)\left|\sum_n \exp(i(\kappa + f)n)\right|^2 \delta\left(\varepsilon_k - \varepsilon_{k'} + \hbar\omega_\beta\right) \\[2ex] (n_\beta + 1)\left|\sum_n \exp(i(\kappa - f)n)\right|^2 \delta\left(\varepsilon_k - \varepsilon_{k'} - \hbar\omega_\beta\right) \end{array}\right. , \tag{11.88}$$

where the delta function with respect to the energy $\delta\left(\Delta\varepsilon + \Delta E_{if}\right)$ in the upper expression in curly brackets corresponds to the phonon absorption, and thus $\Delta\varepsilon = \varepsilon_k - \varepsilon_{k'}$ for particles and $\Delta E_{if} = E_i - E_f = \hbar\omega_\beta$ for the phonon energy [1]. At the same time, the delta function with respect to the energy in the lower expression in curly brackets corresponds to the phonon emission, and hence $\Delta\varepsilon = \varepsilon_k - \varepsilon_{k'}$, but $\Delta E_{if} = E_i - E_f = -\hbar\omega_\beta$ for it [1].

Now we can utilize the formulas (11.22) and (11.66) for the absolute values squared of the lattice sums in Eq. (11.88):

$$\left| \sum_n \exp(i(\kappa + f)n) \right|^2 = N \sum_n \exp(i(\kappa + f)n) = N\delta_{\kappa+f,K} = N\delta_{\kappa-K,-f}$$

$$(11.89)$$

$$\left| \sum_n \exp(i(\kappa - f)n) \right|^2 = N \sum_n \exp(i(\kappa - f)n) = N\delta_{\kappa-f,K} = N\delta_{\kappa-K,f}.$$

$$(11.90)$$

However, we should be accurate here. The point is that the transferred momentum $\kappa = k - k'$ can be arbitrary (lies in any of the Brillouin zones) [1]. At the same time, the phonon wave vector f by the definition (from the periodicity condition for the phonon spectrum) can lie only in the first Brillouin zone. Hence, we have to select such a value of the inverse lattice wave vector K, for which the wave vector:

$$\check{\kappa} = \kappa - K = \pm f \tag{11.91}$$

lies in the first Brillouin zone [1,8,30].

Let us remind, that the summation over $\beta = \{\alpha, f\}$ in Eq. (11.88) is the summation over phonon wave vectors f and over the branches of the phonon spectrum α [1]. For the crystal with one atom on the elementary cell we have only 3 branches of the acoustic spectrum ($\alpha = 1, 2, 3$).

Then performing summation over the wave vectors f in Eq. (11.88) and lifting the Kronecker symbols $\delta_{\kappa-K,-f}$ and $\delta_{\kappa-K,f}$ in (11.89) and (11.90), we get for the doubly differential cross section of the inelastic one-phonon process [1]:

$$\frac{d^2\sigma_{\text{inel.}}^{(1)}(\kappa)}{d\varepsilon_{k'}\Omega_{k'}} \sim \exp\left(-Z(\kappa)\right) N \sum_{\alpha=1}^{3} \frac{\hbar \left| \kappa e_\alpha(\check{\kappa}) \right|^2}{2M\omega_\alpha(\check{\kappa})}$$

$$\times \left\{ n\left(\omega_\alpha(\check{\kappa})\right) \delta\left(\varepsilon_k - \varepsilon_{k'} + \hbar\omega_\alpha(\check{\kappa})\right) + \left[n\left(\omega_\alpha(\check{\kappa})\right) + 1 \right] \delta\left(\varepsilon_k - \varepsilon_{k'} - \hbar\omega_\alpha(\check{\kappa})\right) \right\}, \quad (11.92)$$

where for the phonon frequencies $\omega_\alpha(\check{k})$ in Eq. (11.92) the wave vector $\check{\kappa} = \kappa - K$ lies in the first Brillouin zone.

Note, that the doubly differential cross section of the inelastic one-phonon scattering is proportional to the particle number in crystal [1]:

$$\frac{d^2\sigma_{\text{inel.}}^{(1)}(\kappa)}{d\varepsilon_{k'}d\Omega_{k'}} \propto N. \tag{11.93}$$

Dependence of the doubly differential cross section of the one-phonon scattering on the transferred to the crystal energy $\Delta\varepsilon = \varepsilon_k - \varepsilon_{k'}$ is presented in Fig. 11.9.

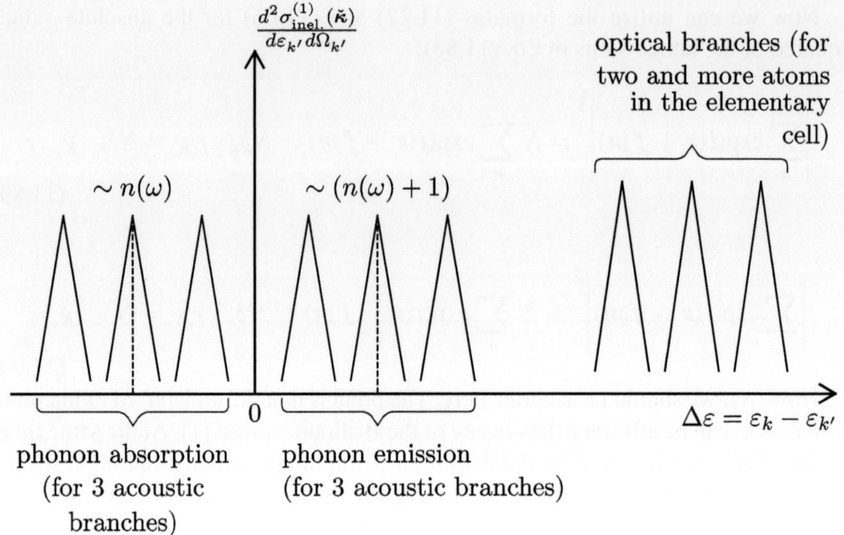

Fig. 11.9 Dependence of the doubly differential cross section of the inelastic one-phonon scattering $\frac{d^2\sigma^{(1)}_{\text{inel.}}(\kappa)}{d\varepsilon_{k'}\,d\Omega_{k'}}$ on the transferred to the crystal energy $\Delta\varepsilon = \varepsilon_k - \varepsilon_{k'}$. For the crystal with one atom on the elementary cell 3 peaks (corresponding to the absorption of the acoustic phonons) emerge at negative values of the transferred energies $\Delta\varepsilon < 0$ in the dependence of the cross section on $\Delta\varepsilon$. They are accompanied by 3 peaks at positive values of the transferred energy $\Delta\varepsilon > 0$ (corresponding to the emission of the acoustic phonons). The absorption peaks intensity is proportional to $n(\omega)$, while the intensity of the emission peaks is proportional to $n(\omega) + 1$. There are no phonons at $T = 0$. They can be emitted only. In this case we have only 3 right peaks [1]. If the crystal is not a monoatomic one, then besides 3 acoustic branches the phonon spectrum contains also optical branches [1]

For the crystal with one atom on the elementary cell 3 peaks (corresponding to the absorption of the acoustic phonons) emerge at negative values of the transferred energies $\Delta\varepsilon < 0$ in the dependence of the cross section on $\Delta\varepsilon$. They are accompanied by 3 peaks at positive values of the transferred energy $\Delta\varepsilon > 0$ (corresponding to the emission of the acoustic phonons) [1]. Let us stress, that the absorption peaks intensity is proportional to $n(\omega)$, while the intensity of the emission peaks is proportional to $n(\omega) + 1$.

There are no phonons at $T = 0$. They can be emitted only. Thus, in this case we have only 3 right peaks [1].

Let us emphasize also that from the intensity peaks of the differential cross section dependence on the transferred energy $\Delta\varepsilon$ we can get the phonon spectrum $\omega_\alpha(\check{k})$ (varying $\Delta\varepsilon$ at fixed values of κ) [1].

If the crystal is not a monoatomic one, then besides 3 acoustic branches the phonon spectrum contains also optical branches [1, 20–23] (see Fig. 11.9).

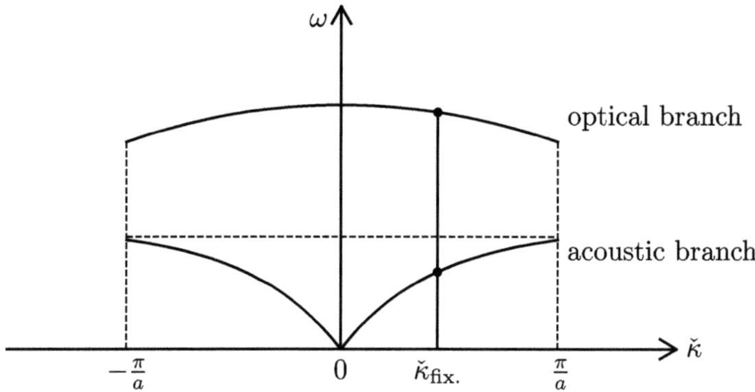

Fig. 11.10 Frequencies of acoustic and optical phonon as the functions of the absolute value of the wave vector $\check{\kappa}$. If we fix the value of $\check{\kappa}_{\text{fix.}}$, then the frequency of the optical phonon for this value of the wave vector is larger than the frequency of the acoustic phonon: $\omega_{\text{opt.}}\left(\check{\kappa}_{\text{fix.}}\right) > \omega_{\text{ac.}}\left(\check{\kappa}_{\text{fix.}}\right)$ [1]

If we fix the absolute value of the wave vector $\check{\kappa}_{\text{fix.}}$, then, as it is shown in Fig. 11.10, the optical phonon frequency for this value of the wave vector is larger than the frequency of the acoustic phonon [1]:

$$\omega_{\text{opt.}}\left(\check{\kappa}_{\text{fix.}}\right) > \omega_{\text{ac.}}\left(\check{\kappa}_{\text{fix.}}\right). \tag{11.94}$$

References

1. Y. Kagan, *Lectures on the Solid-State Theoretical Physics* (Moscow, MEPHI, 1981–1982). Part I. Phonons - Unpublished
2. L.D. Landau, E.M. Lifshitz, *Quantum Mechanics* (Pergamon Press, New York, Non-Relativistic Theory, 1977)
3. M. Born, Z. Angew. Phys. **38**, 803 (1926). (in German)
4. E. Fermi, *Nuclear Physics* (Chicago University Press, 1950)
5. H. Bethe, E. Salpeter, *Quantum mechanics of one- and two-electron atoms* (Springer, Boston MA, 1977)
6. C. Cohen-Tannoudji, B. Diu, F. Laloe, *Quantum Mechanics* (Willey, 1977)
7. W. Pauli, *Wave Mechanics, Pauli Lectures on Physics*, vol. 5 (Dover books on Physics, 2000)
8. E.G. Brovman, Yu. Kagan, *Neutron inelastic scattering*, vol. 1 (IAEA, Vienna, 1968), p.3
9. Y.A. Izumov, N.A. Chernoplekov, *Neutron Spectroscopy* (Springer, Berlin, 1994)
10. S.V. Maleev, Polarized neutron scattering in magnets. Rev. Artic., Phys. Uspekhi **45**, 569 (2002)
11. J.M. Cowley, *Diffraction physics* (North-Holland, Amsterdam, 1975)
12. C.G. Shull, Rev. Mod. Phys. **67**, 753 (1995)
13. R. Mossbauer, Z. Phys. vol. 151 (1958), p. 124; Naturwissenschaften, vol. 45 (1959), p. 538 (in German)
14. R. Mossbauer, Zeit. Naturforsch, vol. 14 (1959) p. 211 (in German)
15. R. Mossbauer, *Recoilless Nuclear Resonance Absorption of Gamma Radiation, Nobel Lecture, December 11, 1961, Nobel Lectures, Physics, 1942–1962* (Elsevier Publishing Company, Amsterdam, 1964)
16. E.G. Brovman, Y. Kagan, Phonons in non-transition metals, in *Lattice Dynamics*, ed. by A. Maradudin, G.K. Horton (Amsterdam, North-Holland, 1974)

17. P. Debye, Ann. Phys. **348**, 49 (1912). ((in German))
18. I. Waller, Zeit. fur Physik A **17**, 398 (1923). ((in German))
19. E.M. Lifshitz, L.P. Pitaevskii, *Statistical Physics Part II: Theory of the Condensed State*, Pergamon. (Elsevier, Pergamon, 2013)
20. O. Madelung, *Introduction to Solid-State Theory*, Springer Series in Solid-State Sciences (1995)
21. C. Kittel, *Quantum Theory of Solids* (Wiley, New York-London, 1963)
22. J.M. Ziman, Electrons and phonons, in *The Theory of Transport Phenomena in Solids* (Clarendon Press, Oxford, 1960)
23. N.W. Ashcroft, N.D. Mermin, *Solid State Physics* (Holt, Rinehart and Winston, 1976)
24. W.H. Bragg, W.L. Bragg, Proc. Royal Soc. London, vol. A88 (1913), p. 428
25. W. Borchardt-Ott, *Crystallography* (Springer, Berlin, 1993)
26. W. Mossa, *Crystal Structure Determination* (Springer, Berlin, 1999)
27. A. Lipson, S.G. Lipson, H. Lipson, *Optical Physics* (Cambridge University Press, Cambridge, 2010)
28. M. Born, E. Wolf, *Principles of Optics* (Cambridge University Press, Cambridge, 1999)
29. G.R. Fowles, *Introduction to Modern Optics* (Dower Publications, New York, 1975)
30. H. Jones, *The Theory of Brillouin Zones and electronic States in Crystals* (North-Holland Publishing Company, Amsterdam, 1962)
31. R.P. Feynman, *Statistical Mechanics, A Set of Lectures (Advanced Books Classics)* (Westview Press, Boulder, 1998)
32. L.D. Landau, E.M. Lifshitz, *Statistical Physics, Part I* (Butterworth-Heinemann, Oxford, 1999)

Lecture 12. Anharmonic Interactions: The Damping of the Phonon Spectrum

12

Abstract

In this Lecture we discuss the anharmonic interactions corresponding to the cubic and quartic terms in the phonon Hamiltonian. These interactions determine the damping of the phonon spectrum (related to the decay and picking up processes), as well as temperature corrections to the energy and specific heat of the crystal. We analyze different temperature dependences of the phonon lifetime in the long wavelength limit at low temperatures.

In this Lecture, we consider the anharmonic interaction of the phonon modes in crystal resulting in the damping of the phonon spectrum [1–3].

The phonon Hamiltonian in crystal with one atom on the elementary cell has the form [1,4]:

$$H = H_0 + H_{\text{int.}}. \tag{12.1}$$

In Eq. (12.1) the unperturbed Hamiltonian

$$H_0 = T + V_{\text{harm.}}, \tag{12.2}$$

which corresponds to the harmonic oscillations of the lattice [1,4–8], contains quadratic in the displacement terms (proportional to u^2) in the operator of the potential energy $V_{\text{harm.}}$.

At the same time, the interaction Hamiltonian:

$$H_{\text{int.}} = H_3 + H_4 \tag{12.3}$$

contains cubic and quartic terms [1] in atom displacements:

$$H_3 = \frac{1}{3!} \sum_{n_1,n_2,n_3} B^{ikl}_{n_1 n_2, n_3} u^i_{n_1} u^k_{n_2} u^l_{n_3}, \tag{12.4}$$

M. Kagan, *Lecture Notes of Professor Yuri Kagan in Theoretical Solid-State Physics*, Lecture Notes in Physics 1048, https://doi.org/10.1007/978-3-032-14621-2_12

203

$$H_4 = \frac{1}{4!} \sum_{n_1,n_2,,n_3,n_4} C^{iklm}_{n_1n_2,n_3n_4} u^i_{n_1} u^k_{n_2} u^l_{n_3} u^m_{n_4}. \tag{12.5}$$

The small parameter of this expansion is $\sqrt{\frac{u^2}{a^2}}$, where a is the interatomic distance. It is important to note that for all the temperatures (even when we come close to the melting temperature):

$$u \ll a. \tag{12.6}$$

Note that even in the melting point the ratio: $\frac{u^2}{a^2} \sim 10\%$ [1,4,5]. Melting takes place in fact due to the presence of the entropy term in the Free energy [1,9]: $F = E - TS$. The point is that the fluid entropy is larger than the solid-state one:

$$S_{\text{liquid}} > S_{\text{solid state}}. \tag{12.7}$$

It means, that the Free energy of the liquid is smaller than the solid-state one

$$F_{\text{liquid}} < F_{\text{solid state}}. \tag{12.8}$$

Hence, the phase transition from the solid state to the liquid phase becomes beneficial.

Let us return back now to the analysis of the cubic and quartic terms in the Hamiltonian (12.3).

Diagonal matrix element [1,10] for the cubic interaction contains the product of three creation and annihilation operators, and equals to zero in the first order of the perturbation theory:

$$\langle\{n\}|H_3|\{n\}\rangle = 0. \tag{12.9}$$

This matrix element is non-zero in the second order [1].

Diagonal matrix element for the quartic term contains the product of 4 creation and annihilation operators, and is non-zero already in the first order of the perturbation theory [1]:

$$\langle\{n\}|H_4|\{n\}\rangle \neq 0. \tag{12.10}$$

That is why the contributions to the energy from H_3 and H_4 prove to be similar.

Let us recollect the expression for the displacement operator from the previous Lectures [1]:

$$\hat{u}^i_{n_1} = \sum_{\beta_1} \sqrt{\frac{\hbar}{2M\omega_{\beta_1}N}} \left[e^i (\beta_1) \exp\left(i f_1 R_{n_1}\right) \hat{a}_{\beta_1} + e^{i*} (\beta_1) \exp\left(-i f_1 R_{n_1}\right) \hat{a}^\dagger_{\beta_1} \right],$$

where the summation index: $\beta_1 = f_1, \alpha$.

Correspondingly, idex $-\beta_1 = -f_1, \alpha$, and under the substitution $f_1 \to -f_1$ for the polarization vectors we get:

$$e^{i*}(-f, \alpha) = e^i(f, \alpha).$$

As a result the displacement operator can be written in the compact form, as [1]:

$$\hat{u}_n^i = \sum_\beta \sqrt{\frac{\hbar}{2M\omega_\beta N}} e^i(\beta) \exp(if R_n) \left[\hat{a}_\beta + \hat{a}^\dagger_{-\beta}\right]. \tag{12.11}$$

Let us introduce a convenient notation:

$$\hat{A}_\beta = \hat{a}_\beta + \hat{a}^\dagger_{-\beta}. \tag{12.12}$$

Then the displacement operator reads [1]:

$$\hat{u}_n^i = \sum_\beta \sqrt{\frac{\hbar}{2M\omega_\beta N}} e^i(\beta) \exp(if R_n) \hat{A}_\beta. \tag{12.13}$$

In terms of the $\hat{A}_\beta$ operator, the anharmonic contributions in Eqs. (12.4) and (12.5) acquire the form [1]:

$$H_3 = \frac{1}{3!} \sum_{\beta_1,\beta_2,\beta_3} B_{\beta_1\beta_2,\beta_3} \hat{A}_{\beta_1} \hat{A}_{\beta_2} \hat{A}_{\beta_3}, \tag{12.14}$$

$$H_4 = \frac{1}{4!} \sum_{\beta_1,\beta_2,\beta_3,\beta_4} C_{\beta_1\beta_2,\beta_3\beta_4} \hat{A}_{\beta_1} \hat{A}_{\beta_2} \hat{A}_{\beta_3} \hat{A}_{\beta_4}. \tag{12.15}$$

The matrix element $B_{\beta_1\beta_2,\beta_3}$ in Eq. (12.14) is written via the initial one $B^{ikl}_{n_1n_2,n_3}$ entering in the expansion (12.4), as follows [1]:

$$B_{\beta_1\beta_2,\beta_3} = \sum_{n_1,n_2,n_3} \left(\frac{\hbar}{2MN}\right)^{3/2} \frac{1}{\sqrt{\omega_{\beta_1}\omega_{\beta_2}\omega_{\beta_3}}} B^{ikl}_{n_1n_2,n_3} e^i(\beta_1) e^k(\beta_2) e^l(\beta_3)$$

$$\times \exp\left(if_1 R_{n_1} + if_2 R_{n_2} + if_3 R_{n_3}\right). \tag{12.16}$$

The matrix elements $B_{\beta_1\beta_2\beta_3}$ poses several important symmetry properties [1]. Firstly, they are symmetric with respect to each pair of indices β.

Secondly, they satisfy the relation [1]:

$$B_{-\beta_1-\beta_2-\beta_3} = B^*_{\beta_1\beta_2,\beta_3}. \tag{12.17}$$

Thirdly, they have interesting behavior at small wave vectors. This behavior follows from the definition of the contribution to the force acting on the atom occupying n_1 site:

$$F_{\text{on the atom } n_1} = -\frac{\partial H_3}{\partial u_{n_1}^i} \tag{12.18}$$

from the anharmonic interaction potential H_3.

In this case the resulting force acting on the center of gravity of the crystal reads:

$$F_{\text{on the center of grav.}} = \sum_{n_1} F_{\text{on the atom } n_1}. \tag{12.19}$$

In the absence of the external fields, this force is zero for the arbitrary values of the indices: n_2, and n_3.

At the translation of all the crystal as a whole, all the sites have equal displacements, and:

$$\sum_{n_1} B_{n_1 n_2 \text{ fix.} n_3 \text{ fix.}}^{ikl} u_{n_2 \text{ fix.}}^k u_{n_3 \text{ fix.}}^l = 0. \tag{12.20}$$

As a result, we get an important relation [1]:

$$\sum_{n_1} B_{n_1 n_2, n_3}^{ikl} = 0. \tag{12.21}$$

Analogous expression for the force matrix in harmonic approximation $\sum_n A_{nn'}^{ik} = 0$ [1,6] is also a consequence of the system invariance with respect to the crystal transfer as a whole.

For the matrix element $B_{\beta_1 \beta_2, \beta_3}$ (expressed via $B_{n_1 n_2, n_3}^{ikl}$ according to Eq. (12.16)) it follows from the condition (12.21) that for $f_1 = 0$ we have [1]:

$$B_{f_1 = 0, \alpha; \beta_2, \beta_3} = 0. \tag{12.22}$$

Correspondingly, for small values of the wave vector, the dependence of $B_{\beta_1 \beta_2, \beta_3}$ on the value of f_1 has a form [1]:

$$B_{f_1, \alpha; \beta_2, \beta_3} \sim \frac{f_1}{\sqrt{\omega_1}} \text{ for } f_1 \to 0. \tag{12.23}$$

But, for the crystal with one atom on the elementary cell we have only 3 acoustic branches with the linear spectrum: $\omega \sim f_1$ [1,4,6]. Hence, for small values of the wave vector f_1 we get [1]:

$$B_{f_1, \alpha; \beta_2, \beta_3} \sim \sqrt{f_1}, \tag{12.24}$$

If now we have small values of all the three wave vectors: f_1, f_2, f_3, then [1]:

$$B_{\beta_1\beta_2,\beta_3} \sim \sqrt{f_1 f_2 f_3} \sim \sqrt{\omega_{\beta_1}\omega_{\beta_2}\omega_{\beta_3}}. \tag{12.25}$$

Finally, one more relation for the matrix elements $B_{\beta_1\beta_2,\beta_3}$ follows from the condition of the translation invariance of the crystal with respect to the transformations [1]:

$$\boldsymbol{R}_n \to \boldsymbol{R}_n + \boldsymbol{a}. \tag{12.26}$$

It is important to note, that the value of the matrix elements $B^{ikl}_{n_1 n_2, n_3}$ itself does not change in the process of these transformations [1], since it depends only on the difference of the sites numbers $(n_1 - n_2) \cdots$ (see Fig. 12.1).

Thus:

$$B = B\left(n_1 - n_2, n_1 - n_3, n_2 - n_3\right). \tag{12.27}$$

Then under the translation transformation (12.26):

$$B \to B \exp\left[i\left(f_1 + f_2 + f_3\right)\boldsymbol{a}\right]. \tag{12.28}$$

The invariance of B means, that:

$$\exp\left[i\left(f_1 + f_2 + f_3\right)\boldsymbol{a}\right] = 1. \tag{12.29}$$

Hence, for the three-body interaction, the momentum conservation law is satisfied with an accuracy of the wave vector of the reciprocal lattice:

$$f_1 + f_2 + f_3 = 0, \boldsymbol{K}. \tag{12.30}$$

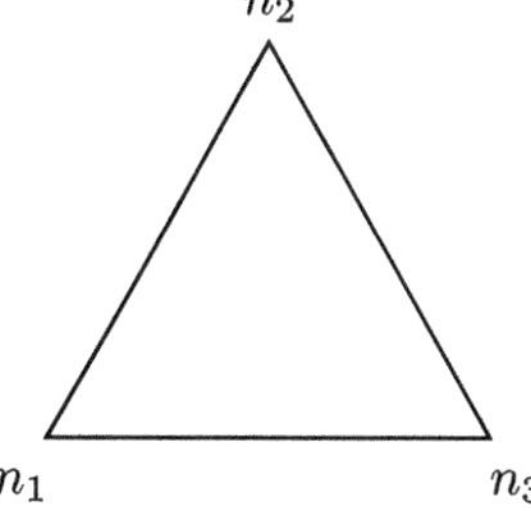

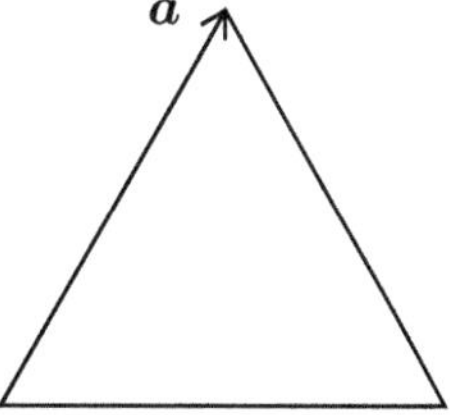

$$B(n_1 - n_2, n_1 - n_3, n_2 - n_3)$$

Fig. 12.1 On the left Figure we show the triangular illustration for the dependence of the matrix element $B\left(n_1 - n_2, n_1 - n_3, n_2 - n_3\right)$ on the difference of the sites numbers $(n_1 - n_2)$, $(n_1 - n_3)$, $(n_2 - n_3)$. On the right, we show the real situation on the lattice with the vector $\boldsymbol{a}$ of the elementary translations [1]

Note that if in (12.30):

$$f_1 + f_2 + f_3 = K, \tag{12.31}$$

then the momentum K is transferred to the lattice. We know however, that all the phonons are determined in the first cell. Hence the reciprocal lattice vector K should return the sum of the vectors $(f_1 + f_2)$ back in the first cell of the Brillouin zone to place there the vector $f_3 = K - (f_1 + f_2)$.

If all the three vectors f_1, f_2, f_3 are small, then $K = 0$. If however f_1, f_2, f_3 are not small, then vector K is fixed, and, $K \neq 0$. In this case, the condition (12.31) corresponds to the Umklapp processes [1–3] (see Fig. 12.2a, b). In the Umklapp processes, shown in Fig. 12.2a, we have triple combinations of the phonon annihilation operators: aaa and creation operators $a^\dagger a^\dagger a^\dagger$. The energy is not conserved in these processes.

The Umklapp processes in Fig. 12.2b correspond to the phonon decay on 2 phonon modes $aa^\dagger a^\dagger$ and phonon annihilation when it picks up another phonon: $aaa^\dagger$. In these processes, the energy conservation law is satisfied exactly.

12.1　Variation of the System Energy in the Perturbation Theory Under the Influence of the Anharmonic Interactions

In the first two orders of the perturbation theory in the interaction $H_{\text{int.}}$ in (12.1) the energy of the system has a form [10]:

$$E = E^0 + \langle i|H_{\text{int.}}|i\rangle + \sum_s \frac{|\langle s|H_{\text{int.}}|i\rangle|^2}{E_i^0 - E_s^0}, \tag{12.32}$$

where $i = \{n\}$ is the set of the occupation numbers of the harmonic problem.

Note, that as we discussed already, the diagonal matrix element $\langle\{n\}|H_3|\{n\}\rangle = 0$, since H_3 contains 3 operators of creation or annihilation. At the same time the average value $\langle\{n\}|H_4|\{n\}\rangle \neq 0$, and according to (12.15) is determined by the following expression:

$$\langle\{n\}|H_4|\{n\}\rangle = \frac{1}{4!} \sum_{1,2,3,4} C_{1,2,3,4}\langle\{n\}|\left(\hat{A}_1\hat{A}_2\right)\left(\hat{A}_3\hat{A}_4\right)|\{n\}\rangle$$

$$= \frac{1}{4!}3\sum C_{1-12-2}\left(2n_1 + 1\right)\left(2n_2 + 1\right), \tag{12.33}$$

where the curly brackets denote pairwise (bilinear) convolutions of the operators $\hat{A}_i$, while the factor 3 in the last line of Eq. (12.33) is related to 3 identical convolutions.

Correspondingly, for the convolution $\hat{A}_1\hat{A}_2$ we get:

$$\hat{A}_1\hat{A}_2 \rightarrow \langle\{n_1, n_2, \ldots\}|\left(a_1 + a^\dagger_{-1}\right)\left(a_2 + a^\dagger_{-2}\right)|\{n_1, n_2, \ldots\}\rangle. \tag{12.34}$$

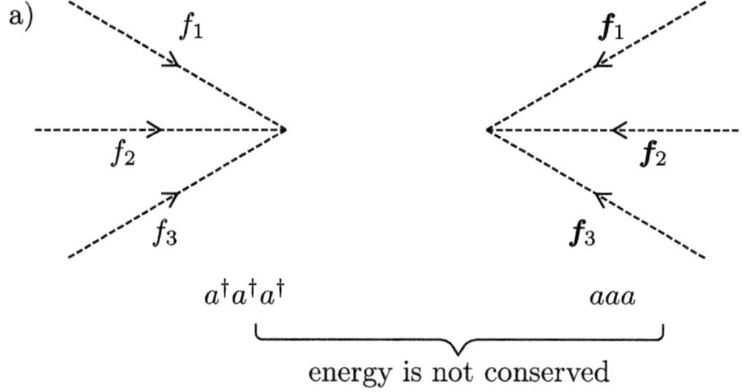

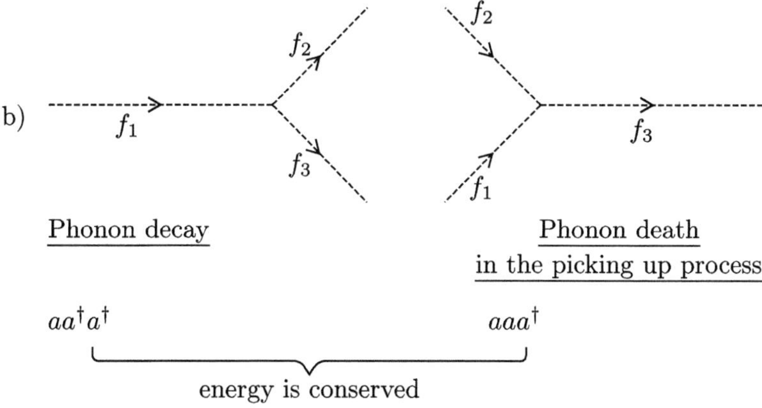

Fig. 12.2 Umklapp processes for the phonon three-body interaction **a** Umklapp processes, where we have triple combinations of the phonon annihilation operators aaa and creation operators $a^\dagger a^\dagger a^\dagger$. The energy is not conserved in these processes **b** Umklapp processes: phonon decay on the two phonon modes $aa^\dagger a^\dagger$ and phonon annihilation when it peaks up another phonon: $aaa^\dagger$. In these processes, the energy conservation law is satisfied exactly [1]

If we put now $1 = -2$, then the diagonal terms in (12.34) acquire the form:

$$\left\langle \left(a_1 + a^\dagger_{-1} \right) \left(a_{-1} + a^\dagger_1 \right) \right\rangle = \left\langle a_1 a^\dagger_1 + a^\dagger_{-1} a_{-1} \right\rangle. \qquad (12.35)$$

But for phonons $\left\langle a^\dagger_{-1} a_{-1} \right\rangle = \left\langle a^\dagger_1 a_1 \right\rangle = n_1(\omega)$, because the phonon frequency is an even function of momentum. Thus, from (12.35) we finally get:

$$\hat{A}_1 \hat{A}_2 \rightarrow \left\langle a_1 a^\dagger_1 + a^\dagger_1 a_1 \right\rangle = 2n_1 + 1. \qquad (12.36)$$

Similarly, the convolution $\hat{A}_3 \hat{A}_4$ yields us the second factor, connected with the phonon occupation number in (12.33):

$$\hat{A}_3 \hat{A}_4 \rightarrow (2n_2 + 1). \tag{12.37}$$

Note, that in (12.33) there are no quartic averages (when $1 = 2 = 3 = 4$), since in analogy with the Wick theorem this convolution contains an extra degree of the inverse system volume $\frac{1}{V} \rightarrow 0$ for $V \rightarrow \infty$.

As a result, the contribution of the quartic terms to the energy reads:

$$E_4 = \frac{1}{4!} 3 \sum_{1,2,} C_{1-12-2} (2n_1 + 1)(2n_2 + 1). \tag{12.38}$$

Let us consider now the contribution to the energy from the second order of the perturbation theory in cubic terms. It is proportional to the product of the matrix elements:

$$E_3 \sim \langle i | H_3 | s \rangle \langle s | H_3 | i \rangle. \tag{12.39}$$

Meanwhile, each matrix element contains the cubic combination of the operators $\hat{A}_i$:

$$\langle s | H_3 | i \rangle \sim \hat{A}_1 \hat{A}_2 \hat{A}_3, \quad \langle i | H_3 | s \rangle \sim \hat{A}_4 \hat{A}_5 \hat{A}_6. \tag{12.40}$$

Moreover, we get $3! = 6$ variants of the convolution, connected with all possible permutations of three numbers in the r.h.s. of this relation:

$$456 = -1 - 2 - 3 \tag{12.41}$$

$(4 = -1, 5 = -2, 6 = -3 \text{ or } 4 = -2, 5 = -1, 6 = -3 \text{ etc.})$

Finally, the contribution of the cubic terms to the energy equals to:

$$\begin{aligned}
E_3 &= \frac{3!}{(3!)^2} \sum_{1,2,3} \left\{ |B_{123}|^2 \frac{n_1 n_2 n_3 - (n_1 + 1)(n_2 + 1)(n_3 + 1)}{\omega_1 + \omega_2 + \omega_3} \right. \\
&\quad \left. + 3 |B_{12-3}|^2 \frac{n_1 n_2 (n_3 + 1) - (n_1 + 1)(n_2 + 1) n_3}{\omega_1 + \omega_2 - \omega_3} \right\} \\
&= \frac{1}{3!} \sum \left\{ |B_{123}|^2 \frac{n_1 n_2 n_3 - (n_1 + 1)(n_2 + 1)(n_3 + 1)}{\omega_1 + \omega_2 + \omega_3} \right. \\
&\quad \left. + 3 |B_{12-3}|^2 \frac{n_1 n_2 (n_3 + 1) - (n_1 + 1)(n_2 + 1) n_3}{\omega_1 + \omega_2 - \omega_3} \right\}.
\end{aligned} \tag{12.42}$$

It contains two terms: first term connected with the creation of 3 phonons and second term related to the annihilation of 2 phonons and creation of 1 phonon.

It is important, that for the matrix element B_{123} the momentum conservation law with the account of the Umklapp processes has the form given by (12.31). At the same time, for the matrix element B_{12-3} in the second term of (12.42) the momentum conservation law reads:

$$f_1 + f_2 - f_3 = K. \tag{12.43}$$

Correspondingly, the sum of the frequencies $\omega_1 + \omega_2 + \omega_3$ in the denominator of the first term transforms to the combination $\omega_1 + \omega_2 - \omega_3$ in the denominator of the second term.

Let us find now the correction to the energy connected with the introduction or removal of the single excitation from the system. In fact, this correction (at $\hbar = 1$) leads to the shift of the phonon frequency:

$$\Delta\omega_1 = \frac{\delta\Delta E}{\delta n_1} = \frac{1}{3!} \sum_{2,3} \left[|B_{123}|^2 \frac{n_2 n_3 - (n_2 + 1)(n_3 + 1)}{\omega_1 + \omega_2 + \omega_3} 3 \right.$$
$$\left. + 3|B_{12-3}|^2 \left\{ 2\frac{n_2(n_3 + 1) - (n_2 + 1)n_3}{\omega_1 + \omega_2 - \omega_3} + \frac{n_2 n_3 - (n_2 + 1)(n_3 + 1)}{\omega_2 + \omega_3 - \omega_1} \right\} \right]$$
$$+ \frac{2 \cdot 2 \cdot 3}{4!} \sum_{2} C_{1-12-2}(2n_2 + 1), \tag{12.44}$$

where $n_1 \equiv n_{\beta_1}$, and we accounted for the contribution of the cubic and quartic terms to the energy.

Note, that factor 3 in the end of the expression for the first term, proportional to $|B_{123}|^2$ in square brackets in (12.44), is related to the blind index n_1 in the initial formula (12.42) for the energy E_3.

Let us emphasize, that in the formula for E_3 the combination $n_1 n_2 n_3 - (n_1 + 1)(n_2 + 1)(n_3 + 1)$ enters under the summation symbol over n_1, n_2, n_3.

Note also, that the second term related to $|B_{12-3}|^2$ in the figure brackets in (12.44) emerges as a result of the substitution $1 \leftrightarrow 3$ in the second term of the initial formula (12.42) for the energy, which is proportional to the combination $n_1 n_2 (n_3 + 1) - (n_1 + 1)(n_2 + 1)n_3$. To get this result we also use the symmetry relation for the matrix elements $B_{12-3} = B^*_{-1-23}$ under the replacement $f_i \rightarrow -f_i$ for all 3 wave vectors which yields:

$$|B_{12-3}|^2 = |B_{-1-23}|^2. \tag{12.45}$$

After the simplification of the expressions in the nominators of the terms related to $|B_{123}|^2$ and $|B_{12-3}|^2$ in (12.44) we get:

$$\Delta\omega_1 = \frac{1}{2} \sum_{2,3} \left[-|B_{123}|^2 \frac{n_2 + n_3 + 1}{\omega_1 + \omega_2 + \omega_3} + |B_{12-3}|^2 \left\{ 2\frac{n_2 - n_3}{\omega_1 + \omega_2 - \omega_3} - \frac{n_2 + n_3 + 1}{\omega_2 + \omega_3 - \omega_1} \right\} \right]$$
$$+ \frac{1}{2} \sum_{2} C_{1-12-2}(2n_2 + 1). \tag{12.46}$$

Further on in the matrix element B_{12-3} we can make a replacement $f_3 \to -f_3$. As a result of this substitution $|B_{12-3}|^2 \to |B_{123}|^2$.

Then after the thermodynamical averaging we get from (12.46):

$$\Delta\omega_1 = -\frac{1}{2}\sum_{2,3}|B_{123}|^2\left\{\frac{\bar{n}_2+\bar{n}_3+1}{\omega_1+\omega_2+\omega_3+i\delta}+2\frac{\bar{n}_3-\bar{n}_2}{\omega_1+\omega_2-\omega_3+i\delta}\right.$$

$$\left.-\frac{\bar{n}_2+\bar{n}_3+1}{\omega_1-\omega_2-\omega_3+i\delta}\right\}+\frac{1}{2}\sum_2 C_{1-12-2}\left(2\bar{n}_2+1\right),\qquad(12.47)$$

where $\bar{n}_2$ and $\bar{n}_3$ are Planck distribution functions for phonons, and we made a replacement $\omega_1 \to \omega_1 + i\delta$.

For the phonon wave function this frequency replacement yields:

$$\exp\left(-i\omega_1 t\right) \to \exp\left(-i\omega_1 t\right)\exp(\delta t).\qquad(12.48)$$

It guarantees adiabatic switching on large time scales $t \to -\infty$ (on these time scales $\exp(\delta t) \to 0$ for positive $\delta = +0$).

Now we can use the well-known substitution:

$$\frac{1}{x+i\delta} \to P\frac{1}{x}-i\pi\delta(x),\qquad(12.49)$$

where P is <u>principal value</u>.

In our problem $P\frac{1}{x}$ yields renormalization of the phonon frequency at $T=0$.

For large $T \geq \omega$ the Planck distribution functions $\bar{n}_i \sim \frac{T}{\omega_i}$ in the nominators of (12.46), and thus, the spectrum renormalization (Re $\Delta\omega$) is linear in T.

Let us analyze now the behavior of the imaginary part of the spectrum Im $\Delta\omega$, which determines the lifetime of phonon. Separating imaginary part in (12.47) using relation (12.49), we get:

$$\text{Im}\,\Delta\omega_1 = \frac{\pi}{2}\sum_{2,3}\left[|B_{123}|^2\left\{(\bar{n}_2+\bar{n}_3+1)\,\delta\left(\omega_1+\omega_2+\omega_3\right)\right.\right.$$

$$+2\left(\bar{n}_3-\bar{n}_2\right)\delta\left(\omega_1+\omega_2-\omega_3\right)-(\bar{n}_2+\bar{n}_3+1)\,\delta\left(\omega_1-\omega_2-\omega_3\right)\}].\quad(12.50)$$

But the first term in figure brackets in (12.50) equals zero, since all the frequencies $\omega_1, \omega_2, \omega_3 > 0$, and as a result the delta function $\delta\left(\omega_1+\omega_2+\omega_3\right)$ does not work (never reaches zero). Then only two terms contribute in the imaginary part Im $\Delta\omega_1$:

$$\text{Im}\,\Delta\omega_1 = -\frac{\pi}{2}\sum_{2,3}|B_{123}|^2\left\{2\left(\bar{n}_2-\bar{n}_3\right)\delta\left(\omega_1+\omega_2-\omega_3\right)\Delta\left(f_1+f_2-f_3\right)\right.$$

$$+(\bar{n}_2+\bar{n}_3+1)\,\delta\left(\omega_1-\omega_2-\omega_3\right)\Delta\left(f_1-f_2-f_3\right)\},\qquad(12.51)$$

where the first term describes the annihilation of phonon when it picks up the other phonon, while the second term is related to the decay of phonon on 2 phonon modes. The symbols $\Delta\left(f_1 + f_2 - f_3\right)$ and $\Delta\left(f_1 - f_2 - f_3\right)$ correspond to the momentum conservation laws within the accuracy of the inverse lattice vector in (12.51).

Note, that the quartic term in (12.47) contributes in $\operatorname{Re}\Delta\omega$, but not in $\operatorname{Im}\Delta\omega_1$. Imaginary part of the spectrum is determined only by the second order with respect to the cubic interaction H_3.

Note also, that the phonon wave function is described by the time-dependent exponent:

$$\exp\left(-i\omega_1 t\right) = \exp\left[-i\left(\omega_1^0 + \operatorname{Re}\Delta\omega_1\right)t\right]\exp\left(\operatorname{Im}\Delta\omega_1 t\right), \tag{12.52}$$

where $\omega_1 = \omega_1^0 + \operatorname{Re}\Delta\omega_1 + i\operatorname{Im}\Delta\omega_1$, and moreover the bare frequency ω_1^0 is purely real. Thus for the imaginary part:

$$\operatorname{Im}\Delta\omega_1 = \operatorname{Im}\omega_1. \tag{12.53}$$

Let us stress, that in delta functions which determine the imaginary part of the spectrum in Eqs. (12.50) and (12.51) enter only the bare and purely real frequencies $\omega_1, \omega_2, \omega_3$.

Thus, the damping of the phonon spectrum arises under the condition:

$$\operatorname{Im}\omega_1 < 0. \tag{12.54}$$

In the opposite case when $\operatorname{Im}\omega_1 > 0$ we have a "swing" (strong increase) of the amplitude of atom fluctuations which leads to the system instability.

For the damping of the phonon spectrum in the long-wavelength limit it is convenient to introduce the phonon lifetime τ_1. Then the inverse lifetime $\frac{1}{\tau_1}$ can be calculated using the formula:

$$\frac{1}{\tau_1} = 2\left|\operatorname{Im}\omega_1\right|. \tag{12.55}$$

Let us emphasize, that the smaller is the frequency, the larger is the occupation number (we have more phonons with small frequencies). In the first term in the figure brackets in (12.51) enters the delta function $\delta\left(\omega_1 + \omega_2 - \omega_3\right)$. It works only if $\omega_3 = \omega_1 + \omega_2 > \omega_2$. Hence $\bar{n}_3 < \bar{n}_2$ and $(\bar{n}_3 - \bar{n}_2) < 0$. Thus, the contribution of the first term really leads to the damping with $\operatorname{Im}\Delta\omega_1 < 0$.

It is interesting, that even at $T = 0$ the imaginary part of the spectrum $\operatorname{Im}\Delta\omega_1 \neq 0$. In this case the contribution of the first term with the picking up of the phonon does not work ($\bar{n}_2 = \bar{n}_3 = 0$, and $\bar{n}_2 - \bar{n}_3 = 0$), and thus only the second term, describing the phonon decay in (12.51), which is proportional to

$$\left(\bar{n}_2 + \bar{n}_3 + 1\right)\delta\left(\omega_1 - \omega_2 - \omega_3\right) = 1 \cdot \delta\left(\omega_1 - \omega_2 - \omega_3\right)$$

is effective (see Fig. 12.3).

$$f_1 = f_2 + f_3 \text{ or } f_3 = f_1 - f_2$$
$$\omega_1 = \omega_2 + \omega_3 \text{ or } \omega_3 = \omega_1 - \omega_2$$

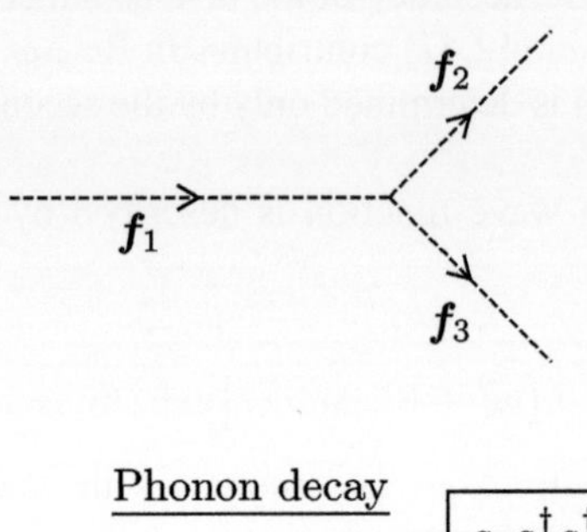

Fig. 12.3 Phonon decay on 2 phonons corresponds to the product of 1 phonon annihilation operator and 2 creation operators $a_1 a_2^\dagger a_3^\dagger$. For these processes the momentum and energy conservation laws have the form: $f_1 = f_2 + f_3$, and $\omega_1 = \omega_2 + \omega_3$. Correspondingly: $f_3 = f_1 - f_2$, and $\omega_3 = \omega_1 - \omega_2$ [1]

As we discussed already, this process describes the decay of one phonon on 2 phonons. In fact, we introduce by hands external phonon on top of the phonon background, which decays on 2 phonons at $T = 0$. We can see, that the contribution of the second term in (12.51) also leads to the damping with the negative imaginary part Im $\Delta\omega_1 < 0$.

12.2 Dying of the Long Wavelength Phonon

Let us analyze now the dependence of damping Im $\Delta\omega_1$ on the wave vector f_1 at $f_1 \to 0$. More exactly we consider small wave vectors satisfying the condition:

$$f_1 a \ll 1, \tag{12.56}$$

where a is the interatomic distance.

12.3 Phonon Decay

Let us consider first the case when $T = 0$. For acoustic branches of the spectrum in crystal with one atom on the elementary cell the frequency $\omega_1 \to 0$ at $f_1 \to 0$. Hence, from the energy conservation law $\omega_1 = \omega_2 + \omega_3$ in the second term in Eq. (12.51) the second and the third phonons also have small frequencies ω_2 and ω_3, and, correspondingly, small wave vectors f_2 and f_3. As a result, (see relation (12.25)), the matrix element $B_{\beta_1\beta_2,\beta_3}$ behaves as: $B_{\beta_1\beta_2,\beta_3} \sim \sqrt{f_1 f_2 f_3} \sim \sqrt{\omega_{\beta_1}\omega_{\beta_2}\omega_{\beta_3}}$, and thus:

$$\left|B_{\beta_1\beta_2,\beta_3}\right|^2 \sim \omega_{\beta_1}\omega_{\beta_2}\omega_{\beta_3}. \tag{12.57}$$

If all 3 wave vectors $\boldsymbol{f}_1$, $\boldsymbol{f}_2$ and $\boldsymbol{f}_3$ are small ones, then the momentum conservation law for the process of the phonon decay is satisfied exactly:

$$\boldsymbol{f}_1 - \boldsymbol{f}_2 - \boldsymbol{f}_3 = 0, \text{ and equivalently }:$$
$$\boldsymbol{f}_3 = \boldsymbol{f}_1 - \boldsymbol{f}_2. \tag{12.58}$$

As a result, in sum over $\beta_2 = \{\alpha_2, \boldsymbol{f}_2\}$ and $\beta_3 = \{\alpha_3, \boldsymbol{f}_3\}$ in (12.51) due to the momentum conservation law one integration on $\boldsymbol{f}_3$ is lifted. Only the summation over the branches α_2 and α_3 and the integration on $\boldsymbol{f}_2$ are remaining.

At $T = 0$ the combination $\bar{n}_2 + \bar{n}_3 + 1 = 1$, and the imaginary part of the spectrum has the form:

$$\text{Im}\,\Delta\omega_1 \sim \sum_{\alpha_2,\alpha_3} \int \frac{d^3 f_2}{(2\pi)^3} \delta\left(\omega_1 - \omega_2 - \omega_3\right)\omega_1\omega_2\omega_3, \tag{12.59}$$

where the frequencies for 3 acoustic branches read:

$$\omega_1 = c_1 f_1,\; \omega_2 = c_2 f_2,\; \omega_3 = c_3 f_3 = c_3 \left|\boldsymbol{f}_1 - \boldsymbol{f}_2\right|. \tag{12.60}$$

Finally with an account of the statistical weight $\frac{d^3 f_2}{(2\pi)^3} \sim f_2^2 df_2$ we get at $T = 0$:

$$\text{Im}\,\Delta\omega_1 \sim \sum_{\alpha_2,\alpha_3} \int f_2^2 df_2 \delta\left(c_1 f_1 - c_2 f_2 - c_3\left|\boldsymbol{f}_1 + \boldsymbol{f}_2\right|\right)$$
$$\times\, c_1 f_1 c_2 f_2 c_3 \left|\boldsymbol{f}_1 + \boldsymbol{f}_2\right| \propto f_1^5. \tag{12.61}$$

If the temperature is finite, then $\bar{n}_2 = \dfrac{1}{\exp\left(\frac{\omega_2}{T}\right) - 1} \approx \dfrac{T}{\omega_2} \gg 1$ and $\bar{n}_3 \approx \dfrac{T}{\omega_3} \gg 1$. As a result, the combination of the occupation numbers in the second term in (12.51) becomes proportional to:

$$\bar{n}_2 + \bar{n}_3 + 1 \approx \bar{n}_2 + \bar{n}_3 \approx \frac{T}{\omega_2} + \frac{T}{\omega_3} = T\left(\frac{\omega_2 + \omega_3}{\omega_2\omega_3}\right) = T\frac{\omega_1}{\omega_2\omega_3} = T\frac{\omega_1^2}{\omega_1\omega_2\omega_3}. \tag{12.62}$$

Meanwhile the imaginary part of the spectrum:

$$\frac{1}{\tau_1} \sim \sum_{\alpha_2,\alpha_3} \int \frac{d^3 f_2}{(2\pi)^3}\delta\left(\omega_1 - \omega_2 - \omega_3\right)\omega_1\omega_2\omega_3 T\frac{\omega_1^2}{\omega_1\omega_2\omega_3}$$
$$\sim T c_1^2 f_1^2 \sum_{\alpha_2,\alpha_3} \int f_2^2 df_2 \delta\left(c_1 f_1 - c_2 f_2 - c_3\left|\boldsymbol{f}_1 - \boldsymbol{f}_2\right|\right) \propto T f_1^4. \tag{12.63}$$

12.4   Phonon Picking Up

But for $T \neq 0$ the first term with phonon picking up in (12.51) also works (see Fig. 12.4). Let the phonon frequency ω_1 to be small in the long wavelength limit:

$$\omega_1 \ll T. \tag{12.64}$$

If the phonon with low energy picks up the phonon also having the low energy, then the contribution of this process to the damping will be the same by order of magnitude, as the process of the phonon decay: $\frac{1}{\tau_1} \propto T f_1^4$.

Important question arises however. Can the phonon with small energy $\omega_1 \ll T$ pick up the phonon with large energy $\omega_2 \sim T$ and also produce the phonon with large energy $\omega_3 \sim T$, satisfying at the same time the energy conservation law: $\omega_3 = \omega_1 + \omega_2$.

(1) Low temperatures $T \ll \theta_D$

To answer this question let us consider first low temperatures with respect to the Debye temperature: $T \ll \theta_D$.

In this case the energy conservation law for the process of phonon picking up with an account of the momentum conservation law $\boldsymbol{f}_1 + \boldsymbol{f}_2 = \boldsymbol{f}_3$ acquires the form:

$$c_1 f_1 + \omega_{\alpha_2}\left(\boldsymbol{f}_2\right) = \omega_{\alpha_3}\left(\boldsymbol{f}_1 + \boldsymbol{f}_2\right), \tag{12.65}$$

where α_2, α_3 are the branches of the phonon spectrum.

Let us expand $\omega_{\alpha_3}\left(\boldsymbol{f}_1 + \boldsymbol{f}_2\right)$ in series, utilizing the assumed smallness of the wave vector $\boldsymbol{f}_1$ with respect to $\boldsymbol{f}_2$. Then:

$$\omega_{\alpha_3}\left(\boldsymbol{f}_1 + \boldsymbol{f}_2\right) \approx \omega_{\alpha_3}\left(\boldsymbol{f}_2\right) + \frac{\partial \omega_{\alpha_3}}{\partial \boldsymbol{f}_2} \boldsymbol{f}_1. \tag{12.66}$$

(a) Coincidence of the phonon branches ($\alpha_2 = \alpha_3$).

If we consider now the case when there is a coincidence between the numbers of the phonon branches: $\alpha_2 = \alpha_3$, then the frequencies $\omega_{\alpha_3}\left(\boldsymbol{f}_2\right) = \omega_{\alpha_2}\left(\boldsymbol{f}_2\right)$, and the energy conservation law in (12.65) is reduced to the simple form:

$$c_1 f_1 \approx \frac{\partial \omega_{\alpha_2}}{\partial \boldsymbol{f}_2} \boldsymbol{f}_1 = \boldsymbol{v}_{\text{gr.}} \boldsymbol{f}_1 = v_{\text{gr.}} f_1 \cos\left(\boldsymbol{f}_1 \boldsymbol{f}_2\right). \tag{12.67}$$

Usually for the spectrum in crystal, the group velocity related to the tangential to the spectrum curve $\omega(f)$, is determined by the smaller tangent of the tilting angle to the horizontal axis, than the tangent of the tilting angle for the sound velocity $v_{\text{gr.}} < c_s$. (see Fig. 12.5).

Then $c_1 f_1 > v_{\text{gr.}} f_1 \cos\left(\boldsymbol{f}_1 \boldsymbol{f}_2\right)$ for all the values of the $\cos\left(\boldsymbol{f}_1 \boldsymbol{f}_2\right)$, and the energy conservation law is not fulfilled.

(b) Different phonon branches ($\alpha_2 \neq \alpha_3$)

$$f_1 + f_2 = f_3$$
$$\omega_1 + \omega_2 = \omega_3$$

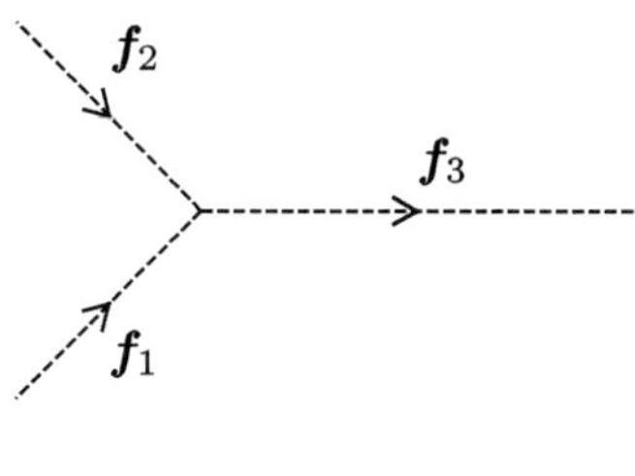

<u>Phonon picking up</u>

$$a_1 a_2 a_3^\dagger$$

Fig. 12.4 Phonon picking up corresponds to the product of 2 phonon annihilation operators and 1 phonon creation operator $a_1 a_2 a_3^\dagger$. Momentum and energy conservation laws for this process have the form: $f_1 + f_2 = f_3$ and $\omega_1 + \omega_2 = \omega_3$ [1]

If, however, the equality $\alpha_2 = \alpha_3$ is not fulfilled and the frequencies $\omega_{\alpha_3}(f_2) \neq \omega_{\alpha_2}(f_2)$, then for the definite value of $\cos(f_1 f_2)$ the energy conservation law:

$$c_1 f_1 + \omega_{\alpha_2}(f_2) \approx \omega_{\alpha_3}(f_2) + v_{gr.} f_1 \cos(f_1 f_2) \tag{12.68}$$

can be satisfied for $\omega_{\alpha_3}(f_2) = c_3 f_2 > \omega_{\alpha_2}(f_2) = c_2 f_2$.

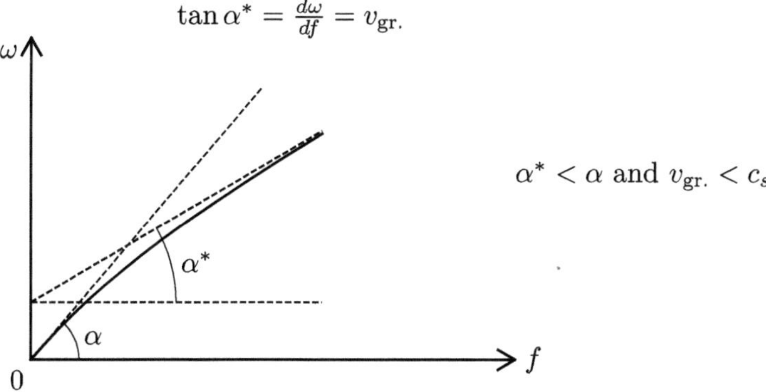

Fig. 12.5 Typical behavior of the phonon spectrum $\omega(f)$ in crystal. The group velocity, related to the tangential to the spectral curve $\omega(f)$, is determined by the smaller tangent of the tilting angle to the horizontal axis, then the tangent of the tilting angle for the sound velocity $\left(\tan\alpha^* = \frac{d\omega}{df} = v_{gr.}\right) < \left(\tan\alpha = \frac{\omega}{f} = c_s\right)$, and $v_{gr.} < c_s$ [1]

Let us recollect now, that the term with the picking up of the phonon is proportional to: $2\,(\bar{n}_2 - \bar{n}_3)\,\delta\,(\omega_1 + \omega_2 - \omega_3)$. Moreover, for the finite temperatures the difference between the average occupation numbers:

$$\bar{n}_2 - \bar{n}_3 \approx \frac{T}{\omega_2} - \frac{T}{\omega_3} = T\left(\frac{\omega_3 - \omega_2}{\omega_2 \omega_3}\right) = T\frac{\omega_1}{\omega_2 \omega_3} = T\frac{\omega_1^2}{\omega_1 \omega_2 \omega_3}, \qquad (12.69)$$

where we utilized again the energy conservation law $\omega_1 + \omega_2 = \omega_3$.

Then the inverse lifetime related to the picking up process:

$$\left(\frac{1}{\tau_1}\right)_{\text{picking up}} \sim T\omega_1^2 \int_0^{f_T} f_2^2 df_2 \int d\cos\theta \frac{\omega_1 \omega_2 \omega_3}{\omega_1 \omega_2 \omega_3}\delta\,(\omega_1 + \omega_2 - \omega_3)$$

$$= T\omega_1^2 \int_0^{f_T} f_2^2 df_2 \int d\cos\theta\,\delta\,(\omega_1 + \omega_2 - \omega_3), \qquad (12.70)$$

where the frequency $\omega_1 = c_1 f_1 \ll T$ (phonon 1 is on top of the phonon background). At the same time, the frequency $\omega_2 = c_2 f_2 \sim T$ (phonon 2 belongs to the phonon background), and the upper limit of the integration on df_2 in (12.70) by order of magnitude equals to:

$$f_T \sim \frac{T}{c_2}. \qquad (12.71)$$

Let us emphasize, that the upper limit of integration distinguishes those wave vectors, for which the classical limit of the Planck distribution function is valid:

$$\bar{n}_2 \approx \frac{T}{\omega_2}.$$

In (12.70) the measure of integration:

$$\frac{d^3 f_2}{(2\pi)^3} \sim f_2^2 df_2 d\cos\theta, \qquad (12.72)$$

where the angle $\theta = \mathbf{f}_1 \mathbf{f}_2$. At the same time, the frequency:

$$\omega_3\left(\left|\mathbf{f}_1 + \mathbf{f}_2\right|\right) \approx \omega_3\left(\mathbf{f}_2\right) + v_{2\,\text{gr.}} f_1 \cos\theta = c_3 f_2 + v_{2\,\text{gr.}} f_1 \cos\theta. \qquad (12.73)$$

As a result, the inverse lifetime due to pick-up processes:

$$\left(\frac{1}{\tau_1}\right)_{\text{picking up}} \sim T\omega_1^2 \int_0^{f_T} f_2^2 df_2 \int d\cos\theta$$

$$\times \delta\left(c_1 f_1 + c_2 f_2 - c_3 f_2 - v_{2\,\text{gr.}} f_1 \cos\theta\right). \qquad (12.74)$$

Let us remind, that, at the same time, we consider low temperatures with respect to Debye temperature, and thus the chain of inequalities arises:

$$\omega_1 \ll (\omega_2 \sim T) \ll \left(\theta_D \sim \frac{c}{a}\right). \tag{12.75}$$

Note that background phonon (phonon 2) is still the long wavelength one:

$$f_T a \ll 1. \tag{12.76}$$

Integration on $d\cos\theta$ in (12.74) is fulfilled elementary, and for $c_3 > c_2$, (when the delta function works) yields:

$$\left(\frac{1}{\tau_1}\right)_{\text{picking up}} \sim T\omega_1^2 \int_0^{f_T} f_2^2 df_2 \frac{1}{v_{2\,\text{gr.}} f_1} \sim \frac{T\omega_1 c_1}{v_{\text{gr.}}} \int_0^{f_T} f_2^2 df_2. \tag{12.77}$$

But $\int_0^{f_T} f_2^2 df_2 \sim T^3$. Hence finally:

$$\left(\frac{1}{\tau_1}\right)_{\text{picking up}} \propto \omega_1 T^4. \tag{12.78}$$

Let us emphasize, that we consider the case of very small frequencies for the phonon 1 which is on top of the background: $\omega_1 \ll T$. That is why the channel, connected with the phonon pick-up, is substantially more effective for the inverse lifetime, than the decay channel:

$$\left(\frac{1}{\tau_1}\right)_{\text{decay}} \propto T f_1^4 \ll \left(\frac{1}{\tau_1}\right)_{\text{picking up}} \propto \omega_1 T^4. \tag{12.79}$$

In a more general case for low temperatures $T \ll \theta_D$ the inverse lifetime behaves according to the formula:

$$\frac{1}{\tau_1} \propto T^n f_1^m, \tag{12.80}$$

where the sum of the powers for the three-dimensional crystal:

$$m + n = 5. \tag{12.81}$$

In particular, the following law can be realized:

$$\frac{1}{\tau_1} \propto T^3 \omega_1^2. \tag{12.82}$$

<u>Comment.</u> If two different spectrum branches ($\alpha_2 \neq \alpha_3$) intersect (see Fig. 12.6), then the conservation laws can be satisfied not only for the small values of the wave

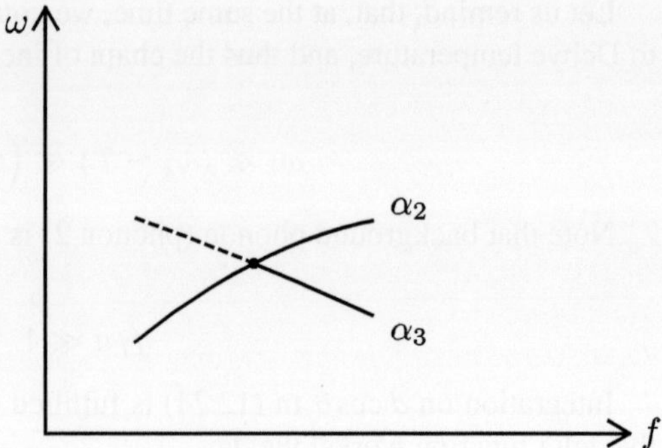

Fig. 12.6 Interception of the two branches of the phonon spectrum $\theta(f)$ in crystal at the finite value of the wave vector f [1]

vector $f_1 \to 0$, but also for the finite values of f_1 close to the branches interception point (depending on the character of the phonon spectrum degeneracy at this point, space dimensionality, number of atoms in the elementary cell and so on).

(2) High temperatures $T > \theta_D$.

Let us consider now the inverse lifetime of the phonon at high temperatures $T > \theta_D$. Planck distribution function in this case is still determined by its classical limit $n_i \sim \frac{T}{\omega_i}$. At the same time, we have no limitations on the region of integration, connected with the energy and momentum delta functions.

The inverse time in this limit occurs to be linear in temperature:

$$\frac{1}{\tau_1} \propto T. \tag{12.83}$$

An interesting question arises: how physically we can understand why the inverse lifetime $\frac{1}{\tau_1} \propto T$ behaves in the marginal way (as we used to say) at high temperatures instead, e.g., of more familiar way for us as $\frac{1}{\tau_1} \propto T^2$? This behavior ($\frac{1}{\tau_1} \propto T^2$) can appear for example due to the product of the occupation numbers $\bar{n}_2 \bar{n}_3$ in the expression for the imaginary part of the spectrum (with an account of the distribution function of the first phonon $\bar{n}_1$ which is on top of the background).

The point is that in our problem there is not only the phonon escape (as in α−decay problem), but also the phonon arrival (which is absent in the decay of the elementary particles). Thus, we have a problem with the phonon reservoir.

In this problem the phonon damping is determined by the balance: escape minus arrival, as in the kinetic equation. It is just this balance which leads to the linear dependence of the inverse lifetime on temperature at high temperatures in (12.83).

Escape in our problem is connected with the creation of phonons in the other states and is proportional to the product: $(\bar{n}_2 + 1)(\bar{n}_3 + 1)$. Arrival is connected with the phonon annihilation in the other states and is proportional to the product: $\bar{n}_2 \bar{n}_3$. As a result, the difference between escape and arrival is proportional to the sum of $\bar{n}_2 + \bar{n}_3$, which determines the linear temperature behavior in (12.83).

12.5 Temperature Dependent Corrections to the Energy and Specific Heat of the Crystal Due to Anharmonic Interactions

Let us determine now the temperature dependent corrections to the energy and specific heat of the system, connected with the anharmonic interactions.

We start with the initial formulas (12.1) for the phonon Hamiltonian in crystal. According to (12.1), operator of the Hamiltonian: $\widehat{H} = \widehat{H}_0 + \widehat{H}_{\text{int.}}$. Then the thermodynamic energy in the formalism of the density matrix can be written as a trace from the convolution of the density matrix with the Hamiltonian:

$$E = \frac{1}{Z} Sp \left(\exp\left(-\frac{\widehat{H}}{T} \right) \cdot \widehat{H} \right), \tag{12.84}$$

where Z is the statistical sum.

In the basis of the eigenfunctions of the system energy E_i the density matrix is diagonal, and thermodynamic energy equals to:

$$E = \frac{1}{Z} \sum_i \exp\left(-\frac{E_i}{T} \right) \cdot E_i, \tag{12.85}$$

where the statistical sum:

$$Z = \sum_i \exp\left(-\frac{E_i}{T} \right). \tag{12.86}$$

If we introduce the standard notation for the inverse temperature:

$$\beta = \frac{1}{T}, \tag{12.87}$$

then expression (12.85) for the thermodynamic energy can be represented as the derivative on β from the logarithm of the statistical sum (taken with the minus sign):

$$E = \frac{1}{Z} \left(-\frac{\partial}{\partial \beta} Sp(-\beta \widehat{H}) \right) = -\frac{1}{Z} \frac{\partial Z}{\partial \beta} = -\frac{\partial}{\partial \beta} \ln Z. \tag{12.88}$$

Thus, the thermodynamic energy:

$$E_{\text{therm.}} = -\frac{\partial}{\partial \beta} \ln Z. \tag{12.89}$$

At the same time the initial Quantum-mechanical energy depends on the set of the occupation numbers $\{n\}$ and can be written as:

$$E(\{n\}) = E_0(\{n\}) + \zeta E_{\text{an.}}(\{n\}), \tag{12.90}$$

where $\zeta = 1$. This multiplier shows, that:

$$E_{\text{an.}} \ll E_0. \tag{12.91}$$

Then the statistical sum:

$$Z = \sum_{\{n\}} \exp\left(-\frac{E(\{n\})}{T}\right) = \sum_{\{n\}} \exp\left(-\frac{E_0(\{n\}) + \zeta E_{\text{an.}}(\{n\})}{T}\right). \tag{12.92}$$

Further on it is convenient to calculate a derivative from the statistical sum on the parameter ζ:

$$\frac{\partial}{\partial \zeta} \ln Z = -\frac{1}{Z} \sum_{\{n\}} \exp\left(-\frac{E(\{n\})}{T}\right) \cdot \frac{E_{\text{an.}}(\{n\})}{T}. \tag{12.93}$$

Within the accuracy till the second order terms $E_{\text{an.}}^2$ we get from (12.93):

$$\frac{\partial}{\partial \zeta} \ln Z \approx -\frac{1}{Z} \sum_{\{n\}} \exp\left(-\frac{E_0(\{n\})}{T}\right) \cdot \frac{E_{\text{an.}}(\{n\})}{T} \approx -\beta \bar{E}_{\text{an.}}^0. \tag{12.94}$$

where $\bar{E}_{\text{an.}}^0$ denotes the averaging of the anharmonic energy with the density matrix of the harmonic approximation.

Now in $E_{\text{an.}}$ we can replace the exact values of the occupation numbers $\{n\}$ on the averaged values $\{\bar{n}\}$, which are determined by the Planck distribution. This is connected with the fact, that the averaging in (12.93) is performed with exponential multiplier $\exp\left(-\frac{E_0(\{n\})}{T}\right)$.

Analogously within the accuracy till the second order terms $E_{\text{an.}}^2$ we can replace the exact statistical sum Z in (12.93) by the statistical sum of the harmonic approximation Z_0:

$$\frac{\partial}{\partial \zeta} \ln Z = -\frac{1}{Z_0} \sum_{\{n\}} \beta E_{\text{an.}}(\{n\}) \exp\left(-\beta E_0(\{n\})\right) = -\beta \bar{E}_{\text{an.}}^0. \tag{12.95}$$

Let us integrate this expression on $d\zeta$. Then:

$$\ln Z(\zeta) = \ln Z_0 - \beta \int_0^\zeta \bar{E}_{\text{an.}}^0 \, d\zeta. \tag{12.96}$$

But $\bar{E}_{\text{an.}}^0$ does not depend on ζ. Thus:

$$\ln Z(\zeta) = \ln Z_0 - \beta \bar{E}_{\text{an.}}^0 \cdot \zeta, \tag{12.97}$$

and for $\zeta = 1$:

$$\ln Z(\zeta = 1) = \ln Z_0 - \beta \bar{E}^0_{\text{an.}}. \tag{12.98}$$

As a result, it follows for the thermodynamic energy in (12.89):

$$E_{\text{therm.}} = -\frac{\partial}{\partial \beta} \ln Z_0 + \frac{\partial}{\partial \beta} \left(\beta \bar{E}^0_{\text{an.}} \right) = E_0 + \frac{\partial}{\partial \beta} \left(\beta \bar{E}^0_{\text{an.}} \right), \tag{12.99}$$

where E_0 is the crystal thermodynamic energy in harmonic approximation.

At temperatures $T = 0$: $\bar{E}^0_{\text{an.}}$ does not depend on β, since in the exponents $\exp\left(-\beta E_0(\{n\})\right)$ entering in the expression for the statistical sum for $\beta = \frac{1}{T} \to \infty$ all the Planck distribution functions $\bar{n}_i = 0$. Hence in harmonic approximation statistical sum in the denominator of (12.95) is just $Z_0 = \exp\left(-\beta E_{00}\right)$. It means that the only one exponent survives having the energy of the ground state E_{00} which is independent of temperature.

Analogously, only one term survives in the sum in the nominator of Eq. (12.95) with the correction to the ground state energy $E_{\text{an. 00}}$. which is independent from β. Correspondingly:

$$\beta \bar{E}^0_{\text{an.}} = \frac{1}{Z_0} \sum_{\{n\}} \beta E_{\text{an.}}(\{n\}) \exp\left(-\beta E_0(\{n\})\right)$$

$$= \frac{1}{Z_0} \beta E_{\text{an. 00}} \exp\left(-\beta E_{00}\right) = \beta E_{\text{an. 00}}. \tag{12.100}$$

Thus, $\beta \bar{E}^0_{\text{an.}} = \beta E_{\text{an. 00}}$, and at $T = 0$ we obtain for the thermodynamic energy in (12.99):

$$E_{\text{therm.}} = E_{00} + \frac{\partial}{\partial \beta} \left(\beta \bar{E}^0_{\text{an.}} \right) = E_{00} + E_{\text{an. 00}}. \tag{12.101}$$

As a result, substituting the condition $n_1 = n_2 = n_3 = 0$ on average occupation numbers in the general expressions (12.38), (12.42) for the energy corrections connected with the cubic and quartic terms, we get at $T = 0$:

$$\bar{E}^0_{\text{an.}} = E_{\text{an. 00}} = E_3\left(n_1 = n_2 = n_3 = 0\right) + E_4\left(n_1 = n_2 = 0\right)$$

$$= -\frac{1}{3!} \sum_{1,2,3} |B_{123}|^2 \frac{1}{\omega_1 + \omega_2 + \omega_3} + \frac{1}{8} \sum_{1,2} C_{1-12-2}. \tag{12.102}$$

High temperatures $T \geq \theta_D$.
Let us consider now the opposite limit of high temperatures $T \gg \theta_D$.

In this case the average occupation numbers $\bar{n}_i = \frac{T}{\omega_i} \gg 1$. Moreover, the bilinear combinations $n_1 n_2 \gg n_1$. Then the energy corrections (with the accuracy till the terms $n_1 + n_2 + n_3$) equal to:

$$\bar{E}^0_{\text{an.}} = -\frac{1}{3!} \sum_{1,2,3} \left\{ |B_{123}|^2 \frac{n_1 n_2 + n_1 n_3 + n_2 n_3}{\omega_1 + \omega_2 + \omega_3} \right.$$

$$\left. +3 |B_{123}|^2 \frac{(n_1 + 1)(n_2 + 1) n_3 - n_1 n_2 (n_3 + 1)}{\omega_1 + \omega_2 - \omega_3} \right\} + \frac{1}{2} \sum_{1,2} C_{1-12-2} n_1 n_2.$$

$$(12.103)$$

Further on we utilize the useful relations for the combinations of the occupation numbers in the nominators of Eq. (12.103):

$$n_1 n_2 = T^2 \frac{1}{\omega_1 \omega_2}, \quad n_1 n_2 + n_1 n_3 + n_2 n_3 = T^2 \frac{\omega_1 + \omega_2 + \omega_3}{\omega_1 \omega_2 \omega_3},$$

$$(n_1 + 1)(n_2 + 1) n_3 - n_1 n_2 (n_3 + 1) \approx n_1 n_3 + n_2 n_3 - n_1 n_2$$

$$= T^2 \frac{\omega_1 + \omega_2 - \omega_3}{\omega_1 \omega_2 \omega_3}.$$

$$(12.104)$$

Hence:

$$\bar{E}^0_{\text{an.}} = -\frac{T^2}{3!} \sum_{1,2,3} |B_{123}|^2 \left\{ \frac{1}{(\omega_1 + \omega_2 + \omega_3)} \cdot \frac{(\omega_1 + \omega_2 + \omega_3)}{\omega_1 \omega_2 \omega_3} \right.$$

$$\left. +\frac{3}{\omega_1 \omega_2 \omega_3} \frac{(\omega_1 + \omega_2 - \omega_3)}{(\omega_1 + \omega_2 - \omega_3)} \right\} + \frac{T^2}{2} \sum_{1,2} C_{1-12-2} \frac{1}{\omega_1 \omega_2}$$

$$= -\frac{T^2}{6} \sum_{1,2,3} |B_{123}|^2 \frac{1}{\omega_1 \omega_2 \omega_3} [1 + 3] + \frac{T^2}{2} \sum_{1,2} C_{1-12-2} \frac{1}{\omega_1 \omega_2}$$

$$= -\frac{2T^2}{3} \sum_{1,2,3} |B_{123}|^2 \frac{1}{\omega_1 \omega_2 \omega_3} + \frac{T^2}{2} \sum_{1,2} C_{1-12-2} \frac{1}{\omega_1 \omega_2}$$

$$\sim T^2 \sim \frac{1}{\beta^2}.$$

$$(12.105)$$

Of course, the correction to the thermodynamic energy $\Delta E_{\text{therm.}}$ due to the anharmonic interactions also has a quadratic temperature dependence at high temperatures:

$$\Delta E_{\text{therm.}} = \frac{\partial}{\partial \beta} \left(\beta \bar{E}^0_{\text{an.}} \right) \sim \frac{\partial}{\partial \beta} \left(\beta \cdot \frac{1}{\beta^2} \right) = \frac{\partial}{\partial \beta} \left(\frac{1}{\beta} \right) = -\frac{1}{\beta^2} \propto T^2. \quad (12.106)$$

Moreover, the sign of the energy correction connected with the interplay of the positive and negative terms both quadratic in temperature in (12.105), is not rigorously determined in the general case.

Let us emphasize, that small corrections to the Free energy $\Delta F_{\text{therm.}}$, and other thermodynamic potentials also depend quadratically on temperature.

Correspondingly, corrections to the specific heat (which is constant: $C_V = \text{const}$ at high temperatures in harmonic approximation according to the Dulong–Petit law) are linear in temperature:

$$\Delta C_V \propto T. \tag{12.107}$$

References

1. Y. Kagan, *Lectures on the Solid-state Theoretical Physics* (Moscow, MEPHI, 1981–1982). Part I. Phonons - Unpublished
2. V.L. Gurevich, *Transport in Phonon Systems* Modern Problems in Condensed Matter Sciences. (Elsevier Science LTD., Oxford, 1988)
3. E.M. Lifshitz, L.P. Pitaevskii, *Physical Kinetics, Course of Theoretical Physics*, vol. 10 (Pergamon Press, Oxford, 1981)
4. J.M. Ziman, Electrons and phonons, in *The Theory of Transport Phenomena in Solids* (Clarendon Press, Oxford, 1960)
5. D. Pines, *Elementary Excitations in Solids* (W.A. Benjamin Inc., New York-Amsterdam, 1963)
6. O. Madelung, *Introduction to Solid-State Theory*, Springer Series in Solid-State Sciences (1995)
7. C. Kittel, *Quantum Theory of Solids* (Wiley, New York-London, 1963)
8. N.W. Ashcroft, N.D. Mermin, *Solid State Physics* (Holt, Rinehart and Winston, 1976)
9. L.D. Landau, E.M. Lifshitz, *Statistical Physics, Part I* (Butterworth-Heinemann, Oxford, 1999)
10. L.D. Landau, E.M. Lifshitz, *Quantum Mechanics, Non-Relativistic Theory* (Pergamon Press, New York, 1977)

Lecture 13. Phonon Thermal Conductivity in Crystals

13

Abstract

We derive Quantum kinetic equation for phonons and reveal the important role of the Umklapp processes. We solve the kinetic equation by Chapman–Enskog method and calculate the phonon relaxation time in τ-approximation. We establish different temperature regimes for the phonon thermal conductivity coefficient in the dielectric crystal including diffusive and ballistic (Knudsen) regime.

As we discussed already in the previous Lectures, at high temperatures $T \geq \theta_D$ mean square of the atomic displacement in crystal and thus the phonon number in the unit volume is proportional to temperature [1–6]. If there is temperature gradient in crystal, then the phonon flux emerges in the system [1,7–11].

Note that in gas the heat flow does not demand the presence of the particle flow. In other words, the mechanism of the thermal conductivity without convection is realized in the system [1,9,10].

In the solid state the heat flow is accompanied by the phonon flow [1,7–9]. In the transport phenomena in crystal, we cannot think in terms of the plane waves but only in terms of the localized wave packets, or quasiparticles [1,9].

There are two important requirements. Firstly, the scale L on which the temperature varies substantially, should be much larger than the phonon wavelength:

$$L \gg \left(\lambda \sim \frac{c}{\omega}\right). \tag{13.1}$$

Secondly, we should satisfy the condition on weak phonon damping (the quasiparticle should have large lifetime [1,7–12]):

$$\omega_\beta \tau_\beta \gg 1, \tag{13.2}$$

where τ_β is the phonon lifetime, and $\beta = f, \alpha$.

© The Author(s), under exclusive license to Springer Nature Switzerland AG 2026   227
M. Kagan, *Lecture Notes of Professor Yuri Kagan in Theoretical Solid-State Physics*,
Lecture Notes in Physics 1048, https://doi.org/10.1007/978-3-032-14621-2_13

From these requirements follows the condition on the length of the mean free path $l \sim v_{\mathrm{gr.}}\tau \sim c\tau$. It means that the free path length should be smaller than the characteristic scale L [1,8–14]:

$$l \ll L. \tag{13.3}$$

In fact, the condition (13.3) should be fulfilled to have quasiequilibrium collisional (hydrodynamic) regime [1,8–21], and not a collisionless (Knudsen) regime [1,8–11,21–23].

We know, that if the system is in quasiequilibrium state, then in each small volume of the sample (and in each time moment) we can introduce the local values of temperature T, density ρ, etc. [1,8,17,19,21]. In other words, the situation is locally an equilibrium one, but we have to take gradients into account [1,17,19,21].

As usual in kinetics, we introduce weakly non-equilibrium phonon distribution function $N(\beta, \boldsymbol{r}, t)$. The equilibrium distribution function meanwhile serves as an initial condition (or as zeroth approximation) and coincides with Planck distribution function:

$$(N(\beta, \boldsymbol{r}, t))_{\mathrm{equil.}} = \bar{n}(\beta). \tag{13.4}$$

Further on we solve quantum kinetic equation [1,9] and find non-equilibrium distribution function:

$$\frac{\partial N_\beta}{\partial t} + \frac{\partial N_\beta}{\partial \boldsymbol{r}} \boldsymbol{v}_{\mathrm{gr.}\beta} = -\sum_\alpha I_{\mathrm{col.}}\left(N_\alpha\right), \tag{13.5}$$

where $\boldsymbol{v}_{\mathrm{gr.}\beta}$ is the group velocity for the given branch of the phonon spectrum.

Note, that collision integral $I_{\mathrm{col.}}$ in the r.h.s. of Eq. (13.5) is related to phonon-phonon collisions in dielectric crystal.

In the stationary case the time derivative $\frac{\partial}{\partial t} = 0$.

Note that while solving the kinetic equation we take into account only the leading term connected with the three-phonon interaction (see Fig. 13.1):

$$\Delta H_3 = \frac{1}{3!} \sum_{1,2,3} B_{1,2,3} \hat{A}_1 \hat{A}_2 \hat{A}_3, \tag{13.6}$$

where again the operator $\hat{A}_\beta = \hat{a}_\beta + \hat{a}^\dagger_{-\beta}$, and we determined the coefficients $B_{1,2,3}$ in the previous Lecture.

Now we can represent the collision integral (which equals the difference of the arrival and departure of phonons) via the scattering probability in Born approximation [1,9]. Then we get for the phonon decay and picking up processes:

Fig. 13.1 We illustrate three-phonon interaction here by the phonon decay process, which yields the most important contribution to the collision integral [1]

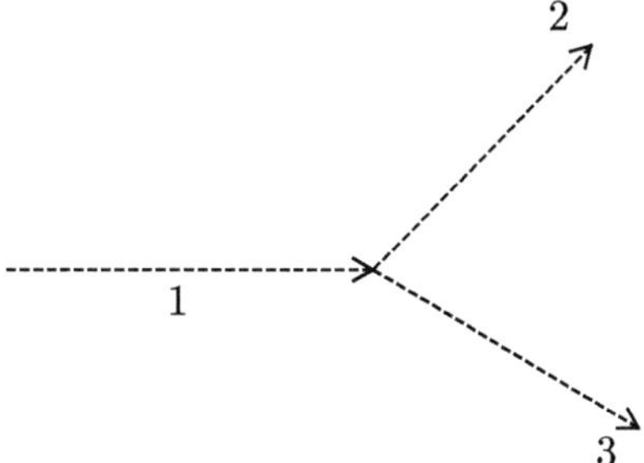

$$I_{\text{col.}} = \text{arrival - departure} \ = \ \frac{1}{2} \cdot \frac{2\pi}{\hbar} \sum_{2,3} \left|B_{1,2,3}\right|^2 \{N_1 \left(N_2 + 1\right) \left(N_3 + 1\right)$$

$$- \left(N_1 + 1\right) N_2 N_3\} \, \delta \left(\omega_1 - \omega_2 - \omega_3\right) \Delta \left(f_1 - f_2 - f_3\right)$$

$$+ \frac{2\pi}{\hbar} \cdot 1 \cdot \sum_{2,3} \left|B_{1,2,3}\right|^2 \{N_1 N_2 \left(N_3 + 1\right) - \left(N_1 + 1\right) \left(N_2 + 1\right) N_3\} \, \delta \left(\omega_1 + \omega_2 - \omega_3\right)$$

$$\cdot \Delta \left(f_1 + f_2 - f_3\right), \tag{13.7}$$

where the symbols $\Delta \left(f_1 - f_2 - f_3\right)$ and $\Delta \left(f_1 + f_2 - f_3\right)$ denote the fulfillment of the momentum conservation law within the accuracy of the inverse lattice vector for the processes of phonon decay and picking up, correspondingly. The factors $\frac{1}{2}$ and 1 in (13.7) are connected with the summation on indices 2 and 3 under the replacement of these indices.

In equilibrium the collision integral $I_{\text{col.}} = 0$. It means, that for the first term, connected with the decay process in (13.7), the balance condition is fulfilled:

$$N_1 \left(N_2 + 1\right) \left(N_3 + 1\right) = \left(N_1 + 1\right) N_2 N_3 \tag{13.8}$$

(and analogously for the second term related to the phonon picking up process).

Let us write the balance condition in Eq. (13.8) in a more convenient way [1,9]:

$$\frac{N_1}{N_1 + 1} = \frac{N_2}{N_2 + 1} \cdot \frac{N_3}{N_3 + 1}. \tag{13.9}$$

If we take the logarithm from the l.h.s. and r.h.s. of (13.9), we get

$$\ln \frac{N_1}{N_1 + 1} = \ln \frac{N_2}{N_2 + 1} + \ln \frac{N_3}{N_3 + 1}. \tag{13.10}$$

This relation should be valid for all the values of N_1, N_2, N_3, satisfying the conservation laws.

Thus, logarithm in (13.10) is an additive one [1,9]. Then it is reasonable to assume (in the spirit of the derivation of the Gibbs distribution in Statistical physics [1,6]), that the logarithm $\ln \frac{N_i}{N_i + 1}$ satisfies the relation:

$$\ln \frac{N_i}{N_i + 1} = -\gamma \omega_i. \tag{13.11}$$

This relation due to the energy conservation law (for the phonon frequencies in (13.7)) automatically guarantees the fulfillment of the additivity condition for the logarithm.

Note that in principle we can add the linear in momentum term $\boldsymbol{a} \boldsymbol{f}_i$ to $-\gamma \omega_i$ in the r.h.s. of (13.11). We should not do it nevertheless since the momentum conservation law is not exactly fulfilled.

As a result, we get from (13.11):

$$\frac{N_i}{N_i + 1} = \exp\left(-\gamma \omega_i\right), \tag{13.12}$$

and thus derive the Planck distribution:

$$N_i = \frac{\exp\left(-\gamma \omega_i\right)}{1 - \exp\left(-\gamma \omega_i\right)} = \frac{1}{\exp\left(\gamma \omega_i\right) - 1}, \tag{13.13}$$

where

$$\gamma = \frac{1}{T} \tag{13.14}$$

is the same for all N.

According to Landau and Gibbs [6], condition (13.13) yields us another way to determine the temperature.

For weakly non-equilibrium situation we can use equilibrium distribution function N_β^0 as zeroth approximation, and calculate the small corrections ΔN_β from the perturbation theory. Then, the non-equilibrium distribution function reads

$$N_\beta = N_\beta^0 + \Delta N_\beta. \tag{13.15}$$

Further on we solve the kinetic equation utilizing the Chapman–Enskog method [1,9,24]. It means that we substitute equilibrium distribution function N_β^0 in the l.h.s. of the kinetic equation calculating the collision integral meanwhile on the corrections to equilibrium distribution function ΔN_β.

Then the l.h.s. of the kinetic equation in the stationary case acquires a form:

$$\frac{\partial N_\beta^0}{\partial \boldsymbol{r}} \boldsymbol{v}_{\text{gr}.\beta} = \frac{\partial N_\beta^0}{\partial T} \frac{\partial T}{\partial \boldsymbol{r}} \boldsymbol{v}_{\text{gr}.\beta} = N_\beta^0 \left(N_\beta^0 + 1\right) \frac{\omega_\beta}{T^2} \boldsymbol{v}_{\text{gr}.\beta} \frac{\partial T}{\partial \boldsymbol{r}}. \tag{13.16}$$

Thus:

$$N_\beta^0 \left(N_\beta^0 + 1\right) \frac{\omega_\beta}{T^2} \boldsymbol{v}_{\text{gr}.\beta} \frac{\partial T}{\partial \boldsymbol{r}} = -I_{\text{col.}} \left(\Delta N_\beta\right), \tag{13.17}$$

where

$$N_\beta^0 \left(N_\beta^0 + 1\right) = \frac{\exp\left(\frac{\omega_\beta}{T}\right)}{\left(\exp\left(\frac{\omega_\beta}{T}\right) - 1\right)^2}. \tag{13.18}$$

Let us remind that the condition for the validity of the Chapman–Enskog method [1,9,24] assumes the smallness of the temperature gradients:

$$\frac{l\nabla T}{T} \ll 1. \tag{13.19}$$

This condition is just equivalent to the condition in Eq. (13.3) according to which the free path length $l \ll L$.

Now we can linearize the collision integral $I_{\text{col.}}\left(\Delta N_\beta\right)$ with respect to small corrections ΔN_β to equilibrium function. Then:

$$N_1 \left(N_2 + 1\right)\left(N_3 + 1\right) - \left(N_1 + 1\right) N_2 N_3 = \Delta N_1 \left[\left(N_2^0 + 1\right)\left(N_3^0 + 1\right) - N_2^0 N_3^0\right]$$
$$+ \Delta N_2 \left[N_1^0 \left(N_3^0 + 1\right) - \left(N_1^0 + 1\right) N_3^0\right] + \Delta N_3 \left[N_1^0 \left(N_2^0 + 1\right) - \left(N_1^0 + 1\right) N_2^0\right]. \tag{13.20}$$

Further on let us rewrite expression for the correction ΔN_1 in (13.20) in a more convenient form using the balance Eq. (13.8):

$$\left(N_2^0 + 1\right)\left(N_3^0 + 1\right) - N_2^0 N_3^0 = \frac{N_1^0 \left(N_2^0 + 1\right)\left(N_3^0 + 1\right)}{N_1^0} - \frac{\left(N_1^0 + 1\right) N_2^0 N_3^0}{\left(N_1^0 + 1\right)}$$
$$= \frac{N_1^0 \left(N_1^0 + 1\right)\left(N_2^0 + 1\right)\left(N_3^0 + 1\right) - N_1^0 \left(N_1^0 + 1\right) N_2^0 N_3^0}{N_1^0 \left(N_1^0 + 1\right)}$$
$$= \frac{N_1^0 \left(N_1^0 + 1\right)\left(N_2^0 + 1\right)\left(N_3^0 + 1\right) - N_1^0 N_1^0 \left(N_2^0 + 1\right)\left(N_3^0 + 1\right)}{N_1^0 \left(N_1^0 + 1\right)}$$
$$= \frac{N_1^0 \left(N_2^0 + 1\right)\left(N_3^0 + 1\right)}{N_1^0 \left(N_1^0 + 1\right)}. \tag{13.21}$$

Afterwords we can make a convenient substitution for the correction ΔN_1 in a linearized collision integral in (13.20):

$$\Delta N_1 = N_1^0 \left(N_1^0 + 1\right) \chi_1. \tag{13.22}$$

In terms of the variable χ_1:

$$\Delta N_1 \left[\left(N_2^0 + 1\right)\left(N_3^0 + 1\right) - N_2^0 N_3^0\right] = N_1^0 \left(N_2^0 + 1\right)\left(N_3^0 + 1\right) \chi_1. \tag{13.23}$$

Analogously we can write the corrections ΔN_2 and ΔN_3 as

$$\Delta N_2 = N_2^0 \left(N_2^0 + 1\right) \chi_2, \quad \Delta N_3 = N_3^0 \left(N_3^0 + 1\right) \chi_3, \tag{13.24}$$

and get from (13.20):

$$N_1 \left(N_2 + 1\right)\left(N_3 + 1\right) - \left(N_1 + 1\right) N_2 N_3 = N_1^0 \left(N_2^0 + 1\right)\left(N_3^0 + 1\right)\left(\chi_1 - \chi_2 - \chi_3\right). \tag{13.25}$$

As a result, after the linearization, the collision integral in Eq. (13.7) can be written as follows:

$$\begin{aligned}
I_{\text{col.}} = \frac{2\pi}{\hbar} \sum_{2,3} |B_{1,2,3}|^2 \Big\{ &\frac{1}{2} N_1^0 \left(N_2^0 + 1\right)\left(N_3^0 + 1\right)(\chi_1 - \chi_2 - \chi_3) \\
&\times \delta\left(\omega_1 - \omega_2 - \omega_3\right) \Delta\left(f_1 - f_2 - f_3\right) \\
&+ N_1^0 N_2^0 \left(N_3^0 + 1\right)(\chi_1 + \chi_2 - \chi_3)\,\delta\left(\omega_1 + \omega_2 - \omega_3\right)\Delta\left(f_1 + f_2 - f_3\right) \Big\}.
\end{aligned} \tag{13.26}$$

We can often use so-called τ-approximation [1,9,10] for the solution of the kinetic equation. In this approximation, in the l.h.s. of the kinetic equation enters equilibrium distribution function N_1^0, and the collision integral in the r.h.s. can be linearized here, as in the Chapman–Enskog method [1,9,10,24].

Moreover, in τ-approximation, the corrections to the distribution functions ΔN_2 and ΔN_3 are assumed to be zero, and thus in the r.h.s. of the kinetic equation survives only the dependence on ΔN_1.

In fact, in τ-approximation we neglect the effect of the excitation of the other modes on the relaxation of the given mode. In other words, we can say, that we calculate the diagonal matrix element on $\beta = f, \alpha$, that is we consider the scattering process of one phonon in non-equilibrium state on the remaining equilibrium phonons.

Then the kinetic equation:

$$N_1^0 \left(N_1^0 + 1\right) \frac{\omega_1}{T^2} v_{\text{gr.}1} \frac{\partial T}{\partial \boldsymbol{r}} = -I_{\text{col.}}\left(\Delta N_1\right) = -\frac{\Delta N_1}{\tau}, \tag{13.27}$$

where $\frac{1}{\tau}$ is the inverse phonon lifetime.

Correspondingly:

$$\frac{\omega_1}{T^2} v_{\text{gr.}1} \frac{\partial T}{\partial \boldsymbol{r}} = -\frac{\chi_1}{\tau}. \tag{13.28}$$

Thus, non-equilibrium correction to the distribution function is proportional to the gradient of temperature:

$$\chi_1 = -\frac{\omega_1 \tau}{T^2} \boldsymbol{v}_{\text{gr.1}} \frac{\partial T}{\partial \boldsymbol{r}} \propto \nabla T. \tag{13.29}$$

Note, that exact inverse lifetime of phonon enters in Eq.'s (13.27)–(13.29):

$$\frac{\left(N_1^0 + 1\right)}{\tau} = \frac{\pi}{\hbar} \sum_{2,3} |B_{1,2,3}|^2 \left(N_3^0 + 1\right) \left\{ \left(N_2^0 + 1\right) \right.$$

$$\times \, \delta\left(\omega_1 - \omega_2 - \omega_3\right) \Delta\left(f_1 - f_2 - f_3\right) + 2N_2^0 \delta\left(\omega_1 + \omega_2 - \omega_3\right) \Delta\left(f_1 + f_2 - f_3\right) \Big\} . \tag{13.30}$$

At $T = 0$ equilibrium (Planck) distribution functions $N_1^0 = N_2^0 = N_3^0 = 0$, and exact expression for the inverse lifetime:

$$\frac{1}{\tau} = \frac{\pi}{\hbar} \sum_{2,3} |B_{1,2,3}|^2 \, \delta\left(\omega_1 - \omega_2 - \omega_3\right) \Delta\left(f_1 - f_2 - f_3\right) \tag{13.31}$$

coincides (for $\hbar = 1$) with expressions (12.51) and (12.55) for the imaginary part of the phonon spectrum Im ω and damping $\frac{1}{\tau} = 2|\operatorname{Im} \omega|$, obtained in previous Lecture in Born approximation for phonon decay.

In particular, the dependence $\frac{1}{\tau} \sim f_1^5$ on the wave vector f_1 is conserved in the long wavelength limit at $T = 0$ in the exact expression for the inverse lifetime in Eq. (13.30).

At finite temperatures in the long wavelength limit (when equilibrium distribution functions $N_1^0 = \frac{T}{\omega_1} \gg 1$, $N_2^0 = \frac{T}{\omega_2} \gg 1$, $N_3^0 = \frac{T}{\omega_3} \gg 1$) expression (13.30) reproduces also the result $\frac{1}{\tau} \sim T f_1^4$ for phonon decay and picking up.

13.1 When τ-Approximation Does Not Work

For the applicability of τ-approximation, it is important not to lose the entropy increase law. The entropy increase law is a consequence of the exact collision integral [1,9,10] and additivity of the logarithm in Eq. (13.10). Its violation can be connected with the presence of the magnetic field, time irreversibility, and so on.

13.2 Properties of the Collision Integral

1. Collision integral is real and self-conjugated. We can prove it utilizing the linearized form of the collision integral in Eq. (13.26), which reads:

$$\langle \chi_k I_{\text{col.}}(\chi_i) \rangle = \langle \chi_i I_{\text{col.}}(\chi_k) \rangle . \tag{13.32}$$

2. If we found an approximate solution of the kinetic equation for the correction to the distribution function, then the exact solution is even more non-equilibrium one. Thus, the exact correction to the equilibrium distribution function is larger in the absolute value. This fact follows from the properties of the operator for the collision integral (it is self-conjugated), and of the exact solution (it corresponds to the maximum of entropy).

Let us return back now to the expression for the non-equilibrium correction to the distribution function obtained in Eq. (13.29) in τ-approximation. It is clear, that even in the most general case, when τ-approximation does not work, we still have

$$\chi_\beta = A_\beta \nabla T. \tag{13.33}$$

Due to space homogeneity and isotropy, in the absence of the crystalline lattice the vector A_β should have the components parallel to the wave vector:

$$A^i_\beta = \varphi\left(f^2\right) f^i, \tag{13.34}$$

where the scalar function φ in the long wavelength limit can be represented as a series expansion in powers of f^2:

$$\varphi\left(f^2\right) = \alpha + \beta f^2 + \gamma f^4 + \cdots. \tag{13.35}$$

Meanwhile the kinetic equation for the correction ΔN_β acquires a form:

$$N^0_\beta\left(N^0_\beta + 1\right) \frac{\omega_\beta}{T^2} v^i_{\text{gr}.\beta} \nabla_i T = -I_{\text{col.}}\left(\Delta N_\beta\right) = -J^i_{\text{col.}}\left\{A_\beta = \varphi\left(f^2\right)f\right\} \nabla_i T, \tag{13.36}$$

where $I_{\text{col.}} = J^i_{\text{col.}} \nabla_i T$.

Dividing the l.h.s. and r.h.s. on the temperature gradient (which is different from zero), we get

$$N^0_\beta\left(N^0_\beta + 1\right) \frac{\omega_\beta}{T^2} v^i_{\text{gr}.\beta} = -J^i_{\text{col.}}\left\{\left(\alpha + \beta f^2 + \gamma f^4 + \cdots\right)f\right\}. \tag{13.37}$$

Further on, multiplying the l.h.s. and r.h.s. of Eq. (13.37) on α, βf^2 and γf^4 respectively, and integrating on momentum, we can find unknown coefficients α, β and γ. Thus, we can calculate the vector A_β.

In fact, we use here the procedure which is well-known in kinetics. It is called the momentum method [1, 9, 10]. This method yields very good approximation to the exact solution if we break somewhere the series expansion in (13.35). We should remember that exact solution of the kinetic equation always gives the larger correction to the equilibrium distribution function by absolute value, than approximate solution.

Note that all these considerations are valid for homogeneous and isotropic space.

There are distinguished directions in crystal, connected with the vectors of the inverse lattice. Hence the determination of the components of vector A_β^i becomes more difficult task. In this case let us use again the τ-approximation.

Then the kinetic equation reads

$$N_\beta^0\left(N_\beta^0+1\right)\frac{\omega_\beta}{T^2}v_{\text{gr}.\beta}^i\nabla_i T = -\frac{\Delta N_\beta}{\tau_\beta},\tag{13.38}$$

and correspondingly the correction to the distribution function is given by

$$\Delta N_\beta = -N_\beta^0\left(N_\beta^0+1\right)\frac{\omega_\beta\tau_\beta}{T^2}v_{\text{gr}.\beta}^i\nabla_i T.\tag{13.39}$$

Now we can utilize the formula for the heat flow connected with phonon modes:

$$Q^i = \sum_\beta N_\beta\omega_\beta v_{\text{gr}.\beta}^i = \sum_\beta\left(N_\beta^0+\Delta N_\beta\right)\omega_\beta v_{\text{gr}.\beta}^i = \sum_\alpha\int\frac{d^3 f}{(2\pi)^3}\Delta N_\beta\omega_\beta v_{\text{gr}.\beta}^i,\tag{13.40}$$

where under the derivation of Eq. (13.40) we take into account that the group velocity $v_{\text{gr}.\beta}^i = \frac{\partial\omega_\beta}{\partial f^i}$ is an odd function of the wave vector f. At the same time, the phonon frequency ω_β is an even function of momentum. Hence the sum:

$$\sum_\beta N_\beta^0\omega_\beta v_{\text{gr}.\beta}^i = 0.\tag{13.41}$$

Let us emphasize that in initial formula (13.40) for the heat flow we do not have inverse system volume (the multiplier $\frac{1}{V}$ is absent). It is connected with the fact, that as always in kinetics, the integral from the distribution function on momentum is normalized on the system density.

Substituting in (13.40) expression for ΔN_β from (13.39), we get

$$Q^i = -\sum_\alpha\int\frac{d^3 f}{(2\pi)^3}N_\beta^0\left(N_\beta^0+1\right)\frac{\omega_\beta^2\tau_\beta}{T^2}v_{\text{gr}.\beta}^i v_{\text{gr}.\beta}^k\nabla_k T,\tag{13.42}$$

where we replaced the summation over the Cartesian index i by the summation over $k : v_{\text{gr}.\beta}^i\nabla_i T \to v_{\text{gr}.\beta}^k\nabla_k T$ in Eq. (13.39).

But by definition the heat flow caused by the temperature gradient equals

$$Q^i = -\kappa^{ik}\frac{\partial T}{\partial x^k} \equiv -\kappa^{ik}\nabla_k T,\tag{13.43}$$

where we introduced the thermal conductivity tensor:

$$\kappa^{ik} = \sum_\alpha\int\frac{d^3 f}{(2\pi)^3}N_\beta^0\left(N_\beta^0+1\right)\frac{\omega_\beta^2\tau_\beta}{T^2}v_{\text{gr}.\beta}^i v_{\text{gr}.\beta}^k.\tag{13.44}$$

Note that as a consequence of the Onsager principle [1,9,25] for the kinetic coefficients: κ^{ik} is symmetric tensor.

Moreover, in the crystal with cubic symmetry the thermal conductivity tensor acquires a diagonal form:

$$\kappa^{ik} = \kappa \delta^{ik}.$$

Then the index contraction in the l.h.s. and r.h.s. of (13.44) yields

$$\kappa = \frac{1}{3} \sum_{\alpha} \int \frac{d^3 f}{(2\pi)^3} N_\beta^0 \left(N_\beta^0 + 1 \right) \frac{\omega_\beta^2 \tau_\beta}{T^2} v_{\mathrm{gr}.\beta}^2. \tag{13.45}$$

Recollecting expression for the contribution to the specific heat from one-phonon mode:

$$C_\beta = N_\beta^0 \left(N_\beta^0 + 1 \right) \frac{\omega_\beta^2}{T^2}, \tag{13.46}$$

and introducing the free path length also for one mode as

$$l_\beta = v_{\mathrm{gr}.\beta} \tau_\beta, \tag{13.47}$$

we get the canonical expression (as in kinetic theory of gases [1,9,10]) for the thermal conductivity coefficient:

$$\kappa = \frac{1}{3} \sum_{\alpha} \int \frac{d^3 f}{(2\pi)^3} C_\beta l_\beta v_{\mathrm{gr}.\beta}. \tag{13.48}$$

Let us stress once more that in the general case the kinetic equation reads according to (13.36):

$$N_\beta^0 \left(N_\beta^0 + 1 \right) \frac{\omega_\beta}{T^2} \boldsymbol{v}_{\mathrm{gr}.\beta} = -\boldsymbol{J}_{\mathrm{col.}} \left(\boldsymbol{A}_\beta \right).$$

Let us multiply the l.h.s. and r.h.s. of this equation on A_β^k and integrate it. Then we get $\left\langle A_\beta^k \boldsymbol{J}_{\mathrm{col.}} \left(A_\beta^i \right) \right\rangle$. But since the collision operator is real and self-conjugated, then as in (13.32):

$$\left\langle A_\beta^k \boldsymbol{J}_{\mathrm{col.}} \left(A_\beta^i \right) \right\rangle = \left\langle A_\beta^i \boldsymbol{J}_{\mathrm{col.}} \left(A_\beta^k \right) \right\rangle. \tag{13.49}$$

Thus, the r.h.s. of the kinetic equation is symmetric. Hence, the l.h.s. is also symmetric. But the l.h.s. is proportional to the thermal conductivity tensor κ^{ik}. Hence, we come to the conclusion that the thermal conductivity tensor κ^{ik} is symmetric in general case (in agreement with the Onsager principle [1,9,10,25]).

13.3   General Behavior of the Thermal Conductivity Coefficient

1. <u>High Temperatures</u> $T > \theta_D$

In this case, the Dulong–Petit law is fulfilled in the main approximation (without an account of the corrections connected with the anharmonic interactions), and correspondingly specific heat $C_\beta = \text{const}$ in the expression for the heat flow in Eq. (13.48).

At the same time, for high temperatures, as we showed in the previous Lecture, the inverse lifetime: $\frac{1}{\tau_\beta} \sim T$. As a result, the free path length in (13.48):

$$l_\beta = v_{\text{gr}.\beta}\tau_\beta \sim \frac{1}{T}, \tag{13.50}$$

since the group velocity $v_{\text{gr}.\beta} = \frac{\partial \omega_\beta}{\partial f}$ does not depend on temperature.

Thus, the thermal conductivity coefficient at high temperatures is inversely proportional to the temperature:

$$\kappa \sim \frac{1}{T}. \tag{13.51}$$

13.4   The Problem of the Long Wavelength Phonons

If the process goes on with small initial and final phonon energies, then the inverse lifetime for $T > \theta_D$:

$$\frac{1}{\tau_\beta} \sim f^4 T,$$

and the integral over the momenta in the expression for the thermal conductivity coefficient in (13.48) becomes divergent on small wave vectors [1]:

$$\kappa \sim \int d^3 f \, \tau \sim \frac{1}{T}\int \frac{d^3 f}{f^4} \sim \frac{1}{T}\int \frac{df}{f^2}. \tag{13.52}$$

The infrared catastrophe emerges [1] with strong power law singularity of the integral for the thermal conductivity coefficient on the lower limit of integration: $\kappa \sim \frac{1}{T f_{\min}} \sim \frac{L}{T}$, where $f_{\min} = \frac{2\pi}{L}$, and L is the system size (the size of the crystal).

Note, that for small f the momentum conservation law is fulfilled exactly, and thus, the decay process, as well as the picking up process for $T \gg \omega_\beta$ yield the same temperature dependence of the inverse lifetime $\frac{1}{\tau_\beta} \sim f^4 T$.

The presence of defects in crystal does not lift the singularity for the thermal conductivity coefficient on the lower limit of the integration [1]. If the size of the inhomogeneity is much less than the wavelength, then the differential of the cross section for phonon scattering on the crystal inhomogeneities (analogously to the

Rayleigh scattering of light [26]) $d\sigma \sim \omega^4$, and the free path length l (as well as the phonon lifetime τ [1,26]):

$$l = \frac{1}{n_{\text{def.}}\sigma} \sim \tau \sim \frac{1}{\omega^4}. \tag{13.53}$$

Hence under the scattering on the small size inhomogeneities (e.g., on the point like defects-impurities or vacancies) the same type of the singularity as in the ideal crystal without defects (see Eq. (13.52)) is conserved on the small wave vectors.

Even the <u>dislocation</u> (large one-dimensional defect) does not entirely lift the singularity on the lower limit [1]. In this case: $l \sim \tau \sim \frac{1}{\omega^3}$, and weaker logarithmic singularity arises instead of the power law one [1]:

$$\kappa \sim \int \frac{d^3 f}{\omega^3} \sim \int \frac{df}{f}. \tag{13.54}$$

Thus, we have the singularity on the lower limit of integration for the thermal conductivity coefficient, which is not so simple to lift.

This singularity can be lifted only due to more intensive processes of the phonon picking up, which can be realized for the crossing of the branches of the phonon spectrum (e.g., for the longitudinal phonons with the maximal sound velocity).

In case of more intensive picking up processes, as we discussed in the previous Lecture, it is possible to get the different behavior of the inverse lifetime, which can lift the singularity in small wave vectors:

$$\frac{1}{\tau_\beta} \sim f^2 T^3.$$

(2) $\underline{T \ll \theta_D}$.
For $\overline{T \ll \theta_D}$ let us write the kinetic equation again as

$$N_\beta^0 \left(N_\beta^0 + 1 \right) \frac{\omega_\beta}{T^2} v_{\text{gr.}\beta}^i = -J_{\text{col.}}^i \left(A_\beta \right),$$

and, moreover, let us consider the simple case:

$$A_\beta^i = b f^i. \tag{13.55}$$

Let us determine the unknown constant b. To do that, as the momentum method [1,9,10,24] advises, let us multiply the l.h.s. and r.h.s. of the kinetic equation on f_i and integrate over $\frac{d^3 f}{(2\pi)^3}$. Then, the l.h.s. of the kinetic equation will be the even function of momentum proportional to the scalar product $f_i v_{\text{gr.}\beta}^i = f_i \frac{\partial \omega_\beta}{\partial f^i}$. Moreover, the l.h.s. will be finite and equal to the const:

$$\sum_\alpha \int \frac{d^3 f}{(2\pi)^3} N_\beta^0 \left(N_\beta^0 + 1 \right) \frac{\omega_\beta^2}{T^2} v_{\text{gr.}\beta}^i f_i = \text{const}. \tag{13.56}$$

At the same time, the r.h.s. occurs to be equal to the scalar product $\langle f_i\, J^i_{\text{col.}}\,(b\mathbf{f})\rangle$. Hence from the equality of the l.h.s. and r.h.s. we get

$$\left\langle f_i\, J^i_{\text{col.}}\,(b\mathbf{f})\right\rangle = \sum_\alpha \int \frac{d^3 f}{(2\pi)^3}\, N^0_\beta \left(N^0_\beta + 1\right) \frac{\omega^2_\beta}{T^2}\, v^i_{\text{gr.}\beta}\, f_i = \text{const}. \qquad (13.57)$$

Note, that the terms with the phonon decay and picking up in the collision integral in Eq. (13.26) are proportional, respectively, to:

$$(\chi_1 - \chi_2 - \chi_3)\, \Delta\left(\mathbf{f}_1 - \mathbf{f}_2 - \mathbf{f}_3\right), \ \text{and} \ (\chi_1 + \chi_2 - \chi_3)\, \Delta\left(\mathbf{f}_1 + \mathbf{f}_2 - \mathbf{f}_3\right).$$

In our case the combinations of the correction functions $(\chi_1 - \chi_2 - \chi_3)$ and $(\chi_1 + \chi_2 - \chi_3)$ depend linearly on the same momentum combinations, which enter in the momentum conservation laws. Thus, in the collision integral enter the terms which are proportional to:

$$b\left(f^i_1 - f^i_2 - f^i_3\right) \Delta\left(\mathbf{f}_1 - \mathbf{f}_2 - \mathbf{f}_3\right), \ \text{and}$$
$$b\left(f^i_1 + f^i_2 - f^i_3\right) \Delta\left(\mathbf{f}_1 + \mathbf{f}_2 - \mathbf{f}_3\right). \qquad (13.58)$$

Correspondingly, if the momentum conservation law is fulfilled exactly, then the combinations $(\chi_1 - \chi_2 - \chi_3) = 0$ and $(\chi_1 + \chi_2 - \chi_3) = 0$.

As a result, the scalar product in Eq. (13.57):

$$\left\langle f_i\, J^i_{\text{col.}}\,(b\mathbf{f})\right\rangle \sim b \cdot 0 = \text{const}. \qquad (13.59)$$

Then, $b = \frac{\text{const}}{0} \to \infty$, and the thermal conductivity coefficient $\kappa^{ik} \sim b \to \infty$. But the exact solution always yields larger by the absolute value correction to the distribution function, than the approximate solution in (13.55) which has simple dependence $A^i_\beta = b f^i$. That is why, in the exact solution the absolute value of coefficient b is even larger, and the exact value of the thermal conductivity tensor κ^{ik} even in larger extent tends to the infinity.

Thus, only Umklapp processes which do not conserve momentum can save us. Umklapp can be connected only with the phonon picking up processes when the first phonon with small energy $\omega_1 \sim T$ picks up the second phonon with larger energy $\omega_2 > T$ (see Fig. 13.2).

For $\omega_2 > T$ Planck (equilibrium) distribution function of the second phonon is exponentially small at low temperatures:

$$N^0_2 = \frac{1}{\exp\left(\frac{\omega_2}{T}\right) - 1} \approx \exp\left(-\frac{\omega_2}{T}\right), \qquad (13.60)$$

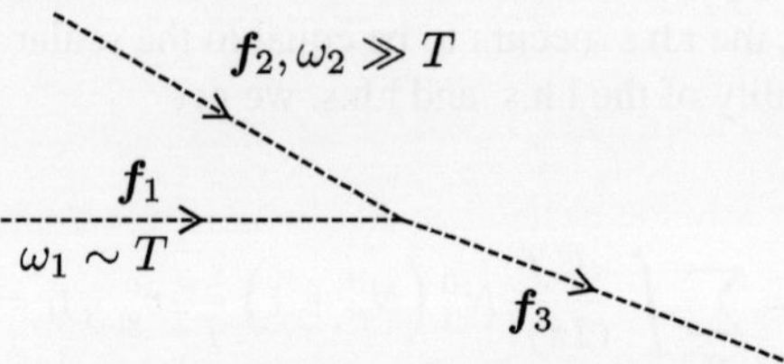

picking up with Umklapp process

Fig. 13.2 Phonon picking up effective for the Umklapp processes when the first phonon with small energy $\omega_1 \sim T$ picks up the second phonon with larger energy $\omega_2 > T$. This process guarantees the finite value of the thermal conductivity coefficient at low temperatures $T < \theta_D$ [1]

But the collision integral in the kinetic equation for the phonon picking up is proportional to the following combination:

$$I_{\text{col.}} \sim N_1^0 N_2^0 \left(N_3^0 + 1\right) (\chi_1 + \chi_2 - \chi_3)\, \Delta \left(f_1 + f_2 - f_3\right) \sim \frac{1}{\tau_\beta}. \tag{13.61}$$

Hence the inverse time for the Umklapp processes is also exponentially small at low temperatures [1,8,9]. Indeed, in τ-approximation:

$$I_{\text{col.}} \sim \frac{1}{\tau_\beta} \sim \frac{1}{\tau_{\text{Umklapp}}} \sim N_2^0 \sim \exp\left(-\frac{\omega_2}{T}\right) \sim \exp\left(-\gamma \frac{\theta_D}{T}\right), \tag{13.62}$$

where γ is the numerical coefficient of order 1.

Of course, at temperatures $T \to 0$:

$$\frac{1}{\tau_{\text{Umklapp}}} \to 0. \tag{13.63}$$

Thus, the thermal conductivity coefficient exponentially depends on temperature in this regime [1]:

$$\kappa \sim \tau_{\text{Umklapp}} \sim \exp\left(\gamma \frac{\theta_D}{T}\right). \tag{13.64}$$

Qualitatively we can explain an importance of the Umklapp processes in the following way. If there are no Umklapp processes, then the initial momentum is conserved. Hence, if we pushed the phonon flow at the initial time moment, then it cannot stop. Correspondingly, thermal resistance equals zero and thermal conductivity equals infinity.

But the heat flow Q^i is proportional to the mean phonon energy $\langle \omega \rangle$ multiplied by the mean flow of phonons $n_{\text{phon.}} v^i_{\text{gr.}} \sim \sum_\alpha \int \frac{d^3 f}{(2\pi)^3} \Delta N_1 v^i_{\text{gr.}\alpha}$. Indeed, as we already discussed, the heat flow in the dielectric crystal always emerges together with the phonon flow. Hence, to stop the heat flow we should stop the flow of phonons.

To stop the phonon flow can only the momentum outflow from the system connected with the Umklapp processes. If there are no Umklapp processes, then the phonon flow stops only due to the finite size L of the crystal [1,8].

But there are no Umklapp processes at $T \to 0$. The free path length in this case reads

$$l_\beta = v_{\text{gr}.\beta}\tau_\beta \leq L, \tag{13.65}$$

and collisionless (Knudsen) regime [1,9,10,22,23] is realized. In this regime, the thermal conductivity coefficient is proportional to the phonon specific heat, and for very low temperatures $T \ll \theta_D$:

$$\kappa \sim C_\beta l_\beta v_{\text{gr}.\beta} \sim C_\beta L_{\text{cryst}.} v_{\text{gr}.\beta} \sim C_\beta \sim T^3. \tag{13.66}$$

If the temperature is slightly higher (but still $T \ll \theta_D$), then instead of the collision less regime for the phonon flight from one crystal boundary to another one, the diffusive regime emerges [1] resembling the random walk problem in Statistical physics [1,6,27] (see Fig. 13.3). In this problem the phonon in between of two scattering acts on the crystal boundaries experiences the multiple scattering on the other phonons.

For the diffusive regime effective phonon scattering time in the kinetic equation can be determined from the condition [1]:

$$\tau_{\text{eff}.} \sim \frac{L^2}{D}, \tag{13.67}$$

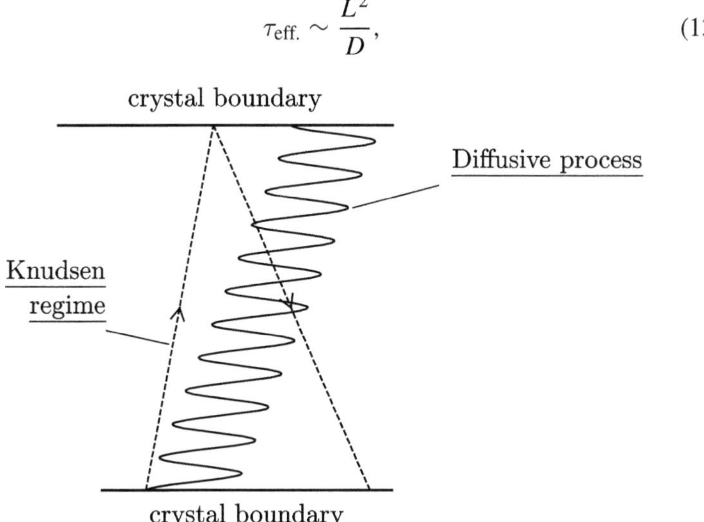

Fig. 13.3 Diffusive regime for the phonon scattering resembling the random walk problem (solid line), and collision less (Knudsen) regime (dotted line) for phonon scattering on the crystal boundaries. In the diffusive regime the phonon in between two scattering acts on crystal boundaries experiences the multiple scattering on the other phonons [1]

where L is the crystal size, D is the diffusion coefficient.

Meanwhile for the diffusion coefficient the following estimate is valid:

$$D \sim v_{\text{gr.}} l_N \sim v_{\text{gr.}}^2 \tau_N, \tag{13.68}$$

where l_N and τ_{N-} are normal (not related to the Umklapp processes) free path length and scattering time corresponding to the phonon decay and picking up.

As we know, the inverse (normal) scattering time for the long wavelength phonons: $\frac{1}{\tau_N} \sim f^m T^n$, where $m + n = 5$. Then, the diffusion coefficient:

$$D \sim \frac{1}{f^m T^n}, \tag{13.69}$$

and effective scattering time:

$$\tau_{\text{eff.}} \sim T^5, \tag{13.70}$$

where we took into account that at low temperatures $T \ll \theta_D$ the phonon energy is $cf \sim T$. Hence, the thermal conductivity coefficient in the diffusive regime [1]:

$$\kappa \sim C_\beta \tau_{\text{eff.}} \sim T^8. \tag{13.71}$$

The dependence of the phonon thermal conductivity coefficient on temperature in different regimes is presented in Fig. 13.4 [1]. At very low temperatures the thermal conductivity coefficient $\kappa \sim T^3$ in collisionless (Knudsen) regime, then $\kappa \sim T^8$ in

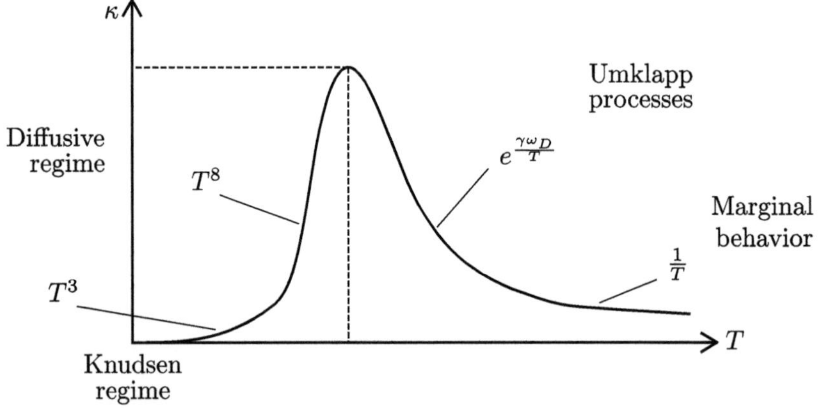

Fig. 13.4 Dependence of the phonon thermal conductivity coefficient on temperature in the dielectric crystal [1]. We present different temperature regimes for the thermal conductivity coefficient in the Figure. At very low temperatures $\kappa \sim T^3$ in Knudsen regime, then $\kappa \sim T^8$ in the diffusive regime. Further on at higher temperatures (but still less than the Debye one) $\kappa \sim \exp\left(\gamma \frac{\theta_D}{T}\right)$ with an account of the Umklapp processes. Finally at higher temperatures $T > \omega_D : \kappa \sim \frac{1}{T}$ due to the marginal behavior of the phonon lifetime [1]

the diffusive regime. Further on at higher temperatures (but still smaller than the Debye one) $\kappa \sim \exp\left(\gamma \frac{\theta_D}{T}\right)$ with an account of the Umklapp processes. Finally at high temperatures $T > \theta_D : \kappa \sim \frac{1}{T}$ due to the marginal behavior of the phonon lifetime.

References

1. Y. Kagan, *Lectures on the Solid-State Theoretical Physics* (Moscow, MEPHI, 1981–1982). Part I. Phonons - unpublished
2. D. Pines, *Elementary Excitations in Solids* (W.A. Benjamin Inc., New York-Amsterdam, 1963)
3. O. Madelung, *Introduction to Solid-State Theory*, Springer Series in Solid-State Sciences (1995)
4. C. Kittel, *Quantum Theory of Solids* (John Wiley and Sons, New York-London, 1963)
5. N.W. Ashcroft, N.D. Mermin, *Solid State Physics* (Holt, Rinehart and Winston, 1976)
6. L.D. Landau, E.M. Lifshitz, *Statistical Physics* (Part I, Butterworth-Heinemann, Oxford, 1999)
7. J.M. Ziman, *Electrons and Phonons: The Theory of Transport Phenomena in Solids* (Clarendon Press, Oxford, 1960)
8. V.L. Gurevich, *Transport in Phonon Systems*, Modern Problems in Condensed Matter Sciences (Elsevier Science LTD, Oxford, 1988)
9. E.M. Lifshitz, L.P. Pitaevskii, *Physical Kinetics, Course of Theoretical Physics*, vol. 10 (Pergamon Press, Oxford, 1981)
10. V.P. Silin, *Kinetic Theory of Gases* (Nauka, Moscow, 1971) (in Russian)
11. S.N. Burmistrov, *Physical Kinetics: Quantum and Classical Problems and Solutions* (Springer, Berlin, 2022)
12. V.F. Gantmakher, Y.B. Levinson, *Carrier Scattering in Metals and Semiconductors* (Elsevier, 2012)
13. M. Apostol, *Physical Kinetics* (Cambridge Scholars Publishing, 2019)
14. Physical and Chemical Kinetics, ed. by R.S. Berry, S.A. Rice, J. Ross (Oxford University Press, 2001)
15. I.D. Boyd, T.E. Scwartzentruber, *Nonequilibrium Gas Dynamics and Molecular Simulations* (Cambridge University Press, 2017)
16. C. Cercignani, *Theory and Application of the Boltzmann Equation* (Scottish Academic Press, 1975)
17. L.D. Landau, E.M. Lifshitz, *Fluid Mechanics* (Pergamon Press, 1986)
18. A.A. Abrikosov, *Fundamentals of the Theory of Metals* (Elsevier, 1988)
19. I.M. Khalatnikov, *An Introduction to the Theory of Superfluidity* (Perseus Publishing, Cambridge, 2000)
20. S.J. Putterman, *Superfluid Hydrodynamics*, North Holand Series in Low Temperature Physics (North-Holland, New York, 1974)
21. M.Y. Kagan, *Modern Trends in Superconductivity and Superfluidity*, Lecture Notes in Physics, vol. 874 (Springer Dordrecht, 2013)
22. L.B. Loeb, *The Kinetic Theory of Gases* (Dover, 2004)
23. J.R. Wilty, C.E. Wicks, R.E. Wilson, G.L. Rorrer, *Fundamentals of Momentum: Heat and Mass Transfer* (Wiley, 2008)
24. S. Chapman, T.G. Cowling, *The Mathematical Theory of Non-Uniform Gases* (Cambridge University Press, 1970)
25. L. Onsager, Reciprocal relations in irreversible processes. Phys. Rev. **37**, 408 (1931)
26. L.D. Landau, E.M. Lifshitz, *Electrodynamics of Continuous Media* (Pergamon Press, 1984)
27. R.P. Feynman, R.B. Leighton, M. Sands, *The Feynman Lectures on Physics*, vol. 4 (Addison-Wesley Publishing Company, Kinetics, Heat, Sound, 1963)

Lecture 14. Spin Waves in Ferromagnets: Exchange Interaction

14

Abstract

We present the foundations of the Heisenberg theory for the exchange interaction and derive the exchange Hamiltonian. We analyze the magnon (spin-wave) spectrum in ferromagnet. We find the temperature dependence of the magnetization and specific heat of the spin subsystem. We consider ferromagnet–paramagnet (FM–PM) phase transition both at the mean-field level and with account of the fluctuations close to the critical point.

At low temperatures many substances acquire the magnetic (spin) order, "working" against the entropy [1–20]. If all the spins are parallel to each other (see Fig. 14.1a), then we get the ferromagnetic (FM) state [1–6].

At the same time, if the checkerboard spin order is realized (see Fig. 14.1b), then we obtain an antiferromagnetism (AFM) in the system [1–6].

At well-defined (critical) temperatures [1–6, 13–25] the magnetic order is destroyed. Sometimes these temperatures can be very large.

14.1   What is the Nature of the Interactions Acting in Magnetic Systems

First of all, we often have a magneto-dipole interaction in magnetic systems [1–6, 8, 26]. However, this interaction usually is rather small. By order of magnitude the magneto-dipole interaction reads:

$$\frac{\mu_1 \mu_2}{r^3} \sim \frac{10^{-40}}{10^{-23}} \sim 10^{-17} \text{erg} \sim (0.1 \div 1)\text{K}, \tag{14.1}$$

where $\mu_1 \sim \mu_2 \sim \mu_B \sim 10^{-20} \frac{\text{erg}}{\text{G}}$ is an electron Bohr magneton.

This weak interaction alone cannot stabilize the magnetic order.

© The Author(s), under exclusive license to Springer Nature Switzerland AG 2026

M. Kagan, *Lecture Notes of Professor Yuri Kagan in Theoretical Solid-State Physics*, Lecture Notes in Physics 1048, https://doi.org/10.1007/978-3-032-14621-2_14

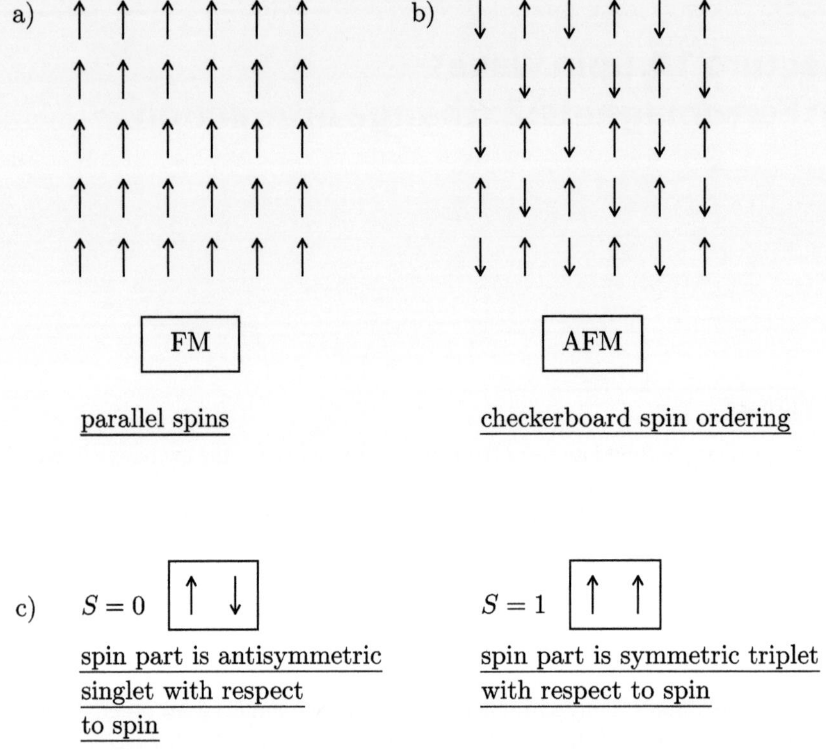

Fig. 14.1 Parallel spin orientation in ferromagnet (**a**) and checkerboard spin order in anti-ferromagnet (**b**). Singlet and triplet state with total spin $S = 0$ and $S = 1$ respectively for two electron spins $\frac{1}{2}$ (**c**). For the singlet state the spin part of the two-particle wave function is antisymmetric, while for the triplet state it is symmetric [1]

It occurs, that magnetic order is stabilized by the exchange interaction (the classical papers on the exchange interaction belong to Heisenberg and Frenkel [27,28]).

Let us consider 2 electrons, occupying two closely located potential wells (see Fig. 14.2). And let us denote the well indices as a and b, and particle indices as 1 and 2 respectively (see Fig. 14.3). Then, particle wave functions in these wells can be written as $\varphi_a(\boldsymbol{r}_1)$ and $\varphi_b(\boldsymbol{r}_2)$. Moreover, if each particle (each electron) has spin $\frac{1}{2}$, then the total spin of two electrons S ($S = S_1 + S_2$) according to Quantum mechanics [29] (see Fig. 14.1c) is either $S = 0$ (singlet $\uparrow\downarrow$) or $S = 1$ (triplet $\uparrow\uparrow$).

Coordinate part of the two-particle wave function has a form [1,29,30]:

$$\Psi_\pm = \frac{1}{\sqrt{2}}\left[\varphi_a(\boldsymbol{r}_1)\,\varphi_b(\boldsymbol{r}_2) \pm \varphi_a(\boldsymbol{r}_2)\,\varphi_b(\boldsymbol{r}_1)\right], \qquad (14.2)$$

where sign + (symmetric coordinate part Ψ_+) according to Pauli principle [2,29, 36,39] corresponds to antisymmetric spin part for the spin singlet $S = 0(\uparrow\downarrow)$. Of course, rigorously speaking spin singlet reads $\frac{1}{\sqrt{2}}(\uparrow\downarrow - \downarrow\uparrow)$.

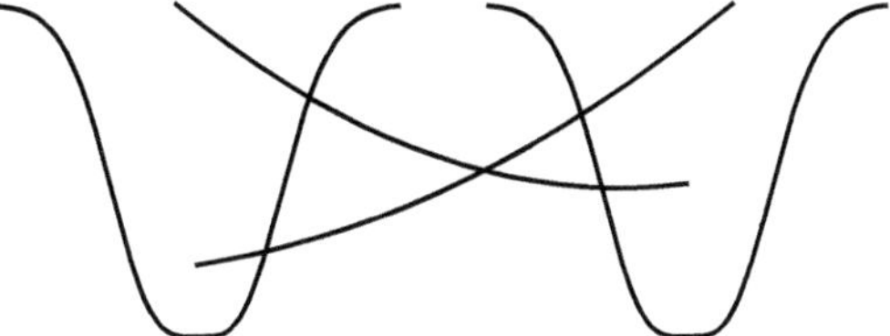

Tails overlapping in the double-well potential

Fig. 14.2 Overlapping of the tails of the electronic wave functions in the double-well potential. If the wells are sufficiently far from each other, then the overlap of the wave functions has an exponential character [1]

$$W_{\text{main int.}} = -\frac{e^2}{|r_1 - R_a|} - \frac{e^2}{|r_2 - R_b|}$$

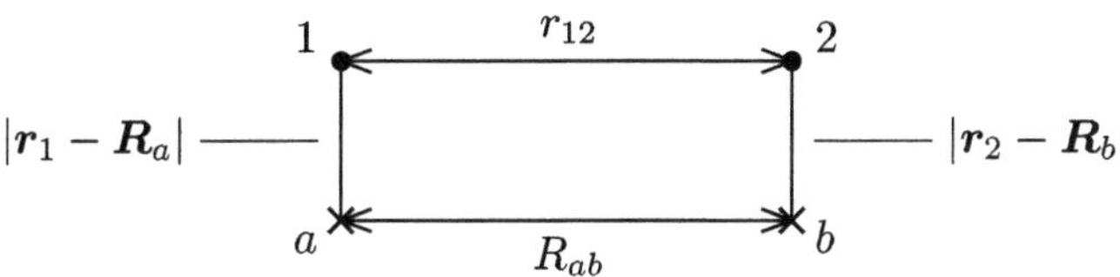

$$\boxed{1,2} \text{ - particle (electron) indices}$$
$$\boxed{a,b} \text{ - potential well (nucleus) indices}$$

Fig. 14.3 Transpositions of particles and wells in the double-well potential. For the hydrogen molecule [30] (1 and 2) are electron coordinates. At the same time (a and b) are nucleus coordinates. The distance between the centers of the wells (between two nuclei in hydrogen molecule) is R_{ab}, and between the electrons r_{12}. In hydrogen molecule r_1, r_2 and R_a, R_b are coordinates of 2 electrons and 2 nuclei respectively. Coordinates r_1 and R_a are related to the first hydrogen atom, while r_2 and R_b—to the second atom. Thus, the main part of the interaction which forms the atoms reads:

$$W_{\text{main int. in mol. } H_2} = -\frac{e^2}{|r_1 - R_a|} - \frac{e^2}{|r_2 - R_b|} \ [1]$$

At the same time, sign-(antisymmetric coordinate part Ψ_-) is related to symmetric spin part for the spin triplet $S = 1(\uparrow\uparrow)$.

Note, that the second term in (14.2), proportional to the product $\varphi_a(r_2)\varphi_b(r_1)$, emerges due to the fact, that potential wells are close to each other, and thus particle 1 can be located not only in the well a, but in the well b as well.

Let us consider the pairwise interaction of two electrons $V(r_1 - r_2)$ in the system. Then the simplest (first order in interaction) Hartree–Fock correction to the energy [1,2,8,9,31–35], presented in Fig. 14.4 on the left, acquires a form:

$$\Delta E = \langle \Psi_\pm | V(r_1 - r_2) | \Psi_\pm \rangle = A \pm J, \tag{14.3}$$

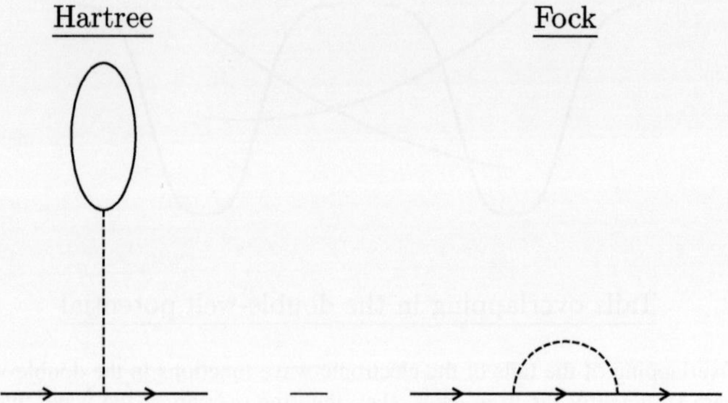

Fig. 14.4 The simplest linear in the interaction corrections, to the energy [1,2,31–35]. On the left, in the figure, we show the Hartree positive correction. On the right we show the Fock exchange correction which is negative [1]

where the integral A of the density-density type diagonal on one-particle wave functions reads:

$$A = \frac{1}{2} \int |\varphi_a(\boldsymbol{r}_1)|^2 \, |\varphi_b(\boldsymbol{r}_2)|^2 \, V(\boldsymbol{r}_1 - \boldsymbol{r}_2) \, d^3 r_1 d^3 r_2, \qquad (14.4)$$

while non-diagonal exchange interaction (exchange integral) J is given by:

$$J = \frac{1}{2} \int \varphi_a(\boldsymbol{r}_1) \, \varphi_b(\boldsymbol{r}_1) \, V(\boldsymbol{r}_1 - \boldsymbol{r}_2) \, \varphi_b^*(\boldsymbol{r}_2) \, \varphi_a^*(\boldsymbol{r}_2) \, d^3 r_1 d^3 r_2. \qquad (14.5)$$

Exchange interaction proves to be proportional to the overlap integral (see Fig. 14.2):

$$J \sim I^2, \qquad (14.6)$$

where the overlap integral:

$$I = \int d^3 r_1 \varphi_a(\boldsymbol{r}_1) \, \varphi_b(\boldsymbol{r}_1). \qquad (14.7)$$

At large distance between the wells the overlap integral is exponentially small. Indeed, if the binding energies of the particle in the wells a and b:

$$E_{\text{bind.}\,a} = \frac{\hbar^2 \kappa_a^2}{2m_a}, \; E_{\text{bind.}\,b} = \frac{\hbar^2 \kappa_b^2}{2m_b}, \qquad (14.8)$$

where κ_a and κ_b are the wave vectors related to the size of the bound state in each well, and R_{ab} is the distance between the wells centers, then one-particle wave functions, determining J in Eq. (14.5), are exponentially small as:

$$\varphi_a \sim \exp\left(-\frac{\kappa_a R_{ab}}{2}\right), \varphi_b \sim \exp\left(-\frac{\kappa_b R_{ab}}{2}\right). \tag{14.9}$$

Hence, the overlap integral:

$$I \sim \exp\left(-\frac{\kappa_a R_{ab}}{2}\right)\exp\left(-\frac{\kappa_b R_{ab}}{2}\right), \tag{14.10}$$

and as a result, exchange interaction can be estimated as follows [1,2,29,30]:

$$J \sim I^2 \sim \exp\left(-\kappa_a R_{ab}\right)\exp\left(-\kappa_b R_{ab}\right) = \exp\left[-\left(\kappa_a + \kappa_b\right) R_{ab}\right]. \tag{14.11}$$

If we assume that the exchange integral is positive ($J > 0$), then $J = |J|$ and the negative correction to the energy $\Delta E_{\min} = A - |J|$ corresponds to the two-particle wave function Ψ_- (i.e., to the triplet with respect to the total electron spin $S = 1$).

However, this is impossible, since the ground state should not have any nodes [1,2,29,30]. Thus, the ground state corresponds to $\Psi_{\mathrm{gr.st.}} = \Psi_+$ (i.e., to the singlet with respect to the total electron spin $S = 0$). In this case the exchange integral is negative:

$$J = -|J| < 0.$$

Indeed, this is a case for the hydrogen molecule H_2 [29,30]. The bound state appears only at $S = 0$ (see Fig. 14.5).

Note, that residual interaction (which does not participate in the formation of 2 hydrogen atoms) for the H_2 molecule, consisting of 2 nuclei (2 protons) and 2 electrons, in the notations of Fig. 14.3 reads:

$$W_{\mathrm{res.\ int.}} = -\frac{e^2}{|r_1 - R_b|} - \frac{e^2}{|r_2 - R_a|} + \frac{e^2}{r_{12}} + \frac{e^2}{R_{ab}}, \tag{14.12}$$

where R_{ab} is the distance between 2 nuclei in hydrogen molecule, and r_{12} is the distance between 2 electrons. Meanwhile $|r_1 - R_b|$ is the distance between the second nucleus and the first electron, whereas $|r_2 - R_a|$ is the distance between the first nucleus and the second electron respectively. First 2 terms in the residual interaction correspond to the Coulomb attraction between electrons and nuclei. At the same time, the third and the fourth term are related to the Coulomb repulsion between 2 electrons and 2 nuclei. Let us stress, that the main part of the interaction, which provides the formation of hydrogen atoms, corresponds to the diagonal Coulomb attraction between electrons and nuclei of the same well (sitting at one center) in Fig. 14.3:

$$W_{\text{main int.}} = -\frac{e^2}{|\boldsymbol{r}_1 - \boldsymbol{R}_a|} - \frac{e^2}{|\boldsymbol{r}_2 - \boldsymbol{R}_b|}. \tag{14.13}$$

It occurs, that the crossed terms in (14.12), related to the interaction between electrons and nuclei of different centers, yield the largest contribution to the first correction to the system energy. They are given by:

$$-\frac{e^2}{|\boldsymbol{r}_1 - \boldsymbol{R}_b|} - \frac{e^2}{|\boldsymbol{r}_2 - \boldsymbol{R}_a|}.$$

As a result, in the Heitler–London theory [1,2,29,30] for the hydrogen molecule the singlet and triplet term in Fig. 14.5 are specified by the total spin I of two atoms. Moreover, the total spin of each atom is formed by a sum of nuclear and electron spins. At the same time, potential energy of singlet and triplet terms is mainly governed by the internuclei distance R_{ab} (see Fig. 14.5) due to adiabatic tracking principle considered in Lecture 6.

Let us return back now to more simple problem with 2 potential wells and 2 electrons (2 nuclei and, correspondingly, 2 nuclear spins are absent).

Then, for the negative exchange integral $J < 0$ the energy correction is positive and maximal for the coordinate wave function Ψ_-:

$$\Delta E_{\max} = A - J = A + |J|. \tag{14.14}$$

At the same time, the energy correction is negative and minimal for the ground state coordinate wave function $\Psi_{\text{gr.st.}} = \Psi_+$:

$$\Delta E_{\min} = A + J = A - |J|. \tag{14.15}$$

Now we can understand qualitatively how the exchange interaction between 2 atoms with large number of electrons is formed. It emerges due to the exchange between all the electron pairs, when one of the electrons of the first atoms exchanges with one of the electrons of the second atom.

In this case the exchange interaction for the transposition of two electron pairs is much smaller than the exchange interaction for the transposition of one pair:

$$J_{\text{transp.of 2 pairs}} = \left(J_{\text{transp.of 1 pair}}\right)^2 \sim \exp\left[-2\left(\kappa_a + \kappa_b\right)R_{ab}\right] \ll J_{\text{transp.of 1 pair}}. \tag{14.16}$$

In general case the exchange interaction between two spins $\frac{1}{2}$ can be written in the following form which is independent of coordinates [1,2,27–30]:

$$V_{\text{ex.}} = \alpha + \beta\left(\boldsymbol{S}_1 \boldsymbol{S}_2\right), \tag{14.17}$$

where the scalar product $(\boldsymbol{S}_1 \boldsymbol{S}_2)$ of 2 pseudovectors $\boldsymbol{S}_1$ and $\boldsymbol{S}_2$ is the only possible scalar combination in the system, and $\alpha = \text{const}$

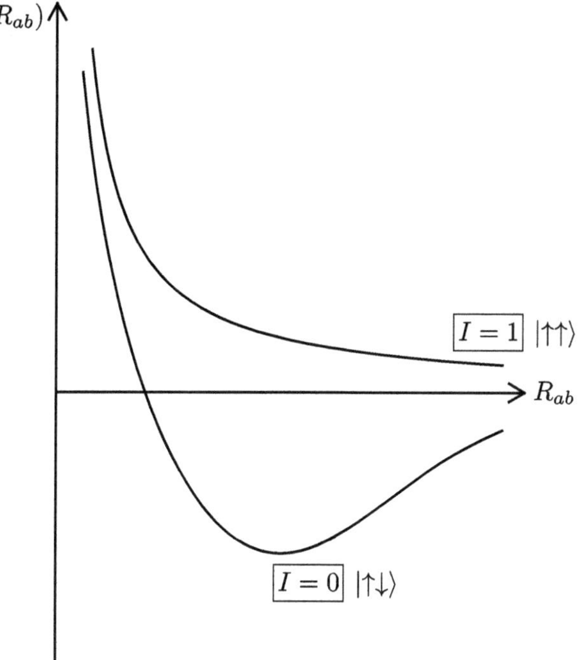

Fig. 14.5 Singlet ($I = 0$) and triplet ($I = 1$) spin terms for the interaction of 2 hydrogen atoms in Heitler–London theory [1,2,29,30]. For singlet term (with respect to total spin I) the bound state of two particles (two atoms in hydrogen molecule) can arise. For triplet term the bound state is absent. Total spin of each atom is formed by a sum of nuclear and electron spin, I is the total spin of two hydrogen atoms, R_{ab} is internuclei distance [1]

We can show meanwhile, that the higher order terms, e.g., quartic product $(S_1 S_2)^2$ for two spins $\frac{1}{2}$, can be reduced again to the sum of the scalar and the scalar product of two spins as in Eq. (14.17):

$$(S_1 S_2)^2 = \gamma + \delta\,(S_1 S_2). \tag{14.18}$$

Now we can calculate the diagonal correction to the energy in (14.3) for the general form of the exchange interaction (14.17). To do that we have to find the average from the scalar product in (14.17).

According to Quantum mechanics [29], we get in the operator form:

$$\hat{S}^2 = \left(\hat{S}_1 + \hat{S}_2\right)^2 = \hat{S}_1^2 + \hat{S}_2^2 + 2\left(\hat{S}_1 \hat{S}_2\right), \tag{14.19}$$

Hence, the scalar product of the spin operators equals to:

$$\left(\hat{\mathbf{S}}_1\hat{\mathbf{S}}_2\right) = \frac{1}{2}\left(\hat{\mathbf{S}}^2 - \hat{\mathbf{S}}_1^2 - \hat{\mathbf{S}}_2^2\right). \tag{14.20}$$

Further on let us find the average from the l.h.s. and r.h.s. of Eq. (14.20) (with respect to the eigenvalues) proceeding to the eigenstates of the operators of the spin squared in the r.h.s.:

$$\left\langle\left(\hat{\mathbf{S}}_1\hat{\mathbf{S}}_2\right)\right\rangle = \frac{1}{2}\left\langle\left(\hat{\mathbf{S}}^2 - \hat{\mathbf{S}}_1^2 - \hat{\mathbf{S}}_2^2\right)\right\rangle, \tag{14.21}$$

where:

$$\left\langle\hat{\mathbf{S}}^2\right\rangle = S(S+1), \left\langle\hat{\mathbf{S}}_1^2\right\rangle = \left\langle\hat{\mathbf{S}}_2^2\right\rangle = \frac{1}{2}\cdot\frac{3}{2} = \frac{3}{4}. \tag{14.22}$$

Then for the spin singlet $S = 0$:

$$\left\langle\left(\hat{\mathbf{S}}_1\hat{\mathbf{S}}_2\right)\right\rangle = \frac{1}{2}\left(0 - \frac{3}{4} - \frac{3}{4}\right) = -\frac{3}{4}. \tag{14.23}$$

At the same time, for the spin triplet $S = 1$:

$$\left\langle\left(\hat{\mathbf{S}}_1\hat{\mathbf{S}}_2\right)\right\rangle = \frac{1}{2}\left(1\cdot 2 - \frac{3}{4} - \frac{3}{4}\right) = \frac{1}{4}. \tag{14.24}$$

As a result, taking into account the normalization condition for the two-particle wave function, the diagonal corrections to the energy acquire a form:

$$\Delta E_{\min} = \alpha - \frac{3}{4}\beta = A - |J| = A + J \tag{14.25}$$

for the spin singlet $S = 0$, and

$$\Delta E_{\max} = \alpha + \frac{1}{4}\beta = A + |J| = A - J \tag{14.26}$$

for the spin triplet $S = 1$.

Solving these 2 equations with respect to A and J, we get:

$$2A = 2\alpha - \frac{1}{2}\beta, 2J = -\beta. \tag{14.27}$$

If we would like to select purely exchange part, then we have to put $A = 0$. In this case:

$$2\alpha - \frac{1}{2}\beta = 0, \text{ and } \alpha = \frac{1}{4}\beta = -\frac{J}{2}. \tag{14.28}$$

Then the exchange interaction in (14.17) can be represented in terms of the exchange integral J in canonical form as:

$$V_{\text{ex.}} = \alpha + \beta\left(S_1 S_2\right) = \beta\left(\frac{1}{4} + (S_1 S_2)\right) = -2J\left(\frac{1}{4} + S_1 S_2\right). \qquad (14.29)$$

Note, that a very important question is connected with the sign of the overlap integral. For hydrogen the sign of the overlap integral is clear.

However, for more complicated atoms the answer to this question is not clear, since the electron wave functions of the inner shells in the complicated atom are strongly oscillating (to get deeper–that is closer to the atomic core, the wave functions of the inner shells should have large quantum numbers, otherwise they will be repelled outward).

14.2 Heisenberg Hamiltonian

Heisenberg Hamiltonian serves as a basic Hamiltonian for the description of magnetic systems [1,2,27–29]. Heisenberg Hamiltonian has a form:

$$\widehat{H} = -\frac{1}{2}\sum_{n,m} J_{n,m}\,\hat{S}_n \hat{S}_m - g\mu_B B \sum_n \hat{S}_n^z, \qquad (14.30)$$

where $\hat{S}_n$ and $\hat{S}_m$ are operators of electron spin $\frac{1}{2}$ on sites n and m, $J_{n,m}$ are exchange integrals between the spins located on sites n and m.

Note that, in the exchange integrals $J_{n,m}$ enter the overlap integrals and that is why $n \neq m$.

In Eq. (14.30) μ_B is electron Bohr magneton, g is gyromagnetic ratio, B is magnetic field directed along z-axis, $\hat{S}_n^Z$ are operators of the projection of the spin (located on site n) on z-axis.

If the main exchange integrals are positive ($J_{n,m} > 0$), then at $T = 0$ the ground state of the system corresponds to the configuration with parallel spin directions.

As a result, the ferromagnetic state (FM) is realized.

In the absence of the magnetic field, we have an infinite degeneracy in the system (there is no distinguished direction), and hence the total spin equals zero, since there is no preferred direction for its orientation.

Thus, before averaging we should violate the symmetry. To do that first of all we should apply very weak magnetic field and only after the averaging put the magnetic field equal zero.

We can utilize meanwhile convenient canonical form for the scalar product of spins in Heisenberg Hamiltonian [1,2,27–29]. Let us write at first the scalar product of spins in the Cartesian coordinates:

$$S_n S_m = S_n^z S_m^z + S_n^x S_m^x + S_n^y S_m^y. \qquad (14.31)$$

Further on, let us introduce according to Quantum mechanics [29] raising and lowering spin operators:

$$S_n^{\pm} = S_n^x \pm i S_n^y. \tag{14.32}$$

Then transverse spin components (relative to the magnetic field direction) can be represented as:

$$S_n^x = \frac{1}{2}\left(S_n^{\dagger} + S_n^{-}\right), S_n^y = \frac{1}{2i}\left(S_n^{\dagger} - S_n^{-}\right). \tag{14.33}$$

Correspondingly for the bilinear combinations of the transverse spin component in Eq. (14.31) we get:

$$S_n^x S_m^x + S_n^y S_m^y = \frac{1}{2}\left(S_n^{\dagger} S_m^{-} + S_n^{-} S_m^{\dagger}\right). \tag{14.34}$$

As a result, the scalar product of spins:

$$\mathbf{S}_n \mathbf{S}_m = S_n^z S_m^z + \frac{1}{2}\left(S_n^{\dagger} S_m^{-} + S_n^{-} S_m^{\dagger}\right), \tag{14.35}$$

and the Heisenberg Hamiltonian reads:

$$\widehat{H} = -\frac{1}{2}\sum_{n,m} J_{n,m}\hat{\mathbf{S}}_n \hat{\mathbf{S}}_m - g\mu_B B \sum_n S_n^z$$

$$= -\frac{1}{2}\sum_{n,m} J_{n,m}\left[S_n^z S_m^z + \frac{1}{2}\left(S_n^{\dagger} S_m^{-} + S_n^{-} S_m^{\dagger}\right)\right] - g\mu_B B \sum_n S_n^z. \tag{14.36}$$

Let us emphasize that spins located on different sites commute with each other. Moreover, according to our assumption, the lattice consists of identical atoms. Hence the exchange integrals are symmetric with respect to the transposition of indices (of the sites n and m):

$$J_{n,m} = J_{m,n}. \tag{14.37}$$

Hence in the second term in round brackets in (14.40) we can transpose the spins (the site numbers) and simplify the Heisenberg Hamiltonian:

$$\widehat{H} = -\frac{1}{2}\sum_{n,m} J_{n,m}\hat{\mathbf{S}}_n \hat{\mathbf{S}}_m - g\mu_B B \sum_n S_n^z$$

$$= -\frac{1}{2}\sum_{n,m} J_{n,m}\left[S_n^z S_m^z + S_n^{-} S_m^{\dagger}\right] - g\mu_B B \sum_n S_n^z. \tag{14.38}$$

14.3 Ground State of the System at $T = 0$

Let us determine the energy of the ground state of the system at $T = 0$.

At $T = 0$ in Quantum mechanics, as in the classical picture, all the spins are oriented parallel to each other, and the operator of the projection of the total spin on z-axis:

$$\hat{S}^Z = \sum_n \hat{S}_n^Z \tag{14.39}$$

commutes with the Hamiltonian of the system:

$$\left[\hat{H}\hat{S}^z\right] = 0. \tag{14.40}$$

This fact allows us to classify the states of the system by the total spin.

Meanwhile the spin wave function of the whole system can be represented as a product of the one-particle spin wave functions:

$$\chi = \prod_n \chi_{S^z=S}^n, \tag{14.41}$$

where $\chi_{S^z=S}^n$ is an eigenfunction of the operator of the spin projection S_n^z on z-axis at the site n, and moreover the eigenvalue of the operator $\hat{S}_n^z$ is maximal: $S^z = S$.

Now we can determine the average of the scalar product $\hat{S}_n \hat{S}_m$ in the ground state. Suppose that the projection of the operator of the total spin on z-axis for two sites n and m:

$$\hat{S}_{\text{tot.}} = \hat{S}_n + \hat{S}_m \tag{14.42}$$

equals to $S_{\text{tot.}}^z$. Then we can calculate the average:

$$\left\langle S_{\text{tot.}}^z | \hat{S}_n \hat{S}_m | S_{\text{tot.}}^z \right\rangle, \tag{14.43}$$

utilizing the scalar product of the operators:

$$\left\langle S_{\text{tot.}}^z | \hat{S}_n \hat{S}_m | S_{\text{tot.}}^z \right\rangle = \frac{1}{2} \left\langle S_{\text{tot.}}^z | \left(\hat{S}_n + \hat{S}_m\right)^2 - \hat{S}_n^2 - \hat{S}_m^2 | S_{\text{tot.}}^z \right\rangle$$

$$= \frac{1}{2} \left\langle S_{\text{tot.}}^z | \hat{S}_{\text{tot.}}^2 - \hat{S}_n^2 - \hat{S}_m^2 | S_{\text{tot.}}^z \right\rangle. \tag{14.44}$$

Further on proceeding to the eigenvalue of the operators of spin squared, we obtain:

$$\left\langle S_{\text{tot.}}^z | \hat{S}_n \hat{S}_m | S_{\text{tot.}}^z \right\rangle = \frac{1}{2} \left[S_{\text{tot.}} \left(S_{\text{tot.}} + 1 \right) - 2S_0 \left(S_0 + 1 \right) \right], \tag{14.45}$$

where $S_0 = \frac{1}{2}$.

We can see, that the maximum of this expression (and thus the minimum of the Hamiltonian (14.38)) is realized for $S_{\text{tot.}} = 1$. In this case:

$$\left\langle S_{\text{tot.}}^z | \hat{\mathbf{S}}_n \hat{\mathbf{S}}_m | S_{\text{tot.}}^z \right\rangle = \frac{1}{2}\left(1 \cdot 2 - 2 \cdot \frac{1}{2} \cdot \frac{3}{2}\right) = \frac{1}{4} = S_0^2. \tag{14.46}$$

At the same time, the minimum of this expression (and maximum of the Hamiltonian (14.38) respectively) corresponds to $S_{\text{tot.}} = 0$, and yields:

$$\left\langle S_{\text{tot.}}^z | \hat{\mathbf{S}}_n \hat{\mathbf{S}}_m | S_{\text{tot.}}^z \right\rangle = \frac{1}{2}\left(0 \cdot 1 - 2 \cdot \frac{1}{2} \cdot \frac{3}{2}\right) = -\frac{3}{4}. \tag{14.47}$$

Now we can calculate the energy of the system at minimum, or in other words to find the ground state energy of the system. To do that in the Hamiltonian (14.38) we proceed from the summation over n and m to the summation over n and $p = n - m$.

Then, taking into account that the exchange integral $J_{n,m} = J(\mathbf{n} - \mathbf{m})$ depends only on the difference $\mathbf{p} = \mathbf{n} - \mathbf{m}$, we get

$$E_0 = -\frac{1}{2} S_0^2 N \sum_{\mathbf{p}} J(\mathbf{p}) - g\mu_B B S_0 N, \tag{14.48}$$

where N is the total number of sites in the system.

The calculation which we perform shows that the ground state of the ferromagnet does not have the fluctuating component. The contribution to the ground state energy connected with the transverse components $S_n^- S_m^\dagger$ equals zero, since all the spins are oriented in the same direction and we have no spin states to raise.

For antiferromagnet the exchange integrals are negative: $J_{n,m} < 0$. Then, it is convenient for the spins to be oriented antiparallel [1–12]. As a result, the emergent system has two sublattices embedded in each other (see Fig. 14.6a).

Let us stress, that in contrast to the ferromagnet, zero oscillations, related to the transverse components $S_n^- S_m^\dagger$ in the Hamiltonian, are allways present in the antiferromagnet. Effectively they permanently transpose the neighboring spins belonging to the different sublattices by raising one of them while lowering the other.

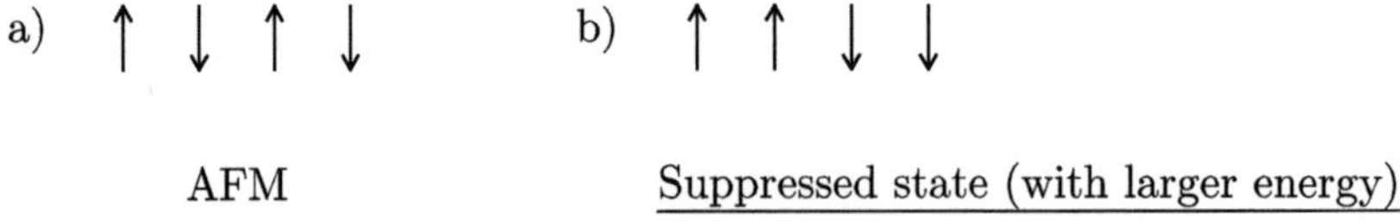

Fig. 14.6 a Checkerboard configuration of spins in the antiferromagnet (AFM). On **b** we show the suppressed spin configuration which corresponds to the larger energy and does not arise in the antiferromagnet even virtually at low temperatures [1]

Hence, we cannot think about the antiferromagnet in terms of two independent sublattices.

Note, that the spin configuration, presented in Fig. 14.6b, is totally suppressed in the antiferromagnet. It cannot arise even virtually with an account of quantum corrections $S_n^- S_m^{\dagger}$ at low temperatures, since it acquires an energy which is larger on $|J|$ by order of magnitude than the checkerboard spin configuration presented in Fig. 14.6a.

14.4 Magnons in Ferromagnet at Low Temperatures

Let us consider now the temperatures different from zero ($T \neq 0$). We assume, nevertheless, that the temperatures are low and thus the excited state is only slightly different from the ground state. In this case we have gas of excitations (with small number of the quasiparticles—which are magnons). In this section we assume the value of spin S in Heisenberg Hamiltonian to be arbitrary (and not necessarily equal to $\frac{1}{2}$)

The main term, describing magnons, is connected with quantum corrections $S_n^- S_m^{\dagger}$. It leads to the running wave for the raising and lowering spin operators similar to the sound wave for the displacement operators in the phonon subsystem. Fourier expansion for these operators yields:

$$S_n^{\pm} = \frac{1}{\sqrt{N}} \sum_q S_q^{\pm} \exp(\pm i\boldsymbol{q}\boldsymbol{n}). \tag{14.49}$$

Hence, the Fourier transformed operators read:

$$S_q^{\pm} = \frac{1}{\sqrt{N}} \sum_n S_n^{\pm} \exp(\mp i\boldsymbol{q}\boldsymbol{n}). \tag{14.50}$$

Note, that the Fourier components of lowering and raising operators represent some spin objects, smeared on the lattice sites and having the same normalization conditions (for the square root from the number of sites: $S_q^{\pm} \sim \frac{1}{\sqrt{N}}$), as the displacement operators.

It means, that if we rotate the smeared spin (the Fourier component $S_q^{\pm}$), then effectively under this translation each spin (spin on each site) rotates only insignificantly. In other words, it does not acquire the value of $-\frac{1}{2}$ instead of $+\frac{1}{2}$, or vice versa, thus varying on 1. It varies only on $\frac{1}{\sqrt{N}}$.

We will show soon, that as a consequence of this small variation, the operators $S_q^{\pm}$ similarly to phonon creation and annihilation operators satisfy bosonic commutation relations.

As a result, expanding bilinear term $S_n^- S_m^{\dagger}$ in the Fourier series, we get:

$$S_n^- S_m^\dagger = \frac{1}{N} \sum_{q,q_1} S_q^- S_{q_1}^\dagger \exp\left(-i q n + i q_1 m\right). \tag{14.51}$$

Correspondingly, the quantum corrections term in the Heisenberg Hamiltonian can be written as:

$$-\frac{1}{2} \sum_{n,m} J_{n,m} S_n^- S_m^\dagger = -\frac{1}{2N} \sum_{n,m} J_{n,m} \sum_{q,q_1} S_q^- S_{q_1}^\dagger \exp\left(-i q n + i q_1 m\right). \tag{14.52}$$

Further on we proceed to the summation over $p = n - m$ and m. To do that we utilize the relation for the exponents:

$$\exp\left(-i q n + i q_1 m\right) = \exp\left[-i q (n - m) - i \left(q - q_1\right) m\right]$$
$$= \exp\left[-i q p - i \left(q - q_1\right) m\right], \tag{14.53}$$

and the fact, that the exchange integral depends only on p:

$$J_{n,m} = J(p).$$

Then we obtain:

$$-\frac{1}{2} \sum_{n,m} J_{n,m} S_n^- S_m^\dagger = -\frac{1}{2N} \sum_{q,q_1} S_q^- S_{q_1}^\dagger \sum_{p,m} J(p) \exp(-i q p) \exp\left(-i \left(q - q_1\right) m\right)$$
$$= -\frac{1}{2} \sum_q \left(\sum_p J(p) \exp(-i q p) \right) S_q^- S_q^\dagger, \tag{14.54}$$

where we took into account, that:

$$\sum_{p,m} J(p) \exp(-i q p) \exp\left(-i \left(q - q_1\right) m\right) = \delta_{q,q_1} N \sum_p J(p) \exp(-i q p). \tag{14.55}$$

14.5 Commutation Relations for Raising and Lowering Operators

Let us establish the commutation relations for the Fourier components S_q^- and $S_{q_1}^\dagger$ and thus demonstrate their bosonic character.

To do that let us utilize the inverse Fourier transformation for the commutator:

$$\left[S_q^\dagger S_{q_1}^-\right] = S_q^\dagger S_{q_1}^- - S_{q_1}^- S_q^\dagger = \frac{1}{N} \sum_{n,m} \left(S_n^\dagger S_m^- - S_m^- S_n^\dagger\right) \exp\left(-i q n + i q_1 m\right). \tag{14.56}$$

Now we can use the commutation relations between raising and lowering spin operators in real space which we know from Quantum mechanics [29]:

$$S_n^\dagger S_m^- - S_m^- S_n^\dagger = \delta_{n,m} \cdot 2S_n^z, \tag{14.57}$$

where we accounted, that raising and lowering operators on different sites (for $n \neq m$) commute with each other, while for the operators on the same site ($n = m$) their commutator is given by:

$$\left[S_n^\dagger S_n^- \right] = 2S_n^z \tag{14.58}$$

Note, that in Quantum mechanics for the generators of the rotation group (for projections of the rotating moment J) always holds the relation [29]:

$$\left[J^\dagger J^- \right] = 2\, J^z.$$

Substituting (14.57) in the expression in the r.h.s. of Eq. (14.56), we get:

$$\left[S_q^\dagger S_{q_1}^- \right] = \frac{1}{N} \sum_n 2S_n^z \exp\left(-i\,(q - q_1)\,n\right) = \frac{2\,S}{N} \sum_n \exp\left(-i\,(q - q_1)\,n\right)$$
$$= 2\,S\delta_{q,q_1}. \tag{14.59}$$

Let us emphasize, that for small vectors q (when $qa \ll 1$, a is intersite distance) the spin phases are equal for large spatial region. Then under the summation over large region in (14.59) the fluctuations in this region are small and weakly excited state is close to the ground state. As a result, we can replace S_n^z on S in (14.59).

Now we can introduce creation and annihilation operators of magnons:

$$S_q^\dagger = \sqrt{2\,S}a_q,\ S_q^- = \sqrt{2\,S}a_q^\dagger, \tag{14.60}$$

and get from Eq. (14.59) bosonic commutation relations for them:

$$\left[a_q a_{q_1}^\dagger \right] = \delta_{q,q_1}. \tag{14.61}$$

Let us stress once more, that the Fourier components do not feel the spin discreteness ($\frac{1}{2}$ or $-\frac{1}{2}$), since they are smeared on all the lattice sites. That is why $S_q^\pm$ effectively varies the spin value not on 1, but only on $\frac{1}{\sqrt{N}}$. Thus, as bosonic operator, it can infinitely raise or low the spin (in fact, it stops working only after $\sqrt{N}$ raisings).

Hence the Heisenberg Hamiltonian can be written as:

$$\widehat{H} = \widehat{H}_{zz} + \widehat{H}_\pm - g\mu_B B \sum_n S_n^z, \tag{14.62}$$

where the diagonal part:

$$\widehat{H}_{zz} = -\frac{1}{2} \sum_{n,m} J_{n,m} S_n^z S_m^z, \tag{14.63}$$

while the term, connected with the quantum corrections, can be expressed via the operator of the magnon number density $a_q^\dagger a_q$, and equls to:

$$\widehat{H}_{\pm} = -S \sum_q \left(\sum_p J(p) \exp(-i\boldsymbol{q}\,\boldsymbol{p}) \right) a_q^\dagger a_q. \tag{14.64}$$

Let us analyze at first the diagonal term $\widehat{H}_{zz}$. To do that let us first write the operator $(\hat{S}_n^z)^2$ using the Pythagorean theorem and spin conservation.

Then, proceeding to the eigenvalues for the operator of spin squared: $\hat{\boldsymbol{S}}^2 = S(S+1)$, we can write:

$$(\hat{S}_n^z)^2 = \hat{\boldsymbol{S}}^2 - (\hat{S}_n^x)^2 - (\hat{S}_n^y)^2 = S(S+1) - (\hat{S}_n^x)^2 - (\hat{S}_n^y)^2. \tag{14.65}$$

Further on introducing the lowering and raising operators on the given site, we obtain:

$$(\hat{S}_n^z)^2 = S(S+1) - \frac{1}{2}\left(S_n^- S_n^\dagger + S_n^\dagger S_n^-\right). \tag{14.66}$$

Now we can proceed in the second term in the r.h.s. of (14.66) to the Fourier components of the lowering and raising operators:

$$(\hat{S}_n^z)^2 = S(S+1) - \frac{1}{2N} \sum_{q,q_1} \left[S_q^- S_{q_1}^\dagger + S_{q_1}^\dagger S_q^- \right] \exp\left(i\left(\boldsymbol{q} - \boldsymbol{q}_1\right)\boldsymbol{n}\right). \tag{14.67}$$

After that we can replace the Fourier components in (14.67) on the operators of magnon creation and annihilation according to Eq. (14.60):

$$(\hat{S}_n^z)^2 = S(S+1) - \frac{S}{N} \sum_{q,q_1} \left[a_q^\dagger a_{q_1} + a_{q_1} a_q^\dagger \right] \exp\left(i\left(\boldsymbol{q} - \boldsymbol{q}_1\right)\boldsymbol{n}\right). \tag{14.68}$$

Now we can utilize the commutation relations for bosonic operators (14.61) and make a substitution:

$$a_{q_1} a_q^\dagger = a_q^\dagger a_{q_1} + \delta_{q,q_1} \tag{14.69}$$

in the second term in the square brackets. As a result:

$$(\hat{S}_n^z)^2 = S(S+1) - \frac{S}{N} \sum_{q,q_1} \left[2a_q^\dagger a_{q_1} + \delta_{q,q_1} \right] \exp\left(i\left(q - q_1\right)n\right)$$

$$= S(S+1) - \frac{S}{N} \sum_q 1 - \frac{2S}{N} \sum_{q,q_1} a_q^\dagger a_{q_1} \exp\left(i\left(q - q_1\right)n\right). \tag{14.70}$$

Further on we can use the fact, that $\sum_q 1 = N$, and thus:

$$S(S+1) - \frac{S}{N} \sum_q 1 = S(S+1) - S = S^2. \tag{14.71}$$

Then finally:

$$(\hat{S}_n^z)^2 = S^2 - \frac{2S}{N} \sum_{q,q_1} a_q^\dagger a_{q_1} \exp\left(i\left(q - q_1\right)n\right). \tag{14.72}$$

If we represent now the operator $(\hat{S}_n^z)^2$ as:

$$\hat{S}_n^z = \sqrt{(\hat{S}_n^z)^2}, \tag{14.73}$$

and utilize the square root expansion in series up to the linear terms $\sqrt{1+x} \approx 1 + \frac{1}{2}x$ for $x \ll 1$, then we get from the expression (14.72):

$$\hat{S}_n^z = \sqrt{(\hat{S}_n^z)^2} \approx S - \frac{1}{N} \sum_{q,q_1} a_q^\dagger a_{q_1} \exp\left(i\left(q - q_1\right)n\right). \tag{14.74}$$

Hence the diagonal part of the Heisenberg Hamiltonian acquires a form:

$$\widehat{H}_{zz} = -\frac{1}{2} \sum_{n,m} J_{n,m} S_n^z S_m^z = -\frac{1}{2} \sum_{n,m} J_{n,m} \left[S - \frac{1}{N} \sum_{q,q_1} a_q^\dagger a_{q_1} \exp\left(i\left(q - q_1\right)n\right) \right]$$

$$\times \left[S - \frac{1}{N} \sum_{q,q_1} a_q^\dagger a_{q_1} \exp\left(i\left(q - q_1\right)m\right) \right] = -\frac{1}{2} \sum_{n,m} J_{n,m} S^2$$

$$+ \sum_{n,m} J_{n,m} \frac{S}{2N} \sum_{q,q_1} a_q^\dagger a_{q_1} \left[\exp\left(i\left(q - q_1\right)n\right) + \exp\left(i\left(q - q_1\right)m\right) \right]$$

$$+ O\left(\frac{1}{N^2}\right). \tag{14.75}$$

Further on using the symmetry condition for the exchange integrals $J_{n,m} = J_{m,n}$, we obtain for the diagonal part of the Hamiltonian:

$$\widehat{H}_{zz} = -\frac{1}{2}\sum_{n,m} J_{n,m} S^2 + \sum_{n,m} J_{n,m} \frac{S}{N} \sum_{q,q_1} a_q^\dagger a_{q_1} \exp\left(i\left(\boldsymbol{q} - \boldsymbol{q}_1\right)\boldsymbol{n}\right). \tag{14.76}$$

Proceeding to the summation on the variables $\boldsymbol{p} = \boldsymbol{n} - \boldsymbol{m}$ and $\boldsymbol{n}$, we can write (14.76) as:

$$\widehat{H}_{zz} = -\frac{1}{2}\sum_{n,m} J_{n,m} S^2 + \frac{S}{N}\sum_{p,n} J(\boldsymbol{p}) \sum_{q,q_1} a_q^\dagger a_{q_1} \exp\left(i\left(\boldsymbol{q} - \boldsymbol{q}_1\right)\boldsymbol{n}\right). \tag{14.77}$$

Calculating the sum over $\boldsymbol{n}$ in the second term (which equals to $N\delta_{q,q_1}$) and lifting the summation over q_1, we can rewrite (14.77) in the canonical form:

$$\widehat{H}_{zz} = -\frac{1}{2}\sum_{n,m} J_{n,m} S^2 + S\sum_{p} J(\boldsymbol{p}) \sum_{q} a_q^\dagger a_q. \tag{14.78}$$

Combining expressions (14.78) for $\widehat{H}_{zz}$ and (14.64) for $\widehat{H}_{\pm}$, we can represent the Heisenberg Hamiltonian as follows:

$$\widehat{H} = \widehat{H}_{zz} + \widehat{H}_{\pm} - g\mu_B B^z \sum_n S_n^z = \left(-\frac{1}{2}\sum_{n,m} J_{n,m} S^2 - g\mu_B B \sum_n \hat{S}_n^z\right)$$
$$+ S\sum_q a_q^\dagger a_q \left\{\sum_p J(\boldsymbol{p})(1 - \exp(-i\boldsymbol{q}\,\boldsymbol{p}))\right\}. \tag{14.79}$$

Further on, substituting expression (14.74) for the operator $\hat{S}_n^z$ in the term violating symmetry in magnetic field $g\mu_B B \sum_n \hat{S}_n^z$:

$$-g\mu_B B \sum_n \hat{S}_n^z = -g\mu_B B \sum_n \left[S - \frac{1}{N}\sum_{q,q_1} a_q^\dagger a_{q_1} \exp\left(i\left(\boldsymbol{q} - \boldsymbol{q}_1\right)\boldsymbol{n}\right)\right], \tag{14.80}$$

and calculating the sums over n and q_1, we get:

$$-g\mu_B B \sum_n \hat{S}_n^z = -g\mu_B B S N + g\mu_B B \sum_q a_q^\dagger a_q. \tag{14.81}$$

Now we can unite the first term in (14.81) with the term $-\frac{1}{2}\sum_{n,m} J_{n,m} S^2$ which does not depend on the creation and annihilation operators in the Heisenberg Hamiltonian, denoting the sum of these terms as:

$$E_0 = -\frac{1}{2}\sum_{n,m} J_{n,m} S^2 - g\mu_B B S N. \tag{14.82}$$

At the same time, it is natural to unite the second term in (14.81). containing the operator of the magnon number density, with the last term in the r.h.s. of the Hamiltonian (14.79). Hence the Heisenberg Hamiltonian acquires a form:

$$\widehat{H} = E_0 + \sum_q a_q^\dagger a_q \left\{ S \sum_p J(\boldsymbol{p})(1 - \exp(-i\boldsymbol{q}\boldsymbol{p})) + g\mu_B B \right\}. \tag{14.83}$$

Now we can introduce the magnon energy (or in fact the energy of the spin wave quanta) as:

$$\varepsilon_q = S \sum_p J(\boldsymbol{p})(1 - \exp(-i\boldsymbol{q}\boldsymbol{p})) + g\mu_B B, \tag{14.84}$$

and represent the Heisenberg Hamiltonian quantized on magnons as follows:

$$\widehat{H} = E_0 + \sum_q \varepsilon_q a_q^\dagger a_q = E_0 + \sum_q \varepsilon_q n_q, \tag{14.85}$$

where $n_q = a_q^\dagger a_q$ is the operator of the magnon number density.

14.6 Magnetic Moment of the Unit Volume (Magnetization of the System)

If we know the magnon spectrum, we can determine the magnetic moment of the unit volume (in other words find the magnetization of the system). Magnetic moment of the unit volume equals to:

$$M = g\mu_B \sum_n \left\langle \hat{S}_n^z \right\rangle = g\mu_B SN - g\mu_B \sum_q a_q^\dagger a_q = g\mu_B SN - g\mu_B \sum_q \left\langle n_q \right\rangle. \tag{14.86}$$

The second term in (14.86) corresponds to the reverse moment (directed oppositely), connected with magnons.

Note, that in the simplest spherically-symmetric case, the gyromagnetic ratio g for the electron with spin $\frac{1}{2}$ according to the results of Quantum mechanics [29] is positive and equals to $2 (g = 2 > 0)$.

Thus, each spin excitation (each spin mode) carries the reversed magnetic moment $(-g\mu_B)$, which for the arbitrary value of spin S is related to the variation of spin on 1.

We can say, that in ferromagnet at low temperatures emerges a small number of reversed spins, which are not localized, but are running along the lattice.

The magnon distribution function has a Planck form:

$$\langle n_q \rangle = \frac{1}{\exp\left(\frac{\varepsilon_q}{T}\right) - 1} \tag{14.87}$$

with zero chemical potential $\mu = 0$ (since the number of excitations is not fixed).

Hence, the ferromagnet magnetization at low temperatures reads:

$$M = M_0 - g\mu_B \int \frac{d^3q}{(2\pi)^3} \frac{1}{\exp\left(\frac{\varepsilon_q}{T}\right) - 1}, \tag{14.88}$$

where at $T = 0$ the magnetic moment $M_0 = \frac{1}{V} g\mu_B S N$ corresponds to all the spins oriented parallel to each other (V is the system volume)

For magnetic field $B = 0$ and small wave vectors q expanding the magnon spectrum in series till the second order inclusively we get:

$$\varepsilon_q = S \sum_p J(p)(1 - \exp(-i\boldsymbol{q}\boldsymbol{p})) \approx S \sum_p J(p) \left(i\boldsymbol{q}\boldsymbol{p} + \frac{1}{2}(\boldsymbol{q}\boldsymbol{p})^2 \right). \tag{14.89}$$

But the exchange integral

$$J(\boldsymbol{p}) = J(-\boldsymbol{p}) \tag{14.90}$$

is an even function of $\boldsymbol{p}$. That is why the linear in $\boldsymbol{p}$ term which equals to $(i\boldsymbol{q}\boldsymbol{p})$ drops out in the sum over $\boldsymbol{p}$ in (14.89).

As a result:

$$\varepsilon_q \approx \frac{S}{2} \sum_p J(\boldsymbol{p})(\boldsymbol{q}\boldsymbol{p})^2, \tag{14.91}$$

and the magnon spectrum for the cubic crystal coincides with the spectrum of a classical particle:

$$\varepsilon_q \approx \frac{S}{2} \cdot \frac{1}{3} \sum_q J(\boldsymbol{p}) p^2 q^2 = \alpha q^2. \tag{14.92}$$

If we introduce now a convenient variable for the dimensionless energy:

$$x = \frac{\varepsilon_q}{T} = \frac{\alpha q^2}{T}, \tag{14.93}$$

and correspondingly the differential:

$$dx = \frac{2\alpha q\, dq}{T},\tag{14.94}$$

then expression for the magnetization (14.88) acquires a form:

$$M = M_0 - g\mu_B \frac{T^{3/2}}{4\pi^2 \alpha^{3/2}} \int \sqrt{x}\, dx \frac{1}{e^x - 1}.\tag{14.95}$$

Thus, we get the Bloch law [36] for the magnetization of the three-dimensional crystal:

$$M = M_0 - \gamma T^{3/2}.\tag{14.96}$$

At the same time, the magnetization in the two-dimensional case:

$$M = M_0 - g\mu_B \frac{T}{4\pi\alpha} \int dx \frac{1}{e^x - 1},\tag{14.97}$$

and the magnon contribution is logarithmically divergent at the lower limit of integration (for $x \to 0$):

$$M - M_0 \sim -\eta T \int \frac{dx}{x} \sim -\eta T \ln x.\tag{14.98}$$

The magnon contribution to the magnetization diverges even stronger at the lower limit for $x \to 0$ in the one-dimensional case:

$$M - M_0 \sim -\frac{1}{4\pi}\sqrt{\frac{T}{\alpha}} \int \frac{dx}{x^{3/2}} \sim -\eta_1 T^{1/2} \frac{1}{x^{1/2}}.\tag{14.99}$$

As a consequence of this infrared singularity, a large amount of reversed magnons appears in one-dimensional and two-dimensional systems even at very low temperatures. We can say, that long wavelength magnons completely destroy ferromagnetic order in the system for the space dimensionalities $D = 2$ and $D = 1$.

Comment 1. The less is space dimensionality, the larger role play fluctuations. Hence it becomes beneficial to proceed to the larger space dimensionality thus reducing fluctuations.

This procedure in the theory of the phase transitions is known as Wilson dimensional regularization [23]. By utilizing Wilson regularization, we should relace in the measure of the integration:

$$d^3q \to d^\varepsilon q,\tag{14.100}$$

where ε is continuous parameter.

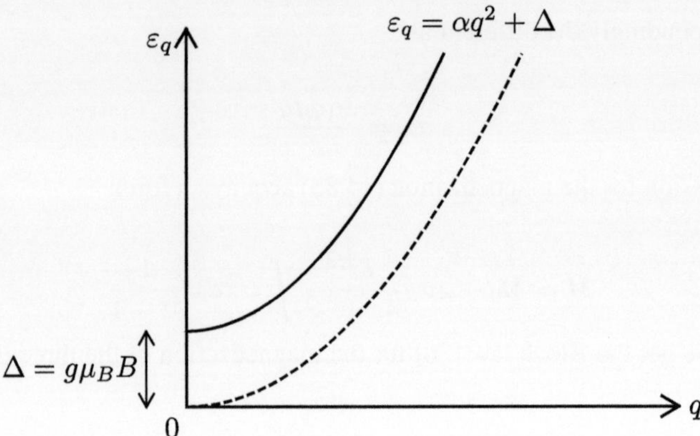

Fig. 14.7 Gapped magnon spectrum $\varepsilon_q = \left(\alpha q^2 + g\mu_B B\right)$ which appears when we apply external magnetic field B to the ferromagnetic system [1]

<u>Comment 2</u>. If we apply an external magnetic field to our system, then it stabilizes the system (due to the emergence of a gap in the excitation spectrum). Indeed, the magnon spectrum in this case (see Fig. 14.7) is given by:

$$\varepsilon_q = \alpha q^2 + g\mu_B B. \tag{14.101}$$

In this case the singularity on large wavelengths does not appear for the magnons with a new spectrum which has a gap. External field stabilizes the system effectively yielding an additional magnetization to the sample.

As a result, for $B \neq 0$ and low temperatures:

$$T < (\Delta = g\mu_B B) \tag{14.102}$$

magnetization (even in systems of reduced dimensionality) behaves as:

$$M = M_0 - \gamma_1 \exp\left(-\frac{\Delta}{T}\right). \tag{14.103}$$

Note, that if we take into account the anisotropy of the real crystal, then the gap appears again in the magnon spectrum.

14.7   Magnon Contribution to the Energy and Specific Heat

Let us determine the magnon contribution to the thermodynamic energy and specific heat of the system at low temperatures. After temperature averaging of the Heisenberg Hamiltonian (quantized on magnons in Eq. 14.85) we get for the thermodynamic energy:

$$E = E_0 + \sum_q \varepsilon_q \langle n_q \rangle = E_0 + V \int \frac{d^3 q}{(2\pi)^3} \frac{\varepsilon_q}{\exp\left(\frac{\varepsilon_q}{T}\right) - 1} = E_0 + \zeta T^{5/2}.$$

$$(14.104)$$

Correspondingly the magnon contribution to the specific heat:

$$C_V \sim T^{3/2}. \tag{14.105}$$

When we increase the temperature, specific heat of the system also increases due to the increase of the number of magnons (we can say that our system can "swallow" more magnons then).

Let us emphasize, that magnons similarly to phonons have the finite lifetime (or finite damping), but their finite lifetime is connected with approximate quantization rules (and not with anharmonic interactions as in the case of phonons).

Let us remind, that exact expression for the commutator $\left[S_q^\dagger S_{q_1}^- \right]$ according to (14.59) has a form:

$$S_q^\dagger S_{q_1}^- - S_{q_1}^- S_q^\dagger = \frac{1}{N} \sum_n 2 S_n^z \exp\left(-i\left(q - q_1\right) n\right),$$

where the operator $\hat{S}_n^z$ (see Eq. (14.74)) equals to:

$$\hat{S}_n^z = S - \frac{1}{N} \sum_{q,q_1} a_q^\dagger a_{q_1} \exp\left(i\left(q - q_1\right) n\right).$$

Note, that we replace the operator $\hat{S}_n^z$ on S in (14.59) (neglecting in fact the second term in (14.74)), and then take the spin S out of the sum symbol.

Hence, to get exact commutation relations, we have to amend the expressions (14.60) for $S_q^\dagger$ and S_q^-. As a result, an exact relation between the commutator $\left[S_q^\dagger S_{q_1}^- \right]$ and the commutator of the magnon creation and annihilation operators $\left[a_q a_{q_1}^\dagger \right]$ (with an account of the so-called Dyson–Maleev corrections [37,38]) acquires a form:

$$\left[S_q^\dagger S_{q_1}^- \right] = S_q^\dagger S_{q_1}^- - S_{q_1}^- S_q^\dagger = 2 S \left(a_q a_{q_1}^\dagger - a_{q_1}^\dagger a_q \right) \left(1 - \frac{1}{2 SN} \sum_{q_2} a_{q_2}^\dagger a_{q_1 + q_2 - q} \right)^2$$

$$\approx 2 S \cdot \delta_{q q_1} \cdot \left(1 - \frac{1}{SN} \sum_{q_2} a_{q_2}^\dagger a_{q_1 + q_2 - q} \right). \tag{14.106}$$

Correspondingly the Fourier components of raising and lowering operators themselves are approximately equal to:

$$S_q^\dagger \approx \sqrt{2S} a_q \cdot \left(1 - \frac{1}{2SN} \sum_{q_2} a_{q_2}^\dagger a_{q_2}\right), \tag{14.107}$$

$$S_q^- \approx \sqrt{2S} a_q^\dagger \cdot \left(1 - \frac{1}{2SN} \sum_{q_2} a_{q_2}^\dagger a_{q_2}\right). \tag{14.108}$$

Let us stress, that due to the finite spin value, in the expressions (14.107)–(14.108) besides linear combinations of the operators $a^\dagger$ and a enter also the cubic combinations ($aa^\dagger a$ and $a^\dagger a^\dagger a$) which contain 3 operators of magnon creation and annihilation.

These cubic combinations are responsible for the magnon damping. It is natural, that the presence of the term related to quantum corrections which is proportional to the bilinear combination $S_n^- S_m^\dagger$ leads to an appearance of the term with the quartic combinations of the magnon creation and annihilation operators in the Heisenberg Hamiltonian:

$$H_4 = -\frac{1}{4!} \sum_{1,2,3,4} \Phi(1, 2, 3, 4) a_1^\dagger a_2^\dagger a_3 a_4. \tag{14.109}$$

As we know, analogous term with the anharmonic interaction, emerges also in the phonon Hamiltonian. At the same time, there is no cubic interaction directly in the magnon Hamiltonian:

$$H_3 = -\frac{1}{3!} \sum_{1,2,3,4} \Phi(1, 2, 3) \left(a_1^\dagger a_2^\dagger a_3 + a_1^\dagger a_2 a_3\right). \tag{14.110}$$

Note, that the cubic interaction H_3 rather seldomly appears in the condensed matter physics. Its emergence in the phonon Hamiltonian is connected with the fact, that the momentum conservation law is not fulfilled exactly (the Umklapp processes on the inverse lattice vector are possible in the system).

14.8   Phase Transition for the Magnetization in the Ferromagnet

In the mean field theory, the phase transition for the magnetization $M(T)$ in the ferromagnet at the critical temperature $T = T_{\mathrm{Curie}}$ is the second order phase transition (see the temperature dependence of the magnetization in Fig. 14.8).

At the Curie point meanwhile, as it is always the case in the second order phase transition, we get the loss of symmetry [1–8].

For temperatures $T > T_{\mathrm{Curie}}$, in the paramagnetic (PM) phase we have full disorder in the magnetic system, and correspondingly the full symmetry. For $T \leq T_{\mathrm{Curie}}$

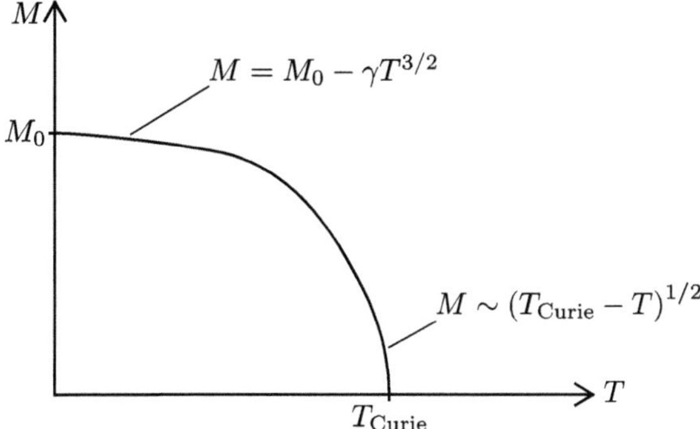

Fig. 14.8 Temperature dependence of magnetization in the ferromagnet in the framework of the mean field theory. At low temperatures $T \ll T_{\text{Curie}}$ the Bloch law is satisfied: $M = M_0 - \gamma T^{3/2}$. For temperatures close to the critical temperature $T \to T_{\text{Curie}}$ the magnetization tends to zero in a square root fashion: $M \sim (T_{\text{Curie}} - T)^{1/2}$ typical for the second order phase transitions [1]

in the ferromagnetic (FM) phase magnetic order appears, and thus we get the lower symmetry in the system.

Besides that, as we know, there is no latent heat in the second order phase transition, the entropy is continuous and we have only the jump in the specific heat (see Fig. 2.3 in Lecture 2). Moreover, the phase transition takes place in all the system volume (in contrast to the first order phase transition, where the transition in the whole volume is not beneficial and we get the seeds of the new phase).

Note, that close to Curie temperature the fluctuations are increasing in the system.

For the description of the ferromagnet thermodynamics the Weiss molecular field approximation [1–8,39,40] exists and is widely used.

In the framework of this approximation the operators $\hat{S}_n$ and $\hat{S}_m$ in the Heisenberg exchange Hamiltonian can be replaced by their average values. In fact, we neglect fluctuations in Weiss approximation [1–8]. Then after the averaging, exchange part of the Hamiltonian equals to:

$$\widehat{H} = -\frac{1}{2} \sum_{n,m} J_{n,m} \left\langle \hat{S}_n \right\rangle \left\langle \hat{S}_m \right\rangle = -\frac{1}{2} \sum_{n,m} J_{n,m} \frac{VM}{g\mu_B N} \cdot \frac{VM}{g\mu_B N}, \qquad (14.111)$$

where the average magnetization $M = g\mu_B \frac{N}{V} \left\langle \hat{S}_n \right\rangle$.

But, from the other hand, if we neglect the fluctuations, then the magnetization can be determined via the mean value of the operator of spin z-component:

$$M = g\mu_B \frac{N}{V} \left\langle \hat{S}_n \right\rangle \approx g\mu_B \frac{N}{V} \left\langle \hat{S}_n^z \right\rangle. \qquad (14.112)$$

As a result, we can represent the Hamiltonian $\widehat{H}$ in (14.111) as effective magnetic field $B_{\text{eff.}}(M)$, acting on the average spin projection $\left\langle \hat{S}_n^z \right\rangle$ (in fact, on the average magnetization M):

$$\widehat{H} = g\mu_B B_{\text{eff.}}(M)\left\langle \hat{S}_n^z \right\rangle, \tag{14.113}$$

where, according to Statistical physics [2], the mean value of the operator of spin z-component can be expressed via the statistical sum:

$$\left\langle \hat{S}_n^z \right\rangle = \frac{1}{Z} \sum_{S^z=-S}^{S} S^z \exp\left(-\frac{g\mu_B B_{\text{eff.}}(M)S^z}{T}\right). \tag{14.114}$$

At the same time, the statistical sum is given by:

$$Z = \sum_{S^z=-S}^{S} \exp\left(-\frac{g\mu_B B_{\text{eff.}}(M)S^z}{T}\right), \tag{14.115}$$

where effective magnetic field (in the energy units) reads:

$$g\mu_B B_{\text{eff.}}(M) = -\frac{1}{2} \sum_{n,m} J_{n,m} \frac{MV}{g\mu_B N} \tag{14.116}$$

Multiplying $\left\langle \hat{S}_n^z \right\rangle$ in (14.111) on $g\mu_B \frac{N}{V}$, we proceed to the self-consistency equation for the magnetization [1–4]:

$$M = g\mu_B \frac{N}{V}\left\langle \hat{S}_n^z \right\rangle = g\mu_B \frac{N}{V}\frac{1}{Z} \sum_{S^z=-S}^{S} S^z \exp\left(-\frac{g\mu_B B_{\text{eff.}}(M)S^z}{T}\right). \tag{14.117}$$

In fact, the self-consistency equation has a form $M = F(M)$. We can say, that the magnetization $M = g\mu_B \frac{N}{V}\left\langle \hat{S}_n^z \right\rangle$ depends on the effective magnetic field $B_{\text{eff.}}$ via its dependence on the average spin component $\left\langle \hat{S}_n^z \right\rangle$. Meanwhile the effective magnetic field $B_{\text{eff.}}$ depends in turn on the magnetization M.

Weiss approximation just corresponds to the mean field theory for the magnetization and leads to its square root temperature dependence typical for the second order phase transitions (see Fig. 14.8):

$$M \sim (T_{\text{Curie}} - T)^{1/2} \tag{14.118}$$

when we approach the Curie temperature $(T \rightarrow T_{\text{Curie}})$ in the ferromagnetic (FM) phase. Of course, in the critical point itself the magnetization $M(T_{\text{Curie}}) = 0$.

At the same time, for temperature $T = 0$, the magnetization goes on the saturation: $M = M_0 = g\mu_B \frac{NS}{V}$. At low temperatures $T \ll T_{\text{Curie}}$ the Bloch law [36] works for the magnetization: $M = M_0 - \gamma T^{3/2}$.

Magnetic susceptibility is divergent for $T \to T_{\text{Curie}}$:

$$\chi = \chi_0 \frac{1}{\left(1 - \frac{T_{\text{Curie}}}{T}\right)} = \chi_0 \frac{T}{(T - T_{\text{Curie}})}, \text{ and}$$

$$\frac{1}{\chi} = \frac{1}{\chi_0} \frac{(T - T_{\text{Curie}})}{T} \approx \frac{1}{\chi_0} \frac{(T - T_{\text{Curie}})}{T_{\text{Curie}}}. \tag{14.119}$$

This result for the magnetic susceptibility (in Statistical physics it is called the Curie–Weiss law [2,36,39,40]) coincides with Landau theory for the second order phase transitions [41].

Let us stress, that in more exact theory [14,21–23,41,42] the magnetization close to the critical point with an account of the fluctuations also tends to zero, but satisfies the different power law:

$$M \sim (T_{\text{Curie}} - T)^{\beta}, \tag{14.120}$$

where the power $\beta \sim \frac{1}{3}$.

Correspondingly, the magnetic susceptibility in the exact theory [14,21–23,41,42] is also divergent close to the critical point, but now as:

$$\chi = \chi_0 \frac{1}{\left(1 - \frac{T_{\text{Curie}}}{T}\right)^{\gamma}}, \tag{14.121}$$

where the power $\gamma \sim \frac{4}{3}$.

References

1. Y. Kagan, *Lectures on the Solid-state Theoretical Physics*, Moscow, MEPHI, 1981–1982. Part I. Phonons (Unpublished)
2. E.M. Lifshitz, L.P. Pitaevskii, *Statistical Physics, Part II, Theory of the Condensed state* (Pergamon, Elsevier, 2013)
3. D.C. Mattis, *The Theory of Magnetism, An Introduction to the Study of Cooperative Phenomena* (Harper and Row Publishers, New York, Evanston, London, 1965)
4. S.V. Vonsovskii, *Magnetism* (Wiley, 1974)
5. A.I. Ahiezer, V.G. Bar'yachtar, S.V. Peletminskii, *Spin Waves* (Nauka Publishing House, Moscow, 1967) (in Russian)
6. S.V. Kusminsky, *Quantum Magnetism. Spin Waves and Optical Cavities* (Springer Briefs in Physics, 2019)
7. D.D. Stansil, A. Prabhakar, *Spin Waves. Theory and Applications.* Springer Business Media (2009)
8. S.V. Tyablikov, *Methods in Quantum Theory of Magnetism* (Springer, Science + Business Media, 1967)
9. Y. A. Izumov, Y. N. Skryabin, *Basic Models in Quantum Theory of Magnetism* (Ekaterinburg, 2002) (in Russian)

10. A.S.T. Pires, *Theoretical Tools to Spin Models in Magnetic Systems* (IOP Publishing, Bristol, UK, 2021)
11. E. Tsymbal, I, Zutic (eds.), *Handbook on Spin Transport and Magnetism* (Tailor and Francis Group, 2020)
12. Y. Takahashi, *Spin Fluctuation Theory of Itinerant Magnetism* (Springer, 2013)
13. R.P. Feynman, *Statistical Mechanics, A Set of Lectures* (W.A. Benjamin Inc., 1972)
14. A.Z. Patashinskii, V.L. Pokrovskii, *Fluctuation Theory of Phase Transitions* (Pergamon Press, 1979)
15. A.M. Tsvelik, *Quantum Field Theory in Condensed Matter Physics* (Cambridge University Press, 1995)
16. S.W. Lowesley, *Frontiers in Physics, Condensed Matter Physics: Dynamic Correlations* (The Benjamin Cummings Publishing Company, 1986)
17. J.W. Negele, H. Orland, *Frontiers in Physics, Quantum Many-Particle Systems* (Addison-Wesley Publishing Company, 1992)
18. G. Parisi, *Frontiers in Physics, Statistical Field Theory* (Addison-Wesley Publishing Company, 1988)
19. A. Abragam, B. Bleaney, *Electron Paramagnetic Resonance of Transition Ions* (Clarendon Press, Oxford, 1970)
20. C.P. Slichter, *Principles of Magnetic Resonance* (Springer, Berlin–Heidelberg–New York, 1980)
21. M. Fisher, *The Nature of Critical Points*. Lectures in Theoretical Physics (University of Colorado Press, Boulder, 1965)
22. L. Kadanoff, *Statistical Physics: Statistics, Dynamics and Renormalization* (World Scientific, 2000)
23. K.G. Wilson, Rev. Mod. Phys. **55**, 583 (1983)
24. C. Itzykson, J.-M. Drouffe, *Statistical Field Theory, v.1, From Brownian Motion to Renormalization and Lattice Gauge Theory* (Cambridge University Press, 1989)
25. E. Fradkin, *Field Theories of Condensed Matter Physics* (Cambridge University Press, 2013)
26. M.Y. Kagan, K.I. Kugel, A.L. Rakhmanov, A.O. Sboychakov, *Electronic Phase Separation in Magnetic and Superconducting Materials. Recent Advances*. Springer Series in Solid-State Sciences, vol. 201 (2024)
27. W. Heisenberg, Zeit. fur Phys. **49**, 619 (1928)
28. Y. Frenkel, Zeit. fur Phys. **49**, 31 (1928)
29. L.D. Landau, E.M. Lifshitz, *Quantum Mechanics, Non-relativistic Theory* (Pergamon Press, New York, 1977)
30. W. Heitler, F. London, Z. Phys. **44**, 455 (1927)
31. A.A. Abrikosov, L.P. Gor'kov, I.E. Dzyaloshinskii, *Methods of Quantum Field Theory in Statistical Physics* (Prentice Hall, Englewood Cliffs, New Jersey, 1963)
32. M.V. Sadovskii, *Diagrammatics*. Lectures on Selected problems in Condensed Matter Theory (World Scientific, Singapore, 2006)
33. L.S. Levitov, *Green's Functions in Problems* (Princeton University Press, 2011)
34. J.C. Slater, *The Self-consistent Field for Molecules and Solids* (McGraw-Hill Book Company, New York, 1974)
35. D. Hartree, *The Calculation of Atomic Structures* (Wiley, New York, 1957)
36. N.W. Ashcroft, N.D. Mermin, *Solid State Physics* (Holt, Rinehart and Winston, 1976)
37. F.J. Dyson, Phys. Rev., **102**, 1217 (1956); ibid. **102**, 1230 (1956)
38. S.V. Maleev, Sov. Phys. JETP **64**, 654 (1958)
39. C. Kittel, *Quantum Theory of Solids* (Wiley, New York–London, 1963)
40. E. Pavarini, E. Koch, U. Schollwock, *Emergent Phenomena in Correlated Matter* (Institute for Advanced Simulations, Julich, 2013)
41. L.D. Landau, E.M. Lifshitz, *Statistical Physics*, Part I (Butterworth-Heinemann, 1980)
42. A.A. Abrikosov, *Fundamentals of the Theory of Metals* (North-Holland, Amsterdam, 1988)

Part II

Electronic Excitations in Normal Metals and Superconductors

Abstract

We consider basic interactions in metals in the second quantization technique. We reveal an essential role of the electroneutrality condition in metals. We introduce important notations of the correlation radius and Wigner–Seitz elementary cell and calculate the average potential energies of electron–ion and electron–electron interactions. In the end of the Lecture, we discuss Fermi gas of non-interacting electrons at zero temperature and calculate the average kinetic energy and Pauli pressure in it.

15.1 Introduction

For electron delocalization (in the situation with the half-filled band typical for metals [1–4], when there is one electron on each site), it is necessary first of all to have the overlap of the wave functions.

Moreover, the overlap of the wave functions should be large enough to overcome the Coulomb repulsion on one site to make it possible to put one more electron on this site in the process of delocalization (and thus to escape Hubbard localization [1,5–10]).

Note that the average kinetic energy of electrons in the delocalized state is minimal $T \propto \frac{1}{r_0^2} \to$ min for large distances between electrons $r_0 \to \infty$ (r_0 is the average distance between electrons) and hence on large distances the total energy on one electron is minimal as well: $E \to$ min.

In the metal physics [1–4, 10–14], we distinguish between transition metals [15] and non-transition metals [1, 16–18].

Transition metal has filled outer electron shell, while inner electron shells are not completely filled [15–17].

Non-transition metal—vice versa has filled inner shells and not completely filled outer shell [1, 18].

In this Lecture, course we will consider only non-transition metals [1, 18–20].

© The Author(s), under exclusive license to Springer Nature Switzerland AG 2026

M. Kagan, *Lecture Notes of Professor Yuri Kagan in Theoretical Solid-State Physics*, Lecture Notes in Physics 1048, https://doi.org/10.1007/978-3-032-14621-2_15

In non-transition metals, ions together with inner electron shells form the periodic structure. At the same time, all the valence electrons are delocalized and collectivize in crystal.

Meanwhile, ions together with inner shells do not overlap with each other (ionic size is smaller than the interatomic distance).

That is why there is only Coulomb interaction acting between ions. Indeed, if $\frac{a_{\text{ion}}}{a_{\text{interat.}}} \ll 1$, then ions feel each other as a whole. In other words, they interact with each other as pointlike particles.

Let us emphasize that ionic charge Z in metals equals to the metal valence. Note also that in metals we have 3 main interactions: ion–ion interaction [1,18,19], electron–ion interaction [1,18,19,21–23] (interaction between valence electrons and nuclei together with electrons of ionic cores) and finally electron–electron interaction [1,9,15–17,20–32].

15.1.1 Hamiltonian

Hamiltonian of the electron and ion subsystems in metal has a form:

$$H = T_e + V_{ee} + V_{ei} + V_{ii} + (T_i), \tag{15.1}$$

where T_e is the operator of the kinetic energy of valence electrons, T_i is the operator of the kinetic energy of ions. We can often neglect the ion kinetic energy since it is usually small in small parameter $\frac{m}{M} \ll 1$ related to the ratio of electron and ion masses.

In the Hamiltonian (15.1), V_{ee} stands for the operator of the potential energy of Coulomb interaction between valence electrons:

$$V_{ee} = \frac{1}{2} \sum \frac{e^2}{|r_\alpha - r_\beta|} = \frac{1}{2} \sum_{\alpha,\beta} \sum_q \frac{4\pi e^2}{q^2 \Omega} \exp\left(iq\left(r_\alpha - r_\beta\right)\right), \tag{15.2}$$

where Ω is the system volume. (Throughout the Lecture course, we use both notations Ω and V for the system volume. We hope it will not lead to any misunderstandings).

At the same time, the operator V_{ei} corresponds to the potential energy of the interaction of the valence electrons and ions (where ions include nuclei and core electrons of the filled inner shells). In the explicit form:

$$V_{ei} = \sum_{\alpha,m} V_{ei}\left(r_\alpha - R_m\right) = \sum_{\alpha,m} \sum_q V_{ei}(q) \exp\left(iq\left(r_\alpha - R_m\right)\right), \tag{15.3}$$

where $V_{ei}(q)$ contains the multiplier $\frac{1}{\Omega}$.

Finally, the operator V_{ii} describes potential energy of ion–ion interaction:

$$V_{ii} = \frac{1}{2} \sum_{m,n} \frac{(Ze)^2}{|R_m - R_n|} = \sum_{m,n} \sum_q \frac{4\pi Z^2 e^2}{q^2 \Omega} \exp\left(iq\left(R_m - R_n\right)\right). \tag{15.4}$$

Further on, we should explicitly account for the electroneutrality condition. To do that we should collect all the terms at $q = 0$. (Note that, due to electroneutrality condition, the sum of zeroth Fourier components of all the potentials V_{ee}, V_{ei}, and V_{ii} should be equal zero).

While collecting the terms, we should remember also that, in metal, we have ZN electrons and N ions.

It is important to have in mind that, for $q \to 0$ (or, respectively, for the wavelength $\lambda \to \infty$), Fourier component of the electron–ion interaction in (15.3) can be represented as a series:

$$V_{ei}(q \to 0) \to -\frac{4\pi Ze^2}{q^2\Omega} + \frac{b}{\Omega} + \alpha q^2 + \cdots, \tag{15.5}$$

where the first Coulomb term of the expansion proportional to $\frac{1}{q^2}$ is connected with the fact that, in the limit $\lambda \to \infty$, the valence electron feels all the ion (with charge Z) as a whole.

The second term of the expansion proportional to $\frac{b}{\Omega}$ does not depend on q for $q \to 0$. It represents non-Coulomb part of the scattering amplitude on angle 0 for the electron–ion interaction, and moreover $b > 0$ [1,18].

Due to the presence of the central symmetry in the system (the system invariance with respect to the replacements $r \to -r, q \to -q$), Eq. (15.5) represents the series expansion in the powers q^2.

Then, collecting the terms at $\frac{1}{q^2}$ and the terms which are independent on q, we get for $q \to 0$

$$\frac{4\pi e^2}{q^2\Omega} \left\{ \frac{NZ(NZ-1)}{2} + Z^2\frac{N(N-1)}{2} - ZNZN \right\} + \frac{b}{\Omega}NZN, \tag{15.6}$$

where the first term in figure brackets corresponds to the number of pairs of the interacting electrons, while the second term—to the number of pairs of the interacting ions. The third term describes the number of electron–ion pairs for the Coulomb part of the electron–ion interaction. Finally, the fourth term (out of the figure brackets) is related to the non-Coulomb part of the amplitude of electron–ion interaction.

After cancelation of the main terms in Eq. (15.6), quadratic in ZN, we are left with the following expression:

$$\frac{4\pi e^2}{q^2\Omega} \left\{ -\frac{NZ}{2} - \frac{NZ^2}{2} \right\}. \tag{15.7}$$

But in the denominator in (15.7) enters the multiplier $q^2\Omega$, where the system volume:

$$\Omega = N\Omega_0, \tag{15.8}$$

and Ω_0 is the volume of the elementary cell.

At the same time, for small wave vectors $q \to 0$ by order of magnitude:

$$q^2 \sim q_{\min}^2, \tag{15.9}$$

where the minimal possible wave vector in the sample with the size L is given by

$$q_{\min} = \frac{2\pi}{L} \sim \Omega^{-1/3} \sim N^{-1/3}. \tag{15.10}$$

As a result, the multiplier

$$q^2 \Omega \sim N^{1/3}, \tag{15.11}$$

and all the expression (15.7) for the non-compensated Coulomb terms by the order of magnitude is proportional to

$$\frac{N}{N^{1/3}} \sim N^{2/3}. \tag{15.12}$$

At the same time, non-Coulomb term in (15.6) is substantially larger, than the residual Coulomb terms in (15.7):

$$\frac{b}{\Omega} NZN \sim N \gg N^{2/3}. \tag{15.13}$$

Hence, we can say that, in main approximation, all the Coulomb terms cancel each other at $q = 0$, and thus the electroneutrality condition is satisfied in metals.

Correspondingly, in the Hamiltonian (15.1) for electron and ion subsystems in metal, and in all the interaction terms in Eqs. (15.2)–(15.5), we can consider only the Fourier harmonics with $q \neq 0$.

15.2   Second Quantization for Electrons

Electron kinetic energy in the Hamiltonian (15.1) in the formalism of the second quantization after proceeding to the momentum space acquires a form [1, 21–23, 25, 26, 33–36]:

$$T_e = \sum_k \varepsilon_k a_k^\dagger a_k, \tag{15.14}$$

where ε_k − is the electron spectrum, $a_k^\dagger$ and a_k are the electron creation and annihilation operators, respectively, index $k = \{\boldsymbol{k}, \sigma\}$, $\boldsymbol{k}$ is wave vector, and σ is the electron spin projection.

Note that creation and annihilation operators are governed by Fermi–Dirac statistics in case of electrons [1,21–23,26,27,33–36] and satisfy the anticommutation relations:

$$\left\{ a_k^\dagger a_k \right\} = 1. \tag{15.15}$$

If electron subsystem is in the external field (or interacts with nonmagnetic impurities or defects), then we should add one more term to the potential energy of the Coulomb interactions in metal:

$$H = \sum_\alpha V\left(r_\alpha\right). \tag{15.16}$$

If the potential energy in (15.16) for the electron in external field does not depend on spin, then in the second quantization technique, an additional term appears when we proceed to the momentum space in the Hamiltonian (15.1):

$$\sum_\alpha \sum_{k_1 k_2 \sigma} \langle k_1 | V\left(r_\alpha\right) | k_2 \rangle \, a_{k_1 \sigma}^\dagger a_{k_2 \sigma}. \tag{15.17}$$

At the same time, potential energy of electron–ion interaction in (15.3) can be written as follows:

$$V_{ei} = \sum_\alpha \sum_q \left\{ \sum_m \exp\left(-iqR_m\right) V_{ei}\left(q\right) \right\} \exp\left(iqr_\alpha\right), \tag{15.18}$$

where all the expression in the figure bracket in (15.18) acts on the exponent $\exp(iqr_\alpha)$, while this exponent in its turn is a function only of electron coordinate r_α.

Further on, we proceed to the momentum space and use the fact that the matrix element of this function, calculated on the basis of plane waves, equals to

$$\langle k_1 | \exp\left(iqr_\alpha\right) | k_2 \rangle = \delta_{k_1, k_2 - q}, \tag{15.19}$$

where:

$$|k_1\rangle = \frac{\exp\left(i k_1 r_\alpha\right)}{\sqrt{\Omega}}. \tag{15.20}$$

Then the potential energy of the electron–ion interaction in the second quantization yields

$$\begin{aligned}
V_{ei} &= \sum_{k,q \neq 0,\sigma} a_{k,\sigma}^\dagger a_{k-q,\sigma} \left(\sum_m V_{ei}(q) \exp\left(-iqR_m\right) \right) \\
&= \sum_{k,q \neq 0,\sigma} U(q) a_{k,\sigma}^\dagger a_{k-q,\sigma},
\end{aligned} \tag{15.21}$$

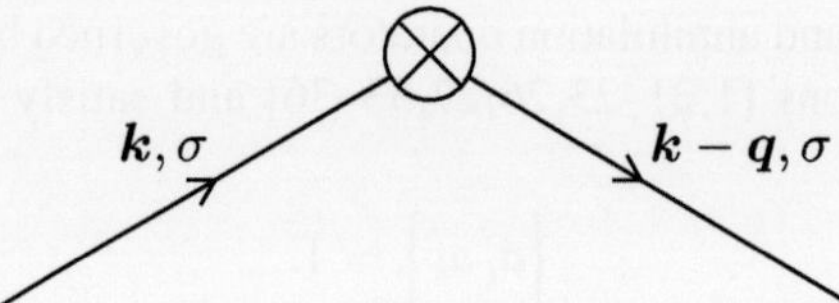

Fig. 15.1 Electron scattering on nucleus (as on the immobile center) with spin projection conservation and momentum non-conservation in the scattering process [1]

where Fourier component of the potential $U(q)$ in the last equality in Eq. (15.21) reads

$$U(q) = \sum_m V_{ei}(q)\exp\left(-i q R_m\right). \tag{15.22}$$

Graphically electron–ion interaction in (15.21)–(15.22) can be represented as electron scattering on the immobile center (on nucleus) with spin projection conservation and momentum non-conservation in the scattering process (see Fig. 15.1).

At the same time, momentum is conserved for electron–electron interaction in Eq. (15.2). This interaction has the following form in the second quantization technique when we proceed to the momentum space:

$$V_{ee} = \frac{1}{2} \sum_{k_1,k_2,q\neq0,\sigma_1,\sigma_2} \frac{4\pi e^2}{q^2\Omega} a^{\dagger}_{k_1,\sigma_1} a^{\dagger}_{k_2,\sigma_2} a_{k_2+q,\sigma_2} a_{k_1-q,\sigma_1}. \tag{15.23}$$

Graphically electron–electron interaction in (15.23) corresponds to electron–electron scattering which conserves both the spin projection and momentum in the scattering process (see Fig. 15.2).

Until now, while deriving expressions for electron–ion and electron–electron interactions, we do not use the crystal periodicity. However, in the expressions (15.21)–(15.22) for electron–ion interaction enters the lattice sum:

$$\sum_m \exp\left(-i q R_m\right) = N\delta_{q,K}, \tag{15.24}$$

where K is the inverse lattice vector.

But as we know, Fourier harmonic $q = 0$ is absent in the sum over q in the expression (15.21) for electron–ion interaction. That is why, in the lattice sum

Fig. 15.2 Electron–electron interaction which conserves both spin projection and momentum in the scattering process [1]

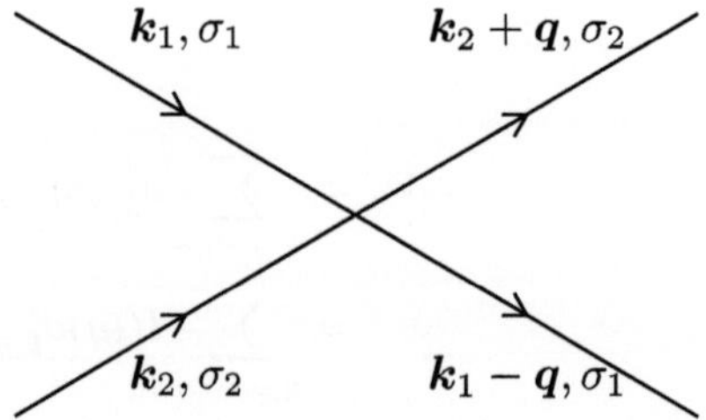

(15.24), work only non-zero values of the inverse lattice vector: $K \neq 0$. Thus, in the expressions (15.21)–(15.22) enters only Fourier component of the potential with large momentum $V_{ei}(K)$, where the absolute value of the momentum:

$$|K| \geq |K_{\min}| = \frac{2\pi}{a}, \tag{15.25}$$

and a is intersite distance.

We know that these Fourier components are strongly oscillating and <u>small</u>. Thus, only due to the lattice periodicity:

$$\frac{V_{ei}(K)}{\varepsilon_F} \sim 0.1 \ll 1, \tag{15.26}$$

where ε_F is the Fermi energy of electrons.

As a result, we have another small parameter in the system, namely: $\frac{V_{ei}(K)}{\varepsilon_F}$.

We can say that, in metals, <u>electron–ion interaction is small</u>.

<u>Comment</u>. It is interesting to note that, in liquid metals (in the absence of the lattice), the ratio $\frac{V_{ei}(K)}{\varepsilon_F} \ll 1$ is still small, since the typical wave vectors, which determine Fourier component of the potential $\left|K_{\text{typ.}}\right| \sim \frac{2\pi}{a_{\text{mean interat. dist.}}}$, are sufficiently large also.

Moreover, in both solid and liquid metals, average electron–ion interaction is zero:

$$\langle |V_{ei}| \rangle = V_{ei}(q = 0) = 0. \tag{15.27}$$

That is why the corrections to the sum of Coulomb and constant (non-Coulomb) term in the expansion (15.5) for electron–ion interaction $V_{ei}(q)$ start from the terms proportional to q^2 for $q \to 0$.

15.3 Electron–Phonon Interaction

Dynamic part of electron–ion interaction, connected with the lattice oscillations in metals, corresponds to electron–phonon interaction [1,18–23,26,27].

As usual, let us represent coordinates of nuclei (of ions) in the following form:

$$R_m = R_m^0 + \hat{u}_m, \tag{15.28}$$

where R_m^0 is an equilibrium ion coordinate, $\hat{u}_m$ is the operator of ion displacement from the equilibrium position, and utilize the fact that, up to the melting temperatures of metal, the mean squared displacement is small in comparison with interatomic distance squared:

$$\frac{\langle u^2 \rangle}{a^2} \ll 1. \tag{15.29}$$

Then exact expression for the exponents in the lattice sum

$$\sum_m \exp\left(-i\boldsymbol{q}\boldsymbol{R}_m\right) = \sum_m \exp\left(-i\boldsymbol{q}\left(\boldsymbol{R}_m^0 + \hat{\boldsymbol{u}}_m\right)\right) \tag{15.30}$$

can be expanded in series till the linear terms:

$$\exp\left(-i\boldsymbol{q}\left(\boldsymbol{R}_m^0 + \hat{\boldsymbol{u}}_m\right)\right) \approx \exp\left(-i\boldsymbol{q}\boldsymbol{R}_m^0\right)\left(1 - i\boldsymbol{q}\hat{\boldsymbol{u}}_m\right). \tag{15.31}$$

In (15.31), we use that the wave vector $|\boldsymbol{q}| \leq \frac{1}{a}$, and correspondingly: $\boldsymbol{q}\hat{\boldsymbol{u}}_m \sim \frac{u}{a} \ll 1$.

Let us write now electron–phonon interaction in the secondly quantized form utilizing expression (15.21):

$$
\begin{aligned}
V_{\text{e-ph}} &= \sum_{\boldsymbol{k},\boldsymbol{q}\neq 0,\sigma} a_{\boldsymbol{k},\sigma}^{\dagger} a_{\boldsymbol{k}-\boldsymbol{q},\sigma} \sum_m V_{ei}(\boldsymbol{q}) \left[\exp\left(-i\boldsymbol{q}\left(\boldsymbol{R}_m^0 + \hat{\boldsymbol{u}}_m\right)\right) - \exp\left(-i\boldsymbol{q}\boldsymbol{R}_m^0\right)\right] \\
&\approx \sum_{\boldsymbol{k},\boldsymbol{q}\neq 0,\sigma} a_{\boldsymbol{k},\sigma}^{\dagger} a_{\boldsymbol{k}-\boldsymbol{q},\sigma} \sum_m V_{ei}(\boldsymbol{q}) \exp\left(-i\boldsymbol{q}\boldsymbol{R}_m^0\right)\left[1 - i\boldsymbol{q}\hat{\boldsymbol{u}}_m - 1\right] \\
&= -i \sum_{\boldsymbol{k},\boldsymbol{q}\neq 0,\sigma} a_{\boldsymbol{k},\sigma}^{\dagger} a_{\boldsymbol{k}-\boldsymbol{q},\sigma} \sum_m V_{ei}(\boldsymbol{q}) \exp\left(-i\boldsymbol{q}\boldsymbol{R}_m^0\right) \boldsymbol{q}\hat{\boldsymbol{u}}_m.
\end{aligned}
\tag{15.32}
$$

Further on, let us substitute in Eq. (15.31) the displacement operator $\hat{\boldsymbol{u}}_m$ expressed via the operators of phonon creation and annihilation (from the first part of the Lecture course):

$$\hat{\boldsymbol{u}}_m = \sum_{\beta} \sqrt{\frac{\hbar}{2M\omega_\beta N}} \left[\boldsymbol{e}_\beta \exp\left(i\boldsymbol{f}\boldsymbol{R}_m^0\right) \hat{b}_\beta + \boldsymbol{e}_\beta^* \exp\left(-i\boldsymbol{f}\boldsymbol{R}_m^0\right) \hat{b}_\beta^{\dagger}\right], \tag{15.33}$$

where index $\beta = \{\boldsymbol{f}, \alpha\}$, $\boldsymbol{f}$ is the wave vector, α is the number of the branch of the phonon spectrum, $\boldsymbol{e}_\beta$ are polarization vectors, $\hat{b}_\beta$ and $\hat{b}_\beta^{\dagger}$ are phonon creation and annihilation operators.

Let us use the symmetry of the polarization vectors with respect to the substitution $\boldsymbol{f} \rightarrow -\boldsymbol{f}$:

$$\boldsymbol{e}_\alpha^*(\boldsymbol{f}) = \boldsymbol{e}_\alpha(-\boldsymbol{f}), \tag{15.34}$$

and write the displacement operator in more compact form:

$$\hat{\boldsymbol{u}}_m = \sum_{\beta} \sqrt{\frac{\hbar}{2M\omega_\beta N}} \boldsymbol{e}_\beta \exp\left(i\boldsymbol{f}\boldsymbol{R}_m^0\right) \left(\hat{b}_\beta + \hat{b}_{-\beta}^{\dagger}\right). \tag{15.35}$$

Further on, let us utilize expression for the lattice sum which we know from the first part of the Course:

$$\sum_m \exp\left(-i q R_m^0 + i f R_m^0\right) = N \Delta(f - q), \tag{15.36}$$

where as usual, symbol $\Delta(f - q)$ means equality of the vectors f and q with an accuracy of the inverse lattice vector K:

$$f - q = K. \tag{15.37}$$

Then electron–phonon interaction acquires the form:

$$\begin{aligned}
V_{\text{e-ph}} &= -i \sum_{k, q \neq 0, \sigma} a_{k,\sigma}^\dagger a_{k-q,\sigma} \sum_m V_{ei}(q) \exp\left(-i q R_m^0\right) q \hat{u}_m \\
&= \sum_{k, q \neq 0, \sigma} a_{k,\sigma}^\dagger a_{k-q,\sigma} A_{q,\alpha} \left(\hat{b}_{\alpha, f} + \hat{b}_{\alpha, -f}^\dagger\right),
\end{aligned} \tag{15.38}$$

where in the terminology of Feynman diagram technique [18,22,37–41] the interaction vertex $A_{q,\alpha}$ is given by

$$\begin{aligned}
A_{q,\alpha} &= i \sqrt{\frac{\hbar}{2M\omega_\alpha(f)N}} \left(q e_{f,\alpha}\right) \sum_m V_{ei}(q) \exp\left(-i q R_m^0 + i f R_m^0\right) \\
&= i \sqrt{\frac{\hbar N}{2M\omega_\alpha(f)}} \left(q e_{f,\alpha}\right) V_{ei}(q) \Delta(f - q).
\end{aligned} \tag{15.39}$$

Graphically electron–phonon interaction for the processes of electron scattering on nucleus (ion) with phonon emission or phonon picking up by electron are illustrated in Fig. 15.3.

Note that secondly quantized interaction in (15.38) is quadratic in electron creation and annihilation operators and at the same time linear in phonon operators. As a result, electron–phonon interaction in Fig. 15.3 conserves the number of electrons, but does not conserve the number of phonons.

Thus, if we have oscillations of nuclei, then we get electron–phonon interaction in the system well known in literature (see [42,43]). In Eqs. (15.38)–(15.39), electron–phonon interaction is described by Frohlich Hamiltonian.

15.4 Wigner–Seitz Lattice

If ,we neglect now the ion oscillations, then we can drop out the Fourier component with $q = 0$ in the ion part. Effectively removing zero Fourier component from ionic part is equivalent to the addition of the homogeneous background to the electron part

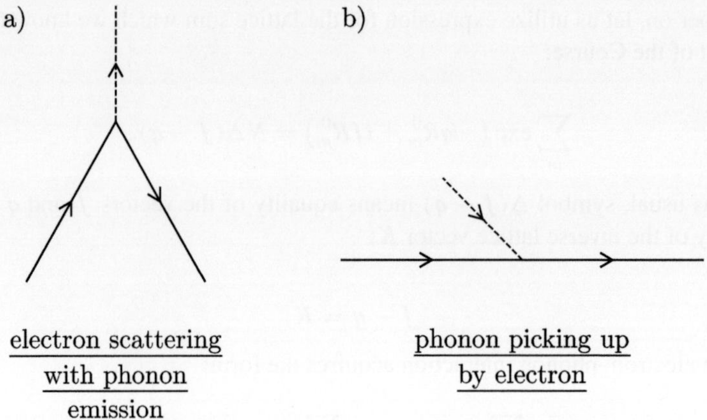

Fig. 15.3 Electron–phonon interaction **a** for the processes of electron scattering on nucleus (ion) with phonon emission by electron **b** phonon picking up by electron. Electron–phonon interaction conserves number of electrons on figures (**a**) and (**b**), but does not conserve the phonon number [1]

Fig. 15.4 Schematic illustration of the Wigner–Seitz lattice [1,24,44]. The lattice is formed by the planes drawn through the midpoints of the distances betweeen the neighboring ions [1]

(to satisfy electroneutrality condition in the system). Then, each elementary cell of the crystal is electroneutral.

Note that, in case of the complicated crystal lattice, we should draw the planes through the midpoints of the distances between neighboring ions. As a result, new lattice splitting appears. This splitting corresponds to the Wigner–Seitz lattice [1,24, 44] (see Fig. 15.4).

In the first approximation, two neighboring cubs in Fig. 15.4 do not interact with each other (there is only quadrupole interaction between them). Then, we can replace each of the Wigner–Seitz polyhedron by the equivalent spheres, drawn around each ion (see Fig. 15.5). Moreover, the volume of the sphere:

$$\Omega_0 = \frac{4\pi}{3} r_0^3 \qquad (15.40)$$

can be determined from the electroneutrality condition:

$$\rho_e \Omega_0 = Ze, \text{ and } \rho_e = \frac{Ze}{\Omega_0}, \qquad (15.41)$$

where Ze is the ion charge, ρ_e is the mean electron density.

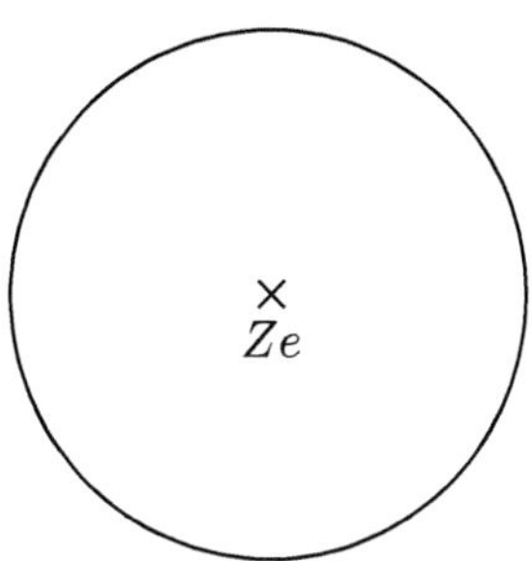

Fig. 15.5 Equivalent Wigner–Seitz sphere with the volume Ω_0, drawn around the ion with the charge Ze [1]

Correspondingly, average electron–ion interaction is given by

$$V_{ei} = \int d\mathbf{r}\,\varphi(\mathbf{r})\rho(\mathbf{r}) = -\int d\mathbf{r}\,\varphi(\mathbf{r})\rho_e = -\int d\mathbf{r}\,\frac{Ze}{r}\frac{Ze}{\Omega_0}$$

$$= -4\pi\frac{Z^2e^2}{\Omega_0}\int_0^{r_0}\frac{r^2 dr}{r} = -4\pi\frac{Z^2e^2}{\Omega_0}\frac{r_0^2}{2} = -\frac{3}{2}\frac{Z^2e^2}{r_0}, \tag{15.42}$$

where $\varphi(\mathbf{r}) = \frac{Ze}{r}$ is the ion potential, $\rho_e = \frac{Ze}{\Omega_0}$ is the average electron density, $\Omega_0 = \frac{4\pi}{3}r_0^3$ is the volume of the Wigner–Seitz sphere, and r_0 is the sphere radius.

The total minus sign in (15.42) accounts for the opposite signs of electron and ion charges and corresponds to attraction.

Further on, it is convenient to express in Rydberg units (Ry) the result for the average electron–ion interaction in Eq. (15.42). We know that 1Ry [1,33]:

$$1\text{Ry} = \frac{e^2}{2a_B} = 13.6\text{eV}, \tag{15.43}$$

where $a_B \approx 0.5\text{Å}$ is the Bohr radius [1,33].

Then multiplying nominator and denominator in the r.h.s. of (15.40), we get

$$V_{ei} = -\frac{3}{2}\frac{Z^2e^2}{r_0} = -\frac{3}{2}\frac{Z^2e^2a_B}{r_0a_B} = -\frac{3Z^2}{(r_0/a_B)}\text{Ry}. \tag{15.44}$$

Now we can introduce a very important notation of the correlation radius for electron plasma in metal [1,26,27]. By the definition the correlation radius r_S equals to

$$r_S = \frac{r_0'}{a_B} = \frac{\left(r_0/\sqrt[3]{Z}\right)}{a_B}. \tag{15.45}$$

Then the average electron–ion interaction in terms of the correlation radius is given by

$$V_{ei} = -\frac{3Z^2}{\sqrt[3]{Z}r_S}\text{Ry} = -\frac{3Z^{5/3}}{r_S}\text{Ry}. \tag{15.46}$$

15.5   Average Electron–Electron Interaction

Average electron–electron interaction in similarity with the average electron–ion interaction can be calculated as follows:

$$V_{ee} = \int d\boldsymbol{r}\,\varphi_e(\boldsymbol{r})\rho_e, \tag{15.47}$$

where $\rho_e = \frac{Ze}{\Omega_0}$ is the average electron density, while electron potential $\varphi_e(\boldsymbol{r})$ outside the shaded sphere in Fig. 15.6 yields

$$\varphi_e(\boldsymbol{r}) = \frac{q(r)}{r} = \frac{\frac{4\pi}{3}r^3\rho_e}{r} = \frac{Zer^2}{r_0^3}. \tag{15.48}$$

Correspondingly:

$$V_{ee} = 4\pi\frac{Z^2e^2}{\Omega_0 r_0^3}\int_0^{r_0} dr\, r^4 = \frac{3Z^2e^2}{r_0^6}\frac{r_0^5}{5} = 0.6\frac{Z^2e^2}{r_0} = 1.2\frac{Z^2e^2 a_B}{2r_0 a_B} = 1.2\frac{Z^2}{(r_0/a_B)}\text{Ry}$$

$$= 1.2\frac{Z^{5/3}}{\left(\dfrac{r_0}{a_B\sqrt[3]{Z}}\right)}\text{Ry} = 1.2\frac{Z^{5/3}}{r_S}\text{Ry}. \tag{15.49}$$

15.6   Average Energy of the Spherical Wigner–Seitz Cell

Average energy of the spherical Wigner–Seitz cell is determined by the sum of average electron–ion and electron–electron interactions, and reads

$$E_i = V_{ei} + V_{ee} = -\frac{3Z^{5/3}}{r_S}\text{Ry} + 1.2\frac{Z^{5/3}}{r_S}\text{Ry} = -1.8\frac{Z^{5/3}}{r_S}\text{Ry}. \tag{15.50}$$

Fig. 15.6 Potential created by the charge of the shaded sphere outside the sphere [1]

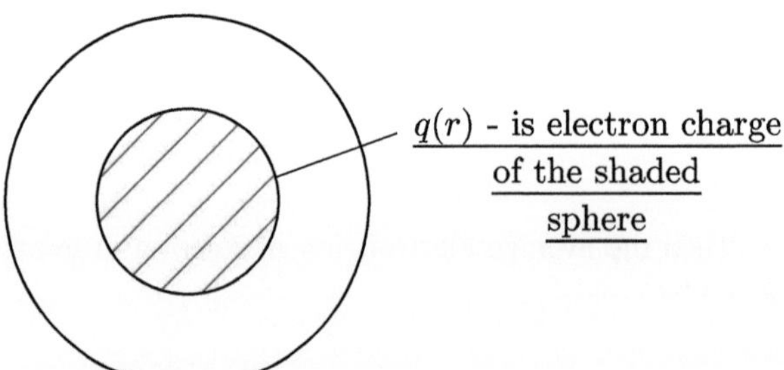

One more contribution to the cell energy yields non-Coulomb part of electron–ion interaction:

$$E = E_i + E_1, \tag{15.51}$$

where non-Coulomb part being recalculated on one cell is given by

$$E_1 = \frac{bZ}{\Omega_0} = \frac{3bZ}{4\pi r_0^3} = \frac{3b}{4\pi r_S^3 a_B^3} \sim \frac{1}{r_S^3}. \tag{15.52}$$

In fact, in the calculation of the average energy on one cell we drop out the Fourier component with $q = 0$. We can say that we consider electrons on top of the positive and homogeneous ion background. In this case the Hamiltonian of the electron subsystem can be written as

$$H = T_e + V_{ee} + V_{ei}. \tag{15.53}$$

15.7 Non-interacting Fermi Gas of Electrons at $T = 0$

Let us consider first the contribution to the kinetic energy from the Fermi gas of non-interacting electrons at $T = 0$ [1,26,27]. The secondly quantized operator of the electron kinetic energy T_e according to (15.14) has a form:

$$T_e = \sum_k \varepsilon_k a_k^\dagger a_k = \sum_k \varepsilon_k \hat{n}_k, \tag{15.54}$$

where $\hat{n}_k = a_k^\dagger a_k$ is the operator of the number of electrons with momentum k and spin projection σ.

At the same time, the secondly quantized operator of the total number of particles:

$$\widehat{N} = \sum_k a_k^\dagger a_k = \sum_k \hat{n}_k. \tag{15.55}$$

It is important to note that the operator of the total number of particles commutes with the total Hamiltonian of the system:

$$\left[\widehat{N}\,\widehat{H}_{\text{tot.}}\right] = 0. \tag{15.56}$$

Let us assume that, at $T = 0$, we have N particles in the system. As we know, non-interacting Fermi particles (electrons) in the ground state at temperature $T = 0$ fill in the Fermi sphere in the momentum space [45] (see Fig. 15.7).

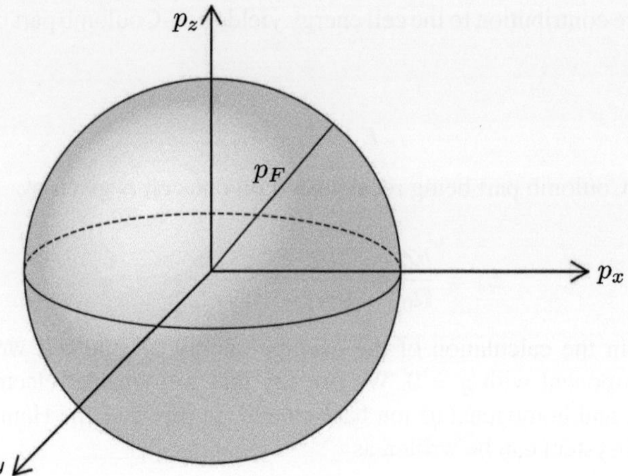

Fig. 15.7   Filled Fermi sphere for the non-interacting gas of electrons at $T = 0$ [1]

The volume of the Fermi sphere equals to

$$V = \frac{4\pi\, p_F^3}{3},$$

(15.57)

where p_F is the Fermi momentum.

Then, according to Quantum statistics [45], the total number of particles (electrons) is given by

$$N_e = 2 \cdot V \cdot \Omega \cdot \frac{1}{(2\pi\hbar)^3} = 2 \cdot \frac{4\pi\, p_F^3}{3(2\pi\hbar)^3} \cdot \Omega = \frac{p_F^3}{3\pi^2\hbar^3} \cdot \Omega = \frac{k_F^3}{3\pi^2} \cdot \Omega,$$

(15.58)

where Ω is the system volume, and factor 2 appears due to the summation over two projections of the electron spin $S = \frac{1}{2}$.

As a result, electron density (particle density with both spin projections):

$$n_e = \frac{N_e}{\Omega} = \frac{k_F^3}{3\pi^2},$$

(15.59)

while Fermi momentum:

$$p_F = \hbar k_F = \hbar \left(\frac{3\pi^2 N_e}{\Omega}\right)^{1/3} = \hbar \left(3\pi^2 n_e\right)^{1/3} \sim n_e^{1/3},$$

(15.60)

where k_F is the Fermi wave vector.

Let us emphasize that, by order of magnitude, the Fermi momentum reproduces the Heisenberg uncertainty principle [1,33,45]:

$$p_F \sim \hbar n_e^{1/3} \sim \frac{\hbar}{r_{\text{between electrons}}}.$$

(15.61)

Correspondingly, the Fermi energy:

$$\varepsilon_F = \frac{p_F^2}{2m} = \frac{\hbar^2 k_F^2}{2m} = \frac{\hbar^2 \left(3\pi^2 n_e\right)^{2/3}}{2m} \sim \frac{\hbar^2 n_e^{2/3}}{m}$$

$$\sim \frac{\hbar^2}{mr_{\text{between electrons}}^2} \propto \frac{1}{r_{\text{between electrons}}^2}. \tag{15.62}$$

Note that electron system is neutral, since it is located on top of the positive ion background.

It is not difficult to calculate kinetic energy of the non-interacting electron gas at $T = 0$. It yields

$$T = \frac{2\Omega}{(2\pi\hbar)^3} \int d^3 p \frac{p^2}{2m} n_p, \tag{15.63}$$

where $n_p = -\theta\left(p - p_F\right) = \begin{cases} 1, \ p < p_F \\ 0, \ p > p_F \end{cases}$ is the Dirac–Fermi-step [1,45].

As a result, the kinetic energy reads

$$\begin{aligned} T &= \frac{2\Omega \cdot 4\pi}{(2\pi\hbar)^3} \int_0^{p_F} \frac{p^2}{2m} \cdot p^2 dp = \frac{\Omega}{2m\pi^2\hbar^3} \int_0^{p_F} dp\, p^4 \\ &= \frac{\Omega}{\pi^2\hbar^3} \cdot \frac{1}{2m} \cdot \frac{p_F^5}{5} = \frac{\Omega p_F^3}{5\pi^2\hbar^3} \cdot \varepsilon_F \\ &= \frac{3}{5} N_e \varepsilon_F. \end{aligned} \tag{15.64}$$

Hence, average kinetic energy on one particle (one electron) equals to

$$\bar{\varepsilon} = \frac{T}{N_e} = \frac{3}{5}\varepsilon_F. \tag{15.65}$$

If the ion charge equals to Z, then, from the electroneutrality condition, it follows that, in the Wigner–Seitz cell, we have Z electrons and one ion. Then the average kinetic energy of electrons on one cell reads

$$E_{\text{kin.on 1 cell}} = Z\bar{\varepsilon} = \frac{3}{5}\varepsilon_F Z = \frac{3Zp_F^2}{10\,m}. \tag{15.66}$$

Let us represent this energy in terms of the correlation radius r_S and let us measure it in Rydberg units again. According to the Quasiclassical Bohr theory [33] (see Eq. (15.43)):

$$1\text{Ry} = \frac{e^2}{2a_B} = \frac{\hbar^2}{2\,ma_B^2}, \tag{15.67}$$

where m is the electron mass, and thus the Bohr radius [33]:

$$a_B = \frac{\hbar^2}{me^2}.$$ (15.68)

Further on, let us utilize the electroneutrality condition (15.41):

$$\rho_e = en_e = \frac{Ze}{\Omega_0}, \text{ or } n_e = \frac{p_F^3}{3\pi^2\hbar^3} = \frac{Z}{\Omega_0} = \frac{3Z}{4\pi r_0^3}.$$ (15.69)

Then we get

$$\frac{4p_F^3}{9\pi\,\hbar^3} = \frac{Z}{r_0^3}.$$ (15.70)

Multiplying the nominators of the l.h.s. and r.h.s. on a_B^3, we obtain

$$\frac{4p_F^3 a_B^3}{9\pi\,\hbar^3} = \frac{Za_B^3}{r_0^3} = \frac{1}{r_S^3}.$$ (15.71)

Thus, we can establish the basic relationship between correlation radius and Fermi momentum for the electron plasma in metals [1,26,27]:

$$r_S = \frac{\hbar}{p_F a_B}\left(\frac{9\pi}{4}\right)^{1/3} \approx 1.92\frac{\hbar}{p_F a_B} = \frac{1.92}{k_F a_B} \sim n_e^{-1/3}.$$ (15.72)

It is interesting to note that, in terms of the radius of the Wigner–Seitz sphere r_0, Fermi wave vector according to the uncertainty principle is given by

$$k_F \approx \frac{1.92\sqrt[3]{Z}}{r_0}.$$ (15.73)

Now we can write the average kinetic energy of electrons on one cell in Rydberg units. To do that we can multiply the nominator and denominator of the last equality in the r.h.s. of Eq. (15.66) on a_B^2. As a result:

$$E_{\text{kin.on 1 cell}} = \frac{3Zp_F^2}{10\,m} = \frac{3Zp_F^2 a_B^2}{10\,ma_B^2} = \frac{3Zk_F^2 a_B^2}{5} \cdot \frac{\hbar^2}{2\,ma_B^2} = \frac{3Zk_F^2 a_B^2}{5}\text{Ry.}$$ (15.74)

Afterwards, it is convenient to multiply the nominator and denominator of the last equality in the r.h.s. of (15.74) additionally on $(1.92)^2$. Then, using the definition of the correlation radius r_S in (15.72), we get for the non-interacting electron gas:

$$E_{\text{kin.on 1 cell}} = \frac{3Zk_F^2 a_B^2}{5}\text{Ry} = \frac{3Zk_F^2 a_B^2 \cdot (1.92)^2}{5 \cdot (1.92)^2}\text{Ry}$$

$$= \frac{3 \cdot (1.92)^2}{5}\frac{Z}{r_S^2}\text{Ry} \approx \frac{2.21Z}{r_S^2}\text{Ry}. \tag{15.75}$$

15.8 Pressure of the Non-interacting Fermi Gas of Electrons at $T = 0$

Let us determine the pressure of the non-interacting Fermi gas of electrons at $T = 0$. To do that let us use the well-known thermodynamic relation [45]:

$$P = -\left(\frac{\partial E_{\text{kin.}}}{\partial \Omega}\right)_{N_e}. \tag{15.76}$$

But the kinetic energy of the electron subsystem is proportional to

$$E_{\text{kin.}} \sim n_e^{2/3} N_e = \left(\frac{N_e}{\Omega}\right)^{2/3} N_e = N_e^{5/3}\Omega^{-2/3}. \tag{15.77}$$

Hence the pressure:

$$P = -\left(\frac{\partial E_{\text{kin.}}}{\partial \Omega}\right)_{N_e} \sim \frac{2}{3}N_e^{5/3}\Omega^{-5/3} = \frac{2}{3}\frac{E_{\text{kin.}}}{\Omega}. \tag{15.78}$$

As a result, equation of state for the Fermi gas of the non-interacting electrons at $T = 0$ acquires a form [1,45]:

$$P\Omega = \frac{2}{3}E_{\text{kin.}}. \tag{15.79}$$

Thus, the system of non-interacting electrons (ideal Fermi gas) in contrast to the ideal Bose gas always has a rigidity due to the Pauli principle [1,33,45].

References

1. Y. Kagan, *Lectures on the Solid-State Theoretical Physics*, Moscow, MEPHI, 1981–1982. Part II. Electrons (Unpublished)
2. I.M. Lifshitz, M.Y. Azbel', M.I. Kaganov, *Electron Theory of Metals* (Consultants Bureau, New York, 1973)
3. A.A. Abrikosov, *Fundamentals of the Theory of Metals* (North-Holland, Amsterdam, 1988)
4. L.D. Landau, E.M. Lifshitz, *Electrodynamics of Continuous Media* (Pergamon Press, 1984)
5. J. Hubbard, Proc. Royal Soc. Lond. **A276**, 238 (1963)
6. J. Hubbard, Proc. Roy. Soc. Lond. **A277**, 237 (1964)

7. J. Hubbard, Proc. Roy. Soc. Lond. **A281**, 401 (1964)
8. J. Hubbard, Proc. Roy. Soc. Lond. **A285**, 542 (1965)
9. P. Fulde, *Electron Correlations in Molecules and Solids*. Springer Series in Solid State Sciences. (Springer, Berlin, Heidelberg, New York, 1993)
10. N.F. Mott, *Metal–Insulator Transitions* (Taylor and Francis LTD, London, 1974)
11. J.M. Ziman (ed.), *The Physics of Metals* (Cambridge University Press, 1960)
12. Y.I. Frenkel, *Introduction to the Theory of Metals* (State Technical and Theoretical Publishing House, Leningrad-Moscow, 1948) (in Russian)
13. Y.G. Dorfman, I.K. Kikoin, *Physics of Metals*. Electric, Optical and Magnetic Properties (State Technical and Theoretical Publishing House, Leningrad-Moscow, 1933) (in Russian)
14. A.H. Wilson, *Quantum Theory of Metals* (Cambridge University Press, New York, 1953)
15. D.I. Khomskii, *Transition Metal Compounds* (Cambridge University Press, 2014)
16. S.V. Vonsovskii, M.I. Kaznelson, *Quantum Solid-State Physics*. Springer Series in Solid-State Sciences (Berlin–Heidelberg–New York, 1989)
17. W.A. Harrison, *Electronic Structure and the Properties of Solids* (W.H. Freeman and Company, San Francisco, 1980)
18. E.G. Brovman, Y. Kagan, *Phonons in Non-transition Metals*. Sov. Phys. Uspekhi. *112*, 369 (1974)
19. E.G. Brovman, Yu. Kagan, A. Kholas, Structure of the metallic hydrogen at zero pressure. Sov. Phys. JETP **34**, 1300 (1972)
20. E.G. Brovman, Y. Kagan, A. Kholas, V. Pushkarev, The role of electron–electron interaction in the formation of the metastable phase of the metallic hydrogen. JETP Lett. **18**, 869 (1973)
21. E.M. Lifshitz, L.P. Pitaevskii, *Statistical Physics, Part II, Theory of the Condensed State* (Pergamon, Elsevier, 2013)
22. A.A. Abrikosov, L.P. Gor'kov, I.E. Dzyaloshinskii, *Methods of Quantum Field Theory in Statistical Physics* (Prentice Hall, Englewood Cliffs, New Jersey, 1963)
23. R.P. Feynman, *Statistical Mechanics: A Set of Lectures*. Advanced Books, Classics (Westview Press, Boulder, 1998)
24. N.W. Ashcroft, N.D. Mermin, *Solid State Physics* (Holt, Rinehart and Winston, 1976)
25. O. Madelung, *Introduction to Solid-State Theory*. Springer Series in Solid-State Sciences (1995)
26. D. Pines, *Elementary Excitations in Solids* (W.A. Benjamin Inc., New York-Amsterdam, 1963)
27. D. Pines, P. Nozieres, *The Theory of Quantum Liquids: Normal Fermi Liquids*, vol. 1 (W.A. Benjamin Inc. New York–Amsterdam, 1966)
28. J.M. Ziman, *Electrons and Phonons: The Theory of Transport Phenomena in Solids* (Clarendon Press, Oxford, 1960)
29. C. Kittel, *Quantum Theory of Solids* (Wiley, New York–London, 1963)
30. J.W. Negele, H. Orland, *Quantum Many-Particle Systems*. Frontiers in Physics (Addison-Wesley Publishing Company, 1992)
31. S.W. Lowesley, *Condensed Matter Physics: Dynamic Correlations*. Frontiers in Physics (The Benjamin Cummings Publishing Company, 1986)
32. W.A. Harrison, *Solid State Theory* (McGraw-Hill Book Company, New-York–London–Toronto, 1970)
33. L.D. Landau, E.M. Lifshitz, *Quantum Mechanics: Non-relativistic Theory* (Pergamon Press, New York, 1977)
34. S. Raimes, *Many-Electron Theory* (North-Holland Publishing Company, Amsterdam–London, 1972)
35. D. Ter Haar, *Introduction to the Physics of Many-Body Systems* (Inter-science Publishers Inc., New York, 1958)
36. F.A. Berezin, *The Method of Second Quantization* (Academic Press, 1966)
37. R.P. Feynman, The Theory of Positrons. Phys. Rev. **76**, 749 (1949)
38. M.V. Sadovskii, *Diagrammatics*. Lectures on Selected problems in Condensed Matter Theory (World Scientific, Singapore, 2006)
39. L.S. Levitov, *Green's Functions in Problems* (Princeton University Press, 2011)
40. R. Mattuck, *A Guide to Feynman Diagrams in the Many-Body Problem* (McGraw-Hill, 1976)

41. C.P. Enz, *A Course on Many-Body Theory Applied to Solid-state Physics*. Lecture Notes in Physics (World Scientific, 1992)
42. H. Frohlich, Adv. Phys. **3**, 325 (1954)
43. J.T. Devreese, Frohlich polarons. Lecture course including detailed theoretical description, presented on Enrico Fermi Summer School, Varenna, Italy 2005, cond.-mat 1012.4576
44. E. Wigner, F. Seitz, Phys. Rev. **43**, 804 (1933)
45. L.D. Landau, E.M. Lifshitz, *Statistical Physics, Part I* (Butterworth-Heinemann, 1980)

Lecture 16. Electron–Electron Interaction: Exchange Hole for the Wave Function of Two Electrons with Parallel Spins

16

Abstract

In this Lecture, we find the first-order Hartree–Fock correction to electron energy. It proves to be negative leading to the formation of the exchange hole for the wave function of two electrons with parallel spins. We calculate the density–density (pairwise) correlation function and find its reduction at small distances.

In the end of the previous Lecture, we considered non-interacting Fermi gas of electrons [1,2]. Let us account now for the electron–electron interaction [1].

16.1 Electron–Electron Interaction

In the second quantization [1], electron–electron interaction according to Eq. (15.23) has a form:

$$
V_{ee} = \frac{1}{2} \sum_{k_1,k_2,q \neq 0,\sigma_1,\sigma_2} \frac{4\pi e^2}{q^2 \Omega} a^\dagger_{k_1\sigma_1} a^\dagger_{k_2,\sigma_2} a_{k_2+q,\sigma_2} a_{k_1-q,\sigma_1}.
$$

Let us determine now the <u>first correction</u> in the interaction to the energy of the electron gas in Eq. (15.74). This correction reads [1]

$$
E^{(1)} = \langle\{n^0\}|V_{ee}|\{n^0\}\rangle \sim \langle\{n^0\}|a^\dagger_{k_1\sigma_1} a^\dagger_{k_2\sigma_2} a_{k_2+q\sigma_2} a_{k_1-q\sigma_1}|\{n^0\}\rangle, \tag{16.1}
$$

where n^0 are the occupation numbers not perturbed by the interaction.

Let us use the Wick theorem [1] and understand what pairwise convolutions of the creation and annihilation operators in (16.1) are different from zero.

For $q = 0$, we can consider the convolutions of the first and fourth operators, and of the second and third operators, respectively (each convolution with its own

M. Kagan, *Lecture Notes of Professor Yuri Kagan in Theoretical Solid-State Physics*, Lecture Notes in Physics 1048, https://doi.org/10.1007/978-3-032-14621-2_16

spin projection σ_1 and σ_2). However, these convolutions are not allowed due to the electroneutrality condition ($\boldsymbol{q} \neq 0$).

The remaining possibility is to use the anticommutation of the second and third operators and to transpose them. Then, after the transposition we can make the convolution of the first and third operators and of the second and the fourth ones. For the convolution of the first and the third operators, the following relations should be fulfilled:

$$\sigma_1 = \sigma_2, \text{ and } \boldsymbol{k}_1 = \boldsymbol{k}_2 + \boldsymbol{q}. \tag{16.2}$$

In this case from (16.2), it follows immediately that

$$\sigma_2 = \sigma_1, \text{ and } \boldsymbol{k}_2 = \boldsymbol{k}_1 - \boldsymbol{q}. \tag{16.3}$$

Note that the relations (16.3) should be satisfied to make the convolution of the second and fourth operators.

Let us emphasize that for both convolutions the condition $\boldsymbol{q} \neq 0$ is valid.

As a result, we get from (16.1)

$$
\begin{aligned}
E^{(1)} &= -\frac{1}{2} \sum_{\boldsymbol{k}_1, \boldsymbol{k}_2, \boldsymbol{q} \neq 0, \sigma_1, \sigma_2} \frac{4\pi e^2}{q^2 \Omega} \left\langle a^{\dagger}_{\boldsymbol{k}_1 \sigma_1} a_{\boldsymbol{k}_2 + \boldsymbol{q}, \sigma_2} \right\rangle \left\langle a^{\dagger}_{\boldsymbol{k}_2, \sigma_2} a_{\boldsymbol{k}_1 - \boldsymbol{q}, \sigma_1} \right\rangle \\
&= -\frac{1}{2} \sum_{\boldsymbol{k}_1, \boldsymbol{q} \neq 0, \sigma} \frac{4\pi e^2}{q^2 \Omega} n_{\boldsymbol{k}_1, \sigma} n_{\boldsymbol{k}_1 - \boldsymbol{q}, \sigma} \\
&= -\frac{1}{2} \sum_{\boldsymbol{k}_1, \boldsymbol{k}_2, \sigma} \frac{4\pi e^2}{(\boldsymbol{k}_1 - \boldsymbol{k}_2)^2 \Omega} n_{\boldsymbol{k}_1 \sigma} n_{\boldsymbol{k}_2 \sigma},
\end{aligned}
\tag{16.4}
$$

where while proceeding to the last equality in the r.h.s. of Eq. (16.4) we used that

$$\boldsymbol{q} = \boldsymbol{k}_1 - \boldsymbol{k}_2. \tag{16.5}$$

Let us emphasize once more that the sign minus in (16.4) is related to the transposition of the second and third operators in (16.1).

It is important to note that first correction in interaction

$$E^{(1)} < 0, \tag{16.6}$$

corresponds to the lowering of energy.

How we can understand this result? It seems from the first sight that electrons repel each other, and the energy with an account of the interaction should increase.

16.2   Explanation of the Effect: Exchange Hole

If we account for the term with $q = 0$ in the interaction, then we would get the positive correction. This correction corresponds to the Hartree term [1] in the diagrammatic expansion [1] for the self-energy [3,5] in Fig. 16.1a.

At the same time, a second term arises in the expansion of the self-energy. It is the Fock term [1] in Fig. 16.1b. It describes purely exchange interaction, connected with the particle transposition, and has a negative sign.

16.3   Symmetry of the Two-Electron Wave Function

Physically the reason for the negative correction to the energy is as follows. The wave function of two electrons is antisymmetric [20]. If two spins are parallel, then the spin part of the wave function is symmetric. It means that coordinate part of the wave function is antisymmetric. Hence, two electrons cannot come close to each other. Hence, the mean distance between electrons is larger than in the case of the homogeneous electron density.

Thus, effective Coulomb energy $E_{\text{eff.}} = \frac{e^2}{r_{\text{eff.}}}$ is reduced. It is just the exchange term with the minus sign which lowers the energy of the homogeneous smearing $V_{ee} = 1.2 \frac{Z^{5/3}}{r_S}$ Ry (calculated in the previous Lecture) till the effective one.

Let us finish now the calculation of the exchange correction in Eq. (16.4) for the case of two parallel spins. Let us utilize the fact that the sum over spin projections:

$$\sum_{\sigma} = 2, \tag{16.7}$$

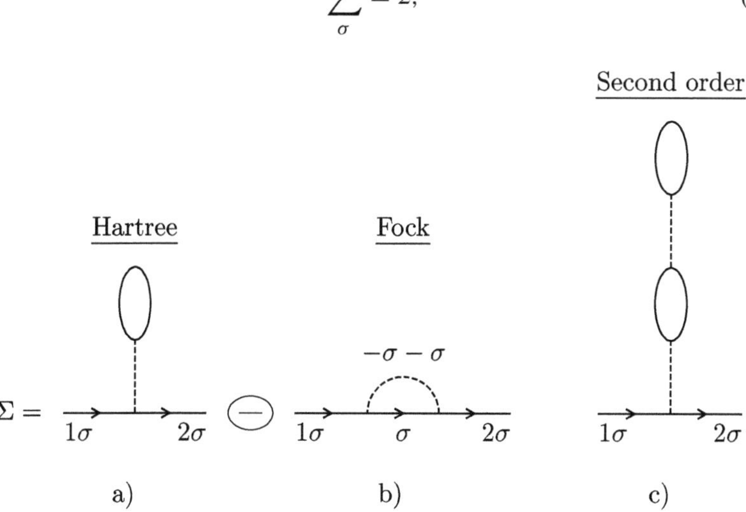

Fig. 16.1 Expansion of the self-energy in the interaction [1]. We show the positive contribution from the Hartree term **a** and the negative contribution from the exchange Fork term **b** in the linear approximation with respect to interaction. On **c**, we illustrate also the simplest quadratic in interaction term [1]

while the Fourier component of the Coulomb potential equals to

$$\frac{4\pi e^2}{(k_1 - k_2)^2 \, \Omega} = \frac{e^2}{\Omega} \int d^3 r \frac{1}{r} \exp\left(i\left(k_1 - k_2\right) r\right). \tag{16.8}$$

Let us proceed also from the summation over wave vectors k_1 and k_2 to the integration by $\Omega \frac{d^3 k_1}{(2\pi)^3}$ and $\Omega \frac{d^3 k_2}{(2\pi)^3}$.

As a result, the first correction to the energy:

$$
\begin{aligned}
E^{(1)} &= -\sum_{k_1, k_2} \frac{4\pi e^2}{(k_1 - k_2)^2 \, \Omega} n_{k_1} n_{k_2} \\
&= -\left(\frac{\Omega}{(2\pi)^3}\right)^2 \cdot \frac{e^2}{\Omega} \int d^3 k_1 n_{k_1} \int d^3 k_2 n_{k_2} \cdot \int d^3 r \frac{1}{r} \exp\left(i\left(k_1 - k_2\right) r\right) \\
&= -\left(\frac{\Omega}{(2\pi)^3}\right)^2 \cdot \frac{e^2}{\Omega} \int d^3 r \frac{1}{r} \cdot \int d^3 k_1 n_{k_1} \exp\left(ik_1 r\right) \int d^3 k_2 n_{k_2} \exp\left(-ik_2 r\right),
\end{aligned}
\tag{16.9}
$$

where we transposed the order of integration.

Now in the second integral, we can replace k_2 on $-k_2$ and utilize the parity of the Fermi–Dirac distribution function $n_{k_2} = -\theta\left(k_2 - k_F\right)$ with respect to this replacement [1,2]. Then, the integrals by $d^3 k_1$ and $d^3 k_2$ equal to each other, and the first correction acquires a form:

$$E^{(1)} = -\left(\frac{\Omega}{(2\pi)^3}\right)^2 \cdot \frac{e^2}{\Omega} \cdot \int d^3 r \frac{1}{r} \cdot \left\{\int d^3 k \, n_k \exp(ikr)\right\}^2. \tag{16.10}$$

Let us calculate the integral in the figure brackets. At $T = 0$:

$$
\begin{aligned}
\int d^3 k \, n_k \exp(ikr) &= 2\pi \int_0^{k_F} dk \, k^2 \int_{-1}^{1} d\cos\theta \, \exp(ikr\cos\theta) \\
&= 2\pi \int_0^{k_F} dk \, k^2 \frac{\exp(ikr) - \exp(-ikr)}{ikr} \\
&= \frac{4\pi}{r} \int_0^{k_F} dk \, k \frac{\exp(ikr) - \exp(-ikr)}{2i} \\
&= \frac{4\pi}{r} \int_0^{k_F} dk \, k \sin kr,
\end{aligned}
\tag{16.11}
$$

where we used the well-known representation of the sine function.

Further on let us introduce dimensionless variable $x = kr$ and calculate by parts the integral in (16.11). Then:

$$
\begin{aligned}
\int d^3k\, n_k \exp(i\mathbf{kr}) &= \frac{4\pi}{r} \int_0^{k_F} dk\, k \sin kr \\
&= \frac{4\pi}{r^3} \int_0^{k_F r} dx\, x \sin x = \frac{4\pi}{r^3} \left\{ -x \cos x \big|_0^{k_F r} + \int_0^{k_F r} dx \cos x \right\} \\
&= \frac{4\pi}{r^3} \left(-k_F r \cos k_F r + \sin x \big|_0^{k_F r} \right) \\
&= \frac{4\pi}{r^3} \left(\sin k_F r - k_F r \cos k_F r \right).
\end{aligned}
\tag{16.12}
$$

Squaring up (16.12) and substituting the result in (16.10), we get

$$
\begin{aligned}
E^{(1)} &= -\left(\frac{16\pi^2 \Omega e^2}{64\pi^6} \right) \int d^3r\, \frac{1}{r} \cdot \frac{1}{r^6} \cdot (\sin k_F r - k_F r \cos k_F r)^2 \\
&= -\frac{\Omega e^2}{4\pi^4} 4\pi \int_0^{\infty} dr\, \frac{r^2}{r^7} \cdot (\sin k_F r - k_F r \cos k_F r)^2 \\
&= -\frac{\Omega e^2}{\pi^3} \int_0^{\infty} \frac{dr}{r^5} \cdot (\sin k_F r - k_F r \cos k_F r)^2 .
\end{aligned}
\tag{16.13}
$$

Let us introduce again the dimensionless variable $y = k_F r$. Then the first correction to the energy:

$$
E^{(1)} = -\frac{\Omega e^2 k_F^4}{\pi^3} \int_0^{\infty} \frac{dx}{x^5} \cdot (\sin x - x \cos x)^2.
\tag{16.14}
$$

Integral in (16.14) is convergent on the lower limit as $\int \frac{dx}{x^5} x^6 = \int dx\, x \sim x^2$ for $x \ll 1$, and can be calculated exactly:

$$
\int_0^{\infty} \frac{dx}{x^5} \cdot (\sin x - x \cos x)^2 = \frac{1}{4}.
\tag{16.15}
$$

As a result, the final expression for the first correction to the energy reads

$$
E^{(1)} = -\frac{\Omega e^2 k_F^4}{4\pi^3}.
\tag{16.16}
$$

Note that the calculation of the integral in the Formulas (16.14)–(16.16) is equivalent to the calculation of the initial sum:

$$
E^{(1)} = -\sum_{k_1, k_2} \frac{4\pi e^2}{(k_1 - k_2)^2 \,\Omega} n_{k_1} n_{k_2} = -\frac{4\pi e^2}{\Omega} \sum_{k_1 < k_F,\, k_2 < k_F} \frac{1}{(k_1 - k_2)^2}.
\tag{16.17}
$$

The detailed calculation of this sum we can find in the classical monograph of Kittel *"Quantum Theory of Solids"* [8].

Let us represent the energy correction in Eq. (16.16) in Rydberg units [20]. To do that let us multiply the nominator and denominator of (16.16) on the Bohr radius a_B and use expression (15.75) for 1 Ry. Then, we obtain

$$E^{(1)} = -\frac{\Omega e^2 k_F^4 a_B}{4\pi^3 a_B} = -\frac{k_F^4 \Omega a_B}{2\pi^3} \cdot \frac{e^2}{2a_B} = -\frac{k_F^4 \Omega a_B}{2\pi^3} \, \text{Ry}. \qquad (16.18)$$

Now we can use the definition (15.72) for the correlation radius (expressed via the Fermi wave vector) and rewrite the first correction $E^{(1)}$ as

$$E^{(1)} = -\frac{k_F^3 \Omega}{2\pi^3} \cdot k_F a_B \, \text{Ry} = -\frac{k_F^3 \Omega}{2\pi^3} \cdot 1.92 \cdot \frac{k_F a_B}{1.92} \, \text{Ry} = -\frac{k_F^3 \Omega}{2\pi^3} \cdot 1.92 \frac{\text{Ry}}{r_S}. \qquad (16.19)$$

Further on, let us substitute in (16.18) expression (15.58) for the total number of particles:

$$E^{(1)} = -\frac{k_F^3 \Omega}{2\pi^3} \cdot 1.92 \frac{\text{Ry}}{r_S} = -\frac{k_F^3 \Omega}{3\pi^2} \cdot \frac{3}{2\pi} \cdot 1.92 \frac{\text{Ry}}{r_S} = -N_e \cdot \frac{3}{2\pi} \cdot 1.92 \frac{\text{Ry}}{r_S}. \qquad (16.20)$$

Then for one particle, we get

$$\frac{E^{(1)}}{N_e} = -\frac{3}{2\pi} \cdot 1.92 \frac{\text{Ry}}{r_S} \approx -\frac{0.916}{r_S} \, \text{Ry}. \qquad (16.21)$$

Note that from the electroneutrality condition, there are Z electrons in one Wigner–Seitz cell [1,27]. Thus the exchange correction to the energy recalculated on one cell yields

$$E_{\text{ex.on 1 cell}} = \frac{ZE^{(1)}}{N_e} \approx -\frac{0.916Z}{r_S} \, \text{Ry}. \qquad (16.22)$$

At the same time, according to (15.75), the kinetic energy of electrons recalculated on one cell is given by

$$E_{\text{kin.on 1 cell}} = \frac{2.21Z}{r_S^2} \, \text{Ry}.$$

We come to the conclusion that since the correlation radius $r_S \sim n_e^{-1/3}$, then when we increase the density (decrease r_S), the kinetic energy $E_{\text{kin.on 1 cell}} \sim \frac{1}{r_S^2}$ grows faster than the absolute value of the exchange energy $|E_{\text{ex.on 1 cell}}| \sim \frac{1}{r_S}$.

Let us emphasize that, in many textbooks on the solid-state theory (see, e.g., [15, 16, 26]), the sum of the kinetic and exchange energies recalculated on one electron

$$E_{HF} = \left(\frac{2.21}{r_S^2} - \frac{0.916}{r_S} \right) \mathrm{Ry} \tag{16.23}$$

is usually called the electron energy in the Hartree–Fock approximation [1,6,7,21,26].

16.4 Correlator of Two Electron Densities

Let us find the expression for the correlator of two electron densities in the second quantization:

$$\langle \hat{\rho}(r)\hat{\rho}(r + R) \rangle.$$

To do that, let us write the operator of electron density

$$\rho(r) = \sum_{\alpha} \delta(r - r_\alpha) \tag{16.24}$$

in the secondly quantized form as

$$\hat{\rho}(r) = \sum_{\alpha} \sum_{k_1,k_2,\sigma} a^{\dagger}_{k_1\sigma} a_{k_2\sigma} \langle k_1 | \delta(r - r_\alpha) | k_2 \rangle. \tag{16.25}$$

In Eq. (16.25), the matrix element:

$$\langle k_1 | \delta(r - r_\alpha) | k_2 \rangle \tag{16.26}$$

is calculated in the basis of the plane waves:

$$|k\rangle = \frac{1}{\sqrt{\Omega}} \exp(ikr), \tag{16.27}$$

Then for the matrix element, we get

$$\delta(r - r_\alpha) \frac{1}{\Omega} \exp(-ik_1 r_\alpha + ik_2 r_\alpha) = \frac{1}{\Omega} \delta(r - r_\alpha) \exp(iqr_\alpha), \tag{16.28}$$

where the transferred momentum:

$$q = k_2 - k_1. \tag{16.29}$$

Further on, replacing the sum over α by the integral $\sum_\alpha \to \int dr_\alpha$, we can represent the density operator in Eq. (16.25) in the following form:

$$\hat{\rho}(\boldsymbol{r}) = \frac{1}{\Omega} \sum_{k_1,k_2,\sigma} a^\dagger_{k_1\sigma} a_{k_2\sigma} \int d\boldsymbol{r}_\alpha \delta\left(\boldsymbol{r} - \boldsymbol{r}_\alpha\right) \exp\left(i\boldsymbol{q}\boldsymbol{r}_\alpha\right)$$

$$= \frac{1}{\Omega} \sum_{k_1,k_2,\sigma} a^\dagger_{k_1\sigma} a_{k_2\sigma} \exp(i\boldsymbol{q}\boldsymbol{r}). \tag{16.30}$$

Now we can proceed from the summation over $\boldsymbol{k}_1$ and $\boldsymbol{k}_2$ in (16.30) to the summation over $\boldsymbol{k}_1$ and $\boldsymbol{q} = \boldsymbol{k}_2 - \boldsymbol{k}_1$. As a result, the secondly quantized density operator in the real space equals to

$$\hat{\rho}(\boldsymbol{r}) = \frac{1}{\Omega} \sum_{k_1,q,\sigma} a^\dagger_{k_1\sigma} a_{k_1+q\sigma} \exp(i\boldsymbol{q}\boldsymbol{r}). \tag{16.31}$$

After that, reversing the order of summation over $\boldsymbol{q}$ and $\boldsymbol{k}_{1,\sigma}$ in (16.31), we can represent the density operator as the Fourier expansion:

$$\hat{\rho}(\boldsymbol{r}) = \sum_q \hat{\rho}(\boldsymbol{q}) \exp(i\boldsymbol{q}\boldsymbol{r}), \tag{16.32}$$

where we introduced the secondly quantized density operator in the momentum space:

$$\hat{\rho}(\boldsymbol{q}) = \frac{1}{\Omega} \sum_{k_1,\sigma} a^\dagger_{k_1\sigma} a_{k_1+q\sigma} = \frac{1}{\Omega} \sum_{k\sigma} a^\dagger_{k\sigma} a_{k+q\sigma}. \tag{16.33}$$

Then the correlator of two densities in (16.24) yields

$$\langle \hat{\rho}(\boldsymbol{r})\hat{\rho}(\boldsymbol{r} + \boldsymbol{R}) \rangle = \frac{1}{\Omega^2} \sum_{kq\sigma k'q'\sigma'} \left\langle 0 | a^\dagger_{k\sigma} a_{k+q\sigma} a^\dagger_{k'\sigma'} a_{k'+q'\sigma'} | 0 \right\rangle$$

$$\cdot \exp\left(i\boldsymbol{q}\boldsymbol{r} + i\boldsymbol{q}'(\boldsymbol{r} + \boldsymbol{R})\right). \tag{16.34}$$

Further on, we can apply the Wick theorem [1,13,14,19] and distinguish two options of the convolutions. The first option is related to the convolutions

$$\left\langle a^\dagger_{k\sigma} a_{k+q\sigma} \right\rangle \left\langle a^\dagger_{k'\sigma'} a_{k'+q'\sigma'} \right\rangle \tag{16.35}$$

for the first and second, and the third and fourth operators. It assumes the fulfillment of the condition $q = q' = 0$. As a result, the following term appears in the density correlator:

$$\frac{1}{\Omega^2} \sum_{k\sigma k'\sigma'} n_{k\sigma} n_{k'\sigma'},$$ (16.36)

where electron occupation numbers for one spin projection read

$$n_{k\sigma} = -\theta \left(k - k_F\right) = \begin{cases} 1, & \text{for } k \leq k_F, \\ 0, & \text{for } k > k_F. \end{cases}$$ (16.37)

The second option is connected with the convolutions of the first and fourth, and of the second and third operators, respectively:

$$\left\langle a_{k\sigma}^{\dagger} a_{k'+q'\sigma'} \right\rangle \left\langle a_{k+q\sigma} a_{k'\sigma'}^{\dagger} \right\rangle.$$ (16.38)

This option requires the fulfillment of the condition:

$$k = k' + q', \sigma = \sigma',$$ (16.39)

for the first convolution in (16.38). At the same time, for the second convolution, we should demand

$$k' = k + q, \sigma' = \sigma.$$ (16.40)

Thus:

$$q' = k - k', q = k' - k, \text{ and } q' = -q.$$ (16.41)

In this case, we get the following term in the density correlator:

$$\frac{1}{\Omega^2} \sum_{k\sigma k'} n_{k\sigma} \left(1 - n_{k'\sigma}\right) \exp\left(i\left(k - k'\right) R\right).$$ (16.42)

Hence, the correlator of two electron densities is given by

$$\left\langle \hat{\rho}(r)\hat{\rho}(r + R) \right\rangle = \frac{1}{\Omega^2} \sum_{k\sigma k'\sigma'} n_{k\sigma} n_{k'\sigma'}$$
$$+ \frac{1}{\Omega^2} \sum_{k\sigma k'} n_{k\sigma} \left(1 - n_{k'\sigma}\right) \exp\left(i\left(k - k'\right) R\right).$$ (16.43)

The correlator depends only on the distance between correlating points. This happens due to space homogeneity.

For $\boldsymbol{R} \neq 0$, we get for the correlator:

$$\langle \hat{\rho}(r)\hat{\rho}(r + \boldsymbol{R})\rangle = n_e^2 + \frac{1}{\Omega^2}\sum_{kk'\sigma} n_{k\sigma} \exp\left(i\left(\boldsymbol{k} - \boldsymbol{k}'\right)\boldsymbol{R}\right)$$

$$- \frac{1}{\Omega^2}\sum_{kk'\sigma} n_{k\sigma}n_{k'\sigma} \exp\left(i\left(\boldsymbol{k} - \boldsymbol{k}'\right)\boldsymbol{R}\right), \qquad (16.44)$$

where electron density (on both spin projections) in the first term in (16.44) equals to

$$n_e = \frac{1}{\Omega}\sum_{k\sigma} n_{k\sigma} = \frac{k_F^3}{3\pi^2}. \qquad (16.45)$$

Further on, let us demonstrate that, for $\boldsymbol{R} \neq 0$, the second term in (16.43) goes to zero.

Indeed, proceeding from the summation to the integration in the second term and utilizing the properties of the Dirac step function in (16.37), we obtain

$$\frac{1}{\Omega^2}\sum_{kk'\sigma} n_{k\sigma} \exp\left(i\left(\boldsymbol{k} - \boldsymbol{k}'\right)\boldsymbol{R}\right) = \sum_\sigma \int_0^{k_F} \frac{d^3k}{(2\pi)^3}\exp(ikR)\int_0^\infty \frac{d^3k'}{(2\pi)^3}\exp\left(-ik'R\right). \quad (16.46)$$

But, by the definition, the integral:

$$\int_0^\infty \frac{d^3k'}{(2\pi)^3}\exp\left(-ik'\boldsymbol{R}\right) = \delta(\boldsymbol{R}) = 0 \text{ for } \boldsymbol{R} \neq 0, \qquad (16.47)$$

and hence the second term in (16.44) equals zero.

As a result, proceeding from the summation to the integration in the residual third term in Eq. (16.44) and performing summation over spin projections $\sum_\sigma = 2$, we can write the correlator as

$$\langle \hat{\rho}(r)\hat{\rho}(r + \boldsymbol{R})\rangle = n_e^2 - 2\int \frac{d^3k}{(2\pi)^3}\int \frac{d^3k'}{(2\pi)^3} n_k n_{k'} \exp\left(i\left(\boldsymbol{k} - \boldsymbol{k}'\right)\boldsymbol{R}\right)$$

$$= n_e^2 - 2\int \frac{d^3k}{(2\pi)^3} n_k \exp(ikR)\int \frac{d^3k'}{(2\pi)^3} n_{k'} \exp\left(-ik'\boldsymbol{R}\right), \qquad (16.48)$$

where n_k and $n_{k'}$ are still the occupation numbers for one spin projection.

Now we can proceed in the similar manner to the calculation of the exchange hole discussed in the first part of this Lecture. Let us replace again in the second integral $k' \to -k'$ and utilize the fact that $n_{k'}$ is an even function of the momentum. As a result, in the integral term in (16.48), the second integral equals to the first one, and hence the correlator:

$$\langle \hat{\rho}(\boldsymbol{r})\hat{\rho}(\boldsymbol{r}+\boldsymbol{R})\rangle = n_e^2 - 2\left\{\int \frac{d^3k}{(2\pi)^3} n_k \exp(i\boldsymbol{k}\boldsymbol{R})\right\}^2.\tag{16.49}$$

After that, let us use the expression (16.12) for the integral in the figure brackets:

$$\int d^3k\, n_k \exp(i\boldsymbol{k}\boldsymbol{r}) = \frac{4\pi}{r^3}\left(\sin k_F r - k_F r \cos k_F r\right)$$

$$= \frac{4\pi k_F^3}{(k_F r)^3}\left(\sin k_F r - k_F r \cos k_F r\right).\tag{16.50}$$

Then the density correlator reads

$$\langle \hat{\rho}(\boldsymbol{r})\hat{\rho}(\boldsymbol{r}+\boldsymbol{R})\rangle = n_e^2 - \frac{2\cdot(4\pi)^2}{(2\pi)^6}k_F^6\left[\frac{\sin k_F r - k_F r \cos k_F r}{(k_F r)^3}\right]^2.\tag{16.51}$$

Now replacing in the second term k_F^6 by $\left(3\pi^2 n_e\right)^2$ and distinguishing n_e^2 as a common multiplier, we finally get for the correlator:

$$\langle \hat{\rho}(\boldsymbol{r})\hat{\rho}(\boldsymbol{r}+\boldsymbol{R})\rangle = n_e^2\left\{1 - \frac{9}{2}\left(\frac{\sin y - y \cos y}{y^3}\right)^2\bigg|_{y=k_F R}\right\},\tag{16.52}$$

or for the dimensionless correlator:

$$\frac{\langle \hat{\rho}(\boldsymbol{r})\hat{\rho}(\boldsymbol{r}+\boldsymbol{R})\rangle}{n_e^2} = 1 - \frac{9}{2}\left(\frac{\sin y - y \cos y}{y^3}\right)^2\bigg|_{y=k_F R}.\tag{16.53}$$

For $R \to \infty$, the correlator

$$\langle \hat{\rho}(\boldsymbol{r})\hat{\rho}(\boldsymbol{r}+\boldsymbol{R})\rangle \to n_e^2,\tag{16.54}$$

and the dimensionless correlator:

$$\frac{\langle \hat{\rho}(\boldsymbol{r})\hat{\rho}(\boldsymbol{r}+\boldsymbol{R})\rangle}{n_e^2} \to 1.\tag{16.55}$$

Note that since there is no interaction in this limit:

$$\langle \hat{\rho}(\boldsymbol{r})\hat{\rho}(\boldsymbol{r}+\boldsymbol{R})\rangle = \langle \hat{\rho}(\boldsymbol{r})\rangle\langle \hat{\rho}(\boldsymbol{r}+\boldsymbol{R})\rangle.\tag{16.56}$$

At the same time for $R \to 0$:

$$\frac{\sin y - y \cos y}{y^3} \approx \frac{y - \frac{y^3}{6} - y\left(1 - \frac{y^2}{2}\right)}{y^3} = -\frac{\frac{y^3}{3}}{y^3} = -\frac{1}{3}. \tag{16.57}$$

Hence, the dimensionless correlator:

$$\frac{\langle \hat{\rho}(r)\hat{\rho}(r+R)\rangle}{n_e^2} = 1 - \frac{9}{2} \cdot \frac{1}{9} = 1 - \frac{1}{2} = \frac{1}{2} \tag{16.58}$$

becomes two times smaller than in the absence of the interaction for $R \to \infty$.

Reduction in 2 times goes on in fact because there is correlation only between electrons with parallel spins.

Graphically, the qualitative behavior of the dimensionless correlator $\frac{\langle \hat{\rho}(r)\hat{\rho}(r+R)\rangle}{n_e^2}$ as a function of the dimensionless distance $k_F R$ between the correlating points is presented in Fig. 16.2.

Pairwise correlation function equals to $\frac{1}{2}$ for $k_F R = 0$ and increases with the increase of the variable $k_F R$. Then finally for $R \to \infty$ this function goes on saturation and becomes equal to 1.

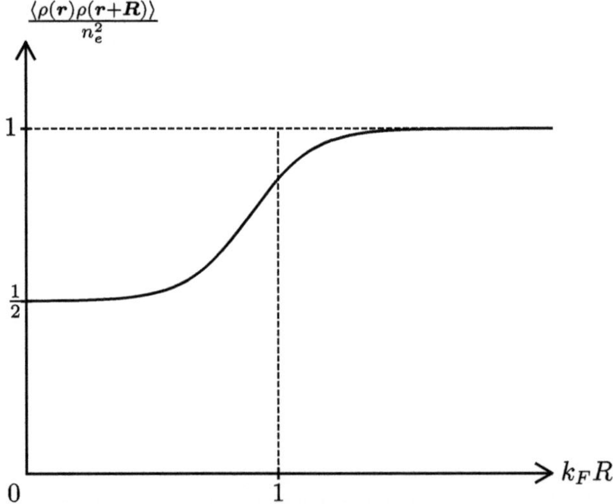

Fig. 16.2 Qualitative behavior of the dimensionless correlator of two electron densities $\frac{\langle \hat{\rho}(r)\hat{\rho}(r+R)\rangle}{n_e^2}$ as a function of the dimensionless distance $k_F R$ between the correlating points. The pairwise correlation function equals to $\frac{1}{2}$ for $k_F R = 0$ (there is correlation only between electrons with parallel spins) and increases with the increase of the variable $k_F R$. Finally, for $k_F R \to \infty$ (in the absence of the interaction), this function goes on saturation and becomes equal to 1 [1]

References

1. Y. Kagan, *Lectures on the Solid-State Theoretical Physics*, Moscow, MEPHI, 1981–1982. Part II. Electrons (Unpublished)
2. L.D. Landau, E.M. Lifshitz, *Statistical Physics, Part I* (Butterworth-Heinemann, 1980)
3. E.M. Lifshitz, L.P. Pitaevskii, *Statistical Physics, Part II, Theory of the Condensed State* (Pergamon, Elsevier, 2013)
4. R.P. Feynman, *Statistical Mechanics: A Set of Lectures*. Advanced Books, Classics (Westview Press, Boulder, 1998)
5. A.A. Abrikosov, L.P. Gor'kov, I.E. Dzyaloshinskii, *Methods of Quantum Field Theory in Statistical Physics* (Prentice Hall, Englewood Cliffs, New Jersey, 1963)
6. D. Pines, *Elementary Excitations in Solids* (W.A. Benjamin Inc., New York–Amsterdam, 1963)
7. D. Pines, P. Nozieres, *The Theory of Quantum Liquids*. Normal Fermi Liquids, vol. 1 (W.A. Benjamin Inc., New York–Amsterdam, 1966)
8. C. Kittel, *Quantum Theory of Solids* (Wiley, New York–London, 1963)
9. J. M. Ziman, *Electrons and Phonons: The Theory of Transport Phenomena in Solids* (Clarendon Press, Oxford, 1960)
10. N.W. Ashcroft, N.D. Mermin, *Solid State Physics* (Rinehart and Winston, Holt, 1976)
11. J.W. Negele, H. Orland, *Quantum Many-Particle Systems*. Frontiers in Physics (Addison-Wesley Publishing Company, 1992)
12. S.W. Lowesley, *Condensed Matter Physics: Dynamic Correlations*. Frontiers in Physics (The Benjamin Cummings Publishing Company, 1986)
13. S. Raimes, *Many-Electron Theory* (North-Holland Publishing Company, Amsterdam–London, 1972)
14. D. Ter Haar, *Introduction to the Physics of Many-Body Systems* (Inter-science Publishers Inc., New York, 1958)
15. J.C. Slater, *The Self-consistent Field for Molecules and Solids* (McGraw Hill, 1974)
16. D. Hartree, *The Calculation of Atomic Structures* (Wiley, London–New York, 1957)
17. S.V. Vonsovskii, M.I. Kaznelson, *Quantum Solid-State Physics*. Springer Series in Solid-State Sciences (Berlin–Heidelberg–New York, 1989)
18. W.A. Harrison, *Solid State Theory* (McGraw-Hill Book Company, New-York–London–Toronto, 1970)
19. F.A. Berezin, *The Method of Second Quantization* (Academic Press, 1966)
20. L.D. Landau, E.M. Lifshitz, *Quantum Mechanics: Non-relativistic Theory* (Pergamon Press, New York, 1977)
21. R.P. Feynman, The theory of positrons. Phys. Rev. **76**, 749 (1949)
22. M.V. Sadovskii, *Diagrammatics*. Lectures on Selected Problems in Condensed Matter Theory (World Scientific, Singapore, 2006)
23. L.S. Levitov, *Green's Functions in Problems* (Princeton University Press, 2011)
24. R. Mattuck, *A Guide to Feynman Diagrams in the Many-Body Problem* (McGraw-Hill, 1976)
25. C.P. Enz, *A Course on Many-Body Theory Applied to Solid-State Physics*. Lecture Notes in Physics (World Scientific, 1992)
26. Y.A. Izumov, Y.N. Skryabin, *Basic Models in Quantum Theory of Magnetism*, Ekaterinburg (2002) (in Russian)
27. E. Wigner, F. Seitz, Phys. Rev. **43**, 804 (1933)

Lecture 17. Correlation Energy for the Interaction of Electrons with Opposite Spins

17

Abstract

In this Lecture, we calculate the second-order correction to the electron energy and determine the correlation energy for the interaction of electrons with opposite spins. We analyze the inhomogeneous part of electron–ion interaction and discuss the self-consistent approximation for the dielectric function in metal. We derive an expression for the polarization operator in the static case defined by the Lindhard function.

In Lecture 16, we calculated linear in electron–electron interaction exchange contribution to the energy [1–13]. This contribution corresponds to attraction and after recalculation on one electron equals to $E_{\mathrm{ex.}} = -\frac{0.916}{r_S} \,\mathrm{Ry} < 0$.

Physically attractive sign of the exchange interaction is connected with the fact, that the mean distance between electrons with parallel spins is larger than the distance between them for the homogeneous smearing. Hence, potential energy of their interaction is smaller than the average one.

As a result, an <u>exchange hole</u> appears in the system [1–13].

We calculated also the correlator of two electron densities and observed the correlation between parallel spins on small distances and its reduction on large distances.

An account of the Coulomb repulsion beyond the first order of the perturbation theory (in this case, we should work with the true wave functions of the Coulomb problem [1, 14]) leads to the emergence of the correlations not only between parallel but also between antiparallel spins [1–13].

As a result, <u>due to the Coulomb interaction</u> electron pushes away electrons with the antiparallel spins as well. The mean distance between electrons with antiparallel spins also becomes larger than the mean distance between them for the homogeneous smearing. This fact leads to the further reduction of the potential energy.

Thus, not only exchange hole but also <u>correlation hole</u> is formed around electron [1–13]. Moreover, the correlation energy $E_{\mathrm{cor.}} < 0$ is negative similar to exchange energy.

M. Kagan, *Lecture Notes of Professor Yuri Kagan in Theoretical Solid-State Physics*,
Lecture Notes in Physics 1048, https://doi.org/10.1007/978-3-032-14621-2_17

Meanwhile, the total energy recalculated on one electron equals to the sum of four terms: the average kinetic energy of electron, exchange energy (due to the first-order correction in electron–electron interaction), correlation energy (the second-order correction in electron–electron interaction), and non-Coulomb part of the amplitude of electron–ion interaction [1–3].

In fact, we get the expansion in the direct and inverse powers of the correlation radius r_S [1–3]:

$$E = E_{\text{kin.el.}} + E_{\text{ex.}} + E_{\text{cor.}} + E_{\text{ei. non-Coul.}} = \frac{3}{5}\varepsilon_F + V_{ee}^{(1)} + V_{ee}^{(2)} + \frac{b}{\Omega_0}$$

$$\sim \frac{1}{r_S^2} - \frac{1}{r_S} + V_{ee}^{(2)} + \frac{1}{r_S^3}, \tag{17.1}$$

where $E_{\text{ex.}} = V_{ee}^{(1)} \sim -\frac{1}{r_S} < 0$ is the exchange energy, $E_{\text{cor.}} = V_{ee}^{(2)} < 0$ is the correlation energy for the interaction between the electrons with the antiparallel spins.

Let us evaluate the contribution to the electron energy from the second order of the perturbation theory in electron–electron interaction. In other words, let us determine the dependence of the correlation energy $E_{\text{cor.}}$ on the correlation radius.

As we know from Quantum mechanics [14], the energy correction in the second order of the perturbation theory:

$$V_{ee}^{(2)} = \sum_{k \neq i} \frac{\langle i \,|V_{ee}|\, k \rangle \, \langle k \,|V_{ee}|\, i \rangle}{E_i - E_k} < 0, \tag{17.2}$$

where the secondly quantized electron–electron interaction [1–3]

$$V_{ee} = \frac{1}{2} \sum_{k_1, k_2, q \neq 0, \sigma_1, \sigma_2} \frac{4\pi e^2}{q^2 \Omega} a_{k_1 \sigma_1}^{\dagger} a_{k_2, \sigma_2}^{\dagger} a_{k_2 + q, \sigma_2} a_{k_1 - q, \sigma_1}$$

does not contain the harmonics with $q = 0$ due to the electroneutrality condition.

But since the Pauli principle [14] allows the transitions only between the electron states close to the Fermi surface [15], we can evaluate $V_{ee}^{(2)}$ as

$$V_{ee}^{(2)} \sim \frac{\left(\frac{e^2}{r_0}\right)^2}{\varepsilon_F} \sim \frac{\left(\frac{e^2}{a_B}\right)^2 \left(\frac{a_B}{r_0}\right)^2}{\frac{e^2}{a_B}\frac{1}{r_S^2}} \sim \frac{e^2}{a_B}\frac{r_S^2}{r_S^2} \sim f\left(r_S^0\right) \text{ Ry.} \tag{17.3}$$

Note that in this estimate, we used the relations: $1 \text{ Ry} \sim \frac{e^2}{a_B}, r_S \sim \frac{r_0}{a_B}, \varepsilon_F \sim \frac{e^2}{a_B}\frac{1}{r_S^2}$.

We can see the contraction of the powers of the correlation radius in this simple estimate. More detailed analysis demonstrates, that in the expression for the interaction $V_{ee}^{(2)}$ enters the function $f\left(r_S^0\right)$ of the zeroth order in the correlation radius. It contains the constant term and the logarithm of the correlation radius:

$$f\left(r_S^0\right) = \alpha + \beta \ln r_S < 0, \tag{17.4}$$

where $\alpha = -0.015$.

In more exact calculation, an appearance of the logarithm in the function $f\left(r_S^0\right)$ is connected with the divergence of the form:

$$\int \frac{dq}{q} = \int \frac{dq/k_F}{q/k_F} \sim \ln \frac{q}{k_F} \tag{17.5}$$

for the dimensionless wave vector q/k_F.

Let us emphasize, that in most of the non-transition metals, the correlation radius equals to [1, 16–18]

$$r_S \sim \frac{r_0}{a_B} = 2 \div 3, \tag{17.6}$$

and generally speaking, an expansion of the electron energy in small r_S (large electron densities) should not work.

It occurs, however, that due to numerical reasons the true small parameter of the theory [1] equals to

$$\frac{r_S}{\gamma}, \tag{17.7}$$

where $\gamma = 4 \div 5$.

Hence, $\frac{r_S}{4 \div 5} < 1$, and in the exact calculation of the correlation energy $\ln\left(\frac{r_S}{4 \div 5}\right) < 0$. This fact explains the negative sign of the correlation energy.

Thus, the small parameter appears in the expansion of the electron energy on the degrees of the correlation radius [1, 16–18].

If vice versa the correlation radius r_S is large (at small electron densities), then we are in the situation when the potential energy $U \sim \frac{e^2}{r}$ is larger than the kinetic one $E_{\text{kin.}} \sim \frac{1}{r^2}$. In this case, the total charge is not compensated. Then with electron subsystem happens just the same as with the ion subsystem in the ordinary crystal. Namely, instead of the ion crystal lattice now electron crystal (or Wigner crystal [19,20]) is formed. It means that electrons are sitting in the knots of the Wigner crystal. Therefore, electron energy becomes equal to the energy of the ion lattice in this case [1, 19, 20]:

$$E_e = E_{\text{ion lattice}} = -\frac{1.8}{r_S} \text{ Ry.} \tag{17.8}$$

17.1   Inhomogeneous Part of the Electron–Ion Interaction Beyond the First Order of the Perturbation Theory

As we demonstrated already in the previous Lectures, the secondly quantized electron–ion interaction has a form:

$$V_{ei} = \sum_{k,q \neq 0,\sigma} a^{\dagger}_{k,\sigma} a_{k-q,\sigma} \left(\sum_m V_{ei}(q) \exp(-i q R_m) \right)$$

$$= \sum_{k,q \neq 0,\sigma} U(q) a^{\dagger}_{k,\sigma} a_{k-q,\sigma}, \tag{17.9}$$

where the Fourier component of the potential:

$$U(q) = \sum_m V_{ei}(q) \exp(-i q R_m). \tag{17.10}$$

We discussed also, that in the first order of the perturbation theory, the correction to the average electron energy related to the electron–ion interaction equals zero:

$$E_{ei}^{(1)} = \langle 0 | V_{ei} | 0 \rangle = 0, \tag{17.11}$$

since interaction harmonic with $q = 0$ is absent in Eq. (17.9) due to the electro-neutrality condition. Note, that we already accounted for the homogeneous part of the amplitude of the electron–ion scattering $\frac{b}{\Omega_0}$ in the expansion of the energy on one electron in powers of the correlation radius r_S in (17.1).

Let us take into account now the inhomogeneous part of electron–ion interaction. In the second order of the perturbation theory [14], the energy correction due to electron–ion interaction yields

$$E_{ei}^{(2)} = \sum_k \frac{\langle 0 | V_{ei} | k \rangle \langle k | V_{ei} | 0 \rangle}{E_0 - E_k}, \tag{17.12}$$

where symbol $|0\rangle$ corresponds to the ground state of the system.

Physically the energy correction $E_{ei}^{(2)}$ describes the situation, when ions create external field, which polarizes electrons (inducing electric multipoles in electron subsystem), and thus interacts with electrons.

Let us emphasize, that the interaction with unpolarized electrons is already accounted by us in the energy of the ion lattice $-\frac{1.8}{r_S}$ Ry and in the non-Coulomb term $\frac{b}{\Omega_0}$ of the scattering amplitude.

Then, we can write the correction to the energy $E_{ei}^{(2)}$ as the interaction of the external (ion) potential with the induced electron density [21]:

$$E_{ei}^{(2)} = \frac{1}{2} \int dr \rho_{e\,\text{ind.}}(r) \varphi_{\text{ext.}}(r) = \frac{e}{2} \int dr n_{\text{ind.}}(r) \varphi_{\text{ext.}}(r), \tag{17.13}$$

where $\rho_{e\,\text{ind.}}(\boldsymbol{r}) = en_{\text{ind.}}(\boldsymbol{r})$ is induced charge density, $\varphi_{\text{ext.}}(\boldsymbol{r})$ is external potential.

Further on, we can use the Poisson equation [21] for the external potential and the density of the external charges:

$$\Delta\varphi_{\text{ext.}} = -4\pi\,en_{\text{ext.}}. \tag{17.14}$$

To solve this equation, it is convenient to expand the external potential $\varphi_{\text{ext.}}$ in the Fourier series in the wave vectors $\boldsymbol{q}$ and the integral on frequencies ω:

$$\varphi_{\text{ext.}}(\boldsymbol{r}, t) = \sum_{q} \int d\omega \exp(i\boldsymbol{q}\boldsymbol{r} - i\omega t)\varphi_{\text{ext.}}(\boldsymbol{q}, \omega). \tag{17.15}$$

Then the solution of the Poisson equation reads

$$-q^2\varphi_{\text{ext.}}(\boldsymbol{q}, \omega) = -4\pi\,en_{\text{ext.}}(\boldsymbol{q}, \omega), \ \text{or}\ \varphi_{\text{ext.}}(\boldsymbol{q}, \omega) = \frac{4\pi e}{q^2}n_{\text{ext.}}(\boldsymbol{q}, \omega), \tag{17.16}$$

Now we can utilize two basic equations of the Electrostatics for the density of the external and total charge in continuous medium [1–3,21]:

$$\operatorname{div}\boldsymbol{D} = 4\pi\,en_{\text{ext.}}, \tag{17.17}$$

and correspondingly

$$\operatorname{div}\boldsymbol{E} = 4\pi\,en_{\text{tot.}}, \tag{17.18}$$

where the total density equals to the sum of induced density and density of the external charges:

$$n_{\text{tot.}} = n_{\text{ext.}} + n_{\text{ind.}}. \tag{17.19}$$

In the momentum space, Electrostatics equations for the vectors of electric induction $\boldsymbol{D}$ and electric field $\boldsymbol{E}$ (17.17)–(17.18) can be written in the following form:

$$i(\boldsymbol{q}\boldsymbol{D}) = 4\pi\,en_{\text{ext.}}(\boldsymbol{q}, \omega), \tag{17.20}$$

$$i(\boldsymbol{q}\boldsymbol{E}) = 4\pi\,en_{\text{tot.}}(\boldsymbol{q}, \omega). \tag{17.21}$$

As a result, we get important relation for the longitudinal components of the vectors $\boldsymbol{D}$ and $\boldsymbol{E}$:

$$\frac{D_{\|}(\boldsymbol{q}, \omega)}{E_{\|}(\boldsymbol{q}, \omega)} = \frac{n_{\text{ext.}}(\boldsymbol{q}, \omega)}{n_{\text{tot.}}(\boldsymbol{q}, \omega)}. \tag{17.22}$$

But the ratio of the longitudinal components:

$$\frac{D_{\|}(\boldsymbol{q}, \omega)}{E_{\|}(\boldsymbol{q}, \omega)} = \varepsilon(\boldsymbol{q}, \omega) \tag{17.23}$$

is by the definition the dielectric permittivity of the medium [1–3,21].

Hence, we get

$$\varepsilon(\boldsymbol{q},\omega) = \frac{n_{\text{ext.}}(\boldsymbol{q},\omega)}{n_{\text{tot.}}(\boldsymbol{q},\omega)} = \frac{n_{\text{ext.}}(\boldsymbol{q},\omega)}{n_{\text{ext.}}(\boldsymbol{q},\omega) + n_{\text{ind.}}(\boldsymbol{q},\omega)} = \frac{1}{1 + \frac{n_{\text{ind.}}(\boldsymbol{q},\omega)}{n_{\text{ext.}}(\boldsymbol{q},\omega)}}. \tag{17.24}$$

It follows from (17.24), that the ratio of the induced and the total charge densities yields

$$\frac{n_{\text{ind.}}(\boldsymbol{q},\omega)}{n_{\text{ext.}}(\boldsymbol{q},\omega)} = \frac{1}{\varepsilon(\boldsymbol{q},\omega)} - 1. \tag{17.25}$$

Now let us return back to Eq. (17.13) for the correction to the energy and expand $n_{\text{ind.}}(\boldsymbol{r})$ and $\varphi_{\text{ext.}}(\boldsymbol{r})$ in this formula in the Fourier series, assuming meanwhile that the frequency $\omega = 0$ (we consider static polarization). Then:

$$E_{ei}^{(2)} = \frac{e}{2}\int d\boldsymbol{r}\left(\sum_{\boldsymbol{q}} n_{\text{ind.}}(\boldsymbol{q})\exp(i\boldsymbol{q}\boldsymbol{r})\right)\left(\sum_{\boldsymbol{q}_1}\varphi_{\text{ext.}}(\boldsymbol{q}_1)\exp\left(i\boldsymbol{q}_1\boldsymbol{r}\right)\right)$$

$$= \frac{e}{2}\sum_{\boldsymbol{q}\boldsymbol{q}_1} n_{\text{ind.}}(\boldsymbol{q})\varphi_{\text{ext.}}(\boldsymbol{q}_1)\int d\boldsymbol{r}\exp\left(i\left(\boldsymbol{q}+\boldsymbol{q}_1\right)\boldsymbol{r}\right). \tag{17.26}$$

But the integral:

$$\int d\boldsymbol{r}\exp\left(i\left(\boldsymbol{q}+\boldsymbol{q}_1\right)\boldsymbol{r}\right) = \Omega\delta_{\boldsymbol{q},-\boldsymbol{q}_1}, \tag{17.27}$$

where Ω is the system volume.

Hence, the energy correction:

$$E_{ei}^{(2)} = \frac{e\Omega}{2}\sum_{\boldsymbol{q}} n_{\text{ind.}}(\boldsymbol{q})\varphi_{\text{ext.}}(-\boldsymbol{q}). \tag{17.28}$$

Further on let us divide and multiply the induced density $n_{\text{ind.}}(\boldsymbol{q})$ in (17.26) on $n_{\text{ext.}}(\boldsymbol{q})$. As a result, we obtain

$$E_{ei}^{(2)} = \frac{e\Omega}{2}\sum_{\boldsymbol{q}}\frac{n_{\text{ind.}}(\boldsymbol{q})}{n_{\text{ext.}}(\boldsymbol{q})}n_{\text{ext.}}(\boldsymbol{q})\varphi_{\text{ext.}}(-\boldsymbol{q}). \tag{17.29}$$

Now according to the formula (17.25), let us express the ratio $\frac{n_{\text{ind.}}(\boldsymbol{q})}{n_{\text{ext.}}(\boldsymbol{q})}$ in (17.29) via the static dielectric permittivity $\varepsilon(\boldsymbol{q},0)$. Then:

$$E_{ei}^{(2)} = \frac{e\Omega}{2}\sum_{\boldsymbol{q}}\left(\frac{1}{\varepsilon(\boldsymbol{q},0)} - 1\right)n_{\text{ext.}}(\boldsymbol{q})\varphi_{\text{ext.}}(-\boldsymbol{q}). \tag{17.30}$$

After that, let us express $n_{\text{ext.}}(\boldsymbol{q})$ via $\varphi_{\text{ext.}}(\boldsymbol{q})$ from the solution of the Poisson equation (17.16):

$$n_{\text{ext.}}(\boldsymbol{q}) = \frac{q^2}{4\pi e}\varphi_{\text{ext.}}(\boldsymbol{q}),\qquad(17.31)$$

and write (17.28) as follows:

$$E_{ei}^{(2)} = \frac{e\Omega}{2}\sum_{\boldsymbol{q}}\left(\frac{1}{\varepsilon(\boldsymbol{q},0)} - 1\right)\varphi_{\text{ext.}}(-\boldsymbol{q})\frac{q^2}{4\pi e}\varphi_{\text{ext.}}(\boldsymbol{q}).\qquad(17.32)$$

Let us remind now, that in our case: $e\varphi_{\text{ext.}}(\boldsymbol{q}) = V_{ei}(\boldsymbol{q})$. Then we get finally:

$$
\begin{aligned}
E_{ei}^{(2)} &= \frac{\Omega}{2}\sum_{\boldsymbol{q}}\left(\frac{1}{\varepsilon(\boldsymbol{q},0)} - 1\right)V_{ei}(-\boldsymbol{q})\frac{q^2}{4\pi e^2}V_{ei}(\boldsymbol{q})\\
&= \frac{\Omega}{2}\sum_{\boldsymbol{q}}\left(\frac{1}{\varepsilon(\boldsymbol{q},0)} - 1\right)\frac{q^2}{4\pi e^2}|V_{ei}(\boldsymbol{q})|^2,
\end{aligned}\qquad(17.33)
$$

where we used the relation $|V_{ei}(\boldsymbol{q})|^2 = V_{ei}(-\boldsymbol{q})V_{ei}(\boldsymbol{q})$.

Now we can substitute in (17.33) well-known expression for the dielectric permittivity $\varepsilon(\boldsymbol{q},\omega)$, derived, e.g., from the diagrammatic technique [8–11] in the Random phase (RPA) approximation [1–3,16–18,22–30]. In this approximation (after the summation of the geometric series of the bubble diagrams with electron–hole loops) we get for the dielectric permittivity [1–3,16–18,22–30] (see more detailed derivation in the next Lecture):

$$\varepsilon(\boldsymbol{q},\omega) = 1 + \frac{4\pi e^2}{q^2}\Pi(\boldsymbol{q},\omega),\qquad(17.34)$$

where $\Pi(\boldsymbol{q},\omega)$ is called the polarization operator [1–3,16–18,22–31] on the language of the diagrammatic technique.

Then from (17.32), we can express the combination:

$$\frac{1}{\varepsilon(\boldsymbol{q},0)} - 1 = \frac{1 - \varepsilon(\boldsymbol{q},0)}{\varepsilon(\boldsymbol{q},0)} = -\frac{\frac{4\pi e^2}{q^2}\Pi(\boldsymbol{q},0)}{\varepsilon(\boldsymbol{q},0)}.\qquad(17.35)$$

Substituting (17.35) in (17.33), we can write the second-order correction to the energy in electron–ion interaction in the canonical form [1–3,16–18]:

$$E_{ei}^{(2)} = -\frac{\Omega}{2}\sum_{\boldsymbol{q}}\left(\frac{\Pi(\boldsymbol{q},0)}{\varepsilon(\boldsymbol{q},0)}\right)|V_{ei}(\boldsymbol{q})|^2,\qquad(17.36)$$

where $\Pi(\boldsymbol{q},0)$ is the static polarization operator, $\varepsilon(\boldsymbol{q},0)$ is the static dielectric permittivity [1–3,16–18,22–32]

17.2   Self-Consistent (Mean Field) Approximation

In the self-consistent field method, the operator of electron density can be written as

$$\rho = \sum_k n_k \, |\Psi_k|^2 , \tag{17.37}$$

where n_k are the occupation numbers, and the wave functions Ψ_k themselves depend on the density ρ.

We will consider electron in the field of other electrons or in the external field. Note that the self-consistent (mean field) approximation works the worse, the less homogeneous is the electron density (in this case, it is bad to consider electron in the mean field of other electrons).

Let us emphasize, that spatial inhomogeneity of electric field statically is governed by the fluctuations. Thus, in the method of the self-consistent field we neglect the fluctuations.

Suppose the electron in the initial state is described by the plane wave with the wave vector $\boldsymbol{k}$ and the energy $\varepsilon(\boldsymbol{k})$:

$$\Psi_k^{(0)}(t) = \frac{1}{\sqrt{\Omega}} \exp\left(i\boldsymbol{k}\boldsymbol{r} - i\frac{\varepsilon(\boldsymbol{k})}{\hbar}t \right). \tag{17.38}$$

Then, under the action on electron of the time-dependent perturbing potential $V_{kk'}(t)$, electron wave function becomes equal to $\Psi_k(t)$, and can be represented as the expansion in series with the functions of the initial states $\Psi_k^{(0)}$:

$$\Psi_k(t) = \sum_k c_k(t)\Psi_k^{(0)}(t) \tag{17.39}$$

with time-dependent expansion coefficients $c_k(t)$. Moreover, if we consider the potential $V_{kk'}(t)$ as weak perturbation, then these coefficients satisfy the equations of motion:

$$\frac{dc_k}{dt} = -\frac{i}{\hbar} \sum_{k'} V_{kk'}c_{k'}^{(0)}(t), \tag{17.40}$$

where according to the initial conditions:

$$c_{k'}^{(0)}(t) = \delta_{kk'}, \tag{17.41}$$

and perturbed wave function:

$$\Psi_k(t) = \Psi_k^{(0)}(t) + \sum_{k'\neq k} c_{k'}(t)\Psi_{k'}^{(0)}(t). \tag{17.42}$$

After substitution of the initial condition (17.41) in (17.40), equation of motion acquires a form:

$$\frac{dc_{k'}}{dt} = -\frac{i}{\hbar} V_{kk'}(t),$$ (17.43)

and correspondingly time-dependent coefficients of the expansion in Eq. (17.42) equal to

$$c_{k'}(t) = -\frac{i}{\hbar} \int_{-\infty}^{t} V_{kk'}(t)dt.$$ (17.44)

Note that time-dependent perturbation $V_{kk'}(t)$ is produced by the Heisenberg operator [14] $\hat{\varphi}_t(\boldsymbol{q}, \omega) \exp(i\boldsymbol{q}\boldsymbol{r} - i\omega t)$ for the effective potential. Thus:

$$V_{kk'}(t) = \int d\boldsymbol{r} \frac{1}{\sqrt{\Omega}} \exp\left(-i\boldsymbol{k'}\boldsymbol{r} + i\frac{\varepsilon\left(\boldsymbol{k'}\right)}{\hbar}t\right) e\hat{\varphi}_t(\boldsymbol{q}, \omega) \exp(i\boldsymbol{q}\boldsymbol{r} - i\omega t)$$

$$\times \frac{1}{\sqrt{\Omega}} \exp\left(i\boldsymbol{k}\boldsymbol{r} - i\frac{\varepsilon(\boldsymbol{k})}{\hbar}t\right) + \left(\text{h.c. where } \varphi \to \varphi^*, \boldsymbol{k} \to -\boldsymbol{k}, \omega \to -\omega\right)$$

$$= e\hat{\varphi}_t(\boldsymbol{q}, \omega) \exp\left(\frac{i}{\hbar}\left(\varepsilon\left(\boldsymbol{k'}\right) - \varepsilon(\boldsymbol{k}) - \hbar\omega\right)t\right) \frac{1}{\Omega} \int d\boldsymbol{r} \exp\left(i\left(-\boldsymbol{k'} + \boldsymbol{k} + \boldsymbol{q}\right)\boldsymbol{r}\right) + \text{h.c.}$$

$$= e\hat{\varphi}_t(\boldsymbol{q}, \omega) \exp\left(\frac{i}{\hbar}\left(\varepsilon\left(\boldsymbol{k'}\right) - \varepsilon(\boldsymbol{k}) - \hbar\omega\right)t\right) \delta_{\boldsymbol{k'},\boldsymbol{k}+\boldsymbol{q}} + \text{h.c.}$$ (17.45)

Actually, in the method of the self-consistent field the operator $\hat{\varphi}_t(\boldsymbol{q}, \omega)$ is connected with the total electron density ρ in the expression (17.37).

Substituting (17.45) in (17.44) and making a replacement $\omega \to \omega \pm i\delta$, where $\delta \to +0$ (for the convergence of the integrals on $-\infty$), we get for the expansion coefficients:

$$c_{k'}(t) = -\frac{i}{\hbar} \left\{ e\hat{\varphi}_t(\boldsymbol{q}, \omega)\delta_{\boldsymbol{k'},\boldsymbol{k}+\boldsymbol{q}} \int_{-\infty}^{t} dt \right.$$

$$\cdot \exp\left(\frac{i}{\hbar}\left(\varepsilon\left(\boldsymbol{k'}\right) - \varepsilon(\boldsymbol{k}) - \hbar(\omega + i\delta)\right)t\right) + e\hat{\varphi}_t^*(\boldsymbol{q}, \omega)\delta_{\boldsymbol{k'},\boldsymbol{k}-\boldsymbol{q}} \int_{-\infty}^{t} dt$$

$$\left. \cdot \exp\left(\frac{i}{\hbar}\left(\varepsilon\left(\boldsymbol{k'}\right) - \varepsilon(\boldsymbol{k}) + \hbar(\omega - i\delta)\right)t\right) \right\} = -\left\{ e\hat{\varphi}_t(\boldsymbol{q}, \omega) \cdot \delta_{\boldsymbol{k'},\boldsymbol{k}+\boldsymbol{q}} \right.$$

$$\cdot \frac{\exp\left(\frac{i}{\hbar}\left(\varepsilon\left(\boldsymbol{k'}\right) - \varepsilon(\boldsymbol{k}) - \hbar(\omega + i\delta)\right)t\right)}{\varepsilon\left(\boldsymbol{k'}\right) - \varepsilon(\boldsymbol{k}) - \hbar(\omega + i\delta)} + e\hat{\varphi}_t^*(\boldsymbol{q}, \omega)\delta_{\boldsymbol{k'},\boldsymbol{k}-\boldsymbol{q}}$$

$$\left. \cdot \frac{\exp\left(\frac{i}{\hbar}\left(\varepsilon\left(\boldsymbol{k'}\right) - \varepsilon(\boldsymbol{k}) + \hbar(\omega - i\delta)\right)t\right)}{\varepsilon\left(\boldsymbol{k'}\right) - \varepsilon(\boldsymbol{k}) + \hbar(\omega - i\delta)} \right\}.$$ (17.46)

Then perturbed wave function in (17.40) can be written as

$$\Psi_k(t) = \Psi_k^{(0)}(t) + \sum_{k' \neq k} c_{k'}(t)\Psi_{k'}^{(0)}(t) = \frac{1}{\sqrt{\Omega}} \exp\left(ikr - i\frac{\varepsilon(k)}{\hbar}t\right)$$

$$- \sum_{k' \neq k} \left\{ e\hat{\varphi}_t(q,\omega)\delta_{k',k+q} \frac{\exp\left(\frac{i}{\hbar}\left(\varepsilon(k') - \varepsilon(k) - \hbar(\omega + i\delta)\right)t\right)}{\varepsilon(k') - \varepsilon(k) - \hbar(\omega + i\delta)} \right.$$

$$\left. + e\hat{\varphi}_t^*(q,\omega)\delta_{k',k-q} \frac{\exp\left(\frac{i}{\hbar}\left(\varepsilon(k') - \varepsilon(k) + \hbar(\omega - i\delta)\right)t\right)}{\varepsilon(k') - \varepsilon(k) + \hbar(\omega - i\delta)} \right\} \frac{1}{\sqrt{\Omega}} \exp\left(ik'r - i\frac{\varepsilon(k')}{\hbar}t\right).$$

$$(17.47)$$

Further on, contracting exponents $\exp\left(-i\frac{\varepsilon(k')}{\hbar}t\right)$ and $\exp\left(i\frac{\varepsilon(k')}{\hbar}t\right)$ and lifting the summation over k' with the help of delta symbols $\delta_{k',k+q}$ and $\delta_{k',k-q}$, we obtain

$$\Psi_k(t) = \frac{1}{\sqrt{\Omega}} \exp\left(ikr - i\frac{\varepsilon(k)}{\hbar}t\right)$$

$$\cdot \left\{ 1 - \left[e\hat{\varphi}_t(q,\omega)\frac{\exp(iqr - i(\omega + i\delta)t)}{\varepsilon(k+q) - \varepsilon(k) - \hbar(\omega + i\delta)} \right.\right.$$

$$\left.\left. + e\hat{\varphi}_t^*(q,\omega)\frac{\exp(-iqr + i(\omega - i\delta)t)}{\varepsilon(k-q) - \varepsilon(k) + \hbar(\omega - i\delta)} \right] \right\}.$$

$$(17.48)$$

Correspondingly, the absolute value squared of the perturbed wave function:

$$|\Psi_k(t)|^2 = \Psi_k(t)\Psi_k^*(t) \approx \frac{1}{\Omega}\left\{ 1 \right.$$

$$-e\hat{\varphi}_t(q,\omega) \cdot \left[\frac{1}{\varepsilon(k+q) - \varepsilon(k) - \hbar\omega} + \frac{1}{\varepsilon(k-q) - \varepsilon(k) + \hbar\omega} \right] \exp(iqr - i\omega t) - \text{h.c.} \left.\right\}. \quad (17.49)$$

Now we can substitute (17.47) in the expression (17.35) for the operator of the electron density. As a result:

$$\rho(q,\omega) = \sum_k n_k |\Psi_k|^2 = \frac{1}{\Omega}\sum_k n_k - \frac{1}{\Omega}e\hat{\varphi}_t(q,\omega)\sum_k n_k \cdot \exp(iqr - i\omega t)$$

$$\cdot \left[\frac{1}{\varepsilon(k+q) - \varepsilon(k) - \hbar\omega} + \frac{1}{\varepsilon(k-q) - \varepsilon(k) + \hbar\omega} \right] - \text{h.c.} = \rho_0 + \rho_{\text{ind.}}(q,\omega).$$

$$(17.50)$$

The first term in (17.50) is related to the unperturbed (average) density:

$$\rho_0 = \frac{1}{\Omega}\sum_k n_k. \qquad (17.51)$$

At the same time, the second term represents the density operator of the induced charges:

$$\rho_{\mathrm{ind.}}(\boldsymbol{q}, \omega) = -\frac{1}{\Omega} e\hat{\varphi}_t(\boldsymbol{q}, \omega) \exp(i\boldsymbol{q}\boldsymbol{r} - i\omega t) \sum_{k} \left[\frac{n_k}{\varepsilon(\boldsymbol{k} + \boldsymbol{q}) - \varepsilon(\boldsymbol{k}) - \hbar\omega} \right.$$
$$\left. + \frac{n_k}{\varepsilon(\boldsymbol{k} - \boldsymbol{q}) - \varepsilon(\boldsymbol{k}) + \hbar\omega} \right] - \mathrm{h.c.} \tag{17.52}$$

Thus, in the method of the self-consistent field the density $\rho(\boldsymbol{q}, \omega)$ in (17.50) as well as in (17.37) is total electron density.

Further on, we introduce the Heisenberg operator for the induced density $\rho_{\mathrm{ind.}t}$ $(\boldsymbol{q}, \omega)$, distinguishing only the exponent $\exp(i\boldsymbol{q}\boldsymbol{r} - i\omega t)$:

$$\rho_{\mathrm{ind.}}(\boldsymbol{q}, \omega) = \rho_{\mathrm{ind.}t}(\boldsymbol{q}, \omega) \exp(i\boldsymbol{q}\boldsymbol{r} - i\omega t). \tag{17.53}$$

Then the Heisenberg operator for the induced density is connected with the Heisenberg operator for electron–ion interaction by the relation:

$$\rho_{\mathrm{ind.}t}(\boldsymbol{q}, \omega) = -\frac{1}{\Omega} e\hat{\varphi}_t(\boldsymbol{q}, \omega) \sum_{k} \left[\frac{n_k}{\varepsilon(\boldsymbol{k} + \boldsymbol{q}) - \varepsilon(\boldsymbol{k}) - \hbar(\omega + i\delta)} \right.$$
$$\left. + \frac{n_k}{\varepsilon(\boldsymbol{k} - \boldsymbol{q}) - \varepsilon(\boldsymbol{k}) + \hbar(\omega - i\delta)} \right]. \tag{17.54}$$

Now in the denominator of the second term in the square bracket in (17.54), we can make the replacement $\boldsymbol{k} \to \boldsymbol{k} + \boldsymbol{q}$, change the signs (distinguishing minus sign between the first and second terms) and get the canonical expression for the induced density:

$$\rho_{\mathrm{ind.}t}(\boldsymbol{q}, \omega) = -\frac{1}{\Omega} e\hat{\varphi}_t(\boldsymbol{q}, \omega) \sum_{k} \frac{n_k - n_{k+q}}{\varepsilon(\boldsymbol{k} + \boldsymbol{q}) - \varepsilon(\boldsymbol{k}) - \hbar\omega}. \tag{17.55}$$

At the same time, the Poisson equation in the method of the self-consistent field is written for the total electron density, which enters in the expression (17.50) and in the initial expression (17.37). Hence:

$$- e\hat{\varphi}_t(\boldsymbol{q}, \omega) = -\frac{4\pi e^2}{q^2} \rho_{\mathrm{tot.}t}(\boldsymbol{q}, \omega). \tag{17.56}$$

In fact, electron–ion interaction in the method of the self-consistent field produces an additional induced density. This additional density in its turn, generates an additional contribution to the effective potential. As a result, effective potential $\hat{\varphi}_t(\boldsymbol{q}, \omega)$ in (17.45–17.50) and in (17.53–17.56) is related to the total density $\rho_{\mathrm{tot.}}(\boldsymbol{q}, \omega)$.

Thus, we get the relationship between $\rho_{\text{ind}.t}(\boldsymbol{q},\omega)$ and $\rho_{\text{tot}.t}(\boldsymbol{q},\omega)$:

$$\rho_{\text{ind}.t}(\boldsymbol{q},\omega) = -\frac{4\pi e^2}{q^2}\rho_{\text{tot}.t}(\boldsymbol{q},\omega)\frac{1}{\Omega}\sum_k \frac{n_k - n_{k+q}}{\varepsilon(\boldsymbol{k}+\boldsymbol{q}) - \varepsilon(\boldsymbol{k}) - \hbar\omega}. \tag{17.57}$$

From (17.57), we can determine the dielectric permittivity as

$$\varepsilon(\boldsymbol{q},\omega) = \frac{\rho_{\text{ext}.t}(\boldsymbol{q},\omega)}{\rho_{\text{tot}.t}(\boldsymbol{q},\omega)} = 1 - \frac{\rho_{\text{ind}.t}(\boldsymbol{q},\omega)}{\rho_{\text{tot}.t}(\boldsymbol{q},\omega)}$$

$$= 1 + \frac{4\pi e^2}{q^2}\frac{1}{\Omega}\sum_k \frac{n_k - n_{k+q}}{\varepsilon(\boldsymbol{k}+\boldsymbol{q}) - \varepsilon(\boldsymbol{k}) - \hbar\omega}. \tag{17.58}$$

But, from the other hand, we know (see Eq. (17.34)) how to express the dielectric permittivity via the polarization operator, namely: $\varepsilon(\boldsymbol{q},\omega) = 1 + \frac{4\pi e^2}{q^2}\Pi(\boldsymbol{q},\omega)$.

As a result, in the method of the self-consistent field we derive the well-known expression for the polarization operator [1,16–18,22,23,31]:

$$\Pi(\boldsymbol{q},\omega) = \frac{1}{\Omega}\sum_k \frac{n_k - n_{k+q}}{\varepsilon(\boldsymbol{k}+\boldsymbol{q}) - \varepsilon(\boldsymbol{k}) - \hbar\omega}. \tag{17.59}$$

At the more accurate derivation of the polarization operator by the methods of the diagrammatic technique [9,10,32–34], we will see that the frequency ω in the denominator of (17.59) should be replaced by $\omega + i\delta$.

Note, that we should understand the sum over k in (17.59) as the sum over the wave vector $\boldsymbol{k}$ and over the spin projection σ. The spin summation in the absence of the magnetic field always leads to the appearance of the additional factor 2.

In the static case, the dielectric function is given by

$$\varepsilon(\boldsymbol{q},0) = 1 + \frac{4\pi e^2}{q^2}\Pi(\boldsymbol{q},0), \tag{17.60}$$

where the static polarization operator reads [1,16–18,31]

$$\Pi(\boldsymbol{q},0) = \frac{1}{\Omega}\sum_k \frac{n_k - n_{k+q}}{\varepsilon(\boldsymbol{k}+\boldsymbol{q}) - \varepsilon(\boldsymbol{k})}. \tag{17.61}$$

17.3 Problem 1. Calculation of the Static Polarization Operator at $T = 0$. Lindhard Function

Let us calculate for the pedagogical reasons the static polarization operator in (17.61) in the three-dimensional case at $T = 0$ on the quadratic spectrum of electrons (see, e.g., [31,35–40]). In this case, the electron energies in the denominator of (17.61):

$$\varepsilon(k + q) = \frac{\hbar^2 (k + q)^2}{2m}, \; \varepsilon(k) = \frac{\hbar^2 k^2}{2m}, \tag{17.62}$$

and correspondingly:

$$\varepsilon(k + q) - \varepsilon(k) = \frac{\hbar^2 \left(q^2 + 2kq\right)}{2m}. \tag{17.63}$$

Let us proceed in (17.61) from the summation to the integration, making as usual the replacement:

$$\Omega \sum_k = \Omega \sum_{k\sigma} = 2\Omega \sum_k \to 2 \int \frac{d^3k}{(2\pi)^3}, \tag{17.64}$$

where factor 2 is related to the summation over the spin projections $\left(\sum_\sigma = 2\right)$.

Then, after the contraction of the system volume Ω, the static polarization operator can be written as

$$\begin{aligned}
\Pi(q, 0) &= 2 \int \frac{d^3k}{(2\pi)^3} \frac{n_k - n_{k+q}}{\varepsilon(k + q) - \varepsilon(k)} \\
&\equiv 2 \int \frac{d^3k}{(2\pi)^3} \left[\frac{n_k}{\varepsilon(k + q) - \varepsilon(k)} - \frac{n_{k+q}}{\varepsilon(k + q) - \varepsilon(k)} \right] \\
&= 4 \int \frac{d^3k}{(2\pi)^3} \frac{n_k}{\varepsilon(k + q) - \varepsilon(k)},
\end{aligned} \tag{17.65}$$

where, in the second term of the second equality in (17.65), we made the inverse replacement $k + q \to k$ and change the signs.

It is important to note, that in (17.65) at $T = 0$, the occupation numbers n_k are described by the Dirac step function for the electrons with one spin projection:

$$n_k = 1 \text{ for } k < k_F. \tag{17.66}$$

Further on, for the quadratic spectrum, it is more convenient to calculate the integral over $\frac{d^3k}{(2\pi)^3}$ in (17.65) in the spherical coordinates [40]. Then, with an account of (17.66), we get

$$\Pi(\boldsymbol{q},0) = 4\int \frac{d^3k}{(2\pi)^3}\frac{n_{\boldsymbol{k}}}{\varepsilon(\boldsymbol{k}+\boldsymbol{q})-\varepsilon(\boldsymbol{k})} = \frac{2m}{\pi^2\hbar^2}\int_0^{k_F} k^2 dk \int_{-1}^1 dx$$

$$\cdot\frac{1}{k^2+2kqx} = \frac{m}{\pi^2\hbar^2}\int_0^{k_F} k^2 dk\frac{1}{qk}\int_{-1}^1 dx\frac{1}{x+\frac{k}{2q}} = \frac{m}{\pi^2\hbar^2 q}\int_0^{k_F} k dk \ln\left|\frac{1+\frac{k}{2q}}{-1+\frac{k}{2q}}\right|$$

$$= \frac{m}{\pi^2\hbar^2 q}\int_0^{k_F} k dk \ln\left|\frac{q+2k}{q-2k}\right|, \tag{17.67}$$

where $x=\cos\theta$, and θ is the angle between wave vectors $\boldsymbol{k}$ and $\boldsymbol{q}$. After that, integrating the last integral in (17.67) by parts, we obtain

$$\Pi(\boldsymbol{q},0) = \frac{m}{\pi^2\hbar^2 q}\left\{\frac{k^2}{2}\left(\ln\left|\frac{q+2k}{q-2k}\right|\right)\Big|_0^{k_F} - \int_0^{k_F}\frac{k^2}{2}\cdot 2\left(\frac{1}{2k+q}-\frac{1}{2k-q}\right)dk\right\}$$

$$= \frac{m}{\pi^2\hbar^2 q}\left\{\frac{k_F^2}{2}\ln\left|\frac{2k_F+q}{2k_F-q}\right| - \int_0^{k_F} dk\left(k^2-\frac{q^2}{4}+\frac{q^2}{4}\right)\left(\frac{1}{2k+q}-\frac{1}{2k-q}\right)\right\}$$

$$= \frac{m}{\pi^2\hbar^2 q}\left\{\frac{k_F^2}{2}\ln\left|\frac{2k_F+q}{2k_F-q}\right| - \frac{1}{2}\int_0^{k_F} dk\left(k-\frac{q}{2}-k-\frac{q}{2}\right) - \frac{q^2}{8}\ln\left|\frac{2k_F+q}{2k_F-q}\right|\right\}$$

$$= \frac{m}{2\pi^2\hbar^2 q}\left\{\left(k_F^2-\frac{q^2}{4}\right)\ln\left|\frac{2k_F+q}{2k_F-q}\right| + qk_F\right\}$$

$$= \frac{mk_F}{2\pi^2\hbar^2}\left\{1+\frac{4k_F^2-q^2}{4k_F q}\ln\left|\frac{2k_F+q}{2k_F-q}\right|\right\}. \tag{17.68}$$

As a result, the static polarization operator $\Pi(\boldsymbol{q},0)$ equals to [31,35–40]

$$\Pi(\boldsymbol{q},0) = N(0)\left\{1+\frac{4k_F^2-q^2}{4k_F q}\ln\left|\frac{2k_F+q}{2k_F-q}\right|\right\}, \tag{17.69}$$

where:

$$N(0) = \frac{mk_F}{2\pi^2\hbar^2} = \frac{mp_F}{2\pi^2\hbar^3} \tag{17.70}$$

is the density of electron states on the Fermi level [41].

Expression in the square brackets of Eq. (17.69) can be rewritten in terms of the well-known Lindhard function $F_{\text{Lind.}}$ [31]. This function depends on the dimensionless variable $z = \frac{q}{2k_F}$ and yields [31]

$$F_{\text{Lind.}}(z) = 1 + \frac{1 - z^2}{2z} \ln \left| \frac{1 + z}{1 - z} \right|. \qquad (17.71)$$

References

1. Y. Kagan, *Lectures on the Solid-state Theoretical Physics* (Moscow, MEPHI, 1981–1982). Part II. Electrons - unpublished
2. D. Pines, *Elementary Excitations in Solids* (W.A. Benjamin Inc., New York-Amsterdam, 1963)
3. D. Pines, P. Nozieres, *The Theory of Quantum Liquids: Normal Fermi Liquids*, vol. 1 (W.A. Benjamin Inc. New York-Amsterdam, 1966)
4. C. Kittel, *Quantum Theory of Solids* (Wiley, New York-London, 1963)
5. J.C. Slater, *The Self-Consistent Field for Molecules and Solids* (McGraw Hill, 1974)
6. D. Hartree, *The Calculation of Atomic Structures* (Wiley, London-New York, 1957)
7. R.P. Feynman, *Statistical Mechanics, A Set of Lectures (Advanced Books, Classics)* (Westview Press, Boulder, 1998)
8. Y.A. Izumov, Y.N. Skryabin, *Basic Models in Quantum Theory of Magnetism* (Ekaterinburg, 2002) (in Russian)
9. M.V. Sadovskii, *Diagrammatics, Lectures on Selected problems in Condensed Matter Theory* (World Scientific, Singapore, 2006)
10. C.P. Enz, *A Course on Many-body Theory Applied to Solid-State Physics*. Lecture Notes in Physics (World Scientific, 1992)
11. S.W. Lowesley, *Frontiers in Physics, Condensed Matter Physics: Dynamic Correlations* (The Benjamin Cummings Publishing Company, 1986)
12. S.V. Vonsovskii, M.I. Kaznelson, *Quantum Solid-State Physics* Springer Series in Solid-State Sciences. (Berlin-Heidelberg-New York, 1989)
13. W.A. Harrison, *Solid State Theory* (McGraw-Hill Book Company, New-York-London-Toronto, 1970)
14. L.D. Landau, E.M. Lifshitz, *Quantum Mechanics* Non-Relativistic Theory. (Pergamon Press, New York, 1977)
15. L.D. Landau, E.M. Lifshitz, *Statistical Physics, Part I* (Butterworth-Heinemann, 1980)
16. E.G. Brovman, Y. Kagan, Phonons in non-transition metals. Sov. Phys. Uspekhi **112**, 369 (1974)
17. E.G. Brovman, Yu. Kagan, A. Kholas, Structure of the metallic hydrogen at zero pressure. Sov. Phys. JETP **34** 1300 (1972)
18. E.G. Brovman, Y. Kagan, A. Kholas, V. Pushkarev, The role of electron–electron interaction in the formation of the metastable phase of the metallic hydrogen. JETP Lett. **18**, 869 (1973)
19. E. Wigner, On the interaction of electrons in metals. Phys. Rev. **46**, 1002 (1934)
20. E.P. Wigner, Effects of the electron interaction on the energy levels of electrons in metals. Trans. Faraday Soc. **34**, 638 (1938)
21. L.D. Landau, E.M. Lifshitz, *Electrodynamics of Continuous Media* (Pergamon Press, 1984)
22. D. Pines, D. Bohm, A collective description of electron interaction. II Collective versus individual particle aspects of the interaction. Phys. Rev. **85**, 338 (1952)
23. D. Bohm, D. Pines, A collective description of electron interactions. III. The Coulomb interaction in a degenerate electron gas. Phys. Rev. **92**, 609 (1953)
24. M. Gell-Mann, K.A. Brueckner, Phys. Rev. **106**, 364 (1957)
25. J. Hubbard, Proc. Roy. Soc. (London) **A240**, 539 (1957)

26. P.W. Anderson, Random phase approximation in the theory of superconductivity. Phys. Rev. **112** (1958)
27. P. Hohenberg, W. Kohn, Inhomogeneous electron gas. Phys. Rev. **136**, B864 (1964)
28. W. Kohn, L.J. Sham, Self-consistent equations including exchange and correlation effects. Phys. Rev. **140**, A1133 (1965)
29. A.L. Fetter, J.D. Walecka, *Quantum Theory of Many-Particle Systems* (McGraw-Hill, S. Francisco, 1971)
30. S. Ring, P. Schick, *The Nuclear Many-body Problem* (Springer, New York, 1980)
31. J. Lindhard, Kgl. Danske Videnskab Elskab, Mat. - Fys. Medd. vol. 28, No. 8 (1954)
32. O.V. Dolgov, D.A. Kirzhnits, E.G. Maksimov, On the admissible sign of the static dielectric function of matter. Rev. Mod. Phys. **53**, 81 (1981)
33. V.A. Markel, Can the imaginary part of the permittivity be negative? Phys. Rev. E **78**, 026608 (2008)
34. L.P. Pitaevskii, On analytical properties of the diamagnetic permittivity in the presence of spatial dispersion. Int. J. Quant. Chem. **112**, 2998 (2012)
35. W. Kohn, J.H. Luttinger, New mechanism of superconductivity. Phys. Rev. Lett. **15**, 524 (1965)
36. D. Fay, A. Layzer, Superfluidity of low-density fermion system. Phys. Rev. Lett. **20**, 187 (1968)
37. M.Y. Kagan, A.V. Chubukov, Possibility of the superfluid transition in slightly non-ideal Fermi-gas with repulsion. JETP Lett. **47**, 614 (1988)
38. A.V. Chubukov, M.Y. Kagan, On the superfluid transition in dense electron system. J. Phys.: Condens. Matt. **1**, 3135 (1989)
39. M.Y. Kagan, *Modern Trends in Superconductivity and Superfluidity*. Lecture Notes in Physics, vol. 874 (Springer, Dordrecht, 2013)
40. M.Y. Kagan, A.V. Ozharovskii, Basic mechanisms of superconductivity in strongly correlated electron systems. Lecture Course, Part II (Moscow, MEPHI, 1999) (in Russian)
41. E.M. Lifshitz, L.P. Pitaevskii, Statistical Physics, Part II, Theory of the Condensed State (Pergamon, Elsevier, 2013)

Lecture 18. Charge Screening in Metals in the Random Phase Approximation

Abstract

We consider charge screening in metals in the random phase approximation (RPA) and derive the plasmon spectrum in metal. We discuss an emergence of acoustic plasmon (Pines demon) in the two-band metal. We introduce important notations of Thomas–Fermi wave vector and Debye radius which define the static screening of the trial charge in metal by electron plasma. We analyze the nature of the Kohn anomaly and Fridel oscillations in the effective interaction related to the non-exponential screening in metals. We briefly comment on Frohlich and Kohn–Luttinger mechanisms of superconductivity.

Summarizing the content of the previous Lecture, let us stress once more, that dielectric permittivity acquires a form (see Eq. (17.34)) [1–5]:

$$\varepsilon(\boldsymbol{q}, \omega) = 1 + \frac{4\pi e^2}{q^2} \Pi(\boldsymbol{q}, \omega),$$

where the polarization operator reads [1–5]:

$$\Pi(\boldsymbol{q}, \omega) = \frac{1}{\Omega} \sum_{k\sigma} \frac{n_{k\sigma} - n_{k+q\sigma}}{\hbar\omega + \varepsilon(\boldsymbol{k} + \boldsymbol{q}) - \varepsilon(\boldsymbol{k}) + i\delta}. \tag{18.1}$$

In the diagrammatic technique [6–13] <u>effective</u> (or screened) Coulomb interaction equals to the bare Coulomb interaction

$$V(\boldsymbol{q}) = \frac{4\pi e^2}{q^2} \tag{18.2}$$

divided on the dielectric permittivity $\varepsilon(\boldsymbol{q}, \omega)$:

$$V_{\text{eff.}}(\boldsymbol{q}, \omega) = \frac{V(\boldsymbol{q})}{\varepsilon(\boldsymbol{q}, \omega)}. \tag{18.3}$$

© The Author(s), under exclusive license to Springer Nature Switzerland AG 2026
M. Kagan, *Lecture Notes of Professor Yuri Kagan in Theoretical Solid-State Physics*,
Lecture Notes in Physics 1048, https://doi.org/10.1007/978-3-032-14621-2_18

Effective interaction corresponds to the summation of the infinite series of the ladder-type diagrams on Fig. 18.1 (in fact to the summation of the infinite geometric progression) [6–13].

In other words, the result (18.3) for the effective interaction corresponds to all the bubble diagrams of the perturbation theory. Let us emphasize, that in diagrammatic technique for fermions [6–8] each closed electron–hole loop made by two solid lines in Fig. 18.1 is related to the polarization operator $\Pi(\boldsymbol{q}, \omega)$ taking with the additional factor (-1) (which appears because of the closed loop). At the same time, each dotted line is connected with the bare Coulomb interaction $V(\boldsymbol{q}) = \frac{4\pi e^2}{q^2}$. The double dotted line corresponds to the effective interaction $V_{\text{eff.}}(\boldsymbol{q}, \omega)$.

As a result, equation for the effective interaction in Fig. 18.1 has a form:

$$V_{\text{eff.}}(\boldsymbol{q}, \omega) = V(\boldsymbol{q}) - V(\boldsymbol{q})\Pi(\boldsymbol{q}, \omega)V_{\text{eff.}}(\boldsymbol{q}, \omega). \tag{18.4}$$

In the diagrammatic technique this equation is called Bethe–Salpeter equation [14,15] Transferring to the l.h.s. the second term in the r.h.s. of the Bethe–Salpeter equation and regrouping the similar terms, we get

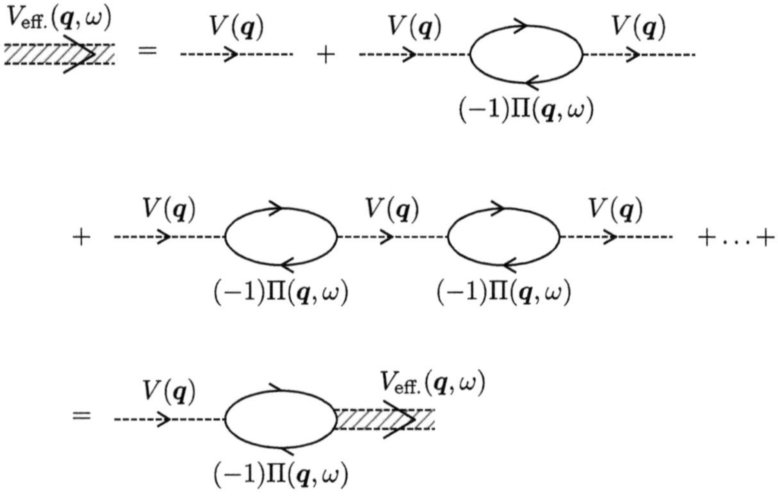

$$V_{\text{eff.}}(\boldsymbol{q}, \omega) = V(\boldsymbol{q}) - V(\boldsymbol{q})\Pi(\boldsymbol{q}, \omega)V_{\text{eff.}}(\boldsymbol{q}, \omega) \Rightarrow$$
$$\Rightarrow V_{\text{eff.}}(\boldsymbol{q}, \omega) = \frac{V(\boldsymbol{q})}{1 + V(\boldsymbol{q})\Pi(\boldsymbol{q},\omega)}$$

Fig. 18.1 Solution of the Bethe–Salpeter equation for the effective Coulomb interaction assumes the summation of the infinite ladder-type series of the bubble diagrams (the infinite geometric progression). In the diagrammatic technique for fermions, to each closed electron–hole loop made by two solid lines on the figure corresponds the polarization operator $\Pi(\boldsymbol{q}, \omega)$ taken with the additional factor (-1) (which appears because of the closed loop). At the same time, each dotted line is related to the bare Coulomb interaction $V(\boldsymbol{q}) = \frac{4\pi e^2}{q^2}$. The double dotted line corresponds to the effective interaction $V_{\text{eff.}}(\boldsymbol{q}, \omega)$ [1]

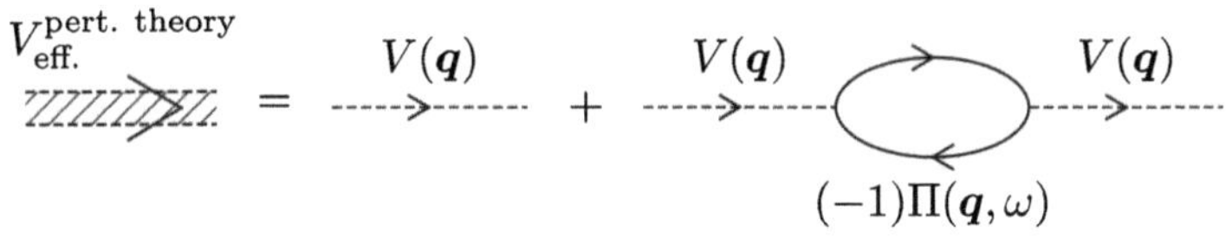

$$V_{\text{eff.}}^{\text{pert. theory}}(q,\omega) = V(q) - V^2(q)\Pi(q,\omega)$$

Fig. 18.2 An account of the first bubble diagram (linear in the polarization operator $\Pi(q,\omega)$ (and quadratic in the bare interaction $V(q)$) for the effective interaction $V_{\text{eff.}}(q,\omega)$ in the framework of the perturbation theory [1]

$$V_{\text{eff.}}(q,\omega) + V(q)\Pi(q,\omega)V_{\text{eff.}}(q,\omega) = V_{\text{eff.}}(q,\omega)(1 + V(q)\Pi(q,\omega)) = V(q). \tag{18.5}$$

From (18.5) we obtain the final expression for the effective interaction, namely:

$$V_{\text{eff.}}(q,\omega) = \frac{V(q)}{1 + V(q)\Pi(q,\omega)}. \tag{18.6}$$

This expression for the effective Coulomb interaction corresponds to the well-known Random phase approximation (RPA) [1–5, 16–23].

Note, that an account of only the first bubble diagram (see Fig. 18.2) would lead us to the replacement of the effective interaction $V_{\text{eff.}}(q,\omega)$ on the bare one $V(q)$ in the second term in the r.h.s. of the Bethe–Salpeter equation (18.4):

$$V_{\text{eff.}}(q,\omega) \approx V(q) - V^2(q)\Pi(q,\omega). \tag{18.7}$$

This result arises in the perturbation theory and corresponds to the expansion of the denominator in the expression for $V_{\text{eff.}}(q,\omega)$ in (18.6) in series up to the linear terms on the polarization operator $\Pi(q,\omega)$ (inclusively).

Let us stress once more, that the result (17.34) for the dielectric permittivity was obtained in the previous Lecture applying the method of the self-consistent field [1–4, 23, 24]. Method of the self-consistent field is valid as we discussed for small fluctuations [25] (that is when the mean-field theory sufficiently well describes the problem).

In other words, the obtained result works well in the limit of high electron density, since in this limit we can neglect the electron discreteness with a good accuracy.

18.1   High Frequency Limit $\omega \to \infty$

Let us consider the behavior of the dielectric permittivity in the high frequency limit at $\omega \to \infty$. Physically the limit $\omega \to \infty$ corresponds to the fulfillment of the relation $\hbar\omega \gg \varepsilon_F$, since the typical electron energies in metal are of the order of the Fermi enery (electron system is degenerate). In this limit the medium does not have time to adjust to the field variations and dielectric permittivity:

$$\varepsilon(\boldsymbol{q}, \omega) \to 1. \tag{18.8}$$

Let us show that in a more rigorous approach for the dielectric permittivity $\varepsilon(\boldsymbol{q}, \omega)$ and polarization operator, the series expansion at large ω in (17.34) and (18.1) leads to the following expression:

$$\varepsilon(\boldsymbol{q}, \omega) \to 1 - \frac{\alpha}{\omega^2}. \tag{18.9}$$

Note, that for $\hbar\omega \gg \varepsilon_F$ in the denominator of the polarization operator (18.1) the condition $\hbar\omega \gg |\varepsilon(\boldsymbol{k} + \boldsymbol{q}) - \varepsilon(\boldsymbol{k})|$ is fulfilled. Thus, polarization operator $\Pi(\boldsymbol{q}, \omega) \approx \frac{1}{\Omega} \sum_{k\sigma} \frac{n_{k\sigma} - n_{k+q\sigma}}{\hbar\omega} = \frac{(N_e - N_e)}{\hbar\omega} \frac{1}{\Omega} = 0$, and the term proportional to $\frac{1}{\omega}$ is absent in the asymptotic expression for $\varepsilon(\boldsymbol{q}, \omega)$ in (18.9).

Let us find the coefficient α in the term proportional to $\frac{1}{\omega^2}$. To do that let us expand in series on large ω the denominator of the polarization operator (18.1), using the relation: $\frac{1}{1+x} \approx 1 - x$ for $x \ll 1$. Then:

$$\varepsilon(\boldsymbol{q}, \omega) \approx 1 - \frac{4\pi e^2}{q^2 \Omega} \sum_k \left(n_k - n_{k+q}\right) \frac{(\varepsilon(\boldsymbol{k} + \boldsymbol{q}) - \varepsilon(\boldsymbol{k}))}{(\hbar\omega)^2}$$

$$= 1 - \frac{4\pi e^2}{q^2 \Omega} \frac{1}{(\hbar\omega)^2} \sum_k \left(n_k - n_{k+q}\right) (\varepsilon(\boldsymbol{k} + \boldsymbol{q}) - \varepsilon(\boldsymbol{k})). \tag{18.10}$$

It is important to note, that if in expressions (18.1) (18.6) the electron energy $\varepsilon(\boldsymbol{k} + \boldsymbol{q}) > \varepsilon(\boldsymbol{k})$, then we can see from the properties of the Dirac step function [25], that $n_k \geq n_{k+q}$ (that is smaller electron energies correspond to the not lesser occupation numbers [1]). Hence the sum over k in expression (18.1) is not negative:

$$\sum_k \left(n_k - n_{k+q}\right) (\varepsilon(\boldsymbol{k} + \boldsymbol{q}) - \varepsilon(\boldsymbol{k})) \geq 0. \tag{18.11}$$

Then the dielectric permittivity can be written in the form:

$$\varepsilon(\boldsymbol{q}, \omega) = 1 - \frac{\widetilde{\omega}^2(\boldsymbol{q})}{\omega^2}, \tag{18.12}$$

where in the nominator of the second term enters the square of the momentum dependent frequency:

$$\alpha = \widetilde{\omega}^2(\boldsymbol{q}) = \frac{4\pi e^2}{q^2 \Omega \hbar^2} \sum_k \left(n_k - n_{k+q}\right) \left(\varepsilon(\boldsymbol{k}+\boldsymbol{q}) - \varepsilon(\boldsymbol{k})\right) \geq 0. \qquad (18.13)$$

Replacing again the sum over $k = \{\boldsymbol{k}, \sigma\}$ by the integral on $\frac{d^3k}{(2\pi)^3}$ and performing summation over spin projections σ $\left(\sum_\sigma = 2\right)$ we get

$$\widetilde{\omega}^2(\boldsymbol{q}) = \frac{4\pi e^2}{q^2 \hbar^2} \cdot 2 \int \frac{d^3k}{(2\pi)^3} \left(n_k - n_{k+q}\right) \left(\varepsilon(\boldsymbol{k}+\boldsymbol{q}) - \varepsilon(\boldsymbol{k})\right), \qquad (18.14)$$

where now n_k and n_{k+q} are the occupation numbers for one spin projection only.

After that we can find an explicit expression for the frequency $\widetilde{\omega}^2(\boldsymbol{q})$ at small values of the vector $\boldsymbol{q}$. To do that let us expand in series on small $\boldsymbol{q}$ the difference of the occupation numbers in (18.14):

$$n_k - n_{k+q} = -\frac{\partial n_k}{\partial \boldsymbol{k}} \boldsymbol{q} = -\frac{\partial n_k}{\partial \varepsilon(\boldsymbol{k})} \cdot \frac{\partial \varepsilon(\boldsymbol{k})}{\partial \boldsymbol{k}} \boldsymbol{q}. \qquad (18.15)$$

But the derivative:

$$\frac{\partial \varepsilon(\boldsymbol{k})}{\partial \boldsymbol{k}} \boldsymbol{q} = \frac{\partial}{\partial \boldsymbol{k}} \left(\frac{\hbar^2 k^2}{2m}\right) \boldsymbol{q} = \frac{\hbar^2 \boldsymbol{k} \boldsymbol{q}}{m}. \qquad (18.16)$$

At the same time, at $T = 0$ the occupation numbers n_k are determined by the Dirac step function and hence:

$$-\frac{\partial n_k}{\partial \varepsilon(\boldsymbol{k})} = \frac{\partial \theta\left(\varepsilon(\boldsymbol{k}) - \varepsilon_F\right)}{\partial \varepsilon(\boldsymbol{k})} = \delta\left(\varepsilon(\boldsymbol{k}) - \varepsilon_F\right). \qquad (18.17)$$

Then the difference of the occupation numbers is given by:

$$n_k - n_{k+q} = \frac{\hbar^2 \boldsymbol{k} \boldsymbol{q}}{m} \delta\left(\varepsilon(\boldsymbol{k}) - \varepsilon_F\right). \qquad (18.18)$$

Analogously the energy difference in (18.14):

$$\varepsilon(\boldsymbol{k}+\boldsymbol{q}) - \varepsilon(\boldsymbol{k}) = \frac{\partial \varepsilon(\boldsymbol{k})}{\partial \boldsymbol{k}} \boldsymbol{q} = \frac{\hbar^2 \boldsymbol{k} \boldsymbol{q}}{m}. \qquad (18.19)$$

As a result, the frequency $\widetilde{\omega}^2(q)$ in (18.14) in spherical coordinates can be written as follows:

$$\widetilde{\omega}^2(q) = \frac{4\pi e^2}{q^2 \hbar^2} \cdot 2 \int \frac{d^3 k}{(2\pi)^3} \left(\frac{\hbar^2 kq}{m}\right)^2 \delta\left(\varepsilon(k) - \varepsilon_F\right)$$

$$= \frac{8\pi e^2}{8\pi^3 q^2} \frac{\hbar^2}{m^2} 2\pi \int k^2 dk \int d\cos\theta (kq\cos\theta)^2 \delta\left(\frac{\hbar^2 k^2}{2m} - \frac{\hbar^2 k_F^2}{2m}\right), \qquad (18.20)$$

where for the polar angle $\theta : kq = kq\cos\theta$.

After taking the integral on $d\cos\theta$

$$\int_{-1}^{1} d\cos\theta \cos^2\theta = \frac{2}{3} \qquad (18.21)$$

and contracting the multipliers q^2 in nominator and denominator of (18.20) we can simplify expression for $\widetilde{\omega}^2(q)$, namely:

$$\widetilde{\omega}^2(q) = \frac{4e^2}{3\pi} \frac{\hbar^2}{m^2} \int k^2 \cdot k^2 dk \delta\left(\frac{\hbar^2 k^2}{2m} - \frac{\hbar^2 k_F^2}{2m}\right). \qquad (18.22)$$

After that performing the change of variables $2kdk \to dk^2$ we obtain:

$$\widetilde{\omega}^2(q) = \frac{2e^2}{3\pi} \frac{\hbar^2}{m^2} \int k^3 \cdot dk^2 \cdot \delta\left(\frac{\hbar^2 k^2}{2m} - \frac{\hbar^2 k_F^2}{2m}\right)$$

$$= \frac{4e^2}{3\pi m} \int k^3 \cdot d\frac{\hbar^2 k^2}{2m} \cdot \delta\left(\frac{\hbar^2 k^2}{2m} - \frac{\hbar^2 k_F^2}{2m}\right) = \frac{4e^2 k_F^3}{3\pi m}. \qquad (18.23)$$

Now recollecting, that electron density (for both spin projections) reads: $n_e = \frac{k_F^3}{3\pi^2}$, we can rewrite (18.23) n the canonical form:

$$\widetilde{\omega}^2(q) \approx \omega_{\mathrm{pl.}}^2 = \frac{4\pi e^2 n_e}{m}, \qquad (18.24)$$

where $\omega_{\mathrm{pl.}}$ is the pasma frequency well known both in Physical kinetics and in physics of electron in metals [1–4, 25–27].

Correspondingly the dielectric permittivity on large frequencies yields:

$$\varepsilon(q,\omega) = 1 - \frac{\omega_{\mathrm{pl.}}^2}{\omega^2}. \qquad (18.25)$$

Note, that in this simple approximation the dielectric permittivity does not depend on the wave vector q.

To get the dependence of $\widetilde{\omega}^2(\boldsymbol{q})$ on $\boldsymbol{q}$ we should expand deeper (till the next terms) the denominator of the polarization operator in (18.1) using for $x \ll 1$ the relation:

$$\frac{1}{1+x} \approx 1 - x + x^2, \tag{18.26}$$

where the dimensionless variable x is given by:

$$x = \frac{\varepsilon(\boldsymbol{k}+\boldsymbol{q}) - \varepsilon(\boldsymbol{k})}{\hbar\omega}. \tag{18.27}$$

In this case the term proportional to $\frac{1}{\omega^3}$ again (similar to the term proportional to $\frac{1}{\omega}$) equals zero. Indeed, under the expansion of the polarization operator this term has the structure:

$$\frac{1}{(\hbar\omega)^3} \sum_k \left(n_k - n_{k+q}\right) \left(\varepsilon(\boldsymbol{k}+\boldsymbol{q}) - \varepsilon(\boldsymbol{k})\right)^2$$

$$= \frac{2}{(\hbar\omega)^3} \int \frac{d^3k}{(2\pi)^3} \left(n_k - n_{k+q}\right) \left(\varepsilon(\boldsymbol{k}+\boldsymbol{q}) - \varepsilon(\boldsymbol{k})\right)^2. \tag{18.28}$$

But the function $(\varepsilon(\boldsymbol{k}+\boldsymbol{q}) - \varepsilon(\boldsymbol{k}))^2$ is an even function under the replacement $\boldsymbol{k} \to \boldsymbol{k}+\boldsymbol{q}$. At the same time, the function $\left(n_k - n_{k+q}\right)$ is an odd function under the replacement $\boldsymbol{k} \to \boldsymbol{k}+\boldsymbol{q}$. Hence the integral over $\frac{d^3k}{(2\pi)^3}$ equals zero.

Only the next term of the expansion proportional to $\frac{1}{\omega^4}$ is different from zero. To determine the structure of this term let us utilize even more deep expansion:

$$\frac{1}{1+x} \approx 1 - x + x^2 - x^3 \text{ for } x \ll 1. \tag{18.29}$$

Then the dielectric permittivity at large frequencies equals to:

$$\varepsilon(\boldsymbol{q},\omega) = 1 - \frac{\omega_{\text{pl.}}^2}{\omega^2} - \frac{8\pi e^2}{q^2 \hbar^4 \omega^4} \int \frac{d^3k}{(2\pi)^3} \left(n_k - n_{k+q}\right) \left(\varepsilon(\boldsymbol{k}+\boldsymbol{q}) - \varepsilon(\boldsymbol{k})\right)^3. \tag{18.30}$$

Substituting expression (18.18) for the difference of the occupation numbers

$$n_k - n_{k+q} = \frac{\hbar^2 \boldsymbol{k}\boldsymbol{q}}{m} \delta\left(\varepsilon(\boldsymbol{k}) - \varepsilon_F\right)$$

in the last term of Eq. (18.30) together with the expression (18.19) $\varepsilon(\boldsymbol{k}+\boldsymbol{q}) - \varepsilon(\boldsymbol{k}) = \frac{\hbar^2 kq}{m}$ for the energy difference, we get for the dielectric permittivity

$$\varepsilon(\boldsymbol{q},\omega) = 1 - \frac{\omega_{\text{pl.}}^2}{\omega^2}$$

$$-\frac{2e^2}{q^2\hbar^4\omega^4\pi}\left(\frac{\hbar^2}{m}\right)^4 \int k^2 dk \int d\cos\theta (kq\cos\theta)^4 \delta\left(\frac{\hbar^2 k^2}{2m} - \frac{\hbar^2 k_F^2}{2m}\right). \tag{18.31}$$

Performing again the angular integration:

$$\int_{-1}^{1} d\cos\theta \cos^4\theta = \frac{2}{5}, \tag{18.32}$$

we can represent (18.31) as follows:

$$\varepsilon(\boldsymbol{q},\omega) = 1 - \frac{\omega_{\text{pl.}}^2}{\omega^2} - \frac{4e^2 q^2}{5\omega^4\pi m^2}\left(\frac{\hbar^2}{m}\right)^2 \int k^4 \cdot k^2 dk \delta\left(\frac{\hbar^2 k^2}{2m} - \frac{\hbar^2 k_F^2}{2m}\right)$$

$$= 1 - \frac{\omega_{\text{pl.}}^2}{\omega^2} - \frac{4e^2 q^2}{5\omega^4\pi m^2}\frac{\hbar^2}{m} \int k^5 \cdot d\frac{\hbar^2 k^2}{2m} \cdot \delta\left(\frac{\hbar^2 k^2}{2m} - \frac{\hbar^2 k_F^2}{2m}\right)$$

$$= 1 - \frac{\omega_{\text{pl.}}^2}{\omega^2} - \frac{4e^2 q^2}{5\omega^4\pi m}\frac{\hbar^2}{m^2}k_F^5 = 1 - \frac{\omega_{\text{pl.}}^2}{\omega^2} - \frac{12\pi e^2 q^2}{5\omega^4 m}\cdot\frac{\hbar^2 k_F^2}{m^2}\cdot\frac{k_F^3}{3\pi^2}$$

$$= 1 - \frac{\omega_{\text{pl.}}^2}{\omega^2} - \frac{12\pi e^2}{5\omega^4 m}\cdot\frac{\hbar^2 k_F^2 q^2}{m^2}\cdot n_e = 1 - \frac{\omega_{\text{pl.}}^2}{\omega^2} - \frac{12\pi e^2 n_e}{5\omega^4 m}(v_F q)^2, \tag{18.33}$$

where $n_e = \frac{k_F^3}{3\pi^2}$ is electron density on both spin projection, while:

$$v_F = \frac{\hbar k_F}{m} \tag{18.34}$$

is Fermi velocity.

Substituting expression (18.24) for the square of plasma frequency $\omega_{\text{pl.}}^2 = \frac{4\pi e^2 n_e}{m}$ in the last term in the r.h.s. of Eq. (18.33) we finally get:

$$\varepsilon(\boldsymbol{q},\omega) = 1 - \frac{\omega_{\text{pl.}}^2}{\omega^2} - \frac{3}{5}\frac{\omega_{\text{pl.}}^2}{\omega^4}\frac{(v_F q)^2}{\omega^4}. \tag{18.35}$$

18.2 Intrinsic Collective Excitations of Electron Subsystem in Metals (Plasmons [1,3,4,16,17,24–26,28])

Intrinsic collective excitations of the electron system (plasmons) on the diagrammatic language [7–13] correspond to the pole in the solution of the Bethe–Salpeter integral equation [14,15] for the effective Coulomb interaction.

$$V_{\text{eff.}}(\boldsymbol{q},\omega) = \frac{V(\boldsymbol{q})}{\varepsilon(\boldsymbol{q},\omega)}.$$

The plasmon spectrum is determined from the condition [1]:

$$\varepsilon(\boldsymbol{q},\omega) = 0. \tag{18.36}$$

In this case the density of the external charges $\rho_{\text{ext.}} = 0$, and thus the induced charge density $\rho_{\text{ind.}}$ equals to the total density:

$$\rho_{\text{ind.}} = \rho_{\text{tot.}}(1 - \varepsilon) = \rho_{\text{tot.}} \text{ for } \varepsilon = 0. \tag{18.37}$$

Let us derive the spectrum of collective excitations on large frequencies from the expression for the dielectric permittivity in (18.35):

$$\varepsilon(\boldsymbol{q},\omega) = 1 - \frac{\omega_{\text{pl.}}^2}{\omega^2} - \frac{3}{5}\frac{\omega_{\text{pl.}}^2 \, (v_F q)^2}{\omega^4} = 0. \tag{18.38}$$

Resolving Eq. (18.38) with respect to ω^2 we get

$$\omega^2 = \omega_{\text{pl.}}^2 + \frac{3}{5}\frac{\omega_{\text{pl.}}^2 \, (v_F q)^2}{\omega^2}. \tag{18.39}$$

For small wave vectors $v_F q \ll \omega$ the second term in the r.h.s. of (18.39) is much less than the first term and we can find the spectrum perturbatively replacing $\omega^2 \to \omega_{\text{pl.}}^2$ in the denominator of the second term. As a result, the plasmon spectrum in the three-dimensional metal acquires a form [1]:

$$\omega^2 = \omega_{\text{pl.}}^2 + \frac{3}{5}(v_F q)^2. \tag{18.40}$$

Let us stress, that plasmons are collective excitations of the bosonic nature. The plasmon spectrum is gapped in the one-band metal. Moreover, the gap in the spectrum equals to the plasma frequency $\omega_{\text{pl.}}$. Physically the origin of the gap is related to the electroneutrality condition in metal and the long-range character of the Coulomb interaction between electrons. The gapped plasmon in the physics of metals and plasma physics is often called Lindhard plasmon [5] (or Vlasov plasmon in Russian literature [26,27]).

In the two-band metal with different Fermi velocities v_{F1} and v_{F2} ($v_{F2} > v_{F1}$) for the two bands, crossing the Fermi level, besides the gapped plasmon [1] with the spectrum (18.40) another plasmon branch appears [4,28], namely the gapless acoustic plasmon with linear spectrum [4,28]

$$\omega = vq, \tag{18.41}$$

where the plasmon velocity:

$$v_{F1} < v < v_{F2}. \tag{18.42}$$

Acoustic plasmon in the two-band metal was predicted by Pines and Nozieres [3,4,28] and is often called the Pines demon [28]. In some sense acoustic plasmon resembles the second sound in the superfluid helium [29]. It is related with the oscillations of the relative charge density in the two bands and not with the oscillations of the total density as for the standard gapped plasmon (and first sound in He). Note that in the second sound we also have the relative oscillations of the superfluid and normal densities (see, e.g., Lecture 4 of the Course).

In the Frohlich mechanism of superconductivity [30] acoustic plasmon plays important role [31–34], replacing acoustic phonon in the BCS mechanism [35,36] as intermediate boson for superconductivity. The search of Pines demon in different strongly correlated electron systems and superconductors is intensified nowadays [37–39].

Note that in two-dimensional metal (in metallic monolayer) in certain frequency window appears another gapless plasmon with the square root character of the spectrum [40,41]:

$$\omega^2 = \gamma q, \text{ and } \omega \sim \sqrt{q}. \tag{18.43}$$

It is necessary to emphasize that under the charged particles passage through the crystal we can see experimentally the peaks of the energy losses [3,4] on the frequencies equal to the multiples of plasma frequencies ($\hbar\Omega = n\hbar\omega_{\mathrm{pl.}}$). These frequencies correspond to the creation of one, two or several plasmons in the system.

Note also that usually the plasmon frequency in metals is of the order of Fermi energy:

$$\hbar\omega_{\mathrm{pl.}} \sim \varepsilon_F. \tag{18.44}$$

However, this is not the case for high and low electron densities, since plasma frequency and Fermi energy depend on density in different way. Namely:

$$\hbar\omega_{\mathrm{pl.}} = \hbar \left(\frac{4\pi e^2 n_e}{m} \right)^{1/2} \sim n_e^{1/2}, \tag{18.45}$$

and at the same time:

$$\varepsilon_F \sim n_e^{2/3}.$$

Let us stress, that individual excitations or electron–hole pairs are also present in the system [3,4].

18.3 Imaginary Part of the Dielectric Permittivity

To calculate the imaginary part of the dielectric permittivity [42–44] let us utilize the well-known formula:

$$\frac{1}{x + i\delta} = P\frac{1}{x} - i\pi\delta(x),$$ (18.46)

where symbol P stands for the principal value.

Then the imaginary part of the dielectric permittivity in the expression (17.34) equals to:

$$\mathrm{Im}\,\varepsilon(\boldsymbol{q}, \omega) = -\pi\frac{4\pi e^2}{q^2\Omega}\sum_{k}\left(n_{\boldsymbol{k}} - n_{\boldsymbol{k}+\boldsymbol{q}}\right)\delta(\hbar\omega + \varepsilon(\boldsymbol{k} + \boldsymbol{q}) - \varepsilon(\boldsymbol{k}))$$

$$= -2\pi\frac{4\pi e^2}{q^2}\int\frac{d^3\boldsymbol{k}}{(2\pi)^3}\left(n_{\boldsymbol{k}} - n_{\boldsymbol{k}+\boldsymbol{q}}\right)\delta(\hbar\omega + \varepsilon(\boldsymbol{k} + \boldsymbol{q}) - \varepsilon(\boldsymbol{k})).$$ (18.47)

Note that the imaginary part $\mathrm{Im}\,\varepsilon(\boldsymbol{q}, \omega)$ is an odd function of frequency $\hbar\omega$ and allways related to the absorption [1,3,4,24].

Let us remind that the spectrum of the electromagnetic wave in crystal has the form [24]:

$$\omega = \frac{ck}{\sqrt{\varepsilon}},$$ (18.48)

where c is speed of light in vacuum.

Correspondingly, the wave vector in medium:

$$k = \frac{\omega}{c}\sqrt{\varepsilon}$$ (18.49)

for small absorption (when $\mathrm{Re}\,\varepsilon = \varepsilon' \gg \mathrm{Im}\,\varepsilon = \varepsilon''$) reads:

$$k = \frac{\omega}{c}\sqrt{\varepsilon'} + i\frac{\varepsilon''}{2\sqrt{\varepsilon'}}\frac{\omega}{c}.$$ (18.50)

Then the electromagnetic plane wave $\exp(ikz)$ incident on crystal from vacuum (from the half-space with $z < 0$) is exponentially decaying in medium:

$$\exp(ikz) \rightarrow \exp\left(i\frac{\omega}{c}\sqrt{\varepsilon'}z\right)\exp\left(-\frac{\varepsilon''}{2\sqrt{\varepsilon'}}\frac{\omega}{c}z\right).$$ (18.51)

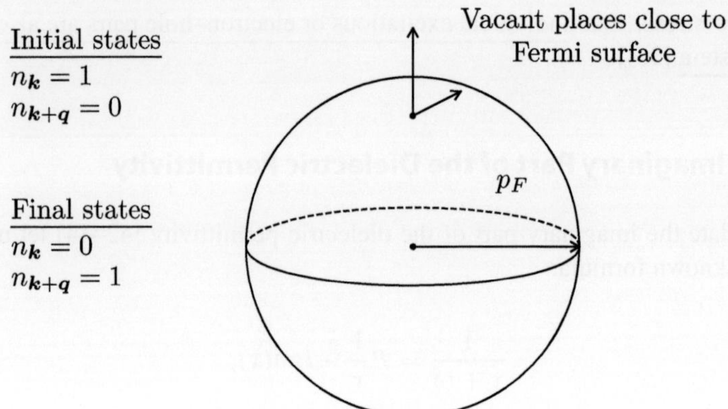

Fig. 18.3 Electron transitions to the vacant places close to the Fermi surface allowed by the Pauli principle. In these transitions, the difference between the occupation numbers is not equal zero $\left(n_k - n_{k+q} \neq 0\right)$ both in initial and final states. Namely, on the Figure in initial state we have electron and thus $n_k = 1, n_{k+q} = 0$, in the final state hole appears and hence vice versa: $n_k = 0, n_{k+q} = 1$ [1]

Note also that in the damping only electrons living close to the Fermi surface work [1,7,8] (see Fig. 18.3). Indeed, as we can see from formula (18.47) the imaginary part of the dielectric permittivity (and hence the damping) is different from zero only in the case when occupation numbers are not equal to each other $n_k \neq n_{k+q}$ (that is do not simultaneously become 0 or 1).

In other words, the presence of the vacant places is necessary close to the Fermi surface in Fig. 18.3.

As a result, electron–hole pair is created. On its creation we have to spend the energy which is analogous to the energy $E = 2mc^2$ required for the creation of electron–positron pair in Quantum relativistic electrodynamics [1]. Thus, imaginary part of the dielectric permittivity is connected with the real process of electron–hole pairs creation in metal [1,46,47].

At the same time, the real part of the dielectric permittivity is related to virtual processes. Let us stress, that the energy is not conserved in the virtual processes, but statistics (Pauli principle) is conserved [1].

18.4   Static Charge Screening in Metal

If in the dielectric permittivity $\varepsilon(q, \omega)$ we consider the energies (the transferred frequencies) which are much less than plasma frequency

$$\hbar\omega \ll \hbar\omega_{\text{pl.}} \sim \varepsilon_F, \tag{18.52}$$

then in fact we work in the static limit for the permittivity $\varepsilon(q, 0)$.

The static limit is often valid for electron–ion interaction since the phonon energies (responsible for the intrinsic vibration motion of ions) usually are small in comparison with the plasmon frequencies:

$$\omega_{\text{phon.}} \ll \omega_{\text{pl.}} \tag{18.53}$$

In the previous Lecture we derived expression (17.69) for the static polarization operator in three-dimensional metal [5]:

$$\Pi(\boldsymbol{q}, 0) = N(0) \left\{ 1 + \frac{4k_F^2 - q^2}{4k_F q} \ln \left| \frac{2k_F + q}{2k_F - q} \right| \right\},$$

where $N(0) = \frac{mk_F}{2\pi^2 \hbar^2}$ is the density of electron states (see (17.70) on the Fermi surface [1,7,8].

Let us consider the case of small wave vectors $q \ll k_F$. Let us expand in this case the logarithm: $\ln \left| \frac{2k_F+q}{2k_F-q} \right|$ in the Lindhard function (17.69) in series up to the cubic terms (inclusively) in the small parameter $\frac{q}{k_F} \ll 1$:

$$\ln \left| \frac{2k_F + q}{2k_F - q} \right| = \ln \left| \frac{1 + \frac{q}{2k_F}}{1 - \frac{q}{2k_F}} \right| \approx \frac{q}{k_F} + \frac{1}{12} \left(\frac{q}{k_F} \right)^3. \tag{18.54}$$

Then, in the linear approximation in $\frac{q}{k_F} \ll 1$, the static polarization operator [1,3–5,12,52,53]:

$$\Pi(q \ll k_F, 0) \approx N(0) \left\{ 1 + \frac{4k_F^2 - q^2}{4k_F q} \left(\frac{q}{k_F} + \frac{1}{12} \left(\frac{q}{k_F} \right)^3 \right) \right\}$$

$$= N(0) \left\{ 1 + \frac{4k_F^2 - q^2}{4k_F^2} \left(1 + \frac{1}{12} \left(\frac{q}{k_F} \right)^2 \right) \right\} = 2N(0) \left(1 - \frac{1}{12} \frac{q^2}{k_F^2} \right). \tag{18.55}$$

Correspondingly, the static dielectric permittivity for small wave vectors $q \ll k_F$ equals to [1–3,7,12,24]:

$$\varepsilon(\boldsymbol{q}, 0) \approx 1 + \frac{4\pi e^2}{q^2} 2N(0) \left(1 - \frac{1}{12} \left(\frac{q^2}{k_F^2} \right) \right) = 1 + \frac{\kappa^2}{q^2} - \frac{\kappa^2}{12 k_F^2}, \tag{18.56}$$

where κ is the Thomas–Fermi wave vector [47,48] playing important role in plasma physics [26,27] and physics of electrons in metals [1–4]. It is convenient to express the square of the Thomas–Fermi wave vector:

$$\kappa^2 = 4\pi e^2 2N(0) = 4\pi e^2 \frac{mk_F}{\pi^2 \hbar^2} = \frac{4e^2 mk_F}{\pi \hbar^2} \tag{18.57}$$

via the ratio of electron density $n_e = \frac{k_F^3}{3\pi^2}$ and Fermi energy $\varepsilon_F = \frac{\hbar^2 k_F^2}{2m}$. In this case multiplying the nominator and the denominator of the last equality in the r.h.s. of Eq. (18.57) on k_F^2, we get the canonical expression for the square of the Thomas–Fermi wave vector [1,3,4]:

$$\kappa^2 = \frac{4e^2 m k_F}{\pi \hbar^2} = \frac{4e^2 m k_F^3}{\pi \hbar^2 k_F^2} = \frac{2e^2 k_F^3}{\pi \varepsilon_F} = \frac{6\pi e^2 k_F^3}{3\pi^2 \varepsilon_F} = \frac{6\pi e^2 n_e}{\varepsilon_F}. \tag{18.58}$$

It is useful also to express the square of the Thomas–Fermi wave vector in the initial formula (18.57) via the Bohr radius $a_B = \frac{\hbar^2}{me^2}$:

$$\kappa^2 = \frac{4e^2 m k_F}{\pi \hbar^2} = \frac{4k_F}{\pi a_B}. \tag{18.59}$$

Hence in dense electron plasma:

$$\frac{\kappa^2}{k_F^2} = \frac{4}{\pi a_B k_F} \sim r_S \ll 1, \tag{18.60}$$

and thus, is the small correction to 1 in the last equality in the r.h.s. of Eq. (18.56). As a result, the dielectric permittivity for $q \to 0$ acquires canonical form:

$$\varepsilon(q) \approx 1 + \frac{\kappa^2}{q^2}. \tag{18.61}$$

At the same time, the effective (screened) Coulomb interaction at small q is given by:

$$V_{\text{eff.}}(q, 0) = \frac{V(q)}{\varepsilon(q, 0)} = \frac{V(q)}{\left(1 + \frac{\kappa^2}{q^2}\right)} = \frac{\frac{4\pi e^2}{q^2}}{\left(1 + \frac{\kappa^2}{q^2}\right)} = \frac{4\pi e^2}{\kappa^2 + q^2}. \tag{18.62}$$

In real space it is described by the Yukawa potential [49] well known in nuclear physics:

$$V_{\text{eff.}}(r) = \int \frac{d^3 q}{(2\pi)^3} V_{\text{eff.}}(q, 0) \exp(iqr) = \int \frac{d^3 q}{(2\pi)^3} \frac{V(q)}{\varepsilon(q, 0)} \exp(iqr) = \frac{e^2}{r} \exp(-\kappa r), \tag{18.63}$$

where the bare Coulomb interaction is expanded in Fourier series as follows:

$$V(r) = \frac{e^2}{r} = \int \frac{d^3 q}{(2\pi)^3} V(q) \exp(iqr) = \int \frac{d^3 q}{(2\pi)^3} \frac{4\pi e^2}{q^2} \exp(iqr). \tag{18.64}$$

Let us stress, that we used expression (18.61) for the effective interaction $V_{\text{eff.}}(\boldsymbol{q}, 0) = \frac{V(\boldsymbol{q})}{\varepsilon(\boldsymbol{q},0)}$ in the momentum space, obtained in the limit of small wave vectors $q \to 0$.

In the solid-state physics the Debye (exponential) charge screening [50] corresponds to Yukawa potential [49]:

$$V_{\text{eff.}}(r) = \frac{e^2}{r} \exp\left(-\frac{r}{r_D}\right), \tag{18.65}$$

where the Debye radius in metal [50] equals to the inverse Thomas–Fermi wave vector:

$$r_D = \frac{1}{\kappa}. \tag{18.66}$$

It is useful also to get the expansion of the static dielectric permittivity in the limit of large wave vectors $q \gg k_F$. In this case it is necessary again to expand $\ln\left|\frac{2k_F + q}{2k_F - q}\right|$ in the Lindhard function in the small parameter $\frac{k_F}{q} \ll 1$ up to the cubic terms (inclusively):

$$\ln\left|\frac{2k_F + q}{2k_F - q}\right| = \ln\left|\frac{1 + \frac{2k_F}{q}}{1 - \frac{2k_F}{q}}\right| \approx \frac{4k_F}{q} + \frac{16}{3}\left(\frac{k_F}{q}\right)^3 = \frac{4k_F}{q}\left(1 + \frac{4}{3}\left(\frac{k_F}{q}\right)^2\right). \tag{18.67}$$

At the same time, expression which stands before the logarithm in the static polarization operator (17.69) can be written as:

$$\frac{4k_F^2 - q^2}{4k_F q} = -\frac{q}{4k_F}\left(1 - 4\left(\frac{k_F}{q}\right)^2\right). \tag{18.68}$$

As a result, the static polarization operator reads in this case:

$$\Pi\left(q \gg k_F, 0\right) \approx N(0)\left(1 - \left(1 - 4\left(\frac{k_F}{q}\right)^2\right)\left(1 + \frac{4}{3}\left(\frac{k_F}{q}\right)^2\right)\right)$$
$$= N(0)\left\{\frac{8}{3}\left(\frac{k_F}{q}\right)^2 + O\left(\left(\frac{k_F}{q}\right)^4\right)\right\} \approx N(0)\frac{8}{3}\left(\frac{k_F}{q}\right)^2. \tag{18.69}$$

Then the static dielectric permittivity for $q \to \infty$ equals to:

$$\varepsilon(\boldsymbol{q}, 0) \approx 1 + \frac{4\pi e^2}{q^2}N(0)\frac{8}{3}\left(\frac{k_F}{q}\right)^2 = 1 + \frac{mk_F}{2\pi^2\hbar^2}\cdot\frac{4\pi e^2}{q^2}\cdot\frac{8}{3}\left(\frac{k_F}{q}\right)^2 = 1 + \frac{\zeta}{q^4}, \tag{18.70}$$

where the constant ζ can be expressed via electron density n_e and the Bohr radius a_B:

$$\zeta = \frac{16\pi m e^2 k_F^3}{3\pi^2 \hbar^2} = \frac{16\pi m e^2 n_e}{\hbar^2} = \frac{16\pi n_e}{a_B}. \tag{18.71}$$

18.5   Macroscopic Description of Charge Screening in Metals

Let us establish the nature of the static charge screening in metals on the macroscopic language. As we discussed in the previous Lecture, the Fourier component of the induced charge density in metal equals to:

$$\rho_{\text{ind.}}(\boldsymbol{q}) = \rho_{\text{tot.}}(\boldsymbol{q}) - \rho_{\text{ext.}}(\boldsymbol{q}) = \rho_{\text{ext.}}(\boldsymbol{q}) \left[\frac{\rho_{\text{tot.}}(\boldsymbol{q})}{\rho_{\text{ext.}}(\boldsymbol{q})} - 1 \right] = \rho_{\text{ext.}}(\boldsymbol{q}) \left[\frac{1}{\varepsilon(\boldsymbol{q},0)} - 1 \right]. \tag{18.72}$$

Then the induced charge density in the real space:

$$\rho_{\text{ind.}}(\boldsymbol{r}) = \int \frac{d^3 q}{(2\pi)^3} \rho_{\text{ind.}}(\boldsymbol{q}) \exp(i\boldsymbol{q}\boldsymbol{r}) = \int \frac{d^3 q}{(2\pi)^3} \rho_{\text{ext.}}(\boldsymbol{q}) \left[\frac{1}{\varepsilon(\boldsymbol{q},0)} - 1 \right] \exp(i\boldsymbol{q}\boldsymbol{r}). \tag{18.73}$$

Thus, the total induced charge in metal:

$$Q_{\text{ind.}} = \int \rho_{\text{ind.}}(\boldsymbol{r}) d^3 r = \int \frac{d^3 q}{(2\pi)^3} \rho_{\text{ext.}}(\boldsymbol{q}) \left[\frac{1}{\varepsilon(\boldsymbol{q},0)} - 1 \right] \int d^3 r \exp(i\boldsymbol{q}\boldsymbol{r})$$
$$= \int \frac{d^3 q}{(2\pi)^3} \rho_{\text{ext.}}(\boldsymbol{q}) \left[\frac{1}{\varepsilon(\boldsymbol{q},0)} - 1 \right] \delta(\boldsymbol{q}) = \rho_{\text{ext.}}(q=0) \left[\frac{1}{\varepsilon(q=0,0)} - 1 \right]. \tag{18.74}$$

But at $q \to 0$ the static dielectric permittivity

$$\varepsilon(q,0) = 1 + \frac{\kappa^2}{q^2} \to \infty. \tag{18.75}$$

Then the square bracket in the last equality in the r.h.s. of Eq. (18.72)

$$\left[\frac{1}{\varepsilon(q=0,0)} - 1 \right] \to -1. \tag{18.76}$$

As a result, the total induced charge:

$$Q_{\text{ind.}} = -\rho_{\text{ext.}}(q = 0) = -\lim_{q \to 0} \int d^3 r \, \rho_{\text{ext.}}(r) \exp(-iqr) = -\int d^3 r \, \rho_{\text{ext.}}(r)$$
$$= -Q_{\text{ext.}}. \tag{18.77}$$

Hence the total charge equals zero in metals:

$$Q_{\text{tot.}} = Q_{\text{ind.}} + Q_{\text{ext.}} = 0. \tag{18.78}$$

Equality of total charge to zero in metal is connected with the fact, that static dielectric permittivity at zero wave vector: $\varepsilon(q = 0) = \infty$. Just as a result of this fact, all the external charges are totally screened in metal, and thus $Q_{\text{ind.}} = -Q_{\text{ext.}}$. As a consequence of that, the total charge: $Q_{\text{tot.}} = 0$.

Let us emphasize, that the situation is absolutely different in dielectrics and semiconductors [1, 24]. In these systems in contrast to metals there is no free electrons, and thus the limit $\varepsilon \to \infty$ at $q \to 0$ is absent for the dielectric permittivity. In other words, the dielectric permittivity ε is finite in dielectrics and semicinductors at $q \to 0$, and hence the total screening of the external charges does not emerge. Correspondingly, external charge in dielectrics is not totally screened and can be seen on large distances [1].

At the same time, due to the presence of free electrons, we get the strong reconstruction of electron subsystem and the induced charge appears which totally screens external (trial) charge, embedded in metal.

18.6 Kohn Anomaly [51]

In the general case, as we already discussed in this Lecture (see Eq. (18.63)) the screened Coulomb interaction in the real space reads:

$$V_{\text{eff.}}(r) = \int \frac{d^3 q}{(2\pi)^3} \frac{V(q)}{\varepsilon(q, 0)} \exp(iqr).$$

If expression under the integral is analytic one, then the integral in (18.61) is effectively sitting on small q.

In general case, the integral sits on small q and on the singularity.

It is important to note, that we have the singular (non-analytical) part [51] in the expression for the static polarization operator (17.69) when the wave vector tends to the diameter of the Fermi sphere $q \to 2k_F$ [51].

In three-dimensional metal the singularity has a form [1, 12]:

$$\Pi_{\text{sing.}}(q \to 2k_F, 0) \sim N(0)\frac{4k_F^2 - q^2}{4k_F q} \ln\left|\frac{2k_F + q}{2k_F - q}\right|$$
$$\approx N(0)\frac{(2k_F - q)}{2k_F} \ln\frac{4k_F}{|q - 2k_F|} = N(0)\frac{(q - 2k_F)}{2k_F} \ln\frac{|q - 2k_F|}{4k_F}. \tag{18.79}$$

Note, that while $\ln\left|\frac{q-2k_F}{4k_F}\right|$ diverges for $q \to 2k_F$, the function $N(0)\frac{(q-2k_F)}{2k_F}\ln\left|\frac{q-2k_F}{4k_F}\right|$ standing in the expression (18.79) for singular part of the polarization operator is convergent (goes to zero) for $q \to 2k_F$.

At the same time, the first derivative on q from $\Pi_{\text{sing.}}$ is divergent:

$$\frac{d\Pi_{\text{sing.}}}{dq} \sim N(0)\frac{1}{2k_F}\ln\left|\frac{q-2k_F}{4k_F}\right| \to \infty \text{ for } q \to 2k_F. \qquad (18.80)$$

The second derivative is divergent as well:

$$\frac{d^2\Pi_{\text{sing.}}}{dq^2} \sim N(0)\frac{1}{2k_F}\cdot\frac{1}{(q-2k_F)} \to \infty \text{ for } q \to 2k_F \qquad (18.81)$$

together with all the higher derivatives on q from the singular part of the polarization operator [1, 12, 51–53].

Physically, emergence of the non-analytical part in the polarization operator is connected with sharp distribution of Fermi particles in momentum space (or with the existence of the Fermi surface for electrons in metals) [1, 12].

Let us emphasize, that qualitative behavior of the static polarization operator $\Pi(q)$ as a function of transferred momentum q has a form, presented in Fig. 18.4. It is important, that $\Pi(q) = 2N(0)$ at $q = 0$.

For small $q \ll k_F$ it decreases as $\Pi(q) = 2N(0)\left(1 - \frac{1}{12}\frac{q^2}{k_F^2}\right)$ [1,12]. At $q = 2k_F$ polarization operator $\Pi(q) = N(0)$ is twice smaller than at $q = 0$. Finally for large $q \gg 2k_F$ the polarization operator is decreasing proportionally to $N(0)\left(\frac{k_F}{q}\right)^2$ [1,12,52] and goes to zero for $q \to \infty$.

At $q = 2k_F$ we can see the change of curvature in the behavior of $\Pi(q)$ as a function of q related to the singularity (the presence of Kohn anomaly [1,12,51–53]).

18.7   Fridel Oscillations [54]

From the Mathematics we know, that the presence of the singularity in momentum space always leads to the oscillations in the real space.

Kohn anomaly (singularity in the static polarization operator and dielectric permittivity for $q \to 2k_F$) leads to the well-known Fridel oscillations [54] in real space.

In other words, we can say, that in metals in contrast with dielectrics and semiconductors, there is non-exponential contribution to the charge screening at large distances. This contribution to the screened interaction is described by the function which is rapidly oscillating in real space and, at the same time, is slowly decreasing (in the power fashion). In the three-dimensional metal it acquires the following form:

$$V_{\text{eff.}}(r) \sim \frac{\cos 2k_F r}{(2k_F r)^3}. \qquad (18.82)$$

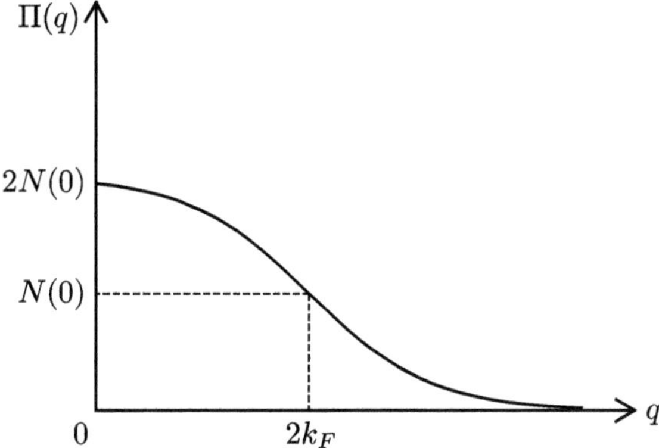

Fig. 18.4 Qualitative behavior of the static polarization operator $\Pi(q)$, as a function of the transferred momentum q [52,53]. Polarization operator $\Pi(q) = 2N(0)$ at $q = 0$. For small $q \ll k_F$ it decreases as $\Pi(q) = 2N(0)\left(1 - \frac{1}{12}\frac{q^2}{k_F^2}\right)$. At $q = 2k_F$ polarization operator $\Pi(q) = N(0)$ is twice smaller than at $q = 0$. For large $q \gg 2k_F$ polarization operator decreases proportionally to $N(0)\left(\frac{k_F}{q}\right)^2$ and tends to zero for $q \to \infty$. At $q = 2k_F$ we can see the change of the curvature in the behavior of $\Pi(q)$ as a function of q connected with the singularity (the presence of Kohn anomaly) [1]

To derive the Fridel oscillations [1,12,52–54] let us return back to the expression (18.63) for the screened Coulomb interaction and let us perform the calculations in the spherical coordinates. Then, after taking the angular integral we get

$$
\begin{aligned}
V_{\text{eff.}}(r) &= \int \frac{d^3\boldsymbol{q}}{(2\pi)^3} \frac{V(q)}{\varepsilon(q,0)} \exp(i\boldsymbol{q}\boldsymbol{r}) \\
&= \frac{2\pi}{(2\pi)^3} \int q^2 dq \frac{V(q)}{\varepsilon(q,0)} \int_{-1}^{1} d\cos\theta \exp(iqr\cos\theta) \\
&= \frac{1}{(2\pi)^2} \int q^2 dq \frac{V(q)}{\varepsilon(q,0)} \frac{\exp(iqr) - \exp(-iqr)}{iqr} = \frac{1}{2\pi^2 r} \int_0^{\infty} \frac{qV(q)}{\varepsilon(q,0)} \sin qr\, dq \\
&= \frac{1}{2\pi^2 r} \int_0^{\infty} dq f(q) \sin qr,
\end{aligned}
\tag{18.83}
$$

where the function:

$$
f(q) = \frac{qV(q)}{\varepsilon(q,0)}.
\tag{18.84}
$$

Further on let us twice integrate expression (18.83) by parts.
After the first integration we obtain for the effective interaction:

$$
V_{\text{eff.}}(r) = -\frac{1}{2\pi^2 r^2} f(q)\cos qr \bigg|_0^{\infty} + \frac{1}{2\pi^2 r^2} \int_0^{\infty} f'(q)\cos qr,
\tag{18.85}
$$

where the derivative:

$$f'(q) = \left(\frac{qV(q)}{\varepsilon(q,0)}\right)'. \tag{18.86}$$

But the bare Coulomb interaction: $V(q) = \frac{4\pi e^2}{q^2}$. At the same time, static dielectric permittivity: $\varepsilon(q,0) = 1 + \frac{\zeta}{q^4}$ for $q \to \infty$ (see Eq. (18.71)). Then the function:

$$f(q \to \infty) = \frac{qV(q)}{\varepsilon(q \to \infty)} \approx qV(q) = \frac{4\pi e^2}{q} \to 0 \text{ for } q \to \infty. \tag{18.87}$$

Analogously the dielectric permittivity $\varepsilon(q,0) = 1 + \frac{\kappa^2}{q^2}$ for $q \to 0$, and the screened interaction:

$$\frac{V(q)}{\varepsilon(q \to 0)} = \frac{4\pi e^2}{\kappa^2 + q^2} \to \frac{4\pi e^2}{\kappa^2}. \tag{18.88}$$

Hence the function:

$$f(q \to 0) = \frac{qV(q)}{\varepsilon(q \to 0)} \approx \frac{4\pi e^2 q}{\kappa^2} \to 0 \text{ for } q \to 0. \tag{18.89}$$

Thus, the first term in the r.h.s. of Eq. (18.85) goes to zero, and screened interaction in real space reads:

$$V_{\text{eff.}}(r) = \frac{1}{2\pi^2 r^2} \int_0^\infty f'(q) \cos qr. \tag{18.90}$$

Let us integrate expression (18.90) by parts again. As a result, we get

$$V_{\text{eff.}}(r) = \frac{1}{2\pi^2 r^3} f'(q) \sin qr \Big|_0^\infty - \frac{1}{2\pi^2 r^3} \int_0^\infty dq \sin qr f''(q). \tag{18.91}$$

Let us first analyze again the first (free) term in (18.91).

For $q \to \infty$ the function $f(q) \to \frac{4\pi e^2}{q}$, and hence the derivative:

$$f'(q) \approx -\frac{4\pi e^2}{q^2} \to 0 \text{ for } q \to \infty. \tag{18.92}$$

At the same time, for $q \to 0$ the function $f(q) \to \frac{4\pi e^2 q}{\kappa^2}$, and its derivative equals to:

$$f'(q) \approx \frac{4\pi e^2}{\kappa^2} = \text{const}. \tag{18.93}$$

However, $\sin qr = 0$ at $q = 0$ on the lower limit in the free term in (18.91). Thus, the free term equals zero again, and the screened interaction after two integrations by parts is given by:

$$V_{\text{eff.}}(r) = -\frac{1}{2\pi^2 r^3} \int_0^\infty dq \sin qr f''(q),$$
(18.94)

where the second derivative:

$$f''(q) = \left(\frac{qV(q)}{\varepsilon(q,0)}\right)'' = \left(\frac{qV(q)}{1 + V(q)\Pi(q)}\right)''.$$
(18.95)

It is important to note, that $f''(q) \approx \frac{8\pi e^2}{q^3} \to 0$ for $q \to \infty$, and $f''(q) \to 0$ for $q \to 0$.

Correspondingly, the integral (18.94) is convergent both on the lower and upper limit. Let us calculate the contribution to the integral from the neighborhood of the singular point (for $q \to 2k_F$). Meanwhile, in all the terms except the singular contribution to the polarization operator in Eq. (18.94), we can put $q = 2k_F$.

Then, for $q \to 2k_F$ the function:

$$f(q \to 2k_F) = \frac{qV(q)}{1 + V(q)\Pi(q)} \approx \frac{2k_F V(q = 2k_F)}{1 + V(q = 2k_F)\Pi(q)},$$
(18.96)

and its first derivative:

$$f'(q \to 2k_F) \approx -\frac{2k_F V^2(2k_F)}{(1 + V(2k_F)\Pi(q))^2}\frac{d\Pi_{\text{sing.}}}{dq} \approx -\frac{2k_F V^2(2k_F)}{(1 + V(2k_F)\Pi(2k_F))^2}\frac{d\Pi_{\text{sing.}}}{dq},$$
(18.97)

where:

$$\Pi(2k_F) = N(0).$$
(18.98)

Hence, for $q \to 2k_F$ the second derivative:

$$f''(q \to 2k_F) \approx -\frac{2k_F V^2(2k_F)}{(1 + V(2k_F)N(0))^2}\frac{d^2\Pi_{\text{sing.}}}{dq^2}.$$
(18.99)

Substituting in (18.99) expression for the second derivative on q of the singular part of the polarization operator $\frac{d^2\Pi_{\text{sing.}}}{dq^2} \sim N(0)\frac{1}{2k_F} \cdot \frac{1}{(q-2k_F)}$ from Eq. (18.81), we obtain finally:

$$f''(q \to 2k_F) \approx -\frac{2k_F V^2(2k_F)}{(1 + V(2k_F)N(0))^2}N(0)\frac{1}{2k_F} \cdot \frac{1}{(q-2k_F)}$$
$$= -\frac{V(2k_F)N(0)}{(1 + V(2k_F)N(0))^2}\frac{1}{(q-2k_F)}.$$
(18.100)

As a result, the screened interaction:

$$V_{\text{eff.}}(r) = \frac{1}{2\pi^2 r^3} \frac{N(0)V^2(2k_F)}{(1 + V(2k_F)N(0))^2} \int_0^\infty dq \sin qr \cdot \frac{1}{(q - 2k_F)}$$

$$= \frac{1}{2\pi^2 r^3} \frac{N(0)V^2(2k_F)}{(1 + V(2k_F)N(0))^2} \int_0^\infty d(qr) \sin qr \cdot \frac{1}{(q - 2k_F)r}. \tag{18.101}$$

Furter on, we can introduce the dimensionless variable:

$$y = (q - 2k_F)r, \tag{18.102}$$

and represent $\sin qr$ in Eq. (18.101) as:

$$\sin qr = \sin\left[(q - 2k_F)r + 2k_F r\right]$$
$$= \sin(y + 2k_F r) = \sin y \cos 2k_F r + \cos y \sin 2k_F r. \tag{18.103}$$

Then, expanding the limits of the integration on the variable y from $-\infty$ till ∞, we can write the screened Coulomb interaction in the form:

$$V_{\text{eff.}}(r) = \frac{1}{2\pi^2 r^3} \frac{N(0)V^2(2k_F)}{(1 + V(2k_F)N(0))^2} \left\{ \cos 2k_F r \int_{-\infty}^\infty \frac{dy}{y} \sin y \right.$$
$$\left. + \sin 2k_F r \int_{-\infty}^\infty \frac{dy}{y} \cos y \right\}. \tag{18.104}$$

Note, that the second integral in (18.104) equals zero being the integral of the odd function over symmetric interval. Hence:

$$V_{\text{eff.}}(r) = \frac{1}{2\pi^2} \frac{N(0)V^2(2k_F)}{(1 + V(2k_F)N(0))^2} \frac{\cos 2k_F r}{r^3} \int_{-\infty}^\infty \frac{dy}{y} \sin y. \tag{18.105}$$

But the last integral in (18.103) is tabulated and yields:

$$\int_{-\infty}^\infty \frac{dy}{y} \sin y = \pi. \tag{18.106}$$

As a result, the Fridel oscillations in screened interaction acquire the canonical form [54]:

$$V_{\text{eff.}}(r) = \frac{1}{2\pi} \frac{N(0)V^2(2k_F)}{(1 + V(2k_F)N(0))^2} \frac{\cos 2k_F r}{r^3} = A\frac{\cos 2k_F r}{(2k_F r)^3}, \tag{18.107}$$

where the constant:

$$A = \frac{4}{\pi^2} \frac{V(2k_F) N(0)}{(1 + V(2k_F) N(0))^2} V(2k_F) k_F^3. \tag{18.108}$$

Substituting in Eq. (18.108) the Fourier component of the bare interaction at $q = 2k_F$:

$$V(2k_F) = \frac{\pi e^2}{k_F^2}, \tag{18.109}$$

we can write the constant A as follows:

$$A = \frac{4}{\pi^2} \frac{\frac{\pi e^2}{k_F^2} N(0)}{\left(1 + \frac{\pi e^2}{k_F^2} N(0)\right)^2} \frac{\pi e^2}{k_F^2} k_F^3. \tag{18.110}$$

Now utilizing expression for the square of the Thomas–Fermi wave vector:

$$\kappa^2 = 8\pi e^2 N(0), \tag{18.111}$$

we can represent the constant as [1]:

$$A = \frac{4}{\pi} \frac{\frac{\kappa^2}{8k_F^2}}{\left(1 + \frac{\kappa^2}{8k_F^2}\right)^2} e^2 k_F = \frac{1}{2\pi} \frac{\frac{e^2 \kappa^2}{k_F}}{\left(1 + \frac{\kappa^2}{8k_F^2}\right)^2}. \tag{18.112}$$

In the limit of large electron density $\frac{\kappa^2}{k_F^2} \sim \frac{1}{k_F a_B} \sim r_S \ll 1$, and expression (18.112) for the constant A can be simplified [53]:

$$A \approx \frac{1}{2\pi} \frac{e^2 \kappa^2}{k_F}. \tag{18.113}$$

As we discussed already, the potential of the Fridel oscillations describes non-exponential charge screening at large distances $k_F r \gg 1$. It slowly decreases with distance (proportionally to $\frac{1}{r^3}$ in the three-dimensional metal). At the same time, it rapidly oscillates with the sign change. The oscillation period equals to the inverse diameter of the Fermi sphere $\frac{1}{2k_F}$.

The sign change of the effective interaction plays very important role in the solid-state physics. In particular, it can lead to the possible realization of the Kohn–Luttinger mechanism of anomalous superconductivity [55] in a large variety of electron systems with repulsive (bare) interaction between the particles and probably even in the superfluid He-3 [55–59].

The mechanism of anomalous (non-phonon) superconductivity of the Kohn–Luttinger type found its further development in papers of D. Fay and A. Layzer, and

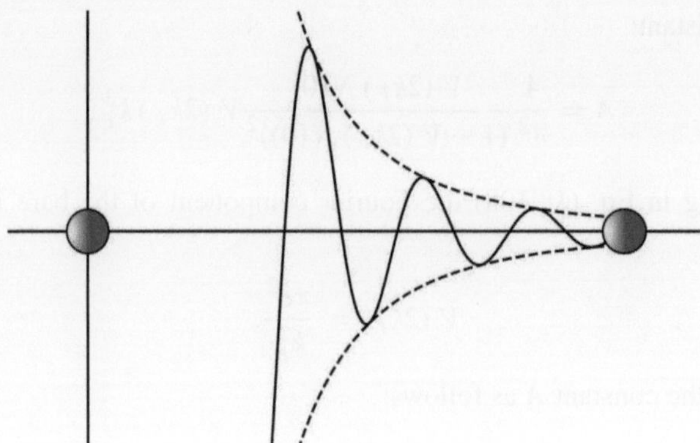

Fig. 18.5 Kohn–Luttinger mechanism of anomalous (non-phonon) superconductivity, based on the presence of Kohn anomaly [51] (Fridel oscillations [54]) in the effective interaction between two electrons forming the Cooper pair [1,52,53]

M. Yu. Kagan and Chubukov [56–59]. These papers were devoted to the possible realization of triplet p-wave pairing in low-density fermion systems with repulsive interaction. In application to the fermionic superfluidity in ultracold quantum gases in magnetic traps the first results on Kohn–Luttinger mechanism were obtained in [60] by M.A. Baranov, M. Yu. Kagan and Yu. Kagan.

Note, that Kohn–Luttinger mechanism is based on the necessity to average the effective interaction between two electrons at large distances on the large number of potential wells and barriers located between them (see Fig. 18.5). When we perform the averaging, it is not evident which contribution to the effective interaction (of the potential wells or of the barriers) is dominant. If the wells contribution is more effective than the barriers contribution, then, in this case, two electrons effectively attract each other forming the Cooper pair [36].

Sign-changing potential of Fridel oscillations in the case of exchange interaction between spins is usually called the RKKY interaction and is well-known in physics of magnetism [61–63].

Let us stress once more, that Fridel oscillations in real space appear due to the presence of Kohn anomaly in effective (screened) interaction in momentum space at the value of the transferred momentum $q = 2k_F$.

Note, that in the two-dimensional metallic monolayer the singular part of the static polarization operator related to the Kohn anomaly is proportional to [12,40,41,53, 64,65,67–69,71–73]:

$$\Pi_{\text{sing.}}(q) \sim N(0) \operatorname{Re} \sqrt{q - 2k_F}, \tag{18.114}$$

. where $N(0) = \frac{m}{2\pi}$ is the two-dimensional density of states, Re stands for the real part. From Eq. (18.114) we can see, that Kohn anomaly in the 2D case has a one-sided (square root) character and works only for the transferred momenta $q > 2k_F$ thus being ineffective for the superconductivity (which sits on the interval from 0 to $2k_F$ for the transferred momenta q [12,52,53,65,67–69,71–73]).

However, the derivative from the singular part of the static polarization operator diverges at $q = 2k_F$ stronger than in the 3D metal:

$$\frac{d\Pi_{\text{sing.}}}{dq} \sim N(0)\frac{1}{\sqrt{q - 2k_F}}. \qquad (18.115)$$

As a result, the Fridel oscillations in the effective interaction decaying on large distances in 2D slower than in the 3D case:

$$V_{\text{eff.}}(r) = B\frac{\cos 2k_F r}{(2k_F r)^2}. \qquad (18.116)$$

Let us stress, that to make strong 2D Kohn anomaly effective for triplet p-wave superconductivity we have to consider either spin-polarized two-dimensional Fermi-gas [58] or two-band (multi-layered) metal [67], or to apply external magnetic field strictly parallel to the metallic monolayer [68].

Finally, in the one-dimensional case the singular part of the static polarization operator behaves as [1,2,73]:

$$\Pi_{\text{sing.}}(q) \sim N(0)\ln\left|\frac{q - 2k_F}{4k_F}\right| \to \infty \text{ for } q \to 2k_F. \qquad (18.117)$$

It means that diverges the polarization operator itself and not only its derivatives. As a result we have the giant Kohn anomaly in 1D. This anomaly leads to Peierls metal–insulator transition [74] in one-dimensional system. We can say, that due to the giant Kohn anomaly the 1D metal cannot exist.

Note, that giant Kohn anomaly in 1D leads to even slower power-fashion decay of Fridel oscillations at large distances $k_F r \gg 1$:

$$V_{\text{eff.}}(r) = C\frac{\cos 2k_F r}{(2k_F r)}. \qquad (18.118)$$

18.8 The Role of Kohn Anomaly in Electron–Phonon Interaction

The presence of Kohn anomaly in the screened Coulomb interaction was firstly confirmed experimentally in the investigations of the phonon spectrums in metals.

Let us emphasize, that the Kohn anomaly is anomaly of the threshold type. The amplitude of the process which contributes to the renormalization of the phonon spectrum has the singularity close to the threshold [1,2,51,73].

The threshold type of the process in Fig.'s 18.6 a and b is connected with the fact, that for $q > 2k_F$ we cannot make real the creation process of electron–hole pair without picking up the phonon with finite energy (which is absent at $T \to 0$). The point is, that for the real process we should fulfill the energy and momentum conservation laws [1]. On the Fermi surface, the energies of electron and hole are equal: $\varepsilon_A\left(\boldsymbol{p}_1\right) = \varepsilon_B\left(\boldsymbol{p}_2\right)$. In other words, if the energy transfer equals zero, then

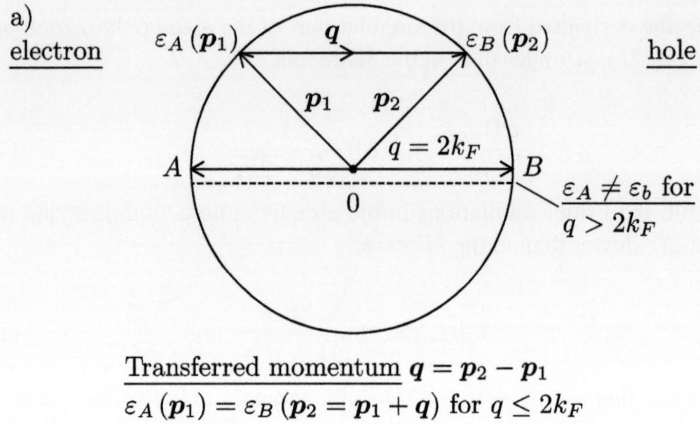

$$\underline{\text{Transferred momentum } \boldsymbol{q} = \boldsymbol{p}_2 - \boldsymbol{p}_1}$$
$$\varepsilon_A(\boldsymbol{p}_1) = \varepsilon_B(\boldsymbol{p}_2 = \boldsymbol{p}_1 + \boldsymbol{q}) \text{ for } q \leq 2k_F$$

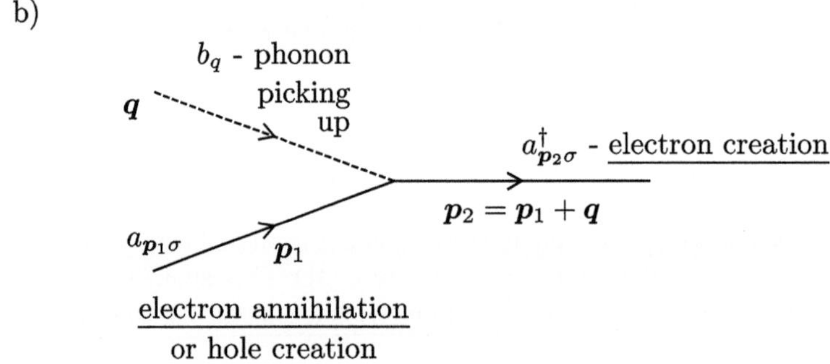

Fig. 18.6 Role of Kohn anomaly in the creation process of the electron–hole pair. In Fig. a we show electron energy $\varepsilon_A(\boldsymbol{p}_1)$ and hole energy $\varepsilon_B(\boldsymbol{p}_2)$ related to the transferred momentum $\boldsymbol{q} = \boldsymbol{p}_2 - \boldsymbol{p}_1$. If both momenta $\boldsymbol{p}_2$ and $\boldsymbol{p}_1$ lie on the Fermi surface for $q \leq 2k_F$, then electron and hole energies are equal: $\varepsilon_A(\boldsymbol{p}_1) = \varepsilon_B(\boldsymbol{p}_2)$. In Fig. b we show the process of phonon picking up by electron with the creation of the electron–hole pair. We can make creation of electron–hole pair in Fig. b to be real for $q \leq 2k_F$. In this case we have elastic process, which does not require the energy transfer [1]

the momentum of electron $\boldsymbol{p}_1$ and hole $\boldsymbol{p}_2 = \boldsymbol{p}_1 + \boldsymbol{q}$ lie on the Fermi surface, and we have elastic process [1] (with phonon energy goes to zero). But to get $\boldsymbol{p}_1$ and $\boldsymbol{p}_2$ lying on the Fermi surface, we have to implement the condition $q \leq 2k_F$ on the transferred momentum. Thus, the value of the transferred momentum $q = 2k_F$ serves as a boundary (as a threshold) between elastic (real) and virtual process [1] (Fig. 18.6).

References

1. Y. Kagan, *Lectures on the Solid-state Theoretical Physics*, Moscow, MEPHI (1981–1982). Part II. Electrons - unpublished
2. E.G. Brovman, Y. Kagan, *Phonons in Non-transition Metals*. Sov. Phys. Uspekhi, **112**, 369(1974)
3. D. Pines, *Elementary Excitations in Solids* (W.A. Benjamin Inc., New York-Amsterdam, 1963)
4. D. Pines, P. Nozieres, *The Theory of Quantum Liquids*, vol. 1, Normal Fermi Liquids (W.A. Benjamin Inc., New York-Amsterdam, 1966)
5. J. Lindhard, Kgl. Danske Videnskab Elskab, Mat. - Fys. Medd. **28**(N8) (1954)
6. R.P. Feynman, The Theory of Positrons. Phys. Rev. **76**, 749 (1949)
7. A.A. Abrikosov, L.P. Gor'kov, I.E. Dzyaloshinskii, *Methods of Quantum Field Theory in Statistical Physics* (Prentice Hall, Englewood Cliffs, New Jersey, 1963)
8. E.M. Lifshitz, L.P. Pitaevskii, *Statistical Physics, Part II, Theory of the Condensed State* (Pergamon, Elsevier, 2013)
9. M.V. Sadovskii, *Diagrammatics, Lectures on Selected problems in Condensed Matter Theory* (World Scientific, Singapore, 2006)
10. C.P. Enz, *A Course on Many-body Theory Applied to Solid-state Physics*. Lecture Notes in Physics (World Scientific, 1992)
11. Y.A. Izumov, Y.N. Skryabin, *Basic Models in Quantum Theory of Magnetism* (Ekaterinburg, 2002) (in Russian)
12. M.Y. Kagan, *Modern trends in Superconductivity and Superfluidity*, Lecture Notes in Physics, vol. 874. (Springer, Dordrecht, 2013)
13. M.Y. Kagan, A.V. Ozharovskii, *Mathematical Methods of the Theory of Superconductivity and Strongly Correlated Electron Systems*. Lecture Course, Part I (Moscow, MEPHI, 1999). (in Russian)
14. E.E. Salpeter, H.A. Bethe, Phys. Rev. **84**, 1232 (1951)
15. E.E. Salpeter, Phys. Rev. **87**, 328 (1952)
16. D. Pines, D. Bohm, A collective description of electron interaction. II Collective vs individual particle aspects of the interaction. Phys. Rev. **85**, 338 (1952)
17. D. Bohm, D. Pines, A collective description of electron interactions. III. The Coulomb interaction in a degenerate electron gas. Phys. Rev. **92**, 609 (1953)
18. M. Gell-Mann, K.A. Brueckner, Phys. Rev. **106**, 364 (1957)
19. J. Hubbard, Proc. Roy. Soc. A (London) **240**, 539 (1957)
20. P.W. Anderson, Random phase approximation in the theory of superconductivity. Phys. Rev. **112**, 1900 (1958)
21. P. Hohenberg, W. Kohn, Inhomogeneous electron gas. Phys. Rev. **136**, B864 (1964)
22. W. Kohn, L.J. Sham, Self-consistent equations including exchange and correlation effects. Phys. Rev. **140**, A1133 (1965)
23. A.L. Fetter, J.D. Walecka, *Quantum Theory of Many-Particle Systems* (McGraw-Hill, S. Francisco, 1971)
24. L.D. Landau, E.M. Lifshitz, *Electrodynamics of Continuous Media* (Pergamon Press, 1984)
25. L.D. Landau, E.M. Lifshitz, *Statistical Physics, Part I* (Butterworth-Heinemann, 1980)
26. E.M. Lifshitz, L.P. Pitaevskii, *Physical Kinetics, Course of Theoretical Physics*, vol. 10 (Pergamon Press, Oxford, 1981)
27. V.P. Silin, *Kinetic theory of gases* (Nauka, Moscow, 1971). (in Russian)
28. D. Pines, Can. Jour. Phys. **34**, 1379 (1956)
29. I.M. Khalatnikov, *An Introduction to the Theory of Superfluidity* (Perseus Publishing, Cambridge, 2000)
30. H. Frohlich, Superconductivity in metals with incomplete inner shells. J. Phys. C **1**, 544, Letter to Editor (1968)
31. H. Rietschel, L.J. Sham, Role of electron Coulomb interaction in superconductivity. Phys. Rev. B **28**, 5100 (1983)
32. M. Grabovski, L.J. Sham, Superconductivity from non-phonon interaction **29**, 6132 (1984)

33. J. Ruhman, P.A. Lee, Superconductivity at very low density: the case of strontium titanate, cond.-mat 1605.01737 (2016)
34. J. Ruhman, P.A. Lee, Pairing from dynamically screened Coulomb repulsion in bismuth. Phys. Rev. B **36**, 235107 (2017)
35. J. Bardeen, L.N. Cooper, J.R. Schrieffer, Phys. Rev. **108**, 1175 (1957)
36. L.N. Cooper, Phys. Rev. **104**, 1189 (1956)
37. A.A. Husain, E.W. Huang, M. Mitrano et al., Pines Demon Observed as 3D Acoustic Plasmon in Sr_2RuO_4. Nature (London) **621**, 66 (2023)
38. A. Balassis, E.V. Chulkov, P. Echenique, V.M. Silkin, First-principles calculations of the dielectric and optical properties of MgB_2. Phys. Rev. **B78**, 224502 (2008)
39. V.M. Silkin, D.V. Efremov, M.Y. Kagan, Doping dependence of Low Energy Charge Collective Excitations in High-T_c cuprates. cond.-mat 2411.1283 (2024). Physica Scripta **100**, 045943 (2025)
40. F. Stern, Phys. Rev. Lett. **18**, 546 (1967)
41. T. Ando, A.B. Fowler, F. Stern, Electronic properties of two-dimensional systems. Rev. Mod. Phys. **54**, 437 (1982)
42. O.V. Dolgov, D.A. Kirzhnits, E.G. Maksimov, On the admissible sign of the static dielectric function of matter. Rev. Mod. Phys. **53**, 81 (1981)
43. V.A. Markel, Can the imaginary part of the permittivity be negative? Phys. Rev. E **78**, 026608 (2008)
44. L.P. Pitaevskii, On analytical properties of the diamagnetic permittivity in the presence of spatial dispersion. Int. J. Quant. Chem. **112**, 2998 (2012)
45. V.B. Berestetskii, E.M. Lifshitz, L.P. Pitaevskii, *Quantum Electrodynamics, v.4 of the Landau-Lifshitz Course on Theoretical Physics* (Pergamon Press, 1982)
46. R.P. Feynman, *QED: The Strange Theory of Light and Matter* (Princeton University Press, 1985)
47. L.H. Thomas, The calculation of atomic fields. Math. Proc. Cambridge Philos. Soc. **23**, 542 (1927)
48. E. Fermi, Un Metodo Statistico per la Determinazione di Alcune Proprieta dell' Atomo. Rend. Accad. Naz. Lincei **6**, 602 (1927). ((in Italian))
49. H. Yukawa, *On the Interaction of Elementary Particles*. Proc. Phys.-Math. Soc. Japan **17**, 48 (1935)
50. P. Debye, E. Huckel, The theory of electrolytes. I. Freezing point depression and related phenomena. Physikalische Zeitschrift **24**, 185 (1923) (originally in German, translated into English in 2019 by M.J. Brauss)
51. W. Kohn, Image of the Fermi Surface in the Vibration spectrum of a metal. Phys. Rev. Lett. **2**, 393 (1959)
52. M.Y. Kagan, A.V. Ozharovskii, *Basic Mechanisms of Superconductivity in Strongly Correlated Electron Systems* (Lecture Course, Part II, Moscow, MEPHI, 1999). (in Russian)
53. M.Y. Kagan, *Physics of the Macroscopic Quantum Systems* (Lectures and Seminars, Moscow, 2014). (in Russian)
54. H. Fridel, Electronic Structure of Primary Solid Solutions in Metals. Adv. Phys. **3**, 446 (1954)
55. W. Kohn, J.H. Luttinger, New mechanism of superconductivity. Phys. Rev. Lett. **15**, 524 (1965)
56. D. Fay, A. Layzer, Superfluidity of low-density fermion system. Phys. Rev. Lett. **20**, 187 (1968)
57. M.Y. Kagan, A.V. Chubukov, Possibility of the superfluid transition in slightly non-ideal Fermi-gas with repulsion. JETP Lett. **47**, 614 (1988)
58. M.Y. Kagan, A.V. Chubukov, Increase of the superfluid transition temperature in spin-polarized Fermi gas with repulsion. JETP Lett. **50**, 483 (1989)
59. A.V. Chubukov, M.Y. Kagan, On the superfluid transition in dense electron system. J. Phys.: Condens. Matt. **1**, 3135 (1989)
60. M.A. Baranov, M.Y. Kagan, Yu. Kagan, On the possibility of the superfluid transition in the Fermi gas of neutral particles at ultralow temperatures. JETP Lett. **64**, 301 (1996)
61. M.A. Ruderman, C. Kittel, Indirect exchange coupling of nuclear magnetic moments by conduction electrons. Phys. Rev. **96**, 99 (1954)

62. T. Kasuya, A theory of metallic Ferro- and Antiferromagnetism on Zener's model. Progress Theoret. Phys. **16**, 45 (1956)
63. K. Yosida, Magnetic properties of Cu−Mn alloys. Phys. Rev. **106**, 893 (1957)
64. M.A. Baranov, M.Y. Kagan, Superconductive pairing in the Hubbard model with a low occupancy. JETP **99**, 1236 (1991)
65. M.A. Baranov, M.Y. Kagan, D-wave pairing in the two-dimensional Hubbard model with low filling. Z. Phys. B: Cond. Matter **86**, 237 (1992)
66. M.A. Baranov, A.V. Chubukov, M.Y. Kagan, Superconductivity and superfluidity in Fermi systems with repulsive interaction. Int. J. Mod. Phys. B **6**, 2471 (1992)
67. M.Y. Kagan, Strong T_c enhancement in the two-dimensional two-band Hubbard model with low filling. Phys. Lett. A **152**, 303 (1992)
68. M.A. Baranov, D.V. Efremov, M.Y. Kagan, The enhancement of the superconductive transition temperature in quasi-2D materials in a parallel magnetic field. Physica C **218**, 75 (1993)
69. A.V. Chubukov, Kohn–Luttinger effect and the instability of a two-dimensional repulsive Fermi-liquid at $T = 0$. Phys. Rev. B **48**, 1097 (1993)
70. M.A. Baranov, M.Y. Kagan, M.S. Mar'enko, Singularity in the quasiparticle interaction function in a 2D Fermi gas. JETP Lett. **58**, 734 (1993)
71. M.Y. Kagan, M.M. Korovushkin, V.A. Mitskan, Anomalous superconductivity and superfluidity in fermionic systems with repulsion. Phys. Uspekhi **58**, 733 (2015)
72. M.Y. Kagan, Unconventional superconductivity in low density electron systems and conventional superconductivity in hydrogen metallic alloys. JETP Lett. **106**, 728 (2016)
73. A.M. Afanasiev, Yu. Kagan, Singularities caused by electron–phonon interaction in the phonon dispersion laws. JETP **16**, 1035 (1962)
74. R.E. Peierls, Annales de I' Institut Henri Poincare. Quelques proprietes typiques des corpses solides **5**, 177 (1935). ((in French))

Lecture 19. Metallic Hydrogen

Abstract

In this Lecture, we start with the second-order correction to the ion–ion interaction and analyze the emergence of sound in metals. We consider the energy functional and determine the volume and the structure (the shape) of elementary cell in metals as a function of the correlation radius. We discuss the energy balance in metallic hydrogen and the critical pressure required for the stabilization of the atomic metal phase. We advocate the vibrational spectrum for the highly anisotropic filamentous metastable phase of metallic hydrogen formed by one-dimensional proton filaments embedded in the three-dimensional electron Fermi liquid. We present the estimate for the lifetime of the metastable phases related to the under-barrier tunneling and growth of the critical seed of the molecular phase in the configurational space.

19.1 Ion–Ion Interaction in the Second Order of the Perturbation Theory

Let us find the correction $E^{(2)}$ to the direct ion–ion interaction (see Eq. (15.4) in Lecture 15 and [1,2])

$$V_{ii} = \frac{1}{2} \sum_{n_1,n_2} V_{ii}\left(\boldsymbol{R}_{n_1} - \boldsymbol{R}_{n_2}\right) = \frac{1}{2} \sum_{n_1,n_2} \sum_q \frac{4\pi(Ze)^2}{q^2\Omega} \exp\left(i\boldsymbol{q}\left(\boldsymbol{R}_{n_1} - \boldsymbol{R}_{n_2}\right)\right)$$

$$(19.1)$$

in the second order of the perturbation theory. This correction describes <u>indirect</u> ion–ion interaction via the electrons. In fact, we would like to consider the process, when one ion perturbs the electronic subsystem in metal while another one reads out this perturbation.

Meanwhile, the energy of the ion–ion interaction via the electrons $E^{(2)}$ is determined by the expression [1,2]:

$$E^{(2)} = -\frac{\Omega}{2} \sum_{n_1,n_2} \sum_q |V_{ei}(q)|^2 \frac{\Pi(q)}{\varepsilon(q,0)} \exp\left(iq\left(R_{n_1} - R_{n_2}\right)\right), \qquad (19.2)$$

where $\Pi(q)$ and $\varepsilon(q,0)$ are static electron polarization operator and dielectric permittivity (see Eq. (17.36) in Lectures 17 and 18 for more details).

For the regular lattice, as we already discussed in the previous Lectures, it is convenient to introduce the potential $U(q)$ equals to [1,2]

$$U(q) = \sum_{n_1} V_{ei}(q) \exp\left(iqR_{n_1}\right) = NV_{ei}(q)\delta_{qK}, \qquad (19.3)$$

where N is the number of the lattice sites (see Eqs. (15.22) and (15.24) in Lecture 15).

Let us remind, that it is just this potential which enters in the secondly quantized expression for electron–ion interaction [1,2]:

$$V_{ei} = \sum_{k,q \neq 0\sigma} U(q)a_{k\sigma}^\dagger a_{k-q,\sigma}. \qquad (19.4)$$

Correspondingly, the second-order correction with respect to electron–ion interaction in Eq. (19.2) in terms of the potential $U(q)$ reads

$$E^{(2)} = -\frac{\Omega}{2} \sum_K |U(K)|^2 \frac{\Pi(K)}{\varepsilon(K)}, \qquad (19.5)$$

where $K \neq 0$ and:

$$U(K) = NV_{ei}(K). \qquad (19.6)$$

In other words, the total ion–ion interaction in (19.1) with an account of the second-order correction equals to the sum of direct and indirect interactions:

$$V_{ii}^{\text{tot.}}\left(R_{n_1} - R_{n_2}\right) = V_{ii}^{\text{dir.}}\left(R_{n_1} - R_{n_2}\right) + V_{ii}^{\text{indir.}}\left(R_{n_1} - R_{n_2}\right), \qquad (19.7)$$

where direct Coulomb interaction between the ions (see Lecture 15):

$$V_{ii}^{\text{dir.}}\left(R_{n_1} - R_{n_2}\right) = \frac{(Ze)^2}{\left|R_{n_1} - R_{n_2}\right|} = \sum_q \frac{4\pi(Ze)^2}{q^2\Omega} \exp\left(iq\left(R_{n_1} - R_{n_2}\right)\right). \quad (19.8)$$

At the same time, the indirect interaction via the electron subsystem in metal:

$$V_{ii}^{\text{indir.}}\left(R_{n_1} - R_{n_2}\right) = -\Omega \sum_q |V_{ei}(q)|^2 \frac{\Pi(q)}{\varepsilon(q)} \exp\left(iq\left(R_{n_1} - R_{n_2}\right)\right). \qquad (19.9)$$

Note that indirect interaction contains the static dielectric permittivity $\varepsilon(q)$ in the denominator and thus oscillates in real space in the same manner as the screened Coulomb interaction between electrons (forming Fridel oscillations [1–3]).

19.2 Sound in Ionic Subsystem of Metal

Let us find the velocity of the longitudinal sound in the ion subsystem of metal, using expressions (19.8)–(19.9) for direct and indirect contributions to the total ion–ion interaction.

For this purpose, we can use the important Formula (8.21) from Lecture 8 on the spectrum of the longitudinal sound in the phonon system with the pairwise interaction between ions:

$$\omega^2(q) = \frac{V_{ii}^{\text{tot.}}(q=0)}{M\Omega_0} q^2,$$ (19.10)

where M is the ion mass, Ω_0 is the volume of the elementary cell of the crystal,

$$V_{ii}^{\text{tot.}}(q=0) = V_{ii}^{\text{dir.}}(q=0) + V_{ii}^{\text{indir.}}(q=0).$$ (19.11)

Fourier component of direct (Coulomb) ion–ion interaction equals to (see Eq. (19.1))

$$V_{ii}^{\text{dir.}}(q) = \frac{4\pi(Ze)^2}{q^2\Omega}.$$ (19.12)

Fourier component of the indirect interaction in Eq. (19.9):

$$V_{ii}^{\text{indir.}}(q) = -\Omega\,|V_{ei}(q)|^2\,\frac{\Pi(q)}{\varepsilon(q)}.$$ (19.13)

Further on, we can use the expansion of electron–ion interaction $V_{ei}(q)$ in series for small values of wave vectors q (see Eq. (15.5) in Lecture 15):

$$V_{ei}(q) \to -\frac{4\pi Ze^2}{q^2\Omega} + \frac{b}{\Omega} \text{ for } q \to 0.$$ (19.14)

Moreover, we can take into account that static polarization operator for $q \to 0$ is given by

$$\Pi(q \to 0) = 2\,N(0)\left[1 - \frac{1}{12}\frac{q^2}{k_F^2}\right],$$ (19.15)

where $N(0)$ is the density of electron states.

At the same time, the static dielectric permittivity for $q \to 0$ yields (see Lectures 17 and 18)

$$\varepsilon(q \to 0) = 1 + \frac{\kappa^2}{q^2} = \frac{\kappa^2 + q^2}{q^2},$$ (19.16)

where the Thomas–Fermi wave vector squared according to Eq. (18.111) [1,4–6]:

$$\kappa^2 = 8\pi e^2 N(0).$$

As a result, utilizing formulas (19.14)–(19.16) and (18.111), we can represent the Fourier component of the indirect interaction for $q \to 0$ as follows:

$$
\begin{aligned}
V_{ii}^{\text{indir.}}(q) &= -\Omega \frac{2\,N(0)q^2}{\left(\kappa^2 + q^2\right)} \left(1 - \frac{1}{12}\frac{q^2}{k_F^2}\right)\left(-\frac{4\pi Z e^2}{q^2 \Omega} + \frac{b}{\Omega}\right)^2 \\[2mm]
&= -\frac{2\,N(0)q^2}{\left(\kappa^2 + q^2\right)\Omega} \left(1 - \frac{1}{12}\frac{q^2}{k_F^2}\right)\left(b^2 + \frac{16\pi^2\left(Ze^2\right)^2}{q^4} - \frac{8\pi Z e^2 b}{q^2}\right) \\[2mm]
&= \frac{2\,N(0)}{\left(\kappa^2 + q^2\right)\Omega} \left(1 - \frac{1}{12}\frac{q^2}{k_F^2}\right)\left(8\pi Z e^2 b - \frac{16\pi^2\left(Ze^2\right)^2}{q^2} - b^2 q^2\right) \\[2mm]
&\approx \frac{2\,N(0)}{\Omega\kappa^2} \left(1 - \frac{q^2}{\kappa^2}\right)\left(1 - \frac{1}{12}\frac{q^2}{k_F^2}\right)\left(8\pi Z e^2 b - \frac{16\pi^2\left(Ze^2\right)^2}{q^2}\right) \\[2mm]
&\approx \frac{2bZ\kappa^2}{\Omega\kappa^2} \left(1 - \frac{q^2}{\kappa^2} - \frac{1}{12}\frac{q^2}{k_F^2}\right) - \frac{4\pi(Ze)^2\kappa^2}{\Omega q^2\kappa^2}\left(1 - \frac{q^2}{\kappa^2} - \frac{1}{12}\frac{q^2}{k_F^2}\right) \\[2mm]
&= \frac{2bZ}{\Omega} - \frac{4\pi(Ze)^2}{\Omega q^2} + \frac{4\pi(Ze)^2}{\Omega}\left(\frac{1}{\kappa^2} + \frac{1}{12}\frac{1}{k_F^2}\right) + O\left(q^2\right).
\end{aligned}
\tag{19.17}
$$

Thus, in total, ion–ion interaction for $q \to 0$ the divergent Coulomb term is canceled and only non-Coulomb contribution in the amplitude of electron–ion scattering survives

$$
\begin{aligned}
V_{ii}^{\text{tot.}}(q \to 0) &= \frac{4\pi(Ze)^2}{q^2\Omega} + \frac{2bZ}{\Omega} - \frac{4\pi(Ze)^2}{\Omega q^2} + \frac{4\pi(Ze)^2}{\Omega}\left(\frac{1}{\kappa^2} + \frac{1}{12}\frac{1}{k_F^2}\right) \\[2mm]
&= \frac{2bZ}{\Omega} + \frac{4\pi(Ze)^2}{\Omega}\left(\frac{1}{\kappa^2} + \frac{1}{12}\frac{1}{k_F^2}\right).
\end{aligned}
\tag{19.18}
$$

Hence, the longitudinal sound velocity squared in metal:

$$
\begin{aligned}
c^2 &= \frac{1}{M\Omega_0}\left[\frac{2bZ}{\Omega} + \frac{4\pi(Ze)^2}{\Omega}\left(\frac{1}{\kappa^2} + \frac{1}{12}\frac{1}{k_F^2}\right)\right] \\[2mm]
&= \frac{1}{MN}\cdot\frac{1}{\Omega_0^2}\left[2bZ + 4\pi(Ze)^2\left(\frac{1}{\kappa^2} + \frac{1}{12}\frac{1}{k_F^2}\right)\right].
\end{aligned}
\tag{19.19}
$$

But in dense Coulomb plasma for the values of the correlation radius $r_S \ll 1$, the ratio $\frac{\kappa^2}{k_F^2} \sim r_S \ll 1$. That is why in Eq. (19.19) $\frac{1}{\kappa^2} \gg \frac{1}{12}\frac{1}{k_F^2}$, and as a result:

$$c^2 \approx \frac{1}{MN} \cdot \frac{1}{\Omega_0^2}\left[2bZ + \frac{4\pi(Ze)^2}{\kappa^2}\right]. \tag{19.20}$$

Substituting expression (18.111) for the Thomas–Fermi [5,6] wave vector squared in Eq. (19.20), we get the famous relation for the sound velocity squared, which was first obtained by John Bardeen and David Pines [7]:

$$c^2 \approx \frac{1}{MN} \cdot \frac{1}{\Omega_0^2}\left[2bZ + \frac{Z^2}{2\,N(0)}\right]. \tag{19.21}$$

Note, that the calculation of the transverse sound velocity (or the shear modulus) in metal [8] is much more complicated task. It demands the evaluation of the indirect ion–ion interaction (via electrons) in the higher orders of the perturbation theory with respect to electron–ion interaction [1,2] together with the electron energy of anisotropy (which curves the Fermi surface in the higher orders of electron–ion interaction).

Let us stress also, that in metallic hydrogen $b = 0$ (ionic core is absent) [9–12]. Hence, the sound velocity squared for the longitudinal sound is given by the simple formula: $c^2 \approx \frac{1}{2MNN(0)\Omega_0^2}$.

19.3 Energy Functional and the Volume of the Elementary Cell in Metal

In the end of Lecture 15 (see Eq. 15.50), we have calculated already the average energy on one elementary cell:

$$E_i = -1.8\frac{Z^{5/3}}{r_S}\,\text{Ry}. \tag{19.22}$$

This energy represents the sum of the average energy of electron–ion and electron–electron interactions.

Additional contribution to the total energy is connected with electron energy $E_{\text{el.}}$ and effective (indirect) interaction of ions via electrons $E_{ii}^{(\text{eff.})}$:

$$E = E_i + E_{\text{el.}} + E_{ii}^{(\text{eff.})}, \tag{19.23}$$

where electron contribution:

$$E_{\text{el.}} = E_{\text{kin.el.}} + E_{1ei} + E_{ee}^{(\text{eff.})} \tag{19.24}$$

equals to the sum of three terms. The first one is kinetic energy of electrons (see Eq. (15.75)):

$$E_{\text{kin.el.}} = \frac{2.21Z}{r_S^2}\,\text{Ry}. \tag{19.25}$$

The second one is non-Coulomb part of electron–ion interaction (see Eq. (15.52)):

$$E_{1ei} = \frac{bZ}{\Omega_0} = \frac{3bZ}{4\pi r_0^3} = \frac{3b}{4\pi r_S^3 a_B^3} \sim \frac{b}{r_S^3}. \tag{19.26}$$

Finally, the third term is related to the effective energy of electron–electron interaction via the ionic lattice. In the second order of the perturbation theory, as it was shown in Eq. (17.4) in Lecture 17, this term is given by

$$E_{ee}^{(2)} = \alpha + \beta \ln r_S. \tag{19.27}$$

Thus, for the construction of the complete theory of the non-transitional metals we have to determine the behavior of the effective amplitude of electron–ion interaction $V_{\text{eff.}}(q)$ with an account of the higher order terms in electron–ion interaction [1,2]. As we discussed in Lecture 15, electron–ion interaction in crystal in the second order of the perturbation theory in the small parameter (see Eq. (15.26)):

$$\frac{V_{ei}(K)}{\varepsilon_F} \ll 1 \tag{19.28}$$

is reduced to the following sum over the wave vectors K of the inverse lattice (see Eq. (19.5)):

$$E^{(2)} = -\frac{\Omega}{2} \sum_{K} |NV_{ei}(K)|^2 \frac{\Pi(K)}{\varepsilon(K)}.$$

Evaluation of the higher order terms in electron–ion interaction, namely of $E^{(3)}$ and $E^{(4)}$ is much more complicated problem, as it was shown by Yuri Kagan and E.G. Brovman in their classical review [2]. The calculation requires consideration not only of the pairwise but also many-particle interactions which are described by so-called many-tails diagrams (see Fig. 19.1). In these diagrams the wavy lines of electron–ion interaction serve as the tails, while the solid lines stand for electrons and holes.

Let us emphasize that an account of the third- and fourth-order terms $\left(E^{(3)} + E^{(4)}\right)$ in indirect ion–ion interaction via electrons as well as in electron–electron interaction via ionic lattice is essential for the determination of the shear modulus and transverse sound velocity in metals.

Moreover, an account of the fourth-order terms $E^{(4)}$ in the indirect electron–electron interaction leads to the emergence of the substantial anisotropy of the electron spectrum and effectively curves the Fermi surface, which is very important for

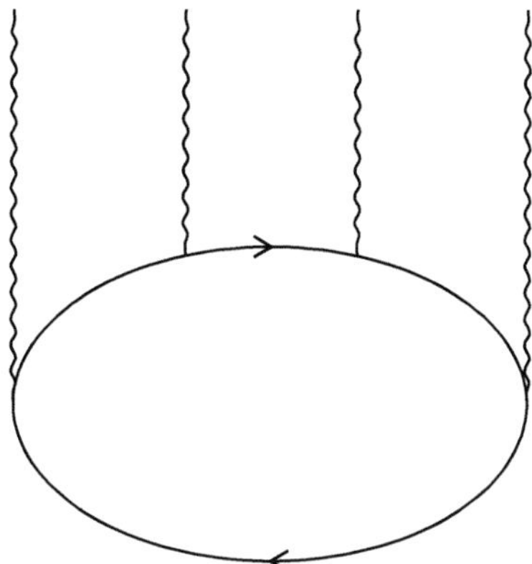

Fig. 19.1 The diagrams with a lot of tails emerging in the higher orders of the perturbation theory in electron–ion interaction [1,2]. The wavy lines describe electron–ion interaction. The solid lines correspond to Green's functions of electrons and holes. These diagrams also contain the Kohn anomaly as an important ingredient [1,13]

the determination of the structure (of the shape) of the elementary cell (whether it is cubic or non-cubic).

All the other ingredients, including the behavior of the static dielectric permittivity $\varepsilon(q)$ and static polarization operator $\Pi(q)$ in all the region of the transferred momenta q in the Random Phase Approximation (RPA [1,2,4,14–16] or even better (with an account of the so-called Hubbard local field corrections [2,17]) we already know (see Lecture 18).

Further on, from the behavior of the effective amplitude $V_{\text{eff.}}(q)$, we can determine $E_{ii}^{(\text{eff.})}$ and find the total energy on one elementary cell E in (19.23). After that from the condition of the absence of the external pressure:

$$P = -\frac{\partial E}{\partial \Omega} = 0 \tag{19.29}$$

we can find the optimal volume of the elementary cell. In the general case, the thermodynamic potential is a function of pressure:

$$\Omega_0 = \Omega_0(P). \tag{19.30}$$

It turns out, that for the evaluation of the optimal volume Ω_0 in simple metals the most important role in the total energy is played by the terms of the zeroth and the first orders, connected with the average energy of the cell:

$$E_i \sim -\frac{1}{r_S},$$

together with the non-Coulomb contribution to the amplitude of electron–ion interaction:

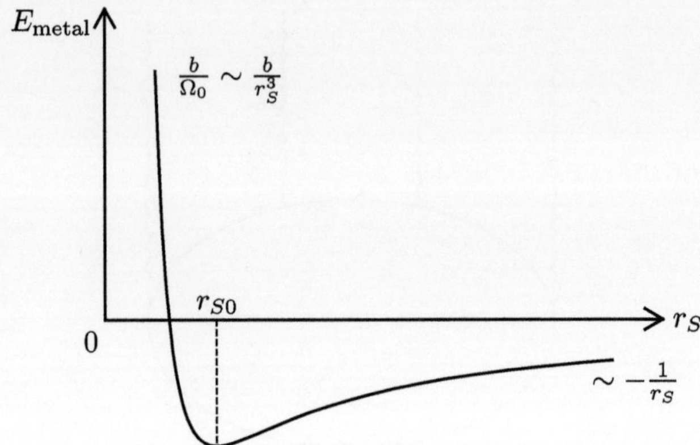

Fig. 19.2 Qualitative behavior of the cell energy $E = E_i + E_{1ei}$ as a function of the correlation radius r_S. For small values of r_S the dominant role is played by the non-Coulomb term E_{1ei} with the excluded volume. For larger values of r_S, the most important is the contribution of the average energy on one cell E_i. The energy minimum for $r_S = r_{S0}$ corresponds to the optimal volume of the elementary cell $\Omega_0 \sim r_{S0}^3$ [1]

$$E_{1ei} = \frac{bZ}{\Omega_0} \sim \frac{b}{r_S^3}.$$

Qualitative behavior of the energy functional of the ion cell $E = E_i + E_{1ei}$ as a function of the correlation radius r_S is shown in Fig. 19.2.

For small values of r_S the dominant role is played by the non-Coulomb term E_{1ei} related to the excluded volume. For larger values of r_S, the most important is the contribution of the average energy of the cell E_i.

The energy minimum for $r_S = r_{S0}$ corresponds to the optimal volume of the elementary cell $\Omega_0 \sim r_{S0}^3$ (see Fig. 19.2).

Note, that the repulsive term proportional to $\frac{1}{\Omega_0}$ appears in the energy functional because the finite radius of the atom does not allow to extra electrons to penetrate into the ionic core. This term is equivalent to the hard-core van der Waals term with excluded volume [18]. It is absent in metallic hydrogen where there is no ionic core and interaction of electron with proton of the nucleus is purely Coulombic one [1,9–12].

19.4   Anisotropy of the Elementary Cell

As we discussed already, for the more detailed determination of the structure and shape of the elementary cell (cubic or non-cubic) besides the energy terms of zeroth- and first-order E_i and E_{1ei}, an important role in the energy functional belongs also to the higher order terms in the energy of indirect ion–ion interaction via electrons $E_{ii}^{(\mathrm{eff.})}$ and in the energy of the electron liquid $E_{ee}^{(\mathrm{eff.})}$.

When we proceed from one structure (one shape) of the ionic cell to another one, the energy difference varies only slightly, whereas the optimal volume of the elementary cell Ω_0 and especially the shear modulus changes much stronger [1,2].

An important fact is that the system is resistant to compression but it is unstable to the shear deformation [1,2,9–12].

It turns out, that ionic lattice tends to make the elementary cell cubic in metals [1,2,9,12].

At the same time, electron liquid with an account of higher order terms in electronion interaction makes the cell non-cubic.

The selection of the optimal structure is determined by the square of the effective amplitude of electron–ion interaction in the energy terms connected with the indirect ion–ion and electron–electron interactions [1,2,9–12]:

$$\delta E \sim |V_{\mathrm{eff.}}(\boldsymbol{K})|^2. \tag{19.31}$$

For purely Coulombic interaction, the electron–electron contribution prevails, leading to the positive sign of the energy of anisotropy:

$$\delta E \sim +\frac{1}{K^4} > 0. \tag{19.32}$$

Due to this contribution to the energy, it becomes more beneficial to produce an anisotropic cell (of the non-cubic shape).

Let us analyze the role of the electron–electron interaction for the simple case comparing the energies of two elementary cells (one isotropic, another one-anisotropic) which have equal areas (see Fig. 19.3).

Suppose one cell has quadratic form with the side a and the area $S = a^2$, while another one has a rectangular form with the sides b and c and the area $S = bc$. We assume also that the areas of the cells are equal:

$$S = a^2 = bc. \tag{19.33}$$

Then in the inverse space the basic vectors for the isotropic cell are given by

$$K_x = K_y = \frac{2\pi}{a}, \tag{19.34}$$

and for rectangular cell:

$$K_x = \frac{2\pi}{c}, \; K_y = \frac{2\pi}{b}. \tag{19.35}$$

Suppose we have strong anisotropy for the rectangular cell and $b \gg c$.

Then for the isotropic cell, the square of the inverse wave vector according to the Pythagorean theorem:

$$K^2 = K_x^2 + K_y^2 = 2\left(\frac{2\pi}{a}\right)^2 = \frac{8\pi^2}{a^2}. \tag{19.36}$$

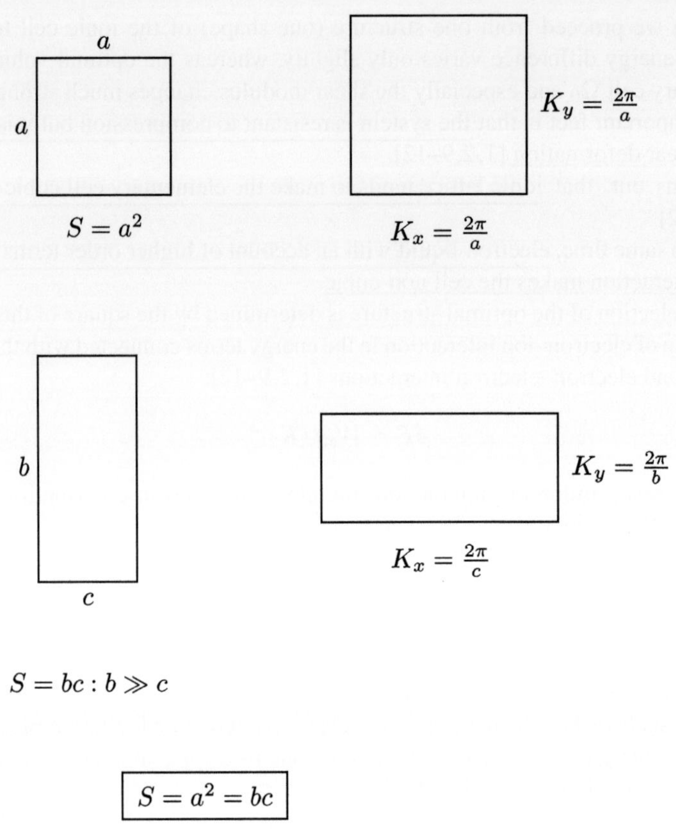

Fig. 19.3 Isotropic and anisotropic (rectangular) cells in real and inverse spaces. The areas of the cells are equal $S = a^2 = bc$. We consider strongly anisotropic case when $b \gg c$ [1]

At the same time, for the strongly anisotropic rectangular cell, we get

$$K^2 = K_x^2 + K_y^2 = 4\pi^2 \left(\frac{1}{b^2} + \frac{1}{c^2} \right) = 4\pi^2 \frac{b^2 + c^2}{b^2 c^2}. \tag{19.37}$$

Hence the contribution to the energy proportional to $\frac{1}{K^4}$ for the isotropic cell yields

$$\frac{1}{K^4} \sim a^4 = S^2, \tag{19.38}$$

while for the rectangular cell this contribution:

$$\frac{1}{K^4} \sim \frac{b^4 c^4}{\left(b^2 + c^2 \right)^2} \approx \frac{b^4 c^4}{b^4} = \frac{S^4}{b^4}. \tag{19.39}$$

As a result, the ratio of the positive contributions to the energy proportional to $\frac{1}{K^4}$ for the anisotropic and isotropic cells equals to

$$\frac{\frac{1}{K^4_{\text{anisotr.}}}}{\frac{1}{K^4_{\text{isotr.}}}} = \frac{\frac{S^4}{b^4}}{S^2} = \frac{S^2}{b^4} = \frac{b^2 c^2}{b^4} = \frac{c^2}{b^2} \ll 1. \tag{19.40}$$

Thus, the energy minimum ($\min \delta E$) corresponds to the anisotropic cell.

19.5 Metallic Hydrogen

All substances at high pressures become metals [19, 20]. Hydrogen at low pressures is not a metal because from the beginning it is more beneficial for it to form the molecule [1, 19, 20]. In molecular phase, we lose in kinetic energy of electrons [19], but we benefit from the binding energy of the molecule.

For simple metals (Na, Li, K), the situation is inverse: electron delocalization is beneficial while the binding energy (for example of the molecular Na) is small.

The difference can be explained by the fact, that hydrogen atom does not have the filled inner shells [1]. That is why the interaction between the atoms in hydrogen molecule is purely Coulombic.

If there are filled inner shells in the atom, then the electron wave function is strongly oscillating one and electron kinetic energy increases in this case. As a result, the valence electron is pushed out of the region close to the nucleus [1].

There was a suspicion, that hydrogen has metastable metallic state (in analogy with graphite, for which the metastable state corresponds to the diamond structure).

Let us stress, that if we heat the diamond, we get the soot [1].

19.6 Energy Balance in the Metallic Hydrogen

The total energy on one cell in the metallic hydrogen according to Eq. (19.24) is determined by the sum of the energies:

$$E = E_i + E_{\text{el.}} + E_{ii}^{(\text{eff.})},$$

where as usual E_i is the sum of the average electron–ion interaction and homogeneously smeared electron–electron interaction on one cell. For the nucleus charge $Z = 1$, it yields

$$E_i = -\frac{1.8}{r_S} \, \text{Ry.}$$

However, the electron contribution in metallic hydrogen [1, 2, 9–12]:

$$E_{\text{el.}} = E_{\text{kin.el.}} + E_{\text{exchange}} + E_{ee}^{(\text{eff.})} \tag{19.41}$$

does not contain non-Coulomb term E_{1ei} related to the ionic core, since the core (the filled shell) is absent in metallic hydrogen and the interaction of electron with nucleus (with proton) is purely Coulombic.

Moreover, the electroneutrality condition in metallic hydrogen practically eliminates all nontrivial contributions, and thus, in the second order of the perturbation theory on electron–ion interaction, both the indirect ion–ion interaction via electrons $E_{ii}^{(2)}$ and electron–electron interaction via ions $E_{ee}^{(2)}$ tend to zero.

At the same time, in electron contribution $E_{\text{el.}}$ in Eq. (19.41) besides the average kinetic energy of electrons

$$T = \frac{2.21}{r_S^2}\, \text{Ry}, \tag{19.42}$$

an important role belongs also to the exchange Fock term [21,22], which lowers the energy (see Eq. (16.22) in Lecture 16):

$$E_{ee}^{(1)} = -\frac{0.916}{r_S}\, \text{Ry}. \tag{19.43}$$

Thus, the total energy reads

$$E = E_i + T + E_{ee}^{(1)} = -\frac{1.8}{r_S}\, \text{Ry} - \frac{0.916}{r_S}\, \text{Ry} + \frac{2.21}{r_S^2}\, \text{Ry}. \tag{19.44}$$

Note, that for large electron density ($r_S \to 0$), the main role in the total energy plays the term with electron kinetic energy $T \sim \frac{1}{r_S^2}$. This term promotes delocalization of electrons and thus the metallization of the system.

In the absence of the external pressure, the condition for the optimal volume of the elementary cell can be written as

$$P_{\min} = 0 = -\left(\frac{\partial \tilde{E}}{\partial V}\right)_N = -\frac{1}{N}\left(\frac{\partial \tilde{E}}{\partial \Omega_0}\right)_N = -\left(\frac{\partial E}{\partial \Omega_0}\right)_N, \tag{19.45}$$

where N is the number of elementary cells in crystal, $V = N\Omega_0$ is the total volume of the system, and $\tilde{E} = NE$ is the total energy.

Correspondingly, the volume of the elementary cell (which equals to the volume of the Wigner–Seitz cell [23]) yields (see Fig. Fig. 15.4 in Lecture 15)

$$\Omega_0 = \frac{4\pi r_0^3}{3} = \frac{4\pi r_S^3 a_B^3}{3}, \tag{19.46}$$

where the correlation radius in metallic hydrogen:

$$r_S = \frac{r_0}{a_B}. \tag{19.47}$$

Hence, the differential of the volume of the spherical Wigner–Seitz cell in Eq. (19.46):

$$d\Omega_0 = 4\pi a_B^3 r_S^2 dr_S.$$ (19.48)

As a result, the condition (19.45) can be represented in the following form:

$$P_{\min} = -\left(\frac{\partial E}{\partial \Omega_0}\right)_N = -\frac{1}{4\pi a_B^3 r_S^2}\frac{\partial}{\partial r_S}\left(-\frac{1.8}{r_S} - \frac{0.916}{r_S} + \frac{2.21}{r_S^2}\right)$$

$$= -\frac{1}{4\pi a_B^3 r_S^2}\left(\frac{2.716}{r_S^2} - \frac{4.42}{r_S^3}\right) \approx -\frac{1}{4\pi a_B^3 r_S^5}(2.72 r_S - 4.42) = 0.$$ (19.49)

Thus, the optimal value of the correlation radius for the metallic hydrogen corresponds to the intermediate electron densities [9–12]:

$$r_{S\,\text{opt.}} \approx \frac{4.42}{2.72} \approx 1.63.$$ (19.50)

It is not difficult also to calculate the compression modulus for metallic hydrogen [9–12]:

$$B = -\frac{\partial P}{\partial V} = \frac{\partial^2 \tilde{E}}{\partial V^2} = \frac{1}{\left(4\pi a_B^3\right)^2 r_S^2 N}\frac{\partial}{\partial r_S}\left(\frac{2.716}{r_S^4} - \frac{4.42}{r_S^5}\right)$$

$$\approx \frac{1}{\left(4\pi a_B^3\right)^2 r_S^2 N}\left(-\frac{2.72\cdot 4}{r_S^5} + \frac{4.42\cdot 5}{r_S^6}\right)$$

$$= \frac{1}{\left(4\pi a_B^3\right)^2 r_S^8 N}(-2.72\cdot 4 r_S + 4.42\cdot 5).$$ (19.51)

Substituting in Eq. (19.51), the optimal value of the correlation radius $r_{S\,\text{opt.}} \approx \frac{4.42}{2.72}$ from Eq. (19.50) we make sure that the compression modulus is positive [9–12]:

$$B = \frac{1}{\left(4\pi a_B^3\right)^2 r_{S\,\text{opt.}}^8 N}(4.42\cdot 5 - 4.42\cdot 4) = \frac{4.42}{\left(4\pi a_B^3\right)^2 r_{S\,\text{opt.}}^8 N}$$

$$= \frac{0.49}{\Omega_{0\,\text{opt.}}^2 r_{S\,\text{opt.}}^2 N} > 0.$$ (19.52)

Hence the optimal value of the correlation radius corresponds to the minimum of energy ($B > 0$). As a result, the optimal volume of the elementary cell in metallic hydrogen yields [9]

$$\Omega_{0\,\text{opt.}} = \frac{4\pi r_{S\,\text{opt.}}^3 a_B^3}{3} \approx 18.18 a_B^3.$$ (19.53)

However, we have to determine also the structure (the optimal shape) of the cell and investigate the stability of the cell toward different perturbations.

It proves, that it is more beneficial for us to produce strongly anisotropic structure with the filaments of charged protons embedded in the Fermi liquid of conductivity electrons. The proton filaments form the triangular lattice in the transverse direction perpendicular to the filaments [1,9] (similar to the Abrikosov vortex lattice [24] in type-II superconductors or Feynman–Onsager vortex lattice [25,26] in the superfluid helium considered in Lecture 4).

At the same time, the filaments move freely relative to each other in the longitudinal direction similar to the linear vortices in helium.

It is interesting to analyze the vibrational spectrum of this system. In fact, in this phase we are dealing with crystal in the transverse direction and with liquid in the longitudinal direction similar to the vortex crystal in rotating helium [1,9,27–29].

The vibrational spectrum of the system proves to be linear in transverse direction (transverse sound mode) and has quadratic dependence on the wave vector (bending oscillations) in the longitudinal direction. In the general case the spectrum (see the pioneering paper of Brovman et al. [9]) is given by

$$\omega^2 (q_\perp, q_z) = \alpha q_\perp^2 + \beta q_z^4, \tag{19.54}$$

where $q_\perp = (q_x, q_y)$.

The character of the spectrum in Eq. (19.54) is related to the absence of the term proportional to q_z^2. Thus, it describes the reduction to zero of the transverse sound velocity and the shear modulus, connected with the displacements of filaments relative to each other along z-axis [9].

Correspondingly, the mean square displacement of ions along z-axis (along proton filaments):

$$\langle u_z^2 \rangle \sim \int \frac{dq_z}{q_z^2} \tag{19.55}$$

diverges at temperatures $T \neq 0$ as $\frac{1}{q_z}$ for small $q_z \to 0$ [9].

Note that at $T = 0$ the mean square displacement along z-axis diverges logarithmically [9].

This behavior of the mean square displacement (see Lecture 9) is typical for the one-dimensional system (for the set of individual filaments).

At the same time, the mean square displacement in the transverse direction (perpendicular to the filaments) $\langle u_x^2 \rangle$ is finite both for $T = 0$ and for $T \neq 0$. Thus, the fluctuations do not destroy the crystal structure in the direction perpendicular to the filaments [9].

The critical pressure required to stabilize this anisotropic phase as global minimum of the thermodynamic potential is very large [9–12]:

$$P = (3 \div 4) \text{ Mbar.}$$

For smaller pressures, the filamentous phase corresponds to the metastable state (to the local minimum of energy or the Gibbs thermodynamic potential [9,18]). Let us tress, that for $P = T = 0$ the energy of the molecular phase of the solid hydrogen is one order of magnitude lower than the energy of the anisotropic filamentous phase of metallic hydrogen. Nevertheless, as it was shown by Burmistrov and Dubovski [31], the anisotropic phase serves as a local minimum till rather low pressures

$$P = 0.1\,\text{Mbar}.$$

It is important to answer the question: how long does the metastable phase live? [9,31] In all the textbooks the lifetime of the metastable state at temperature $T = 0$ equals to infinity.

For finite temperatures the temperature fluctuations facilitate the transition from metastable to the stable state [18].

However, even for $T = 0$, the lifetime of the metastable state is finite due to the under-barrier tunneling and growth of a critical nucleus of a new (stable) phase in the configurational space (see the classical paper of Lifshitz and Kagan [30] on the Quantum kinetics of the decay of the metastable state at temperature $T = 0$).

In our case, the decay of the metastable metallic phase is fulfilled via the virtual emergence and growth of the nuclei of the stable molecular phase in the process of the quasiclassical under-barrier tunneling. The tunneling goes on in the effective anharmonic potential, which includes the difference of the surface contributions of both phases (proportional to R^2) and the bulk contribution proportional to R^3 (see Fig. 19.4 and [30]). Namely:

$$U(R) = 4\pi\sigma R^2 - \frac{4\pi R^3 n_2\,(\mu_1 - \mu_2)}{3}, \tag{19.56}$$

where R is the radius of the spherical nucleus of the molecular phase, σ is the surface tension coefficient, μ_1 and μ_2 are the chemical potentials of the metallic and molecular phase, n_2 is the density of the molecular phase.

If the probability of the decay of the metastable phase (with the emergence of the nucleus of the stable phase) $W = \omega_0 \exp\left(-\frac{S_{\text{eff.}}}{\hbar}\right)$ is small for large values of the quasiclassical action $\beta = \frac{S_{\text{eff.}}}{\hbar} \gg 1$, then, correspondingly, we get the large lifetime of the metastable state $\tau \sim \frac{1}{W} \sim \frac{1}{\omega_0}\exp(\beta)$ [30]. Note that recent experimental discovery of metallic hydrogen on the synchrotron facility in Toulouse (France) at high pressures of $5\,\text{Mbar}$ [32] together with the discovery of superconductivity in the metallic hydrides (H_2S, H_3S, LaH_{10}) [33,34] strongly intensified both experimental and theoretical investigations on the route from high-temperature to room temperature superconductivity.

For electron subsystem of metallic hydrogen Eliashberg mechanism of superconductivity based on strong electron–phonon interaction is likely to be dominant. In [36,37] M. Yu. Kagan, E.A. Mazur and R. Sh. Ikhsanov from the solution of the generalized system of Eliashberg equations on the imaginary axis (with an account of the correction to the chemical potential) got $T_c = 221\,\text{K}$

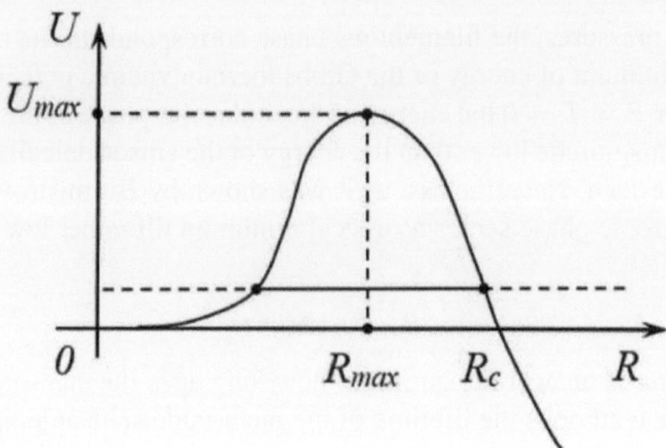

Fig. 19.4 Metastable (metallic) and stable (molecular) state of solid hydrogen serving as local and global extrema of the effective interaction. We show on the Figure quasiclassical under-barrier tunneling of the nucleus of the molecular phase. The tunneling goes on in the effective anharmonic potential which includes the difference of the surface contributions of both phases (proportional to R^2) and the bulk contribution (proportional to R^3), where R is the radius of the spherical nucleus of molecular phase [1,30]

for the critical temperature in 141/AMD phase of metallic hydrogen at pressures 5 Mbar. This estimate is in good agreement with the previous calculations of N.A. Kudryashov, A.A. Kutukov, and E.A. Mazur based on the solution of Elishberg equations on the real axis in which the authors obtained $T_c = 217$ K for the critical temperature [38,39].

Moreover, according to the ideas of Aschcroft [40,41], we can get very interesting coexistence and interplay of two Bose condensates in the system (see also the development of these ideas in [29,42]).

Let us stress, that the first condensate corresponds to the Cooper pairs in electron subsystem. At the same time, the second one is bi-proton condensate promoting superfluidity in ion (proton) subsystem of metallic hydrogen (similar to bi-neutron and bi-proton superfluidity in neutron stars).

Note also, that, as we already discussed in this Lecture, there are interesting analogies between quantum (or vortex) crystals and strongly anisotropic phases (such as filamentous phase) of metallic hydrogen [29,42–45].

Finally, very important for our understanding is the analysis of the phase diagram of metallic hydrogen in the whole region of temperatures and pressures (see nice review article of Silvera et al. [46] and earlier papers of Abrikosov [47–50] and Schneider et al. [51]).

Let us emphasize, that at low temperatures, we get the insulating crystalline phase of molecular hydrogen (H_2) at low pressures and solid metallic phase of atomic hydrogen (H) at high pressures. At the same time, at high temperatures and low pressures, we obtain liquid molecular phase of H_2 transforming into the liquid metallic phase (H) at higher pressures (see Fig. 19.5 and [46]).

Very interesting are the investigations of the parameter region close to the first-order phase boundary for the metal–insulator transition (MIT). In this region, in the shock wave experiments (first performed for deuterium [52]), close to the phase

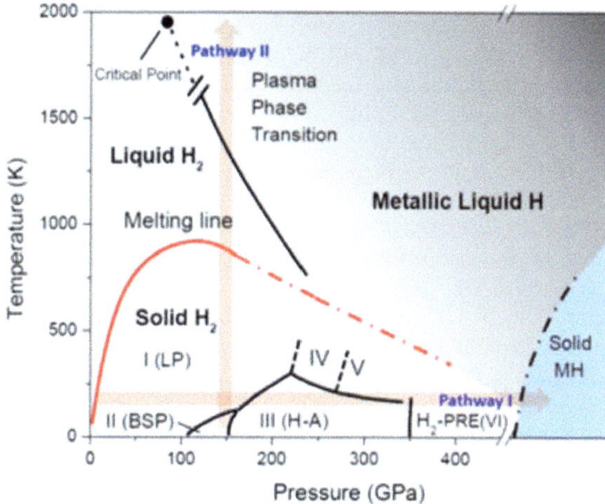

Fig. 19.5 $T - P$ phase diagram of molecular and atomic hydrogen at high pressures. Phase diagram contains four important phases of molecular (H_2) and metallic (H) hydrogen. Namely, depending on pressure, we get two phases of solid (crystalline) molecular and metallic hydrogen at low temperatures and two phases of liquid molecular and metallic hydrogen at high temperatures (from the review article of R.P. Dias and I.F. Silvera [46], see also [1,43–45])

boundary, we get the inhomogeneous state with metallic droplets in the insulating matrix (as it happens often for the first-order phase transitions).

References

1. Y. Kagan, *Lectures on the Solid-state Theoretical Physics*, Moscow, MEPHI, 1981–1982. Part II. Electrons (Unpublished)
2. E.G. Brovman, Y. Kagan, Phonons in non-transition metals. Sov. Phys. Uspekhi **112**, 369 (1974)
3. H. Fridel, *Electronic Structure of Primary Solid Solutions in Metals*. Advances in Physics, vol. 3 (1954), p. 446
4. D. Pines, P. Nozieres, *Theory of Quantum Liquids* (Benjamin, 1966)
5. L.H. Thomas, The calculation of atomic fields. Math. Proc. Cambr. Philos. Soc. **23**, 542 (1927)
6. E. Fermi, Un Metodo Statistico per la Determinazione di Alcune Proprieta dell' Atomo. Rend. Accad. Naz. Lincei **6**, 602 (1927) (in Italian)
7. J. Bardeen, D. Pines, Electron–phonon interaction in metals. Phys. Rev. **99**, 1140 (1955)
8. L.D. Landau, E.M. Lifshitz, *Theory of Elasticity* (Butterworth-Heinemann, 1986)
9. E.G. Brovman, Y. Kagan, A. Kholas, Structure of the metallic hydrogen at zero pressure. Sov. Phys. JETP **34**, 1300 (1972)
10. E.G. Brovman, Y. Kagan, A. Kholas, V.V. Puhkarev, Role of electron–electron interaction in the formation of metastable state of metallic hydrogen. JETP Lett. **18**, 869 (1973)
11. Yu. Kagan, V.V. Pushkarev, A. Kholas, Sov. Phys. JETP **46**, 511 (1977)
12. Yu. Kagan, E.G. Brovman, Problem of metallic hydrogen. Physics Uspekhi **105**, 777 (1971)
13. W. Kohn, Image of the Fermi surface in the vibration spectrum of a metal. Phys. Rev. Lett. **2**, 393 (1959)
14. D. Pines, D. Bohm, A collective description of electron interaction. II Collective vs individual particle aspects of the interaction. Phys. Rev. **85**, 338 (1952)

15. D. Bohm, D. Pines, A collective description of electron interactions. III. The Coulomb interaction in a degenerate electron gas. Phys. Rev. **92**, 609 (1953)
16. M. Gell-Mann, K.A. Brueckner, Phys. Rev. **106**, 364 (1957)
17. J. Hubbard, Proc. Roy. Soc. (Lond.) **A240**, 539 (1957)
18. L.D. Landau, E.M. Lifshitz, *Statistical Physics*, Part I (Butterworth-Heinemann, 1980)
19. E. Wigner, H.B. Huntington, Jour. Chem. Phys. **3**, 764 (1935)
20. R. Kronig, J. de Boer, I. Korringa, Physica (Utrecht) **12**, 245 (1946)
21. D. Hartree, *The Calculation of Atomic Structures* (Wiley, New York, 1957)
22. V.A. Fock, Näherungsmethode zur Lösung des quantenmechanischen Mehrkörperproblems. Z. Angew. Phys. **61**, 126 (1930). ((in German))
23. E. Wigner, F. Seitz, Phys. Rev. **43**, 804 (1933)
24. A.A. Abrikosov, Sov. Phys. JETP **5**, 1174 (1957)
25. R.P. Feynman, *Progress in Low Temperature Physics*, vol. 1, Chap. 2 (1955), p. 17
26. L. Onsager, Nuovo Cimento. **6**(Suppl. 2), 249 (1949)
27. A.F. Andreev, M.Y. Kagan, Sov. Phys. JETP **59**, 318 (1984)
28. M.Y. Kagan, *Modern Trends in Superconductivity and Superfluidity*. Lecture Notes in Physics, vol. 874. (Springer, Dordrecht, 2013)
29. M.Y. Kagan, A. Bianconi, Fermi-Bose mixtures and BCS-BEC crossover in high-T_C superconductors. Condens. Matter (Switzerland) **4**, 51 (2019)
30. I.M. Lifshitz, Yu. Kagan, Quantum kinetics of phase transitions at temperature close to absolute zero. Sov. Phys. JETP **35**, 206 (1972)
31. S.N. Burmistrov, L.B. Dubovski, On the lifetime of metastable metallic hydrogen, cond-mat 1611.025938 Nov. 2016. Low Temp. Phys. **43**, 1152 (2017)
32. P. Loubeyre, F. Occelli, P. Dumas, Nature **577**, 631 (2020)
33. D. Duan, Y. Liu, F. Tian et al., Sci. Rep. **4**, 6968 (2014)
34. A.P. Drozdov, M.I. Eremets, I.A. Troyan et al., Nature **525**, 73 (2015)
35. G.M. Eliashberg, Sov. Phys. JETP **11**, 696 (1960)
36. E.A. Mazur, R.S. Ikhsanov, M.Y. Kagan, J. Phys.: Conf. Ser. **2036**, 012019 (2021)
37. R.S. Ikhsanov, E.A. Mazur, M.Y. Kagan, Software complex for the numerical solution of the isotropic imaginary-axis Eliashberg equations. cond-mat (2022), arXiv:2202.01452
38. N.A. Kudryashov, A.A. Kutukov, E.A. Mazur, JETP Lett. **104**, 488 (2016)
39. N.A. Kudryashov, A.A. Kutukov, E.A. Mazur, JETP Lett. **105**, 430 (2017)
40. N.W. Aschcroft, Phys. Rev. Lett. **21**, 1748 (1961)
41. N.W. Aschcroft, Jour. Phys. A: Math Gen. **36**, 6137 (2003)
42. M.Y. Kagan, Unconventional superconductivity in low density electron systems and conventional superconductivity in hydrogen metallic alloys. JETP Lett. **106**, 728 (2016)
43. M.Y. Kagan, S.V. Aksenov, A.V. Turlapov et al., Formation of droplets of the order parameter and superconductivity in inhomogeneous Fermi-Bose mixtures. JETP Lett. **117**, 754 (2023)
44. M.Y. Kagan, R.S. Ikhsanov, I.A. Kovalev, A.V. Krasavin et al., Superconductivity and inhomogeneous states in metallic hydrogen and electron systems with attraction. ZHETP **106**, 89 (2024)
45. M.Y. Kagan, K.I. Kugel, A.L. Rakhmanov, A.O. Sboychakov, *Electronic Phase Separation in Magnetic and Superconducting Materials*. Recent Advances, Springer Series in Solid-State Sciences, vol. 201 (2024)
46. R.P. Dias, I.F. Silvera, Science **355**, 715 (2017)
47. A.A. Abrikosov, Astron. Jour. **31**, 112 (1954)
48. A.A. Abrikosov, Sov. Phys. JETP **12**, 1254 (1961)
49. A.A. Abrikosov, Sov. Phys. JETP **14**, 401 (1962)
50. A.A. Abrikosov, Sov. Phys. JETP **18**, 1399 (1964)
51. T. Schneider, E. Stoll, Preprint AF-SSR-34 (1969)
52. M. Houtput, J. Tempere, I.F. Silvera, Phys. Rev. B **100**, 134106 (2019)

Lecture 20. One-Electron Excitations in Metals

20

Abstract

In the first part of this Lecture, we evaluate the damping of electron excitations near the Fermi surface. We consider one-electron problem in metals and introduce the basic ideas of Landau Fermi liquid theory. We derive the spectrum of electronic excitations in metals and establish its interconnection with the microscopic theory. We provide the definition of Bloch waves and obtain the quasiclassical expressions for the operators of electron velocity and acceleration. We illustrate the emergence of the electronic bands in metals both in weak-coupling and strong-coupling approximation.

20.1   One-Electron Excitations

Important role in Landau Fermi liquid theory [1–7] and in metal physics [8–15] play one-electron excitations (quasiparticles obeying Fermi statistics [16, 17]).

The energy of one-electron excitation ε_k with momentum k is determined as the <u>variational derivative</u> from the total electron energy E on the occupation number of this electron state n_k:

$$\varepsilon_k = \frac{\delta E}{\delta n_k}. \tag{20.1}$$

The physical meaning of this formula according to Landau [2,3] is related to the transition from electron liquid of strongly interacting particles to weakly interacting gas of the quasiparticles.

Let us remind, that the average energy of electron–ion and electron–electron interaction for the whole crystal is given by [1]:

$$NE_i = N\left(E_{ei} + E_{ee}\right) = -3\frac{Z^{5/3}N}{r_S} + 1.2\frac{Z^{5/3}N}{r_S}, \tag{20.2}$$

where E_i is the energy on one elementary cell, and N is the total number of cells in crystal.

M. Kagan, *Lecture Notes of Professor Yuri Kagan in Theoretical Solid-State Physics,*
Lecture Notes in Physics 1048, https://doi.org/10.1007/978-3-032-14621-2_20

First term in (20.2) corresponds to electron–ion interaction, while the second one—to electron–electron interaction.

However, there is substantial difference in these two terms (which at first sight have similar structure). Namely, electron–ion interaction is linear in electron density:

$$n_e = \frac{\sum n_k}{\Omega},\tag{20.3}$$

where n_k are the occupation numbers of different electronic states, and Ω is total volume of the system.

Note that total number of electrons:

$$N_e = n_e \Omega = \sum n_k.\tag{20.4}$$

Let us remind that electron–electron interaction is proportional to the square of electron density:

$$n_e^2 = \frac{\sum n_k}{\Omega}\frac{\sum n_k}{\Omega}.$$

As we discussed already, the electroneutrality condition has a form $n_e = Z n_i$ where n_i is ion density, and Z is ion charge.

Correspondingly, the total number of electrons $N_e = \sum n_k = ZN_i = ZN$ for simple metals with one ion on the elementary cell. Then, the average energy of electron–ion interaction in Eq. (20.2) can be written as [1]:

$$NE_{ei} = -3\frac{Z^{5/3}N}{r_S} = -3\frac{Z^{2/3}N_e}{r_S} = -3\frac{Z^{2/3}\sum n_k}{r_S}.\tag{20.5}$$

At the same time, the average energy of electron–electron interaction yields [1]:

$$NE_{ee} = 1.2\frac{Z^{5/3}N}{r_S} = 1.2\frac{Z^2N^2}{Z^{1/3}Nr_S} = 1.2\frac{N_e^2}{Z^{1/3}Nr_S} = 1.2\frac{\left(\sum n_k\right)^2}{Z^{1/3}Nr_S}.\tag{20.6}$$

As a result, the contribution of these terms in the quasiparticle energy is given by:

$$\varepsilon_k^i = \frac{\delta\left(NE_i\right)}{\delta n_k} = \frac{\delta\left(NE_{ei}+NE_{ee}\right)}{\delta n_k} = -3\frac{Z^{2/3}}{r_S} + 2.4\frac{\sum n_k}{Z^{1/3}Nr_S} = -3\frac{Z^{2/3}}{r_S}$$

$$+ 2.4\frac{N_e}{Z^{1/3}Nr_S} = -3\frac{Z^{2/3}}{r_S} + 2.4\frac{ZN}{Z^{1/3}r_S N} = -3\frac{Z^{2/3}}{r_S} + 2.4\frac{Z^{2/3}}{r_S} = -0.6\frac{Z^{2/3}}{r_S}.$$
$$\tag{20.7}$$

Analogously non-Coulomb part of electron–ion interaction for the whole crystal (for ions with the core of filled inner shells):

$$NE_{ei}^{(1)} = \frac{bZN}{\Omega_0} = \frac{bN_e}{\Omega_0} = \frac{b\sum n_k}{\Omega_0},\tag{20.8}$$

and thus, the contribution of this term to the quasiparticle energy equals to:

$$\varepsilon_k^{(1)} = \frac{b}{\Omega_0}. \tag{20.9}$$

Let us emphasize, that the energy sum $\varepsilon_k^i + \varepsilon_k^{(1)} = -0.6\frac{Z^{2/3}}{r_s} + \frac{b}{\Omega_0}$ is the potential energy of electron in the potential well of constant depth in metal.

If we distinguish now purely electronic contribution to the total energy, then the energy of the electronic subsystem in main approximation:

$$E^{(0)} = \sum \varepsilon_{k0} n_k + E_{\text{exchange}} = \sum \varepsilon_{k0} n_k - \frac{1}{2} \sum_{k_1,k_2,\sigma} \frac{4\pi e^2 n_{k_1} n_{k_2}}{|k_1 - k_2|^2 \Omega}, \tag{20.10}$$

where in the first term $\varepsilon_{k0} = \frac{\hbar^2 k^2}{2m}$ is the bare kinetic energy of electrons (with the bare electron mass m), while the second term represents the exchange Fock term [18,19] related to the exchange hole between electrons with parallel spins.

For the whole crystal with Z electrons on the cell the exchange term equals to:

$$E_{\text{exchange}} = -\frac{1}{2} \sum_{k_1,k_2,\sigma} \frac{4\pi e^2 n_{k_1} n_{k_2}}{|k_1 - k_2|^2 \Omega}. \tag{20.11}$$

Then the quasiparticle energy in this approximation:

$$\varepsilon_k^{(0)} = \frac{\delta E^{(0)}}{\delta n_k} = \varepsilon_{k0} - \sum_{k_2} \frac{4\pi e^2 n_{k_2}}{|k - k_2|^2 \Omega} = \frac{\hbar^2 k^2}{2m} - \sum_{k_2} \frac{4\pi e^2 n_{k_2}}{|k - k_2|^2 \Omega}. \tag{20.12}$$

Let us correct the exchange term in (20.11) and thus the second term in quasiparticle energy (20.12) taking into account the screening of the Coulomb interaction [1,6,7]. If the Fourier component of the bare Coulomb interaction had a form (see Lectures 17–18):

$$V(q = k_1 - k_2) = \frac{4\pi e^2}{|k_1 - k_2|^2 \Omega},$$

then the screened Coulomb interaction reads:

$$V_{\text{eff.}}(q, \omega) = \frac{V(q)}{\varepsilon(q, \omega)},$$

where $\varepsilon(q, \omega)$ is the dielectric permittivity depending not only on transferred momentum $q = k_1 - k_2$ (see Fig. 20.1), but also on the transferred frequency $\omega = \varepsilon_{k_1} - \varepsilon_{k_2}$ which equals to the difference of the bare kinetic energies of electrons.

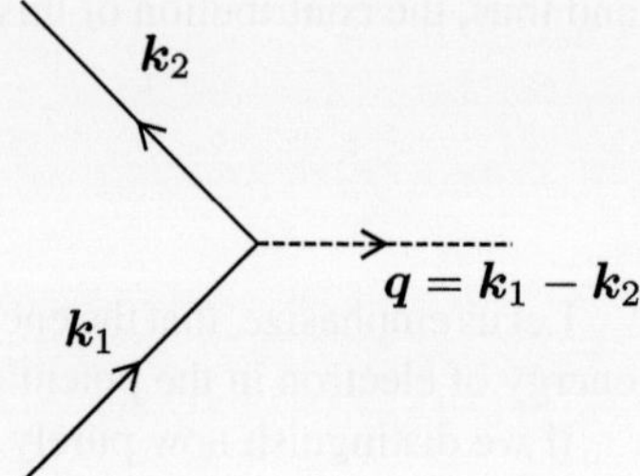

Fig. 20.1 Bare Coulomb interaction in momentum space $V(q = k_1 - k_2) = \frac{4\pi e^2}{|k_1 - k_2|^2 \Omega}$ between two electrons with wave vectors k_1 and k_2 [1]

Correspondingly the quasiparticle energy with an account of screening reads:

$$\varepsilon_k^{(0)} = \frac{\hbar^2 k^2}{2m} + \frac{\partial E_{\text{exchange}}}{\partial n_k} = \frac{\hbar^2 k^2}{2m} - \sum_k \frac{4\pi e^2 n_{k_2}}{|k - k_2|^2 \Omega} \frac{1}{\varepsilon \left(k - k_2, \varepsilon_k - \varepsilon_{k_2} \right)}.$$

$$(20.13)$$

Generally speaking, as we discussed already, the dielectric permittivity is a complex quantity:

$$\varepsilon(q, \omega) = \varepsilon'(q, \omega) + i\varepsilon''(q, \omega).$$

However, its real and imaginary parts behave differently under the replacement $k_1 \rightarrow k_2$ which is equivalent to the replacement:

$$(q, \omega) \rightarrow (-q, -\omega).$$

Real part of the permittivity $\varepsilon(q, \omega)$ proves to be even function of (q, ω). At the same time, imaginary part is odd function.

Let us substitute in the corrected exchange term in (20.11) and correspondingly in the second term of Eq. (20.13) the full expression for the inverse dielectric permittivity:

$$\frac{1}{\varepsilon} = \frac{1}{\varepsilon' + i\varepsilon''} = \frac{\varepsilon' - i\varepsilon''}{(\varepsilon')^2 + (\varepsilon'')^2}.$$

Let us keep in mind that the bare Coulomb interaction

$$V(q = k - k_2) = \frac{4\pi e^2}{|k - k_2|^2 \Omega}$$

does not change under the replacement $k \rightarrow k_2$. The product of the occupation numbers $n_k n_{k_2}$ in (20.11) does not change as well. That is why, in the nominator of the full expression for the inverse dielectric permittivity in the sum over k and k_2 in (20.11) only real part of the permittivity survives. Thus, in corrected exchange term (20.11), and hence in quasiparticle energy (20.13), enters only the combination:

$$\frac{\varepsilon'}{(\varepsilon')^2 + (\varepsilon'')^2} = \text{Re}\,\frac{1}{\varepsilon}.$$

$$(20.14)$$

As a result, the quasiparticle spectrum in (20.13) is purely real and yields [1]:

$$\varepsilon_k^{(0)} = \frac{\hbar^2 k^2}{2m} + \Delta\varepsilon_{\text{exchange}} = \frac{\hbar^2 k^2}{2m} - \sum_k \frac{4\pi e^2 n_{k_2}}{|k - k_2|^2 \, \Omega} \, \text{Re} \, \frac{1}{\varepsilon\left(k - k_2, \varepsilon_k - \varepsilon_{k_2}\right)}.$$

(20.15)

20.2 Effective Mass and Quasiparticle Spectrum

From expression (20.15) for the quasiparticle spectrum we can easily calculate the group velocity of electrons:

$$V_{\text{gr.}} = \frac{1}{\hbar} \frac{\partial \varepsilon_k^{(0)}}{\partial k} = \frac{\hbar k}{m} + \frac{1}{\hbar} \frac{\partial \Delta\varepsilon_{\text{exchange}}}{\partial k}.$$

(20.16)

On the Fermi surface $\hbar k = p_F$, and the group velocity equals to the Fermi velocity [1]:

$$V_F = \frac{p_F}{m^*} = \frac{p_F}{m} - \frac{1}{\hbar} \frac{\partial}{\partial k} \left(\sum_{k_2} \frac{4\pi e^2 n_{k_2}}{|k - k_2|^2 \, \Omega} \, \text{Re} \, \frac{1}{\varepsilon\left(k - k_2, \varepsilon_k - \varepsilon_{k_2}\right)} \right) \Bigg|_{\hbar k = p_F},$$

(20.17)

where m^* is effective mass of electron in metal.

It proves, that the second exchange term in (20.17) depends weakly on the wave vector k and the electron mass is weakly renormalized due to this term in simple metals [1].

Note that electron correlations considered in the framework of the Hubbard model and related basic models (such as t-J or Kondo-lattice models) lead to strong mass enhancement in transition and rare-earth metal compounds (including heavy-fermion systems and high-T_C cuprates) [15,21–25].

At the same time, in simple metals electron–ion interaction yields important contribution to the effective mass [1–20]. Namely, in the second order of the perturbation theory on electron–ion interaction we should add the term related to the sum over non-zero vectors of the inverse lattice ($K \neq 0$) to the electron energy functional in (20.10) and correspondingly to (20.15).

As we discussed in the previous Lecture, this term acquires the form:

$$E^{(2)} = -\frac{\Omega}{2} \sum_{K \neq 0} |N V_{ei}(K)|^2 \frac{\Pi(K)}{\varepsilon(K)},$$

where the dielectric permittivity on the inverse lattice vector:

$$\varepsilon(K) = 1 + \frac{4\pi e^2}{K^2} \Pi(K),$$

and the static polarization operator depends on the occupation numbers as follows:

$$\Pi(\boldsymbol{K}) = \frac{1}{\Omega} \sum_k \frac{n_k - n_{k+K}}{\varepsilon_{k+K} - \varepsilon_k} = \frac{1}{\Omega} \sum_k \left(\frac{n_k}{\varepsilon_{k+K} - \varepsilon_k} - \frac{n_{k+K}}{\varepsilon_{k+K} - \varepsilon_k} \right). \tag{20.18}$$

Further on we can make the replacement $\boldsymbol{k} + \boldsymbol{K} \to \boldsymbol{k}$ and $\boldsymbol{k} \to \boldsymbol{k} - \boldsymbol{K}$ in nominator and denominator of the second term in the r.h.s. of Eq. (20.18). Then we can rewrite expresion (20.18) as:

$$\Pi(\boldsymbol{K}) = \frac{1}{\Omega} \sum_k \left(\frac{n_k}{\varepsilon_{k+K} - \varepsilon_k} - \frac{n_k}{\varepsilon_k - \varepsilon_{k-K}} \right). \tag{20.19}$$

Correspondingly, the multiplier connected with screening reads:

$$\frac{\Pi(\boldsymbol{K})}{\varepsilon(\boldsymbol{K})} = \frac{\varepsilon(\boldsymbol{K}) - 1}{\varepsilon(\boldsymbol{K})} \frac{K^2}{4\pi e^2} = \left(1 - \frac{1}{\varepsilon(\boldsymbol{K})} \right) \frac{1}{\frac{4\pi e^2}{K^2}}. \tag{20.20}$$

Now we can calculate the electron–ion contribution specified by $E^{(2)}$ in the quasiparticle energy:

$$\varepsilon^{(2)}(\boldsymbol{k}) = \frac{\delta\left(E^{(2)}\right)}{\delta n_k} = -\frac{\Omega}{2} \sum_{K \neq 0} |NV_{ei}(\boldsymbol{K})|^2 \frac{1}{\varepsilon^2(\boldsymbol{K})} \frac{\frac{4\pi e^2}{K^2}}{\frac{4\pi e^2}{K^2}} \frac{\delta\Pi(\boldsymbol{K})}{\delta n_k}. \tag{20.21}$$

Substituting in (20.21) the variational derivative from the expression for polarization operator in the r.h.s. of (20.19):

$$\frac{\delta\Pi(\boldsymbol{K})}{\delta n_k} = \frac{1}{\Omega} \sum_k \left(\frac{1}{\varepsilon_{k+K} - \varepsilon_k} - \frac{1}{\varepsilon_k - \varepsilon_{k-K}} \right), \tag{20.22}$$

we get for the contribution to the quasiparticle energy:

$$\frac{\delta\left(E^{(2)}\right)}{\delta n_k} = -\frac{1}{2} \sum_{K \neq 0} \left| \frac{NV_{ei}(\boldsymbol{K})}{\varepsilon(\boldsymbol{K})} \right|^2 \left(\frac{1}{\varepsilon_{k+K} - \varepsilon_k} - \frac{1}{\varepsilon_k - \varepsilon_{k-K}} \right). \tag{20.23}$$

Afterwards we can use the fact that expression (20.23) is an even function under the replacement: $\boldsymbol{K} \to -\boldsymbol{K}$. Taking this fact into account, the variational derivative equals to:

$$\varepsilon^{(2)}(\boldsymbol{k}) = \frac{\delta\left(E^{(2)}\right)}{\delta n_k} = \sum_{K \neq 0} \left| \frac{NV_{ei}(\boldsymbol{K})}{\varepsilon(\boldsymbol{K})} \right|^2 \frac{1}{\varepsilon_k - \varepsilon_{k+K}} = \sum_{K \neq 0} |NV_{\text{eff.}}(\boldsymbol{K})|^2 \frac{1}{\varepsilon_k - \varepsilon_{k+K}}, \tag{20.24}$$

where $V_{\text{eff.}}(\boldsymbol{K}) = \frac{V_{ei}(\boldsymbol{K})}{\varepsilon(\boldsymbol{K})}$ is screened electron–ion interaction.

Let us collect all the terms now including potential energy of electrons in the potential well of the constant depth, kinetic electron energy, screened exchange contribution and second order term in screened electron–ion interaction. Then the quasiparticle spectrum is given by:

$$\varepsilon_k = -0.6\frac{Z^{2/3}}{r_S} + \frac{b}{\Omega_0} + \frac{\hbar^2 k^2}{2m} + \Delta\varepsilon_{\text{exchange}}(\boldsymbol{k}) + \varepsilon^{(2)}(\boldsymbol{k}). \tag{20.25}$$

20.3 Quasiparticle Damping

Let us consider electron living above the Fermi surface and let us find out how fast it scatters with the loss of energy and goes down in the energy scale.

The Hamiltonian of the secondly-quantized electron–electron interaction equals to:

$$H_{\text{int.}} = \frac{1}{2} \sum_{k,k_2,q} \frac{4\pi e^2}{q^2\Omega} a^{\dagger}_{k+q} a^{\dagger}_{k_2-q} a_k a_{k_2}, \tag{20.26}$$

where $k = \{\boldsymbol{k}, \sigma\}$.

Let us consider the process, when the initial particle with the energy lying above the Fermi surface interacts with another particle with the energy lying below the Fermi surface. Suppose two particles with the energies above the Fermi surface emerge as a result of this interaction.

In fact, we analyze the process in which two electrons and one hole are created.

The probability of this process is determined from the Golden rule of Quantum mechanics [16] and yields:

$$W_k = \frac{\hbar}{\tau}$$

$$= \frac{2\pi}{\hbar} \sum_{k_2,q} \left| \frac{4\pi e^2}{q^2\varepsilon(q)} \right|^2 n_{k_2} \left(1 - n_{k+q}\right) \left(1 - n_{k_2-q}\right) \delta\left(\varepsilon_k + \varepsilon_{k_2} - \varepsilon_{k+q} - \varepsilon_{k_2-q}\right), \tag{20.27}$$

where τ is the damping time (or lifetime) of quasiparticles. For better convergence on small wave vectors q and for simplicity of the calculations we can substitute in Eq. (20.27) the screened Coulomb interaction by:

$$V_{\text{eff.}}(q = 0) = \frac{4\pi e^2}{q^2\varepsilon(q)} \approx \frac{4\pi e^2}{\kappa^2} = \text{const.}$$

Further on, to determine the inverse lifetime of quasiparticles we should accurately analyze the double integral of delta function related to the energy conservation:

$$\frac{1}{\tau} \sim \text{const} \iint \frac{d^3 k_2}{(2\pi)^3} \frac{d^3 q}{(2\pi)^3} \delta\left(\varepsilon_k + \varepsilon_{k_2} - \varepsilon_{k+q} - \varepsilon_{k_2-q}\right). \tag{20.28}$$

Let us proceed to the more convenient integration on $\frac{d^3k_2}{(2\pi)^3}\frac{d^3(k+q)}{(2\pi)^3}$ and make the change of variables $k + q = k_3$ according to the textbook of Abrikosov, Gor'kov, Dzyaloshinskii (see p. 35 in their classical monograph "*Methods of Quantum Field Theory in Statistical Physics*" [4]).

Then for k_4 we get:

$$k_4 = k_2 - q = k + k_2 - k_3. \tag{20.29}$$

As a result, the inverse lifetime:

$$\frac{1}{\tau} \sim \text{const} \iint \frac{d^3k_2}{(2\pi)^3} \frac{d^3k_3}{(2\pi)^3} \delta\left(\varepsilon_k + \varepsilon_{k_2} - \varepsilon_{k_3} - \varepsilon_{k_1+k_2-k_3}\right), \tag{20.30}$$

where conservation law for the energies is fulfilled:

$$\varepsilon_k + \varepsilon_{k_2} = \varepsilon_{k_3} + \varepsilon_{k_1+k_2-k_3}. \tag{20.31}$$

Hence for the energies counted from the Fermi energy:

$$(\varepsilon_k - \varepsilon_F) + \left(\varepsilon_{k_2} - \varepsilon_F\right) = \left(\varepsilon_{k_3} - \varepsilon_F\right) + \left(\varepsilon_{k_1+k_2-k_3} - \varepsilon_F\right) \geq 0. \tag{20.32}$$

Note that in Eqs. (20.31)–(20.32) the absolute values of the wave vectors:

$$k > k_F, k_2 < k_F, k_3 > k_F, |k + k_2 - k_3| > k_F. \tag{20.33}$$

Let us assume that initial particle lies above the Fermi surface but close to it (see Fig. 20.2) and thus for $\hbar = 1$ its energy can be written as:

$$\varepsilon_k = \frac{k^2}{2m} = \varepsilon_F + \omega, \text{ where } \omega \ll \varepsilon_F. \tag{20.34}$$

Correspondingly for the energy counted from the Fermi energy:

$$\omega = \varepsilon_k - \varepsilon_F \approx V_F (k - k_F), \tag{20.35}$$

and the difference of the wave vectors is small:

$$k - k_F \approx \frac{\omega}{V_F} \ll k_F. \tag{20.36}$$

Since both energies ε_{k_3} and $\varepsilon_{k_1+k_2-k_3}$ lie above the Fermi level, the r.h.s. of Eq. (20.32) is positive.

Thus, the l.h.s. of (20.32) is also positive:

$$(\varepsilon_k - \varepsilon_F) + \left(\varepsilon_{k_2} - \varepsilon_F\right) \approx V_F (k - k_F + k_2 - k_F) = \omega + \varepsilon_{k_2} - \varepsilon_F \geq 0. \tag{20.37}$$

It means that the second initial particle lies in the ω-neighborhood below the Fermi surface:

$$\xi_2 = \varepsilon_{k_2} - \varepsilon_F \geq -\omega. \tag{20.38}$$

Moreover, for the absolute value of its wave vector k_2 we obtain:

$$-\frac{\omega}{V_F} \leq k_2 - k_F \leq 0. \tag{20.39}$$

From Eq. (20.37) also follows, that the sum of the absolute values of the wave vectors:

$$k + k_2 \geq 2k_F. \tag{20.40}$$

At the same time, since $k_2 < k_F$:

$$k + k_2 \leq k + k_F. \tag{20.41}$$

Thus, the sum of the absolute values of these wave vectors:

$$2k_F \leq k + k_2 \leq k + k_F \approx 2k_F + \frac{\omega}{V_F} \tag{20.42}$$

lies close to the diameter of the Fermi sphere $2k_F$.

On the other hand, in the r.h.s. of (20.32) enters the sum of positive energies:

$$\left(\varepsilon_{k_3} - \varepsilon_F\right) + \left(\varepsilon_{k_1 + k_2 - k_3} - \varepsilon_F\right) \approx V_F \left(k_3 - k_F + |k + k_2 - k_3| - k_F\right) \geq 0. \tag{20.43}$$

Then, equating the l.h.s. to the r.h.s. in Eq. (20.32) we get:

$$k + k_2 - 2k_F = (k_3 - k_F) + (|k + k_2 - k_3| - k_F) \geq k_3 - k_F, \tag{20.44}$$

where we used that $|k + k_2 - k_3| - k_F \geq 0$.

Hence, the difference of the absolute values of the wave vectors $k_3 - k_F$ is limited by inequalities:

$$0 \leq k_3 - k_F \leq k + k_2 - 2k_F \leq \frac{\omega}{V_F}. \tag{20.45}$$

As a result, the energy of the emergent particle ε_{k_3} also lies in ω-neighborhood above the Fermi surface:

$$0 \leq \xi_3 = \varepsilon_{k_3} - \varepsilon_F \leq \omega. \tag{20.46}$$

For the quadratic spectrum expression in the argument of the delta function equals to [4]:

$$\frac{k^2 + k_2^2 - 2k_3^2 - (k_1 + k_2)^2 + 2(k_1 + k_2)k_3}{2m}$$
$$= \frac{k^2 + k_2^2 - 2k_3^2 - (k_1 + k_2)^2 + 2|k_1 + k_2|k_3\cos\theta}{2m}. \tag{20.47}$$

Correspondingly delta function in Eq. (20.30) works when cosine of the angle θ between wave vector k_3 and sum of the vectors $k + k_2$ is given by [4]:

$$\cos\theta = -\frac{k^2 + k_2^2 - 2k_3^2 - (k + k_2)^2}{2|k + k_2|k_3}. \tag{20.48}$$

At the same time, cosine of the angle χ between k and k_2 is arbitrary and can take any value in the interval from -1 to 1. Since particles with momenta k_2 and k_3 lie close to the Fermi energy, we can replace the measure of integration in (20.30) according to the standard procedure for the theory of electron Fermi liquid in metal [1,4,5]:

$$\frac{d^3k_2}{(2\pi)^3}\frac{d^3k_3}{(2\pi)^3} \approx N^2(0)d\xi_2\frac{d\cos\chi}{2}d\xi_3\frac{d\cos\theta}{2}, \tag{20.49}$$

where $\xi_2 = \varepsilon_{k_2} - \varepsilon_F, \xi_3 = \varepsilon_{k_3} - \varepsilon_F, N(0) = \frac{mk_F}{2\pi^2}$ is the density of states at the Fermi surface, χ is the angle between k and k_2. Then lifting the integration on $d\cos\theta$ at the expense of the delta function, we obtain from Eq. (20.30) [4]:

$$\frac{1}{\tau} \sim [V_{\text{eff.}}(q=0)N(0)]^2 \int_{-\omega}^0 d\xi_2 \int_{-1}^1 \frac{d\cos\chi}{2} \int_0^\omega d\xi_3 \frac{m}{|k_1 + k_2|k_3}. \tag{20.50}$$

Now, in Eq. (20.50) we can replace:

$$k_3 \to k_F, \quad |k_1 + k_2| \approx \sqrt{2k_F^2(1 + \cos\chi)} = \sqrt{2}k_F\sqrt{1 + \cos\chi}. \tag{20.51}$$

As a result, the last angular integral in (20.50) is calculated elementary [4]:

$$\int_{-1}^1 \frac{d\cos\chi}{2}\frac{1}{\sqrt{1 + \cos\chi}} = \sqrt{2}. \tag{20.52}$$

After taking this integral the expression for the inverse lifetime yields:

$$\frac{1}{\tau} \sim [V_{\text{eff.}}(q=0)N(0)]^2 \frac{m}{k_F^2}\int_{-\omega}^0 d\xi_2 \int_0^\omega d\xi_3 \sim \frac{[V_{\text{eff.}}(q=0)N(0)]^2}{\varepsilon_F}\omega^2 \propto g^2\frac{\omega^2}{\varepsilon_F}. \tag{20.53}$$

It is interesting, that in Eq. (20.53) enters the square of the dimensionless coupling constant related to the screened Coulomb interaction:

$$g = \frac{4\pi e^2}{\kappa^2} N(0). \tag{20.54}$$

Note that in the RPA approximation [1,6,7] the square of the Thomas–Fermi wave vector $\kappa^2 = 8\pi e^2 N(0)$ [26,27], and the coupling constant g is universal (does not depend on density) and is of the order of $g \sim \frac{1}{2}$ in simple metals.

Thus, even for sufficiently strong interaction, the quasiparticles are weakly damped on the Fermi surface. Correspondingly, the real part of the quasiparticle spectrum $\mathrm{Re}\,(\varepsilon - \varepsilon_F) \sim \omega$ is always much larger than the imaginary part $\mathrm{Im}\,(\varepsilon - \varepsilon_F) \sim \frac{\omega^2}{\varepsilon_F}$ [1–9].

Physically this fact is connected with very small allowed phase volume. Particles close to the Fermi surface live for a long time due to the Pauli principle [1–5, 16, 17]. Actually, the factor $\frac{\omega^2}{\varepsilon_F}$ in the damping is related to the fact, that at small energies in the scattering process in Fig. 20.2 due to Pauli principle participates only small fraction (proportional to $\frac{\omega}{\varepsilon_F}$) from the total number of particles. Moreover, each particle of this fraction has an energy of order of ω. At finite temperatures the averaging of the expression for the inverse lifetime with the Fermi–Dirac distribution function yields standard damping for Landau Fermi liquid theory:

$$\frac{1}{\tau} \propto g^2 \frac{T^2}{\varepsilon_F} \ll T, \tag{20.55}$$

where the temperature $T \ll \varepsilon_F$ is effectively the measure of smearing of the Dirac step function in metal.

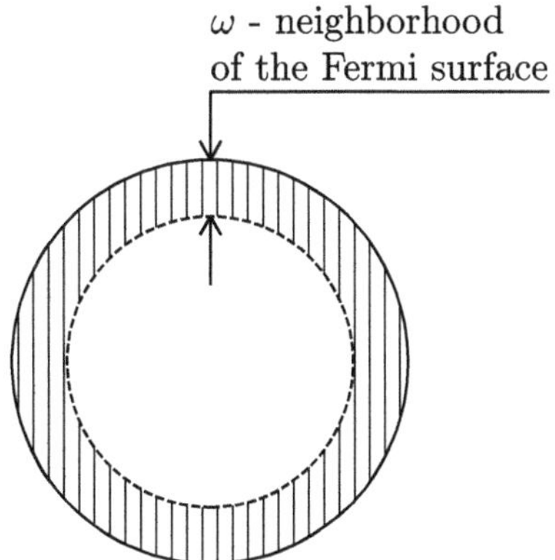

Fig. 20.2 Damping of the quasiparticle with energy $\varepsilon_k - \varepsilon_F = \omega$ lying slightly higher than Fermi level. We show on the Figure an allowed statistical volume for the evaluation of the inverse lifetime $\frac{1}{\tau} \sim g^2 \frac{\omega^2}{\varepsilon_F}$. Small statistical volume is related to small ω—neighborhood of the Fermi surface in agreement with Pauli principle [1]

Let us stress, that just the smallness of damping close to the Fermi surface allows in Landau theory to proceed from the strongly interacting Fermi liquid of real particles to the weakly interacting Fermi gas of quasiparticles.

Note also that in advanced theories of high temperature superconductivity theorists introduced different concepts connected with possible violations of Landau Fermi liquid theory in strongly correlated matter specified, e.g., by Marginal Fermi Liquid (with damping linear in temperature) [28] or Luttinger Fermi Liquid [29] with even larger quasiparticle damping and zero value of the jump (of Z-factor on diagrammatic language [4,5,23,25,30]) in the fermionic distribution of real particles on the Fermi surface.

All of these theories, however satisfy Luttinger theorem [4,5,30–32] according to which the volume of the Fermi surface is conserved, or in other words, the total number of real particles equal to the total number of quasiparticles.

20.4   One-Particle Problem

In one-particle problem we can consider Schrodinger equation [1,8,9,16] for electrons in the field of periodic screened ion potential. In momentum space effective interaction reads:

$$V_{\text{eff.}}(k) = \frac{V(k)}{\varepsilon(k)},$$

where $\varepsilon(k)$ is the static dielectric permittivity. Effective interaction takes full account of the effects of electron liquid in this approach.

In fact, in real space we should write the Schrodinger equation for the electron wave function $\psi(r)$ in the periodic potential $U = V_{\text{eff.}}(r)$:

$$-\frac{\hbar^2}{2m_0}\Delta\psi + U\psi = \varepsilon\psi, \tag{20.56}$$

where m_0 is the bare mass of electron.

Let us expand the wave function $\psi(r)$ in Fourier series:

$$\psi(r) = \sum_q c_q \exp(iqr). \tag{20.57}$$

Then we can rewrite the Schrodinger equation as follows:

$$\sum_q \left(\frac{\hbar^2 q^2}{2m_0} - \varepsilon\right) c_q \exp(iqr) + \sum_q c_q U \exp(iqr) = 0. \tag{20.58}$$

Further on, multiplying on $\exp\left(-i\boldsymbol{q}_1\boldsymbol{r}\right)$ and integrating over $d^3\boldsymbol{r}$ we obtain:

$$\left(\frac{\hbar^2 q_1^2}{2m_0} - \varepsilon\right) c_{q_1} + \sum_q c_q \int d^3\boldsymbol{r} \exp\left(-i\boldsymbol{q}_1\boldsymbol{r}\right) U \exp(i\boldsymbol{q}\boldsymbol{r})$$

$$= \left(\frac{\hbar^2 q_1^2}{2m_0} - \varepsilon\right) c_{q_1} + \sum_q c_q \langle q_1|U|q\rangle = 0, \tag{20.59}$$

where in the second term we introduced the notation:

$$\langle q_1|U|q\rangle = \int d^3\boldsymbol{r} \exp\left(-i\boldsymbol{q}_1\boldsymbol{r}\right) U \exp(i\boldsymbol{q}\boldsymbol{r}). \tag{20.60}$$

We also used the relation

$$\int d^3\boldsymbol{r} \exp\left(-i\boldsymbol{q}_1\boldsymbol{r}\right) \exp(i\boldsymbol{q}\boldsymbol{r}) = \Omega\delta\left(\boldsymbol{q} - \boldsymbol{q}_1\right) = \delta_{\boldsymbol{q}\boldsymbol{q}_1}, \tag{20.61}$$

which lifts summation over q in the first term.

After that we can make a replacement $q \to q_1$ and represent the Schrodinger equation in momentum space in canonical form:

$$\left(\frac{\hbar^2 q^2}{2m_0} - \varepsilon\right) c_q + \sum_{q_1} c_{q_1} \langle q|U|q_1\rangle = 0, \tag{20.62}$$

where:

$$\langle q|U|q_1\rangle = U\left(\boldsymbol{q}_1 - \boldsymbol{q}\right). \tag{20.63}$$

It follows from the periodicity of the potential that:

$$U\left(\boldsymbol{q}_1 - \boldsymbol{q}\right) = U(\boldsymbol{K}),$$

where $\boldsymbol{K} = \boldsymbol{q}_1 - \boldsymbol{q}$ is the wave vector of the inverse lattice.

Then, $\boldsymbol{q}_1 = \boldsymbol{q} + \boldsymbol{K}$ and from the summation over q_1 in the second term in Eq. (20.9) we can proceed to the summation over K. As a result we obtain the infinite system of the homogeneous equations on the coefficients c_q:

$$\left(\frac{\hbar^2 q^2}{2m_0} - \varepsilon\right) c_q + \sum_{K\neq 0} c_{q+K} U(K) = 0. \tag{20.64}$$

Correspondingly, the eigenvalues of the problem read:

$$\varepsilon = \varepsilon(\boldsymbol{q}, n), \tag{20.65}$$

where n is the number of the solution. We can find the eigenvalues from the condition on zero determinant of the linear system of Eq. (20.62). Solutions (20.65) determine the electron spectrum.

In the periodic structure the sheets of the solutions appear corresponding to the energy bands. The bands do not intersect in momentum space at the given value of q. Thus n is the number of the electron band.

Let us make the replacement $q \to q + K$ in the system of equations (20.62). As a result, we get the same system of equations but with redefined coefficients:

$$\left(\frac{\hbar^2 (q + K)^2}{2m_0} - \varepsilon \right) c_{q+K} + \sum_{K_1 \neq 0} c_{q+K+K_1} U(K_1) = 0. \tag{20.66}$$

Determinant of this system remains the same (it still equals zero), and thus its eigenvalues will not change. Hence electron energy in metals as we expected is the periodic function with the period equals to the inverse lattice vector:

$$\varepsilon(q, n) = \varepsilon(q + K, n). \tag{20.67}$$

Thus, the eigenvectors of the problem at q and $q + K$ also coincide:

$$\psi_\varepsilon = \psi_{q,n} = \psi_{q+K,n}. \tag{20.68}$$

That is why we can consider only the first cell of the inverse lattice as in the case of phonons. We should remember, however, that the value of n is limited by the number of the degrees of freedom in the phonon case.

Let us use the condition (20.68) and determine the form of the wave functions from the expansion (20.57):

$$\psi_{q,n} = \psi_{q+K,n} = \sum_{q+K} c^{(n)}_{q+K} \exp(i(q + K)r) = \exp(iqr) \sum_{K} c^{(n)}_{q+K} \exp(iKr),$$

$$\tag{20.69}$$

where we proceed from the summation over $q + K$ to the summation over K.

Let us emphasize, that in expression (20.57) expansion coefficients of the eigenvectors $c^{(n)}_{q+K}$ correspond to eigenvalues $\varepsilon(q, n)$.

Now we can introduce the Bloch amplitudes [1, 8–10] of the wave function:

$$u_{qn}(r) = \sum_{K} c^{(n)}_{q+K} \exp(iKr), \tag{20.70}$$

and represent $\psi_{q,n}$ as:

$$\psi_{q,n} = \exp(iqr) u_{qn}(r). \tag{20.71}$$

The amplitudes $u_{qn}(r)$ in Condensed matter physics usually referred to as Bloch wave functions [1,8–10]. The Bloch wave function is periodic function in real (coordinate) space:

$$u_{qn}(r + a) = u_{qn}(r), \tag{20.72}$$

where a is the lattice period and by the definition: $\exp(iKa) = 1$.

Correspondingly the eigenfunction $\psi_{q,n}$ under the shift on the lattice period is transformed as follows:

$$\psi_{qn}(r + a) = \exp(iqa)\psi_{qn}(r). \tag{20.73}$$

The Bloch wave functions and their amplitudes form the full orthonormalized basis:

$$\int dr\,\psi_{qn}^*(r)\Psi_{q'n'}(r) = \delta_{qq'}\delta_{nn'}, \tag{20.74}$$

$$\int dr\,u_{qn}^*(r)u_{qn'}(r) = \delta_{nn'}. \tag{20.75}$$

Let us stress, that $\hbar q$ is quasimomentum (and not momentum). It is defined with the accuracy of the inverse lattice vector. Consequently in Eq. (20.57):

$$\psi_{q,n} = \psi_{q+K,n} = \sum_{q+K} c_{q+K}^{(n)} \exp(i(q + K)r) = \sum_{K} c_{q+K}^{(n)} \exp(i(q + K)r),$$

and the matrix element of the momentum operator equals to:

$$\left\langle \psi_{qn}^* | \hat{p} | \psi_{qn} \right\rangle = \sum_{K} \left| c_{q+K} \right|^2 \hbar(q + K) = \hbar q + \sum_{K} \left| c_{q+K} \right|^2 \hbar K. \tag{20.76}$$

Note that deriving (20.76) we used the normalization condition:

$$\sum_{K} \left| c_{q+K} \right|^2 = \sum_{K+q} \left| c_{q+K} \right|^2 = 1. \tag{20.77}$$

Thus, the wave function $\psi_{q,n}$ is not the eigenfunction of the momentum operator (the function $\psi_{q,n}$ represents the superposition of plane waves). As a result, momentum is not defined exactly, and hence is not conserved [1].

At the same time, the quasimomentum $p = \hbar q$ is conserved for the electron motion in the periodic field.

20.5   Strong-Coupling and Weak-Coupling Approximation

Let us ask an important question: how exactly we can define electron velocity?

In all the textbooks [8,9,11,15] we can read about electron spectrum and group velocity of electron in two cases: in the approximation of weak periodic potential and in the strong-coupling case, when electron practically sits in one cell.

Weak potential corresponds to the case, when all the coefficients of Fourier expansion c_{q+K} are small for $K \neq 0$.

Strong potential arises in the situation of weakly overlapping potential wells. In this case we get electron energy level in the potential of one cell and correspondingly we have degeneracy. As a result, in strong coupling the wave function in the first approximation acquires the form:

$$\psi = \sum_m c_m \Phi \left(r - R_m \right). \tag{20.78}$$

In strong coupling approach one cell does not differ from the other one and it is physically equivalent to find the particle in each cell. Hence in Eq. (20.78)

$$|c_m|^2 = \text{const, and } c_m = \exp \left(iqR_m \right). \tag{20.79}$$

Then, we can get the Bloch wave functions from the strong-coupling wave functions $\Phi \left(r - R_m \right)$:

$$\psi \left(r \right) = \sum_m \exp \left(iqR_m \right) \Phi \left(r - R_m \right). \tag{20.80}$$

Indeed, the function $\psi \left(r \right)$ in (20.80) has the property of translational symmetry of the Bloch functions:

$$\psi \left(r + a \right) = \sum_m \exp \left(iqR_m \right) \Phi \left(r + a - R_m \right)$$

$$= \exp(iqa) \sum_m \exp \left(iq \left(R_m - a \right) \right) \Phi \left(r - R_m + a \right)$$

$$= \exp(iqa) \sum_{m'} \exp \left(iqR_{m'} \right) \Phi \left(r - R_{m'} \right) = \exp(iqa)\psi(r), \tag{20.81}$$

where we used the fact that $R_m - a = R_{m'}$ is the lattice vector as well.

Let us stress that actually the function $\Phi \left(r - R_m \right)$ corresponds to some atomic level [1,8,9,11,15].

20.6 Operator of Electron Velocity

According to Quantum mechanics [16], operator of the velocity can be expresses via the commutator of the Hamiltonian and coordinate of the particle:

$$\hat{v} = \frac{i}{\hbar}[H, \hat{r}].$$
(20.82)

Let us take advantage now from the fact, that eigenvalues of the Hamiltonian in (q, n) representation equal to $\varepsilon(q, n)$.

Moreover, in non-periodic momentum space the operator of coordinate yields:

$$\hat{r} = i\frac{\partial}{\partial q}.$$
(20.83)

But we have periodic crystal. Thus, we have to find the operator of coordinate in (q, n) space.

Note that the wave function represents the wave packet. If we proceed in the Fourier expansion of the wave function from the summation over q to the integration, then from Eq. (20.4) we can represent the wave packet in the form:

$$\psi(r) = \sum_n \int \frac{d^3q}{(2\pi)^3} c_{qn}\psi_{qn}(r).$$
(20.84)

Correspondingly, the operator of coordinate acts on the wave function as:

$$r\psi = \sum_n \int \frac{d^3q}{(2\pi)^3} c_{qn}r\psi_{qn}(r).$$
(20.85)

If we take now the derivative $i\frac{\partial}{\partial q}$ from $\psi_{qn}(r)$ and utilize the definition of the wave function via the Bloch amplitude in Eq.(20.71), then we obtain:

$$i\frac{\partial\psi_{qn}(r)}{\partial q} = -r\psi_{qn}(r) + i\,\exp(iqr)\frac{\partial u_{qn}}{\partial q}.$$
(20.86)

As a result:

$$r\psi_{qn}(r) = -i\frac{\partial\psi_{qn}(r)}{\partial q} + i\,\exp(iqr)\frac{\partial u_{qn}}{\partial q}.$$
(20.87)

Substituting (20.87) in the expression (20.85) for $r\psi$ we get:

$$r\psi = \sum_n \int \frac{d^3q}{(2\pi)^3} c_{qn}\left(-i\frac{\partial\psi_{qn}(r)}{\partial q} + i\,\exp(iqr)\frac{\partial u_{qn}}{\partial q}\right).$$
(20.88)

Further on let us represent the derivative $\frac{\partial u_{qn}}{\partial q}$ from the Bloch amplitude with number n in Eq. (20.88) in the form of a series of Bloch amplitudes u_{qm} with different numbers m:

$$\frac{\partial u_{qn}}{\partial q} = \sum_m f_{mn} u_{qm}. \tag{20.89}$$

From the orthonormality of Bloch amplitudes the vector coefficients f_{mn} in this expansion read:

$$f_{mn} = \int dr u_{qm}^* \frac{\partial u_{qn}}{\partial q}. \tag{20.90}$$

With an account of the expansion (20.84) $r\psi$ in Eq. (20.88) equals to:

$$r\psi = \sum_n \int \frac{d^3q}{(2\pi)^3} c_{qn} \left(-i \frac{\partial \psi_{qn}(r)}{\partial q} + i \exp(iqr) \sum_m f_{mn} u_{qm} \right). \tag{20.91}$$

After that let us integrate by parts the first term $-c_{qn} i \frac{\partial \psi_{qn}}{\partial q}$ in the r.h.s. of Eq. (20.91):

$$- c_{qn} i \frac{\partial \psi_{qn}}{\partial q} = -i \frac{\partial \left(c_{qn} \psi_{qn} \right)}{\partial q} + i \frac{\partial c_{qn}}{\partial q} \psi_{qn}. \tag{20.92}$$

Then, let us proceed to the surface integral in the first term with full derivative in the r.h.s. of Eq. (20.92) (see Fig. 20.3). Herewith we take into account that the surface elements in Fig. 20.3 have the oppositely directed normal vectors, while all the quantities do not change under the shift on the inverse lattice vector.

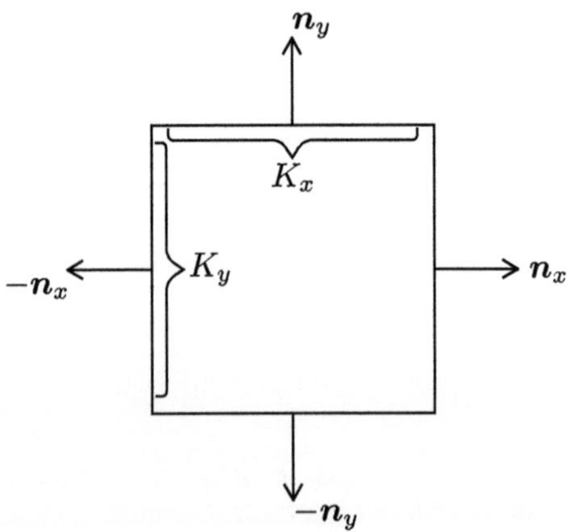

Fig. 20.3 Unit normal vectors n_x, n_y to the sides K_y, K_x of the rectangular cell of the inverse space [1]

Correspondingly the term with the full derivative yields zero contribution in the integral (20.91), and hence:

$$
\begin{aligned}
\boldsymbol{r}\psi &= \sum_n \int \frac{d^3\boldsymbol{q}}{(2\pi)^3}\left(i\frac{\partial c_{qn}}{\partial \boldsymbol{q}}\psi_{qn} + i c_{qn}\exp(i\boldsymbol{q}\boldsymbol{r})\sum_m \boldsymbol{f}_{mn}u_{qm}\right) \\
&= \sum_n \int \frac{d^3\boldsymbol{q}}{(2\pi)^3}\left(i\frac{\partial c_{qn}}{\partial \boldsymbol{q}}\psi_{qn} + i c_{qn}\sum_m \boldsymbol{f}_{mn}\psi_{qm}\right).
\end{aligned}
\tag{20.93}
$$

Now we can interchange the blind indices m and n in the second term in the r.h.s. of Eq. (20.93) and reverse the order of summation over m and n. Then:

$$
\begin{aligned}
\boldsymbol{r}\psi &= \sum_n \int \frac{d^3\boldsymbol{q}}{(2\pi)^3} i\frac{\partial c_{qn}}{\partial \boldsymbol{q}}\psi_{qn} + i\sum_m \int \frac{d^3\boldsymbol{q}}{(2\pi)^3}c_{qm}\sum_n \boldsymbol{f}_{nm}\psi_{qn} \\
&= \sum_n \int \frac{d^3\boldsymbol{q}}{(2\pi)^3}\psi_{qn}\left(i\frac{\partial c_{qn}}{\partial \boldsymbol{q}} + i\sum_m \boldsymbol{f}_{nm}c_{qm}\right) \\
&= \sum_n \int \frac{d^3\boldsymbol{q}}{(2\pi)^3}\psi_{qn}\left(i\frac{\partial c_{qn}}{\partial \boldsymbol{q}} + \widehat{\boldsymbol{\Omega}}c_{qn}\right),
\end{aligned}
\tag{20.94}
$$

where we introduced the vector operator $\widehat{\boldsymbol{\Omega}}$ according to the relation:

$$
\widehat{\boldsymbol{\Omega}}c_{qn} = \sum_m \boldsymbol{\Omega}_{nm}c_{qm} = i\sum_m \boldsymbol{f}_{nm}c_{qm}.
\tag{20.95}
$$

Correspondingly in matrix representation operator:

$$
\boldsymbol{\Omega}_{nm} = i\boldsymbol{f}_{nm}.
\tag{20.96}
$$

Hence coordinate operator acting in $(\boldsymbol{q}, n)$ space on the coefficients c_{qn} equals to:

$$
\hat{\boldsymbol{r}} = \hat{I}i\frac{\partial}{\partial \boldsymbol{q}} + \widehat{\boldsymbol{\Omega}},
\tag{20.97}
$$

where $\hat{I}$ is the unit matrix.

When the interaction is decreased the Bloch amplitudes tend to unity, and operator $\widehat{\boldsymbol{\Omega}}$ goes to zero.

Now let us return back again to the definition of the velocity operator via the commutator of the Hamiltonian and coordinate operator in Eq. (20.82). Let us write the commutator in the $(\boldsymbol{q}, n)$ space. Then the matrix element of the velocity operator can be expressed via electron spectrum and matrix elements of coordinate operator in (20.97):

$$(\hat{v})_{qn,qm} = \frac{i}{\hbar}\{\varepsilon(q,n)(r)_{nm} - (r)_{mn}\varepsilon(q,m)\}$$

$$= \delta_{nm}\frac{1}{\hbar}\frac{\partial \varepsilon(q,n)}{\partial q} + \frac{i}{\hbar}\mathbf{\Omega}_{nm}(\varepsilon(q,n) - \varepsilon(q,m)). \qquad (20.98)$$

Further on we can find the average velocity. To do that we can put $n = m$. Then we get expression for the group velocity of electron which we intuitively expect:

$$\hat{v}_n(q) = \frac{1}{\hbar}\frac{\partial \varepsilon(q,n)}{\partial q}. \qquad (20.99)$$

Note, that if we have electron interaction with light in the system, then the Lagrangian [33] of this interaction reads in the operator form [1,34]:

$$\mathcal{L}_{\text{int.}} = \frac{e}{c}\hat{v}\hat{A} = \frac{1}{c}\hat{j}\hat{A}, \qquad (20.100)$$

where $\hat{j}$ is current operator (for one charge), $\hat{A}$ is vector potential operator, c is speed of light.

In the second quantization Eq. (20.100) in particular describes the process of photon absorption by electron. Let us stress that since we have crystal (actually the third body) we can satisfy the conservation laws for the processes of photon emission and absorption.

In case of light at fixed photon energy $\varepsilon_{\text{phot.}}$, the photon momentum $p_{\text{phot.}} = \frac{\varepsilon_{\text{phot.}}}{c}$ is small. Hence the momentum q is almost conserved in the photon absorption process. Then in the matrix element for the velocity operator in Eq. (20.98) works the second term which equals to:

$$(\hat{v})_{qn,qm} = \frac{i}{\hbar}\mathbf{\Omega}_{nm}(\varepsilon(q,n) - \varepsilon(q,m)). \qquad (20.101)$$

This term corresponds to the electron transition from one band to the other. In this process the photon is absorbed.

20.7   Acceleration Operator

Electron Hamiltonian in the homogeneous electric field has the following form in (q, n) representation [16]:

$$H = H_0 - e\hat{\mathbf{E}}\hat{r} = \varepsilon(q,n) - e\hat{\mathbf{E}}\hat{r}, \qquad (20.102)$$

where $\hat{\mathbf{E}}$ is electric field operator, $\hat{r}$ is coordinate operator in (q, n) representation.

Then acceleration is determined by the commutator of the Hamiltonian and quasi-momentum:

$$\frac{d\boldsymbol{q}}{dt} = \frac{i}{\hbar}[H, \hat{\boldsymbol{q}}] = -\frac{ie}{\hbar} E^\alpha \left[r^\alpha \boldsymbol{q}\right].$$

(20.103)

But from the uncertainty principle:

$$\left[r^\alpha q^\beta\right] = i\delta^{\alpha\beta}.$$

(20.104)

Hence for the acceleration we obtain:

$$\frac{dq^\beta}{dt} = \frac{1}{\hbar} e E^\beta, \text{ and } \hbar\frac{d\boldsymbol{q}}{dt} = e\boldsymbol{E}.$$

(20.105)

Thus, here we get full similarity with Newton law [1,33], but for quasimomentum now. Then using Eqs. (20.100) and (20.105) for time derivative from group velocity, we can write:

$$\frac{dv_n^\alpha}{dt} = \frac{\partial v_n^\alpha}{\partial q^\beta}\frac{dq^\beta}{dt} = \frac{1}{\hbar}\frac{\partial^2\varepsilon(\boldsymbol{q}, n)}{\partial q_\alpha \partial q_\beta}\frac{dq^\beta}{dt} = \frac{1}{\hbar^2}\frac{\partial^2\varepsilon(\boldsymbol{q}, n)}{\partial q_\alpha \partial q_\beta}e E^\beta.$$

(20.106)

But the second Newton law in classical mechanics has the form [1,33]:

$$\frac{d\boldsymbol{v}}{dt} = \frac{1}{m}\boldsymbol{F},$$

(20.107)

where $\boldsymbol{F} = e\boldsymbol{E}$ is the force, acting on the particle with mass m.

Analogously we can introduce the tensor of inverse effective masses $\left(\frac{1}{m^*}\right)^{\alpha\beta}$, and write the second Newton law as follows [1,8,9,11–15]:

$$\frac{dv_n^\alpha}{dt} = \left(\frac{1}{m^*}\right)^{\alpha\beta} e E^\beta,$$

(20.108)

where the tensor:

$$\left(\frac{1}{m^*}\right)^{\alpha\beta} = \frac{1}{\hbar^2}\frac{\partial^2\varepsilon(\boldsymbol{q}, n)}{\partial q_\alpha \partial q_\beta}.$$

(20.109)

Formulas (20.108), (20.109) show, that in crystal we apply the force in one direction but electron moves in the other direction. In fact, electron moves in crossed fields of electric field and periodic crystal [1].

20.8   The Meaning of the Band Spectrum

If the spectrum is limited in energy, then it always has minimum (see Fig. 20.4). In this case, close to the minimum, we can expand electron energy in series in q. If close to the minimal value of the quasimomentum $q_{\min} = 0$ there is no degeneracy, then electron energy close to the minimum in the main axis can be reduced to the elliptic form [1,8,9]:

$$\varepsilon_n(q) = \varepsilon_0 + \sum_{i=1,2,3} \gamma^i q_i q_k = \varepsilon_0 + \sum_i \frac{\hbar^2 q_i^2}{2m_i^*}, \qquad (20.110)$$

where we used the positive definiteness of the coefficients γ^i close to the energy minimum.

Note, that, generally speaking, the bare electron mass in vacuum (m_0) and positively defined bare band masses $m_i^* > 0$ (which do not account for electron–electron interactions) are quite different quantities. They are not equal to each other even in the isotropic case for the cubic crystal (when all the band masses m_i^* are equal).

Similarly in the neighborhood of maximum electron spectrum in main axis is given by [1,8,9]: $\varepsilon_n(q) = \varepsilon_0 + \sum_i \frac{\hbar^2 q_i^2}{2m_i^*}$, where all the band masses $m_i^* < 0$ are negative.

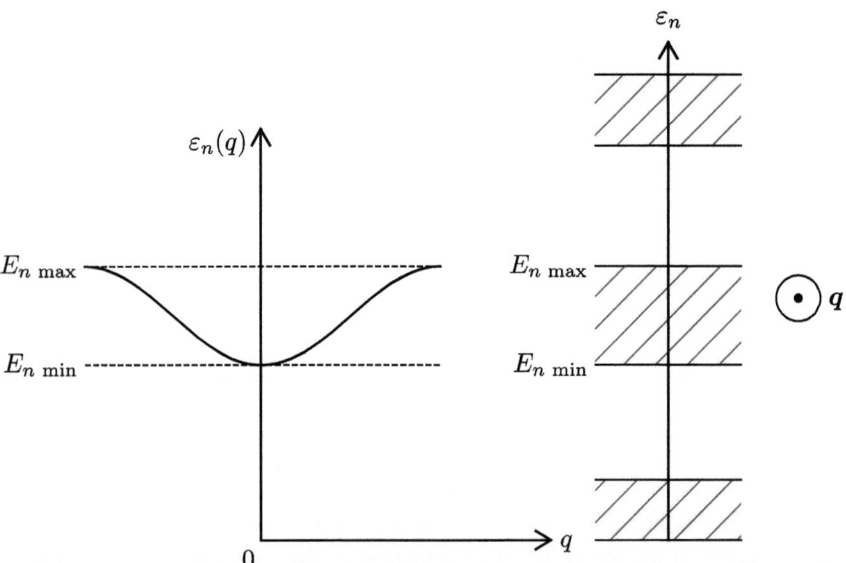

Fig. 20.4 Band energy spectrum of electron in crystal. Each band corresponds to the stripe of energy levels lying in the interval between maximal and minimal values $E_{n\,\min} < \varepsilon_n(q) < E_{n\,\max}$ [1]

We can say, that electron in external field close to the energy maximum moves in opposite direction (as positron) [1].

References

1. Y. Kagan, *Lectures on the Solid-State Theoretical Physics*, Moscow, MEPHI, 1981–1982. Part II. Electrons (Unpublished)
2. L.D. Landau, Sov. Phys. JETP **3**, 920 (1957)
3. L.D. Landau, Sov. Phys. JETP **5**, 101 (1957)
4. A.A. Abrikosov, L.P. Gor'kov, I.E. Dzyaloshinskii, *Methods of Quantum Field Theory in Statistical Physics* (Prentice Hall, Englewood Cliffs, New Jersey, 1963)
5. E.M. Lifshitz, L.P. Pitaevskii, *Statistical Physics, Part II, Theory of the Condensed State* (Pergamon, Elsevier, 2013)
6. D. Pines, *Elementary Excitations in Solids* (W.A. Benjamin Inc., New York–Amsterdam, 1963)
7. D. Pines, P. Nozieres, *The Theory of Quantum Liquids*. Normal Fermi Liquids, vol. 1 (W.A. Benjamin Inc. New York–Amsterdam, 1966)
8. A.A. Abrikosov, *Fundamentals of the Theory of Metals* (Elsevier Science Publishers, North-Holland, Amsterdam, 1988)
9. I.M. Lifshitz, M.Y. Azbel, M.I. Kaganov, *Electron Theory of Metals* (Consultants Bureau, New York, 1973)
10. J. M. Ziman, *The Calculation of Bloch Functions, Solid State Physics*. Advances in Research and Applications (Academic Press, New York and London, 1971)
11. C. Kittel, *Quantum Theory of Solids* (Wiley, New York–London, 1963)
12. W.A. Harrison, *Solid State Theory* (McGrow-Hill Book Company, New York–London–Toronto, 1970)
13. J.M. Ziman (ed.), *Physics of Metals*. Electrons, vol. 1 (Cambridge at the University Press, 1969)
14. N.W. Aschcroft, N.D. Mermin, *Solid State Physics* (Holt, Rinehart and Winston, 1976)
15. S.V. Vonsovskii, M.I. Katsnelson, *Quantum Solid-State Physics*. Springer Series in Solid-State Sciences (Berlin—Heidelberg–New York, 1989)
16. L.D. Landau, E.M. Lifshitz, *Quantum Mechanics, Non-relativistic Theory* (Pergamon Press, New York, 1977)
17. L.D. Landau, E.M. Lifshitz, *Statistical Physics, Part I* (Butterworth-Heinemann, 1980)
18. D. Hartree, *The Calculation of Atomic Structures* (Wiley, New York, 1957)
19. V.A. Fock, Näherungsmethode zur Lösung des quantenmechanischen Mehrkörperproblems. Z. Angew. Phys. **61**, 126 (1930) (in German)
20. P. Fulde, *Electron Correlations in Molecules and Solids*. Springer Series in Solid-State Sciences (1993)
21. D.I. Khomskii, *Transition Metal Compounds* (Cambridge University Press, 2014)
22. M.Y. Kagan, *Modern Trends in Superconductivity and Superfluidity*. Lecture Notes in Physics, vol. 874. (Springer, Dordrecht, 2013)
23. G. Bashkaran, A.E. Ruckenstein, E. Tosatti, Y. Lu (eds.) *Strongly-Correlated Electron Systems*. Progress in High Temperature Superconductivity (World Scientific, 1990)
24. M.Y. Kagan, K.I. Kugel, A.L. Rakhmanov, A.O. Sboychakov, *Electronic Phase Separation in Magnetic and Superconducting Materials*. Recent Advances, Springer Series in Solid-State Sciences, vol. 201 (2024)
25. E.G. Brovman, Y. Kagan, Phonons in non-transition metals. Sov. Phys. Uspekhi **112**, 369 (1974)
26. L.H. Thomas, The calculation of atomic fields. Math. Proc. Cambridge Philos. Soc. **23**, 542 (1927)
27. E. Fermi, Un Metodo Statistico per la Determinazione di Alcune Proprieta dell' Atomo. Rend. Accad. Naz. Lincei **6**, 602 (1927) (in Italian)

28. C.M. Varma, P.B. Littlewood, S. Schmitt-Rink, E. Abrahams, A.E. Ruckenstein, Phenomenology of the normal state of Cu-O high temperature superconductors. Phys. Rev. Lett. **63**, 1996 (1989)
29. P.W. Anderson, Luttinger liquid behavior of the normal state of the 2D Hubbard model. Phys. Rev. Lett. **64**, 1839 (1990)
30. V.M. Galitskii, Sov. Phys. JETP **7**, 104 (1958)
31. J.M. Luttinger, J.C. Ward, Ground state energy of a Many-Fermion system. Phys. Rev. **118**, 1417 (1960)
32. J.M. Luttinger, Fermi surface and some simple equilibrium properties of a system of interacting fermions. Phys. Rev. **119**, 1153 (1960)
33. L.D. Landau, E.M. Lifshitz, *Mechanics* (Butterworth-Heinemann, 1976)
34. V.B. Berestetskii, E.M. Lifshitz, L.P. Pitaevskii, *Quantum Electrodynamics* (Butterworth-Heinemann, 1982)

Abstract

In the first part of this Lecture, we analyze electron motion and effective mass of electron in magnetic field. We consider also quantum statistics of electrons in metal and electron density of states. We discuss the band filling of metals in uncorrelated case. We introduce the double-well potential for the description of the metal–insulator transition in strongly correlated Mott–Hubard regime. We find specific heat of electron gas in metal and compare electron and phonon contributions. In the end of the Lecture, we describe Lifshitz topological phase transition.

21.1 Electron Motion in Magnetic Field

Hamiltonian of electron in magnetic field has the form [1–4]:

$$ H = \frac{1}{2m} \left(\hat{p} - \frac{e}{c} A \right)^2 + V(r), \tag{21.1} $$

where A is vector potential.

In Landau gauge [2]:

$$ A = \frac{1}{2}[Hr], \tag{21.2} $$

and, correspondingly, the magnetic field:

$$ H = \operatorname{curl} A. \tag{21.3} $$

Let us emphasize, that magnetic field is practically always a quasiclassical one [1–4]. Indeed, the wave packet has a width of the order of the interatomic distance, while all the wave vectors are of the order of the Fermi momentum. Thus:

$$ q \sim q_F \sim \frac{1}{a}. \tag{21.4} $$

© The Author(s), under exclusive license to Springer Nature Switzerland AG 2026

397

M. Kagan, *Lecture Notes of Professor Yuri Kagan in Theoretical Solid-State Physics*, Lecture Notes in Physics 1048, https://doi.org/10.1007/978-3-032-14621-2_21

Hence, the term with vector potential in the electron kinetic energy in the Hamiltonian (21.1) yields by order of magnitude:

$$\frac{e}{c}A \sim \frac{e}{c}Ha. \tag{21.5}$$

At the same time the momentum:

$$p \sim \frac{\hbar}{a}, \text{ and } a \sim \frac{\hbar}{p}. \tag{21.6}$$

Then the ratio of these two terms in Eq. (21.1):

$$\frac{\hbar e}{c}Ha\frac{1}{\hbar p} \sim \frac{\hbar eH}{c}\frac{1}{p^2} \sim \hbar\frac{eH}{m^*c}\frac{m^*}{p^2} \sim \frac{\hbar\omega_c}{\varepsilon_F} \ll 1, \tag{21.7}$$

where the typical value of the cyclotron frequency in metal:

$$\hbar\omega_c = \hbar\frac{eH}{m^*c} \sim 1\text{K}. \tag{21.8}$$

Note that the typical value of the Fermi energy:

$$\varepsilon_F \sim \left(10^3 \div 10^4\right)\text{K}. \tag{21.9}$$

Thus, vector potential varies weakly on the size a. Hence in zero approximation we can consider magnetic field to be constant and its space variation can be accounted in the next approximation only.

Then in zero approximation we can replace the momentum p on the generalized momentum $p - \frac{e}{c}A$, and write for electron spectrum [1–6]:

$$\varepsilon_n(p) \to \varepsilon_n\left(p - \frac{e}{c}A\right). \tag{21.10}$$

Let us stress, that from Eq. (21.10) we can get the Lorentz force in crystal [1–8]:

$$\dot{p} = -\frac{\partial\varepsilon_n\left(p - \frac{e}{c}A\right)}{\partial r} = -\frac{\partial\varepsilon_n\left(p - \frac{e}{c}A\right)}{\partial p}\frac{\partial\left(p - \frac{e}{c}A\right)}{\partial r} = \frac{e}{c}[vH], \tag{21.11}$$

where:

$$v = \frac{\partial\varepsilon_n}{\partial p} \tag{21.12}$$

is the group velocity.

In fact, Eq. (21.11) describes the motion of the wave packet with the group velocity $v = v_{\text{gr.}}$.

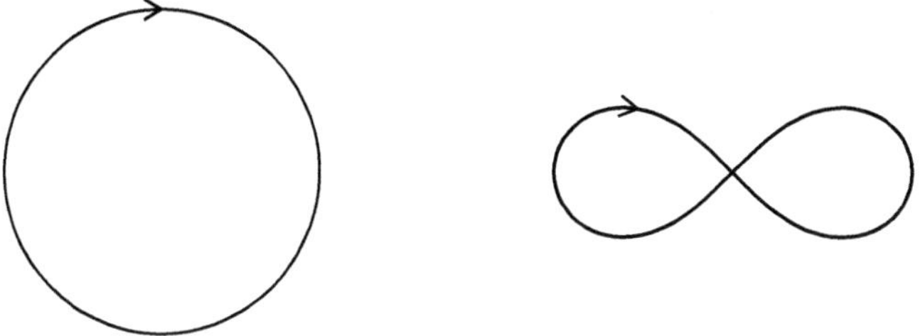

Fig. 21.1 Different trajectories for electron motion in the magnetic field in crystal [1]

Two important consequences follow from this equation. Firstly:

$$\dot{\boldsymbol{p}}H = 0, \tag{21.13}$$

and correspondingly the momentum projection on the direction of the magnetic field is conserved:

$$p_z = \text{const}, \tag{21.14}$$

where $z - axis \| \boldsymbol{H}$.

Secondly:

$$\dot{\boldsymbol{p}}\boldsymbol{v} = 0, \tag{21.15}$$

and thus, the energy ε_n is conserved.

Hence, we get:

$$\varepsilon_n \left(p_x, p_y, p_z = \text{const} \right) = \varepsilon = \text{const}. \tag{21.16}$$

Geometrically Eq. (21.16) describes the curve, which is related to the trajectory of the particle motion in magnetic field (see Fig. 21.1):

$$p_y = f\left(p_x\right)\big|_{p_z,\varepsilon}. \tag{21.17}$$

These trajectories can be quite complex and even non-closed [1,6]. Note, that if the trajectory is non-closed in momentum space, then it is non-closed in real space as well.

21.2 Effective Mass in Magnetic Field

To calculate effective mass let us use equation of motion for the Lorenz force (21.11) in plane perpendicular to magnetic field. Then for electron motion along the trajectory, the time derivative from the tangential component of the momentum p_τ acquires the form (see Fig. 21.2a):

$$\frac{dp_\tau}{dt} = \frac{e}{c}v_\perp H, \tag{21.18}$$

where $v_\perp$ is normal (to the trajectory) component of the group velocity.

By dividing the variables, we get:

$$dp_\tau = \frac{e}{c} v_\perp H dt, \ \text{or} \ \frac{dp_\tau}{\frac{e}{c} v_\perp H} = dt. \tag{21.19}$$

Hence in agreement with the Classical mechanics [8], rotation period of electron:

$$T = \frac{c}{eH} \oint \frac{dp_\tau}{v_\perp}. \tag{21.20}$$

But since rotation frequency equals to the cyclotron frequency (21.8) in magnetic field [1–8], we obtain:

$$\omega_C = \frac{eH}{m^* c} = \frac{2\pi}{T}. \tag{21.21}$$

Hence, effective mass is given by:

$$m^* (p_Z, \varepsilon) = \frac{1}{2\pi} \frac{eH}{c} T = \frac{1}{2\pi} \oint \frac{dp_\tau}{v_\perp}. \tag{21.22}$$

Generally speaking, effective mass depends in magnetic field on two conserving variables p_z, ε (for each of the cross sections and electron trajectory in this cross section). Thus, in Eq. (21.22):

$$p_z = \text{const} \ \text{and} \ \varepsilon = \text{const}.$$

Let us consider two close trajectories lying in the same plane perpendicular to magnetic field (see Fig. 21.2b). These trajectories correspond to the same value of $p_z = \text{const}$ and close values of $\varepsilon = \text{const}$ and $\varepsilon + d\varepsilon = \text{const}$. Each of these trajectories bounds the area $S (\varepsilon, p_z)$ and $S (\varepsilon + d\varepsilon, p_z)$. Smaller area bounded by the trajectory with $p_z = \text{const}$ and $\varepsilon = \text{const}$ is shaded by the dotted line in the figure.

Larger area contains additional unshaded belt. It is convenient to represent the area element for this belt as follows:

$$dp_\tau dp_\perp = dp_\tau \frac{dp_\perp}{d\varepsilon} d\varepsilon = \frac{dp_\tau}{v_\perp} d\varepsilon. \tag{21.23}$$

Then the total area of the belt is given by the formula:

$$dS = S (\varepsilon + d\varepsilon, p_z) - S (\varepsilon, p_z) = d\varepsilon \oint \frac{dp_\tau}{v_\perp}. \tag{21.24}$$

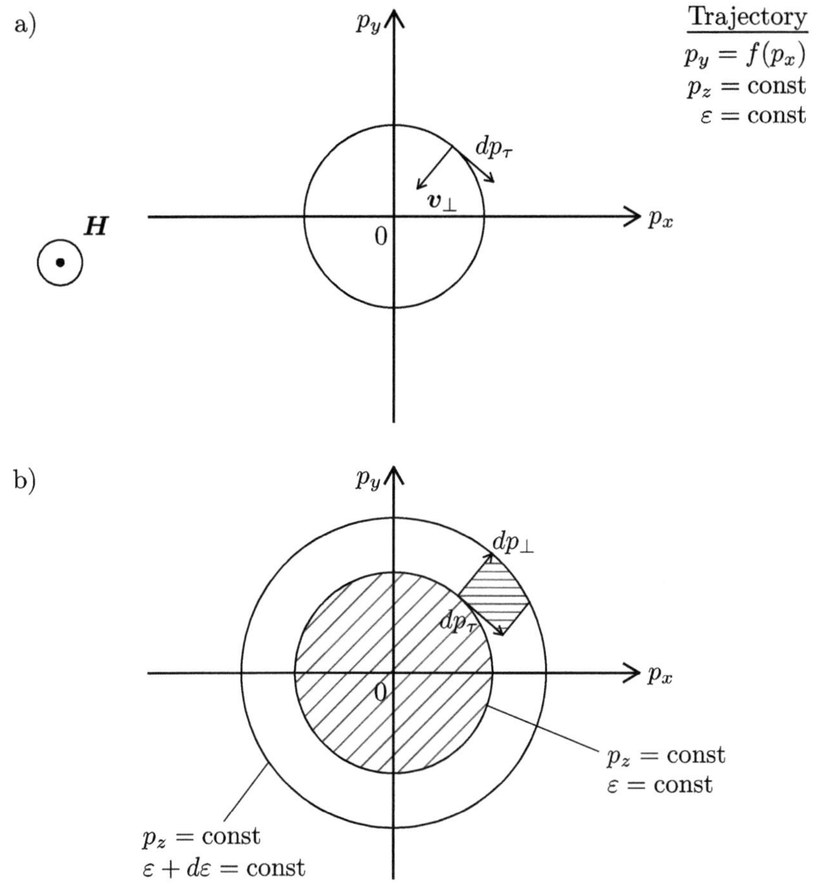

Fig. 21.2 Electron trajectory in magnetic field in crystal in the plane perpendicular to the magnetic field $p_y = f(p_x)$ for $p_z = \text{cost}$ and $\varepsilon = \text{const}$ (**a**) and the area bounded by the trajectory in this plane (**b**). On **a** we show normal (to the trajectory) component of the group velocity $v_\perp$ and differential of the tangential momentum component dp_τ. On **b** we illustrate the increase of the area bounded by the trajectory with $p_z = \text{const}$ and $\varepsilon + d\varepsilon = \text{const}$. This increase corresponds to the unshaded belt. We show also the rectangular element of the belt area $dp_\tau dp_\perp = \frac{dp_\tau}{v_\perp} d\varepsilon$ [1]

Comparing the Formulas (21.22) and (21.24), we can find one more very useful expression for electron effective mass in magnetic field [1–4]:

$$m^* (p_z, \varepsilon) = \frac{1}{2\pi} \frac{\partial S}{\partial \varepsilon}. \tag{21.25}$$

Equation (21.25) shows, that if the area is increased, then $m^* > 0$ and we get the standard cyclotron mass.

If, however, the area is decreased, then electron (and not a hole) starts circling as positron.

Fig. 21.3 The surfaces of constant energy in magnetic field in the elementary cell of the inverse crystal lattice. The area under the trajectory at first increases with the increase of electron filling of the cell and then starts to decrease since the belt perimeter $L = \oint dp_\perp$ decreases [1]

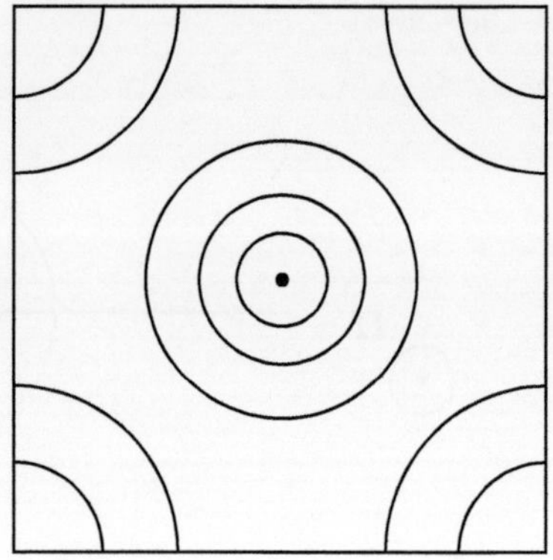

In the limit of one cell of the inverse lattice, the area under the trajectory at first increases with the increase of electron filling, and then starts to decrease since the belt perimeter $L = \oint dp_\perp$ is decreasing (see Fig. 21.3).

21.3   Quantum Statistics of Electrons in Metal

At temperature $T = 0$ the condition for determining the boundary electron energy in metal has the form [1–4,9]:

$$N = \frac{2V}{(2\pi)^3} \sum_n \int d^3q\, \theta\, (\mu - \varepsilon_n(\boldsymbol{q})), \qquad (21.26)$$

where the Dirac theta function [9]:

$$\theta(x) = \begin{cases} 1, & \text{for } x > 0, \\ 0, & \text{for } x < 0. \end{cases} \qquad (21.27)$$

Factor 2 in Eq. (21.26) is related to the summation over spin projections, while the chemical potential in non-interacting electron gas equals to the Fermi energy:

$$\mu = \varepsilon_F. \qquad (21.28)$$

In fact, equation

$$\varepsilon_n(p) = \mu = \varepsilon_F \qquad (21.29)$$

determines the Fermi surface in momentum space for electrons in metal.

In general case, the Fermi surface represents the complex and often even multiconnected surface in momentum space (see for example the classical monograph of Lifshitz et al. [4]).

All the thermodynamics and kinetics of electrons in metal is determined by the neighborhood of the Fermi surface [1–4]. If the spectrum is ordinary (as in the isotropic crystal), then the Fermi surface is a sphere [9].

Let us see what maximal number of electrons can be in one band. Let us consider the simplest case of one atom on elementary cell. In this case, after integration in momentum space over the first cell of the inverse lattice, we get:

$$V \int \frac{d^3 q}{(2\pi)^3} = \frac{V}{\Omega_0} = N_{\text{elem. cells}}.$$ (21.30)

Correspondingly, according to the Pauli principle [2,9], the maximal number of electrons in the band yields [1–4]:

$$N_{\text{max.el.}} = 2 N_{\text{elem. cells}}.$$ (21.31)

Thus, if the number of electrons on elementary cell is odd, then definitely the lower bands are not completely filled [1–4, 10–13].

If, however, the number of electrons is even, then either all the bands are completely filled or some of them are completely empty.

If all the bands are filled, then this substance is insulator [14–16]. Note that breakdown in a dielectric is associated with electron knocking out of the zone.

If the number of electrons is odd or if the bands are overlapping, and thus are not completely filled, then we get a metal [1–4, 10–13] (at least in weak-coupling approximation).

Semiconductor [1, 17, 18] is dielectric [10–13] with small forbidden band.

The valency of the element equals to the number of collectivized electrons in the cell (if there is one atom in the cell).

The maximal number of electrons in one band in case of one atom on the cell is given by:

$$2 N_{\text{elem. cells}} = 2 N_{\text{atoms}}.$$ (21.32)

If each atom gives one electron in collective usage, then the total number of collectivized electrons reads:

$$N_{\text{el.}} = N_{\text{atoms}} = \frac{N_{\text{max.el.}}}{2},$$ (21.33)

and as a result, the band is half filled. This is the case of the valency equal to 1.

At the same time, if the valency equals to 2, then the band is completely filled:

$$N_{\text{el.}} = 2 N_{\text{atoms}} = N_{\text{max.el.}}$$ (21.34)

21.4 Strong-Coupling Approach to the Formation and Filling of the Bands: Mott–Hubbard Metal–Insulator Transition

If we consider now the double-well problem [1, 2, 19–22], then in the strong-coupling approximation we obtain the splitting of electron level on 2 levels which equals to Δ_0 (see Fig. 21.4). This splitting is related to the tunneling process. If the number

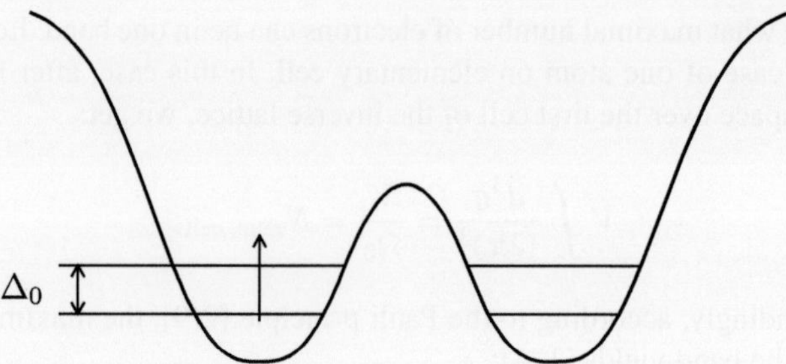

Fig. 21.4 Splitting Δ_0 of electron level on two levels related to the tunneling [1]

of the nearest neighbors on the lattice equals to z, then the level splitting (actually conduction band half-width) yields [19–22]:

$$\Delta = z\Delta_0. \tag{21.35}$$

Note, that in the strong-coupling case electron has a quasi-atomic function [1,10–13]. In principle, we can put one more electron with opposite spin projection in the potential well (there is no contradiction with the Pauli principle in this case). However, the strong onsite Coulomb repulsion (from the second electron) prevents this opportunity.

Thus, on each level there is only one electron and we get the metal.

If we forget about Coulomb repulsion and put 2 electrons in the elementary cell, then the Coulomb repulsion throws them above the conduction band. It will be a dielectric. Correspondingly, effective gap arises in the system. This gap by order of magnitude equals to the onsite (intracell) Coulomb interaction U [23,24].

Let us increase the width of the conduction band till the values:

$$\Delta > 2\varepsilon + U, \tag{21.36}$$

where ε is kinetic energy of each of two electrons. Then electrons remain in the conduction band and according to Hubbard [23,24] we get the insulator–metal transition.

Let us stress, that the condition of Hubbard localization (the condition for the dielectric behavior of the half-filled band) has the form [23,24];

$$\Delta/|U| \sim 1. \tag{21.37}$$

Hubbard localization (or Mott–Hubbard localization [23–26] as it is commonly known) plays the crucial role in the modern physics of superconductivity and strongly correlated electron systems [27–29].

21.5   Work of Electron Escape from Metal

As we already discussed in this Lecture, the Fermi surface for electrons in metal can be found from the condition [1,3,4]:

$$\varepsilon(p) = \mu,$$

where the chemical potential μ is related to the energy variation when we add one particle to the system.

At the temperature $T = 0$ this particle cannot sit lower than the Fermi surface. Hence, the chemical potential counted from zero energy, is given by: $\mu = \varepsilon_F$. At the same time, electron spectrum:

$$\tilde{\varepsilon}(\boldsymbol{p}) = \varepsilon_0 + \varepsilon(\boldsymbol{p}), \tag{21.38}$$

where $\varepsilon_0 < 0$ is negative constant contribution to the energy (which confines electrons in metal), while electron energy counted from ε_0 yields:

$$\varepsilon(\boldsymbol{p}) = \frac{p^2}{2m_*}. \tag{21.39}$$

Correspondingly, the chemical potential related to the spectrum (21.39) reads:

$$\tilde{\mu} = \varepsilon_0 + \varepsilon_F. \tag{21.40}$$

Thus, the difference between electron energy and chemical potential still has the standard form:

$$\tilde{\varepsilon}(\boldsymbol{p}) - \tilde{\mu} = \frac{p^2}{2m_*} - \varepsilon_F. \tag{21.41}$$

Work of electron escape from metal (the work to extract electron sitting at the Fermi level out of metal) equals to electron binding energy [1,12,30] (see Fig. 21.5):

$$A = |\varepsilon_0| - \varepsilon_F. \tag{21.42}$$

We can say, that electron excitations in metal have the constant energy padding (constant energy shell) [1].

Thermoelectric emission (the probability of electron escape from metal) is determined by the thermo-activation exponent [1]:

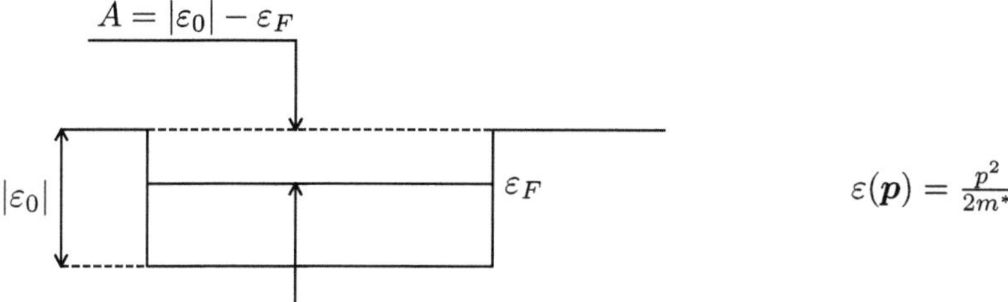

Fig. 21.5 Fermi energy ε_F and the work of electron escape from metal: $A = |\varepsilon_0| - \varepsilon_F$ [1]

$$\exp\left(-\frac{A}{T}\right), \tag{21.43}$$

where the work: $A = |\varepsilon_0| - \varepsilon_F$.

21.6   Thermodynamics of Electron Gas in Metal

To find electron contribution to the specific heat in metal we should solve in the first approximation the thermodynamic problem for the degenerate Fermi gas in the absence of the interaction, but with arbitrary electron dispersion law.

For temperatures much lower than Fermi energy $T \ll \varepsilon_F$, the chemical potential weakly depends on temperature, and if we neglect small corrections (quadratic in temperature) $\mu \approx \varepsilon_F$.

Note that in metal for all the temperatures up to melting temperature the degeneracy condition is fulfilled for the Fermi gas of electrons:

$$\frac{T}{\varepsilon_F} \ll 1. \tag{21.44}$$

As a result, the thermodynamic potential equals to:

$$\begin{aligned}
\Omega(\mu, T, V) &= -VT \sum_n \sum_\sigma \int \frac{d^3 \boldsymbol{p}}{(2\pi\hbar)^3} \ln\left(1 + \exp\left(\frac{\mu - \varepsilon_n(\boldsymbol{p})}{T}\right)\right) \\
&= -VT \sum_{\sigma n} \int d\varepsilon g_n(\varepsilon) \ln\left(1 + \exp\left(\frac{\mu - \varepsilon}{T}\right)\right),
\end{aligned} \tag{21.45}$$

where we introduced important quantity $g_n(\varepsilon)$ for electron density of states and made the replacement:

$$\frac{d^3 \boldsymbol{p}}{(2\pi\hbar)^3} \to g_n(\varepsilon)d\varepsilon. \tag{21.46}$$

Further on performing summation over spin projections and using notation:

$$\sum_n g_n(\varepsilon) = g(\varepsilon) \tag{21.47}$$

for the total density of states, we can write the thermodynamic potential in the following form:

$$\Omega(\mu, T, V) = -2VT \int_0^\infty d\varepsilon g(\varepsilon) \ln\left(1 + \exp\left(\frac{\mu - \varepsilon}{T}\right)\right), \tag{21.48}$$

where $\sum_\sigma = 2$.

21.7 Density of States of Electron Gas

In the first part of the Course, we introduced the density of states for various branches of the phonon spectrum according to the relation:

$$\Omega_0 \frac{d^3 p}{(2\pi \hbar)^3} = g_n(\varepsilon) d\varepsilon, \tag{21.49}$$

where Ω_0 is the volume of the elementary cell.

Analogously the density of states of the unit volume for the electron spectrum according to Eq. (21.46) reads:

$$\frac{d^3 p}{(2\pi \hbar)^3} = g_n(\varepsilon) d\varepsilon.$$

It is convenient to write electron density of states similarly to the phonon density of states as:

$$g_n(\varepsilon) = \int \frac{d^3 p}{(2\pi \hbar)^3} \delta \left(\varepsilon - \varepsilon_n\right). \tag{21.50}$$

The density of states for electron gas is normalized on the number of elementary cells (on the maximal number of electrons with one spin projection in the band):

$$V \int d\varepsilon g_n(\varepsilon) = N. \tag{21.51}$$

Hence the total density of states in agreement with Eq. (21.47):

$$\sum_n g_n(\varepsilon) = g(\varepsilon) = \frac{1}{V} \frac{\partial N_{\text{el.}}}{\partial \varepsilon} = \frac{\partial n_{\text{el.}}}{\partial \varepsilon}, \tag{21.52}$$

where $N_{\text{el.}}$ and $n_{\text{el.}}$ are the total electron number and total density on one spin projection (in all the bands).

Note, that the complex law of electron dispersion sits just in the density of states. In particular, electron density of states as well as the phonon one contains Van Hove singularities [31] for the saddle points of the spectrum.

We can say, that Van Hove singularities are transferred from phonons to electrons since in both cases these singularities are related to the crystal [1,31].

Note also, that in Lectures 17 and 18 we used another notation instead of $g(\varepsilon)$ for the density of states in case of the quadratic spectrum in free space, namely [32]:

$$N(0) = g(\varepsilon_F) = \frac{mk_F}{2\pi^2 \hbar^2}, \tag{21.53}$$

where the energy $\varepsilon - \varepsilon_F$ in the argument of N ($\varepsilon - \varepsilon_F = 0$) is counted from ε_F.

This notation is also quite common for the description of the Fermi systems.

21.8 Specific Heat of Electron Gas

Let us return back now to the expression for the thermodynamic potential in Eq. (21.48) and perform the forthcoming calculations using integration by parts. Then:

$$\Omega(\mu, T, V) = -2\,VTZ(\varepsilon)\ln\left(1 + \exp\left(\frac{\mu - \varepsilon}{T}\right)\right)\bigg|_0^\infty$$

$$+ 2\,VT\int_0^\infty d\varepsilon\, Z(\varepsilon)\frac{\exp\left(\frac{\mu-\varepsilon}{T}\right)}{1 + \exp\left(\frac{\mu-\varepsilon}{T}\right)}\left(-\frac{1}{T}\right), \tag{21.54}$$

where:

$$Z(\varepsilon) = \int_0^\varepsilon d\varepsilon'\, g\left(\varepsilon'\right). \tag{21.55}$$

But for $\varepsilon = 0 : Z(0) = 0$. At the same time, for $\varepsilon \to \infty : \exp\left(\frac{\mu-\varepsilon}{T}\right) \to 0$, and $\ln\left(1 + \exp\left(\frac{\mu-\varepsilon}{T}\right)\right) \to \ln(1 + 0) = 0$. Hence the first term in the r.h.s. of Eq. (21.54) equals zero. As a result:

$$\Omega(\mu, T, V) = -2V\int_0^\infty d\varepsilon\, Z(\varepsilon)\frac{\exp\left(\frac{\mu-\varepsilon}{T}\right)}{1 + \exp\left(\frac{\mu-\varepsilon}{T}\right)}$$

$$= -2V\int_0^\infty d\varepsilon\, Z(\varepsilon)\frac{1}{1 + \exp\left(\frac{\varepsilon-\mu}{T}\right)}$$

$$= -2V\int_0^\infty Z(\varepsilon)d\varepsilon\, n_F(\varepsilon), \tag{21.56}$$

where:

$$n_F(\varepsilon) = \frac{1}{1 + \exp\left(\frac{\varepsilon-\mu}{T}\right)} \tag{21.57}$$

is the Fermi–Dirac distribution function for electrons. At temperatures different from zero, the distribution function close to the Fermi level is smeared in the energy interval of order of T. Thus, $n_F(\varepsilon) \to 0$ for $\varepsilon - \mu \geq T$, and at the same time $n_F(\varepsilon) \to 1$ for $\mu - \varepsilon \geq T$ (see Fig. 21.6).

Correspondingly, we get the important relation for the integrals with the distribution function at low temperatures $\mu \gg T$ (see, for example, the similar calculation in paragraph 58 of the Landau–Lifshitz textbook *Statistical Physics. Part 1* [9]):

$$\int_0^\infty f(\varepsilon)d\varepsilon\, n_F(\varepsilon) = \int_0^\mu f(\varepsilon)d\varepsilon + \frac{\pi^2}{6}T^2 f'(\mu). \tag{21.58}$$

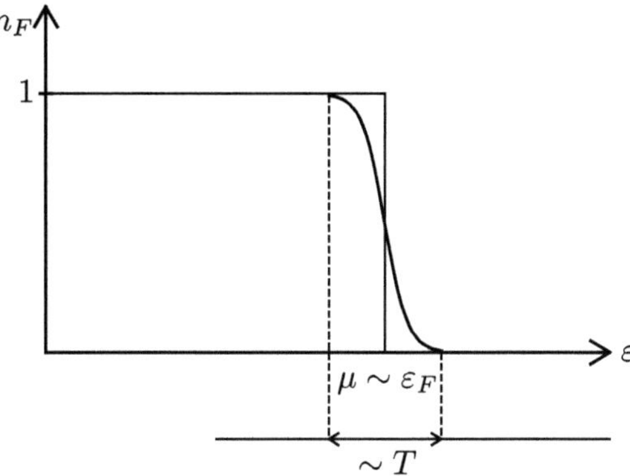

Fig. 21.6 Qualitative behavior of the Fermi–Dirac distribution function at finite temperatures. In this case the Dirac step function is smeared close to the Fermi level in the interval of the order of T. Thus, $n_F(\varepsilon) \to 0$ for $\varepsilon - \mu \geq T$, and at the same time $n_F(\varepsilon) \to 1$ for $\mu - \varepsilon \geq T$ [1]

Let us stress, that deriving Eq. (21.58) we took into account exact cancelation of the terms linear in T.

Consequently, the thermodynamic potential is given by:

$$\Omega(\mu, T, V) = -2V \int_0^\mu Z(\varepsilon)d\varepsilon - V \frac{\pi^2}{3} T^2 Z'(\mu), \qquad (21.59)$$

where:

$$Z'(\varepsilon) \equiv \frac{dZ}{d\varepsilon} = g(\varepsilon). \qquad (21.60)$$

Then finally:

$$\Omega(\mu, T, V) = -2V \int_0^\mu Z(\varepsilon)d\varepsilon - V \frac{\pi^2}{3} T^2 g(\mu). \qquad (21.61)$$

Hence the entropy of the electron gas in metal reads:

$$S = -\left(\frac{\partial \Omega}{\partial T}\right)_{\mu, V} = V \frac{2\pi^2}{3} T g(\mu). \qquad (21.62)$$

Correspondingly, specific heat yields:

$$C = T \frac{dS}{dT} = V \frac{2\pi^2}{3} T g(\mu). \qquad (21.63)$$

We get very important relation according to which electron specific heat linearly depends on temperature for arbitrary dispersion law. We should demand only the degeneracy of electron gas.

Actually, everything is determined by the Fermi surface where the damping is absent for electron excitations. Thus, close to the Fermi surface one-particle description works pretty well.

Let us remind, that phonon specific heat in the three-dimensional crystal $C_{\text{ph.}} \sim T^3$ is proportional to the third power of temperature. Hence, specific heat for sufficiently low temperatures is determined by electrons.

At the same time at high temperature (comparing to Debye temperature):

$$C_{\text{ph.}} = 3N_{\text{nucl.}} = \text{const} \quad (k_B = 1).$$

Let us stress that electron specific heat is proportional to the number of electrons in the energy shell of the order of temperature (see Fig. 21.7). Since their number $N_{\text{eff.el.}} \sim N_{\text{el.}} \frac{T}{\mu} \sim N_{\text{nucl.}} \frac{T}{\mu} \ll N_{\text{nucl.}}$, then as a result:

$$C_{\text{el.}} \sim N_{\text{el.}} \frac{T}{\mu} \ll C_{\text{ph.}} \tag{21.64}$$

for $\theta_D < T \ll \varepsilon_F$. In this temperature region everything is determined by phonons.

Problem to the Lecture. Let us find the temperature at which electron and phonon specific heat become equal to each other.

In the Debye approximation the energy of the phonon gas for the three-dimensional crystal with the density of states $g(\omega) = \frac{3\omega^2}{\omega_D^3}$ (see Lecture 9 in the first part of the Course) yields:

$$E_{\text{phonon}}(T) = 3N \int_0^\infty d\omega \frac{3\omega^2}{\omega_D^3} \frac{\hbar\omega}{\exp\left(\frac{\hbar\omega}{T}\right) - 1} = 3N \frac{3}{(\hbar\omega_D)^3} T^4 \int_0^\infty \frac{dx\, x^3}{\exp(x) - 1}, \tag{21.65}$$

where the tabulated integral in (21.65) equals to the product of the Gamma function and Riemann zeta function [1–9]:

$$\int_0^\infty \frac{dx\, x^3}{\exp(x) - 1} = \Gamma(3)\zeta(3) = 2 \cdot 1.202 = 2.404. \tag{21.66}$$

Fig. 21.7 Specific heat of electron gas is proportional to the number of electrons living in the energy shell of the order of temperature close to the Fermi surface [1]

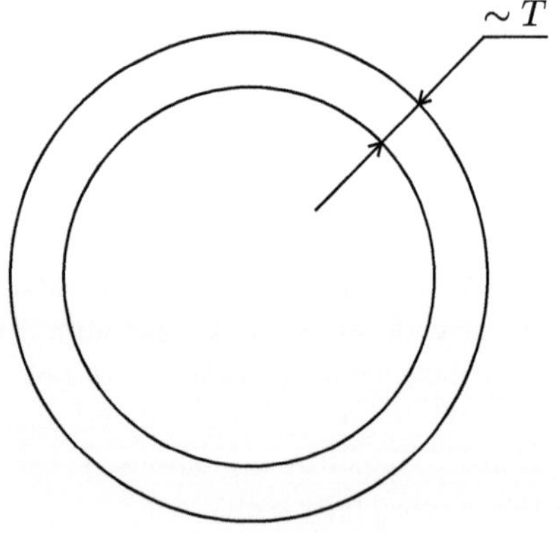

Hence the energy of the phonon gas:

$$E_{\text{phonon}}(T) = 2.404 \cdot \frac{9N}{(\hbar\omega_D)^3} T^4, \tag{21.67}$$

and the phonon specific heat:

$$C_{\text{ph.}} = 9.62 \cdot 9N \cdot \frac{T^3}{(\hbar\omega_D)^3} = 86.6\,N \cdot \frac{T^3}{(\hbar\omega_D)^3}. \tag{21.68}$$

At the same time electron specific heat for the quadratic spectrum equals to: (see Eqs. (21.52)–(21.53)):

$$C_{\text{el.}} = V \frac{2\pi^2}{3} T g(\mu) = V \frac{2\pi^2}{3} T \frac{\partial n_{\text{el.}}}{\partial \varepsilon}\bigg|_{\varepsilon=\mu}, \tag{21.69}$$

where electron density on one spin projection $n_{\text{el.}}\,(\mu = \varepsilon_F) = \frac{k_F^3}{6\pi^2}$, while the Fermi energy $\varepsilon_F = \frac{\hbar^2 k_F^2}{2m_*}$. Correspondingly, $n_{\text{el.}}\,(\mu = \varepsilon_F) \sim \mu^{3/2}$, and:

$$\frac{\partial n_{\text{el.}}}{\partial \varepsilon}\bigg|_{\varepsilon=\mu} = \frac{3n_{\text{el.}}}{2\varepsilon_F}. \tag{21.70}$$

Thus, electron specific heat:

$$C_{\text{el.}} = V \frac{2\pi^2}{3} T \frac{3n_{\text{el.}}}{2\varepsilon_F} = \frac{2\pi^2}{3} T \frac{3N_{\text{el.}}}{2\varepsilon_F} = \pi^2 N_{\text{el.}} \frac{T}{\varepsilon_F}. \tag{21.71}$$

For metal with monovalent ion on the cell:

$$N_{\text{el.}} = N_{\text{nucl.}} = N. \tag{21.72}$$

Hence, $C_{\text{el.}} = C_{\text{ph.}}$ under the fulfillment of the condition:

$$86.6\,N \cdot \frac{T^3}{(\hbar\omega_D)^3} = \pi^2 N_{\text{el.}} \frac{T}{\varepsilon_F}, \ \text{ or } 86.6 \frac{T^3}{(\hbar\omega_D)^3} = \pi^2 \frac{T}{\varepsilon_F}. \tag{21.73}$$

Correspondingly:

$$\frac{86.6\varepsilon_F T^2}{\pi^2 (\hbar\omega_D)^3} = 8.8 \frac{\varepsilon_F T^2}{(\hbar\omega_D)^3} = 1. \tag{21.74}$$

As a result, electron and phonon specific heat become equal to each other at the temperature:

$$T_0 = \sqrt{0.11 \frac{(\hbar\omega_D)^3}{\varepsilon_F}} = 0.33\hbar\omega_D \sqrt{\frac{\hbar\omega_D}{\varepsilon_F}} \equiv 0.33\theta_D \sqrt{\frac{\theta_D}{\varepsilon_F}} \ll \theta_D. \tag{21.75}$$

At higher temperatures the phonons yield the main contribution to specific heat.

21.9   Phase Transition of the Order 2.5 (Topological Lifshitz Phase Transition)

In analogy with Lecture 9 of the Course (where we analyzed the phonon gas contribution to the coefficient of the thermal expansion) let us write the volume of the system and the entropy as the derivatives of the Gibbs thermodynamic potential [1,9], correspondingly, with respect to the pressure and temperature:

$$V = \frac{\partial \Phi}{\partial P}, \quad S = -\frac{\partial \Phi}{\partial T}. \tag{21.76}$$

Let us utilize now the equality of the second (crossed) derivative of the thermodynamic potential by temperature and pressure with respect to the reversal of the derivatives order [9]. Namely:

$$\frac{\partial V}{\partial T} = \frac{\partial^2 \Phi}{\partial P \partial T} = -\frac{\partial S}{\partial P}. \tag{21.77}$$

Let us substitute expression (21.63) for the entropy of electron gas in metal:

$$S = V \frac{2\pi^2}{3} T g(\mu)$$

in this relation.

Then, for the derivative of the volume by temperature:

$$\begin{aligned}
\frac{\partial V}{\partial T} &= -\frac{\partial S}{\partial P} = -\frac{\partial V}{\partial P}\frac{2\pi^2}{3} T g(\mu) - \frac{2\pi^2}{3} T V \frac{\partial g(\mu)}{\partial P} \\
&= -\frac{2\pi^2}{3} T g(\mu)\frac{\partial V}{\partial P} - \frac{2\pi^2}{3} T V \frac{\partial g(\mu)}{\partial \mu}\frac{\partial \mu}{\partial P},
\end{aligned} \tag{21.78}$$

where the chemical potential with an account of the electron binding energy in metal:

$$\mu = \varepsilon_0 + \varepsilon_F. \tag{21.79}$$

At the same time, the derivative of the chemical potential by pressure in Eq. (21.78):

$$\frac{\partial \mu}{\partial P} = \frac{\partial \mu}{\partial V}\frac{\partial V}{\partial P}. \tag{21.80}$$

Hence, we get for the electron contribution to the thermal expansion:

$$\frac{\partial V}{\partial T} = -\frac{2\pi^2}{3} T \frac{\partial V}{\partial P}\left(g(\mu) + V \frac{\partial \mu}{\partial V}\frac{\partial g(\mu)}{\partial \mu}\right), \tag{21.81}$$

where $\left(-\frac{\partial V}{\partial P}\right)$ is inverse compressibility of the system.

As we already discussed, electron density of states similarly to phonon density of states contains Van Hove singularity [1,31].

In the three-dimensional crystal Van Hove singularity in the electron density of states has a square root character [1,31] (see Lecture 7 in the first part of the Course). If the density of states:

$$g(\mu) \sim \sqrt{|\mu - \varepsilon_0|}, \tag{21.82}$$

then its derivative diverges root-wise:

$$\frac{\partial g(\mu)}{\partial \mu} \sim \frac{1}{\sqrt{|\mu - \varepsilon_0|}}. \tag{21.83}$$

Correspondingly, the derivative $\frac{\partial V}{\partial T}$ experiences a jump for $\mu = \varepsilon_0$.

The root character of singularity in the electron density of states is associated with the opening of a new electron band (see Fig. 21.8):

$$g(\varepsilon) \sim \sqrt{\varepsilon - \varepsilon_1}, \tag{21.84}$$

where the energy level ε_1 is related to the bottom of a new band.

This type of singularity in the density of states corresponds to the phase transition of the order of 2.5. In modern terminology this phase transition is often referred to as Lifshitz topological phase transition [1–4,33,34].

We can say, that the emergence or the disappearance of the new band leads to the singularity in the density of states of the type of the one given by Eq. (21.84), that is to the phase transition. Thus, creation of the new band is equivalent to the pressure increase or to the emergence of the band overlap.

In the experiment it is the increase of pressure which usually leads to Lifshitz topological phase transition [1,4,33]. In particular, as a result of the new band opening, we can get the insulator–metal phase transition.

Note that since the thermal expansion as we know from Lecture 8 is proportional to the specific heat of the system:

$$\frac{\partial V}{\partial T} = \left(-\frac{\partial V}{\partial P}\right)\frac{\beta C_{\text{el.}}}{V} \sim C_{\text{el.}}, \tag{21.85}$$

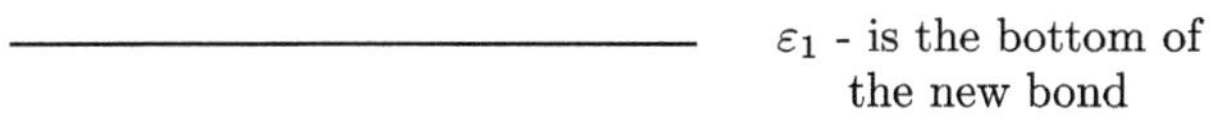

-- ε_1 - is the bottom of
the new bond

Fig. 21.8 The bottom of the new band ε_1 in Lifshitz topological phase transition [1]

then the electron specific heat at the constant volume also diverges under the opening of the new band.

Thus, electron specific heat can change essentially with pressure change (even though it would seem that the pressure is connected only with the lattice).

References

1. Y. Kagan, *Lectures on the Solid-State Theoretical Physics*, Moscow, MEPHI, 1981–1982. Part II. Electrons (Unpublished)
2. L.D. Landau, E.M. Lifshitz, *Quantum Mechanics: Non-relativistic Theory* (Pergamon Press, New York, 1977)
3. A.A. Abrikosov, *Fundamentals of the Theory of Metals* (Elsevier Science Publishers, North-Holland, Amsterdam, 1988)
4. I.M. Lifshitz, M.Y. Azbel, M.I. Kaganov, *Electron Theory of Metals* (Consultants Bureau, New York, 1973)
5. R. Peierls, On the theory of diamagnetism of conduction electrons. Z. Phys. **80**, 763 (1933)
6. D.R. Hofstadter, Energy levels and wave functions of Bloch electrons in rational and irrational magnetic fields. Phys. Rev. B **14**, 2239 (1976)
7. L.D. Landau, E.M. Lifshitz, *The Classical Theory of Fields* (Butterworth-Heinemann, 1994)
8. L.D. Landau, E.M. Lifshitz, *Mechanics* (Butterworth-Heinemann, 1976)
9. L.D. Landau, E.M. Lifshitz, *Statistical Physics, Part I* (Butterworth-Heinemann, 1980)
10. L.D. Landau, E.M. Lifshitz, *Electrodynamics of Continuous Media* (Pergamon, Oxford, 1984)
11. J.M. Ziman (ed.), *The Physics of Metals*. Electrons, vol. 1 (Clarendon at the University Press, 1969)
12. N.W. Ashcroft, N.D. Mermin, *Solid State Physics* (Rinehart and Winston, Holt, 1976)
13. W.H. Harrison, *Electronic Structure and the Properties of Solids* (W.H. Freeman and Company, San Francisco, 1980)
14. H. Frohlich, *Theory of Dielectrics* (Clarendon, Oxford, 1958)
15. M. Davies (ed.), *Dielectric and Related Molecular Processes*, vol. 1 (The Chemical Society, London, 1972)
16. B.K.P. Scaife, *Principles of Dielectrics* (Clarendon, Oxford, 1989)
17. P.Y. Yu, M. Cardona, *Fundamentals of Semiconductors*. Physics and Materials Properties (Springer, 2002)
18. K. Seeger, *Semiconductor Physics* (Springer, Wien-New York, 1973)
19. J. Kondo, A. Yoshimori (eds.), *Fermi Surface Effects*. Springer Series in Solid-State Sciences (1987)
20. E. Burstein, S. Lundkquist (eds.), *Tunneling Phenomena in Solids* (Plenum Press, New York, 1969)
21. Y. Kagan, N.V. Prokof'ev, Electron polaron effect and quantum diffusion of heavy particle in metals. Sov. Phys. JETP **63**, 1276 (1986)
22. Y. Kagan, N.V. Prokof'ev, Quantum tunneling in metals and the heavy fermion problems. Sov. Phys. JETP **66**, 211 (1987)
23. J. Hubbard, Proc. Royal Soc. Lond. **A276**, 238 (1963)
24. J. Hubbard, Proc. Roy. Soc. Lond. **A277**, 237 (1964)
25. N.F. Mott, *Metal–Insulator Transitions* (Taylor and Francis LTD, London, 1974)
26. N.F. Mott, E.A. Davis, *Electronic Processes in Non-crystalline Materials* (Oxford University Press, 1979)
27. P. Fulde, *Electron Correlations in Molecules and Solids*. Springer Series in Solid-State Sciences (1993)

28. P.W. Anderson, Luttinger liquid behavior of the normal state of the 2D Hubbard model. Phys. Rev. Lett. **64**, 1839 (1990)
29. G. Bashkaran, A.E. Ruckenstein, E. Tosatti, Y. Lu (eds.), *Strongly-Correlated Electron Systems. Progress in High Temperature Superconductivity* (World Scientific, 1990)
30. C. Kittel, *Quantum Theory of Solids* (Wiley, 1963)
31. L. Van Hove, The occurrence of singularities in the elastic frequency distribution of a crystal. Phys. Rev. **89**, 1189 (1953)
32. M.Y. Kagan, *Modern Trends in Superconductivity and Superfluidity*. Lecture Notes in Physics, vol. 874. (Springer, Dordrecht, 2013)
33. I.M. Lifshitz, Anomalies of electron characteristics of a metal in the high pressure region. Sov. Phys. JETP **11**, 1130 (1960)
34. G.E. Volovik, Quantum phase transitions from topology in momentum space. Springer Lect. Notes Phys. **718**, 31 (2007)

Lecture 22. Magnetic Oscillations in Metals

22

Abstract

In the beginning of this Lecture, we consider Pauli paramagnetism in metals related to the interaction of electron spin with magnetic field. We analyze the comb of electron levels in magnetic field leading to quantum oscillations of magnetic susceptibility and magnetic moment. We present Lifshitz–Kosevich theory for De Haas–Van Alphen effect and describe the quantum oscillations in bismuth. In the end of the Lecture, we discuss Bohr–Sommerfeld quasiclassical quantization and Landau diamagnetism connected with the orbital electron motion in magnetic field. We compare Pauli and Landau contributions to magnetic susceptibility in metal.

22.1 Pauli Paramagnetism

In this Lecture we will learn in detail about interesting phenomena arising in electron gas when we switch on magnetic field [1–14].

We will start with the description of Pauli paramagnetism [1–6] related to the interaction of electron spin with magnetic field in metal.

We know, that in the absence of magnetic field the number of electrons with spins up (parallel to the field) and down (antiparallel to the field) equal to each other, and chemical potentials for distribution functions of electrons with both spin-projections-coincide (see Fig. 22.1a):

$$\mu_1 = \mu_2. \tag{22.1}$$

When we switch on external magnetic field, we would like to think that the chemical potentials become equal to $\mu_1 = \mu - \mu_B H$ for spins down and $\mu_2 = \mu + \mu_B H$ for spins up, where μ_B is electron Bohr magneton [15] (see Fig. 22.1b).

In fact, this is not the case. The chemical potentials in the equilibrium for both components of electron spin should be equal in the presence of the field.

© The Author(s), under exclusive license to Springer Nature Switzerland AG 2026

M. Kagan, *Lecture Notes of Professor Yuri Kagan in Theoretical Solid-State Physics*, Lecture Notes in Physics 1048, https://doi.org/10.1007/978-3-032-14621-2_22

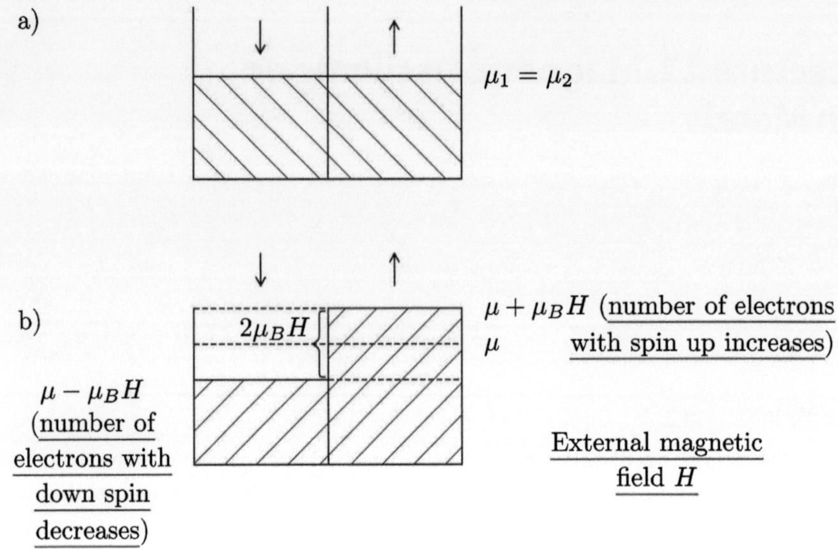

Fig. 22.1 **a** Equal chemical potentials of electrons with spins up and down in the absence of magnetic field **b** Switching on the magnetic field results in larger number of electrons with spin up. Correspondingly the number of electrons with spin down decreases [1]

Only the bottoms of the bands are shifted: down on $\mu_B H$ for spins up and up on $\mu_B H$ for spins down (see Fig. 22.2). At the same time both components of electron spins have the common level of the chemical potential μ.

It is evident from Fig. 22.2, that when we switch on magnetic field the number of electrons with spin up becomes larger than the number of electrons with spin down. As a result, Pauli paramagnetism arises.

Let us introduce convenient notations:

$$\mu' = \mu + \mu_B H, \quad \mu'' = \mu - \mu_B H. \tag{22.2}$$

Then the total number of electrons with spin up in meal:

$$N_\uparrow = V \int d\varepsilon g(\varepsilon) \frac{1}{1 + \exp\left(\frac{\varepsilon - \mu_B H - \mu}{T}\right)} = V \int_0^{\mu'} d\varepsilon g(\varepsilon) + O\left(\frac{T^2}{\varepsilon_F^2}\right). \tag{22.3}$$

Analogously the total number of electrons with spin down:

$$N_\downarrow = V \int d\varepsilon g(\varepsilon) \frac{1}{1 + \exp\left(\frac{\varepsilon + \mu_B H - \mu}{T}\right)} = V \int_0^{\mu''} d\varepsilon g(\varepsilon) + O\left(\frac{T^2}{\varepsilon_F^2}\right). \tag{22.4}$$

Hence the magnetization:

$$M = \mu_B \left(N_\uparrow - N_\downarrow\right) = 2\mu_B^2 H g\left(\varepsilon_F\right) V, \tag{22.5}$$

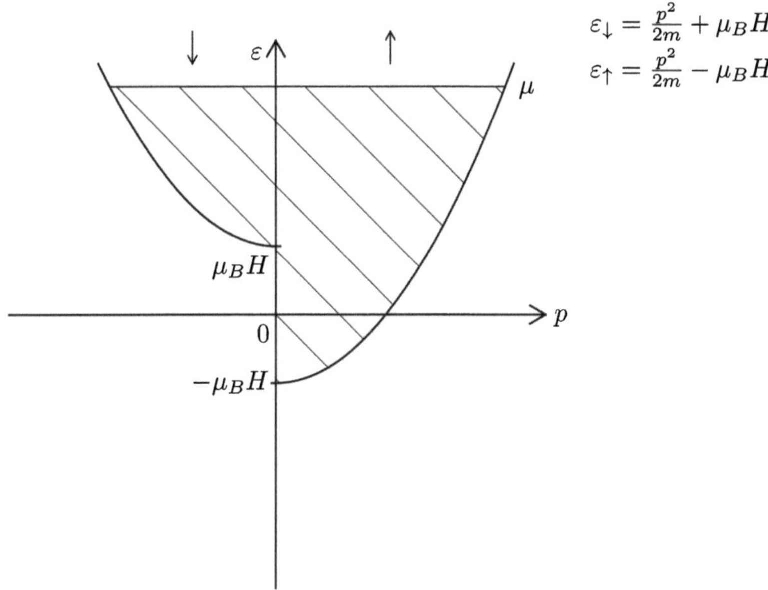

Fig. 22.2 The downward shift of the bottom of the band on $\mu_B H$ for spins up and upward shift on $\mu_B H$ for spins down. At the same time, both components of electron spins get the common level of the chemical potential μ [1]

where we took into account that in magnetic field according to Eqs. (22.3–22.4) the number of up and down spins read:

$$N_\uparrow = N_0 + \mu_B H g(\mu), \; N_\downarrow = N_0 - \mu_B H g(\mu). \tag{22.6}$$

In Eq. (22.6) N_0 is the total number of electrons for each of electron spin components in the absence of the field.

Note that in weak magnetic fields the chemical potential (up to the terms quadratic in magnetic field):

$$\mu \approx \varepsilon_F. \tag{22.7}$$

Correspondingly, magnetic susceptibility:

$$\chi = \frac{\partial M}{\partial H} = 2\mu_B^2 g\left(\varepsilon_F\right) V + O\left(\frac{T^2}{\varepsilon_F^2}\right). \tag{22.8}$$

This is quantum result for degenerate electron gas.

At the same time, for the classical gas we get the Curie law for the susceptibility [1–6]:

$$\chi = \frac{\mu_B^2 n V}{3T} = \frac{\mu_B^2 N}{3T}. \tag{22.9}$$

We can see, that the density of states:

$$g\left(\varepsilon_F\right) \sim \frac{n}{\varepsilon_F}, \text{ and } g\left(\varepsilon_F\right) V \sim \frac{N}{\varepsilon_F}, \tag{22.10}$$

where $N \sim N_0$ is of the order of the electrons number of each component. Hence, quantum and classical result coincide by order of magnitude at very high temperatures $T \sim \varepsilon_F \sim \left(10^3 \div 10^4\right)$ K.

Let us stress that for low and room temperatures the ratio $\frac{T}{\varepsilon_F} \ll 1$, and quantum susceptibility is much smaller than the classical one [1,2]:

$$\chi_{\text{quant.}} \sim \chi_{\text{clas.}} \frac{T}{\varepsilon_F} \ll \chi_{\text{clas.}} \tag{22.11}$$

This result is connected with the fact, that in quantum regime reorientation of spins in magnetic field affects only the region close to the Fermi surface, that is only the narrow shell with the width much smaller than the Fermi energy [1,2]:

$$\frac{\mu_B H}{\varepsilon_F} \ll 1. \tag{22.12}$$

The same small parameter determines the region of weak magnetic fields in degenerate electron gas [1,2]. Note also that expression (22.8) for magnetic susceptibility of degenerate gas is valid for the arbitrary dispersion law of electrons in metal.

22.2   Quantum Oscillations of Magnetic Susceptibility and Magnetic Moment: Energy Level Comb

In Quantum mechanics [15] it is well known the solution of the problem for Landau levels [7] (for Landau quantization of the transverse motion) in constant magnetic field. Let us emphasize, that if the motion is finite in classical picture, then this motion is quantized in Quantum mechanics.

Hamiltonian of the Landau levels problem has the form:

$$H = \frac{\left(\hat{\boldsymbol{p}} - \frac{e}{c}\boldsymbol{A}\right)^2}{2m}, \tag{22.13}$$

where vector potential in Landau gauge reads:

$$A_x = -Hy, \ A_y = A_z = 0. \tag{22.14}$$

In this case we have the free motion for the momentum components p_x, p_z (p_x, p_z are the quantum numbers or the integrals of motion). The wave function of the problem according to Landau–Lifshitz textbook on Quantum mechanics [15] can be written as:

$$\psi_{p_x,p_z,n} = \exp\left(\frac{ip_x x + ip_z z}{\hbar}\right)\varphi_n\left(y - y_0\right), \qquad (22.15)$$

where we used the relation:

$$(p_x + \frac{e}{c}Hy)^2 = (y - y_0)^2(\frac{e}{c}H)^2; \; y_0 = -\frac{cp_x}{eH}. \qquad (22.16)$$

As a result, electron spectrum is degenerate with respect to p_x (the center of the classical circle is arbitrary). It is determined as a sum of the spectrum for the harmonic oscillator [15] in the transverse (to magnetic field) direction and the spectrum of the free motion along the direction of magnetic field:

$$\varepsilon_{p_z,\omega_c} = \hbar\omega_c(n + \frac{1}{2}) + \frac{p_z^2}{2m}, \qquad (22.17)$$

where $\omega_c = \frac{|e|H}{mc}$ is the cyclotron frequency.

22.3 The Density of States at Landau Levels

The number of states for the problem of Landau levels is given by:

$$\sum_{n\sigma}\int \frac{L_x dp_x}{2\pi\hbar}\int \frac{L_z dp_z}{2\pi\hbar}. \qquad (22.18)$$

Since electron spectrum is degenerate with respect to p_x, we can perform the integration over this variable in all the expressions containing (or implicitly dependent) on $\varepsilon_{p_z,\omega_c}$.

Let us consider the problem in the box (see Fig. 22.3). Then from the definition of the circle center in Eq. (22.16) it follows that:

$$p_x = -\frac{y_0 eH}{c}, \qquad (22.19)$$

where:

$$-\frac{L_y}{2} < y_0 < \frac{L_y}{2}. \qquad (22.20)$$

Hence after the integration of the number of states in Eq. (22.18) over p_x (which is equivalent to the integration over the center of the circle due to relation (22.19) or over L_y), we get:

$$\int \frac{L_x dp_x}{2\pi\hbar}\sum_{n\sigma}\int \frac{L_z dp_z}{2\pi\hbar} = \frac{L_x L_y L_z}{(2\pi\hbar)^2}\frac{eH}{c}\sum_{n\sigma}\int dp_z = \frac{V}{(2\pi\hbar)^2}\frac{eH}{c}\sum_{n\sigma}\int dp_z, \qquad (22.21)$$

where $V = L_x L_y L_z$ is the box volume, and e is elementary charge.

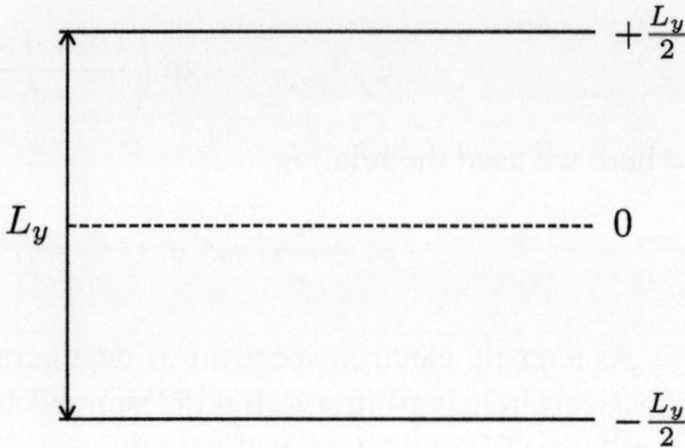

Fig. 22.3 The problem in the box for the determination of the electron density of states at Landau levels [1]

Now we can calculate the number of states which is contained in the given volume with the energies smaller than ε. From equation for the spectrum (22.17) we can see, that the maximal value for the momentum projection p_z reads:

$$p_z^{\max} = \sqrt{2m\left(\varepsilon - \hbar\omega_c\left(n + \frac{1}{2}\right)\right)}. \qquad (22.22)$$

Thus, performing the integration over p_z in Eq. (22.21) and summing up over the spin projections, we obtain:

$$\frac{2eH}{c}\frac{V}{(2\pi\hbar)^2}\sum_n{}'\sqrt{2m\left(\varepsilon - \hbar\omega_c\left(n + \frac{1}{2}\right)\right)}, \qquad (22.23)$$

where the dash at the upper limit of the sum means, that we only take the sum of those n, for which the expression under the square root is positive.

Hence the density of states:

$$
\begin{aligned}
g(\varepsilon) = \frac{1}{V}\frac{\partial N(\varepsilon)}{\partial\varepsilon} &= \frac{2eH}{c}\frac{V}{(2\pi\hbar)^2}\sum_{n=0}{}'\frac{\sqrt{2m}}{\sqrt{\varepsilon - \hbar\omega_c(n + \frac{1}{2})}}\frac{1}{2} \\
&= \frac{eH}{c}\frac{V}{(2\pi\hbar)^2}\sum_{n=0}{}'\frac{\sqrt{2m}}{\sqrt{\varepsilon - \hbar\omega_c(n + \frac{1}{2})}}. \qquad (22.24)
\end{aligned}
$$

Thus, the density of states $g(\varepsilon)$ has square root singularities (see Fig. 22.4a). These singularities in the density of states itself (and not in its derivative) we found in $g(\varepsilon)$ in the one-dimensional case. We have seen these singularities in the phonon spectrum [1]. Now effectively we get the one-dimensional (free) motion of electrons along the field direction (see Fig. 22.4b).

Let us study more attentively the dependence of the density of states on the magnetic field H. We know that the cyclotron frequency $\omega_c = \frac{eH}{mc}$ is proportional to the magnetic field. Hence if we increase the field, then the frequency is also increased. Correspondingly, when we change the frequency, the number of states $N(\varepsilon)$ is also changed.

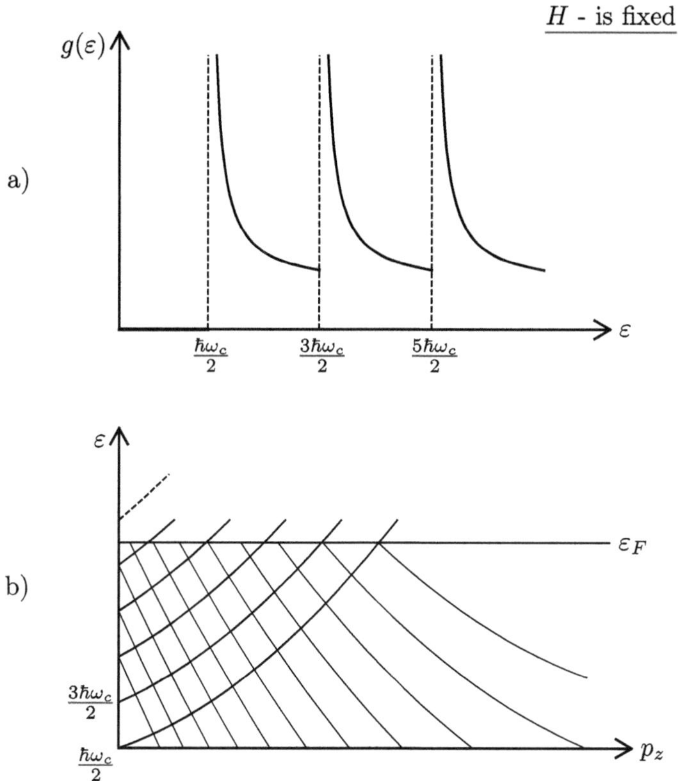

Fig. 22.4 **a** Square root singularities in the density of states for Landau levels of electron in magnetic field **b** Dependence of electron energy on the magnitude of the momentum projection p_z on z-axis parallel to the magnetic field H [1]

Thus, when we increase magnetic field, the number of states for the given energy changes experiencing a jump. As a result, all the physical quantities as the functions of magnetic field start to vary periodically with magnetic field.

To get this periodical behavior we should fulfill the following conditions [1,2]:

(1) The particles should not scatter strongly during the rotation period:

$$\omega_c \tau_{\text{scat.}} \ll 1.$$

(2) Landau levels should not be washed out strongly by temperature:

$$\frac{T}{\hbar \omega_c} \ll 1.$$

22.4   De Haas–Van Alphen Effect

For the analysis of the oscillations of the thermodynamic quantities [1–18], let us consider the behavior of the thermodynamic potential of electron gas in magnetic field with the measure of integration given by Eq. (22.21):

$$\Omega = -\frac{VT}{(2\pi\hbar)^2}\frac{eH}{c}\sum_{n=0,\sigma}^{\infty}\int_{-\infty}^{\infty} dp_z \ln\left(1 + \exp\left(\frac{\mu - \varepsilon(p_z, n)}{T}\right)\right), \qquad (22.25)$$

where according to Eq. (22.17) electron spectrum reads $\varepsilon_{p_z,\omega_c} = \hbar\omega_c(n + \frac{1}{2}) + \frac{p_z^2}{2m}$.

Let us proceed in Eq. (22.25) from the summation to the integration over n using the relation:

$$\sum_{n=0}^{\infty}\varphi(n) = \sum_{l=-\infty}^{\infty}\int_{-\frac{1}{2}}^{\infty}\delta(x - l)dx\,\varphi(x) = \int_{-\frac{1}{2}}^{\infty} f(x)dx\,\varphi(x), \qquad (22.26)$$

where the function:

$$f(x) = \sum_{-\infty}^{\infty}\delta(x - l) \qquad (22.27)$$

is the periodic function with the period 1 (see Fig. 22.5).

But as we know from the first part of this Course, the formula (22.27) and Fig. 22.5 determine the lattice sum in the one-dimensional crystal chain. Thus:

$$f(x) = \sum_{-\infty}^{\infty}\delta(x - l) = \sum_{k=-\infty}^{\infty}\exp(i2\pi kx), \qquad (22.28)$$

and correspondingly:

$$\sum_{n=0}^{\infty}\varphi(n) = \sum_{k=-\infty}^{\infty}\int_{-\frac{1}{2}}^{\infty}\exp(i2\pi kx)dx\,\varphi(x). \qquad (22.29)$$

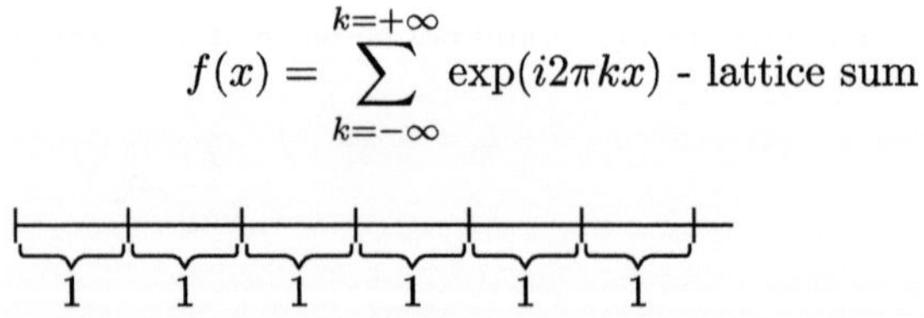

Fig. 22.5 Periodical function $f(x)$, which determines the lattice sum of the one-dimensional crystal chain [1]

$$f(x) = \sum_{k=-\infty}^{k=+\infty}\exp(i2\pi kx) \text{ - lattice sum}$$

Let us distinguish in this sum the term with $k = 0$, and for negative k let us make the substitution $k \to -k$. Then finally:

$$\sum_{n=0}^{\infty} \varphi(n) = \int_{-\frac{1}{2}}^{\infty} dx\, \varphi(x) + 2\,\mathrm{Re} \sum_{k=1}^{\infty} \int_{-\frac{1}{2}}^{\infty} \exp(i 2\pi k x)\, dx\, \varphi(x), \qquad (22.30)$$

where Re stands for the real part.

Thus we derived the well-known Poisson formula [1,2]. We can find the derivation of the relations (22.29–22.30) and the Poisson formula based on them, for example, in Landau–Lifshitz textbook *Statistical Physics, Part I*, paragraph 60 [2]. The derivation assumes the fulfillment of the condition $\frac{T}{\hbar\omega_c} \ll 1$ on the weak temperature smearing of Landau levels.

Note, that the first term with $k = 0$ which equals to $\int_{-\frac{1}{2}}^{\infty} dx\, \varphi(x)$ does not depend on magnetic field. At the same time, electron spectrum depends now on the continuous (and not discrete) variable x:

$$\varepsilon_{p_z,\omega_c}(x) = \hbar\omega_c \left(x + \frac{1}{2}\right) + \frac{p_z^2}{2m}. \qquad (22.31)$$

Hence formula (22.29) for the thermodynamic potential after summation over spin projections can be written as:

$$\Omega = \Omega_{\text{nonmag.}}(k = 0) + 2\,\mathrm{Re} \sum_{k=1}^{\infty} \Omega_{\text{mag.}}(k), \qquad (22.32)$$

where the nonmagnetic contribution:

$$\Omega_{\text{nonmag.}}(k = 0)$$

$$= -\frac{2VT}{(2\pi\hbar)^2} \frac{eH}{c} \int_{-\infty}^{\infty} dp_z \int_{-\frac{1}{2}}^{\infty} dx \ln\left(1 + \exp\left(\frac{\mu - \hbar\omega_c\left(x + \frac{1}{2}\right) - \frac{p_z^2}{2m}}{T}\right)\right). \qquad (22.33)$$

Substituting the variables y and dx in the nonmagnetic contribution:

$$y = \hbar\omega_c\left(x + \frac{1}{2}\right), \, dx = \frac{dy}{\hbar\omega_c} \qquad (22.34)$$

we get:

$$\Omega_{\text{nonmag.}}(k = 0) = -\frac{2VT}{(2\pi\hbar)^2} \frac{eH}{c\hbar\omega_c} \int_{-\infty}^{\infty} dp_z \int_{0}^{\infty} dy \ln\left(1 + \exp\left(\frac{\mu - y - \frac{p_z^2}{2m}}{T}\right)\right)$$

$$= \frac{2VTm}{\hbar(2\pi\hbar)^2} F(\mu, m, T). \qquad (22.35)$$

Thus, since the function $F(\mu, m, T)$ does not explicitly depend on magnetic field, then $\Omega_{\text{nonmag.}}(k=0)$ - does not depend on magnetic field as well (at least in not very strong fields, where the chemical potential $\mu \approx \varepsilon_F$ weakly changes with the field variation).

At the same time, for the magnetic contributions:

$$\Omega_{\text{mag.}}(k)$$

$$= -\frac{2VT}{(2\pi\hbar)^2}\frac{eH}{c}\int_{-\infty}^{\infty}dp_z\int_{-\frac{1}{2}}^{\infty}dx\ln\left(1+\exp\left(\frac{\mu-\hbar\omega_c\left(x+\frac{1}{2}\right)-\frac{p_z^2}{2m}}{T}\right)\right)\exp(i2\pi kx).$$

$$(22.36)$$

Let us analyze magnetic contributions more thoroughly and let us take at first the integral over dx by integrating expression (22.36) by parts. Then:

$$\int_{-\frac{1}{2}}^{\infty}dx\ln\left(1+\exp\left(\frac{\mu-\varepsilon_{p_z,\omega_c}(x)}{T}\right)\right)\exp(i2\pi kx)$$

$$=\frac{1}{i2\pi k}\exp(i2\pi kx)\ln\left(1+\exp\left(\frac{\mu-\varepsilon_{p_z,\omega_c}(x)}{T}\right)\right)\Bigg|_{-\frac{1}{2}}^{\infty}$$

$$-\frac{1}{i2\pi k}\int_{-\frac{1}{2}}^{\infty}dx\exp(i2\pi kx)\frac{\exp\left(\frac{\mu-\varepsilon_{p_z,\omega_c}(x)}{T}\right)}{1+\exp\left(\frac{\mu-\varepsilon_{p_z,\omega_c}(x)}{T}\right)}\left(-\frac{1}{T}\frac{\partial\varepsilon_{p_z,\omega_c}(x)}{\partial x}\right). \quad (22.37)$$

First term in Eq. (22.37) equals zero on the upper limit since:
$\varepsilon_{p_z,\omega_c}(x\to\infty)\to\infty$, and $\ln\left(1+\exp\left(\frac{\mu-\varepsilon_{p_z,\omega_c}(x\to\infty)}{T}\right)\right)\to\ln(1+0)=0$.

At the same time, on the lower limit: $\varepsilon_{p_z,\omega_c}\left(x\to-\frac{1}{2}\right)\to\frac{p_z^2}{2m}$, and

$$\ln\left(1+\exp\left(\frac{\mu-\hbar\omega_c\left(x+\frac{1}{2}\right)-\frac{p_z^2}{2m}}{T}\right)\right)\to\ln\left(1+\exp\left(\frac{\mu-\frac{p_z^2}{2m}}{T}\right)\right).$$

$$(22.38)$$

But we should keep in mind that for $x\to-\frac{1}{2}$:

$$\exp(i2\pi kx)\to\exp(-i\pi k)=(-1)^k,\text{ and }\frac{1}{i2\pi k}\exp(i2\pi kx)\to\frac{(-1)^k}{i2\pi k}. \quad (22.39)$$

As a result, the first term proves to be purely imaginary one on the lower level. Hence its real part entering in Eq. (22.32) equals zero.

Thus, in Eq. (22.37) only the second term remains. It can be transformed also using the definition of the Fermi–Dirac distribution function:

$$\frac{\exp\left(\frac{\mu-\varepsilon_{p_z,\omega_c}(x)}{T}\right)}{1+\exp\left(\frac{\mu-\varepsilon_{p_z,\omega_c}(x)}{T}\right)} = \frac{1}{\exp\left(\frac{\varepsilon_{p_z,\omega_c}(x)-\mu}{T}\right)+1} = n_F\left(\varepsilon_{p_z,\omega_c}(x)\right). \qquad (22.40)$$

Correspondingly, the integral (22.37) can be written as:

$$\int_{-\frac{1}{2}}^{\infty} dx \ln\left(1+\exp\left(\frac{\mu-\varepsilon_{p_z,\omega_c}(x)}{T}\right)\right)\exp(i2\pi kx)$$

$$= \frac{1}{i2\pi kT}\int_{-\frac{1}{2}}^{\infty} dx \exp(i2\pi kx) n_F\left(\varepsilon_{p_z,\omega_c}(x)\right)\left(\frac{\partial \varepsilon_{p_z,\omega_c}(x)}{\partial x}\right)_{p_z}. \qquad (22.41)$$

Further on let us change the variables again:

$$dx \to d\varepsilon = \frac{\partial \varepsilon(x)}{\partial x}dx, \qquad (22.42)$$

where the energy variable:

$$\varepsilon(x) = \hbar\omega_c(x+\frac{1}{2}) + \frac{p_z^2}{2m}.$$

Then:

$$x = \frac{1}{\hbar\omega_c}\left[\varepsilon(x) - \frac{p_z^2}{2m}\right] - \frac{1}{2}. \qquad (22.43)$$

Note that on the lower limit: $\varepsilon\left(x=-\frac{1}{2}\right) = \frac{p_z^2}{2m}$, while on the upper limit: $\varepsilon(x \to \infty) \to \infty$. As a result:

$$\int_{-\frac{1}{2}}^{\infty} dx \ln\left(1+\exp\left(\frac{\mu-\varepsilon_{p_z,\omega_c}(x)}{T}\right)\right)\exp(i2\pi kx)$$

$$= \frac{1}{i2\pi kT}\int_{\frac{p_z^2}{2m}}^{\infty} d\varepsilon \exp\left(\frac{i2\pi k}{\hbar\omega_c}\left(\varepsilon - \frac{p_z^2}{2m}\right) - i\pi k\right)n_F(\varepsilon). \qquad (22.44)$$

At the same time the thermodynamic potential $\Omega_{\text{mag.}}(k)$ is given by:

$$\Omega_{\text{mag.}}(k)$$

$$= -\frac{2VT}{(2\pi\hbar)^2}\frac{eH}{c}\frac{1}{i2\pi kT}\int_{-\infty}^{\infty} dp_z \int_{\frac{p_z^2}{2m}}^{\infty} d\varepsilon \exp\left(\frac{i2\pi k}{\hbar\omega_c}\left(\varepsilon - \frac{p_z^2}{2m}\right) - i\pi k\right)n_F(\varepsilon)$$

$$= -\frac{V}{(2\pi\hbar)^2}\frac{eH}{c}\frac{1}{i\pi k}\int_{-\infty}^{\infty} dp_z \int_{\frac{p_z^2}{2m}}^{\infty} d\varepsilon \exp\left(\frac{i2\pi k}{\hbar\omega_c}\left(\varepsilon - \frac{p_z^2}{2m}\right) - i\pi k\right)n_F(\varepsilon). \qquad (22.45)$$

Let us analyze more attentively the double integral in Eq. (22.45). The integration region over $d\varepsilon$ is shown in Fig. 22.6a. Let us rearrange the order of integration over $d\varepsilon$ and dp_z. Then after this rearrangement the first integration over dp_z, as we can see in Fig. 22.6a, is performed on the interval from 0 to $\sqrt{2m\varepsilon}$, while the second integration over $d\varepsilon$ can be expanded now on the interval from 0 to ∞. Thus:

$$\Omega_{\text{mag.}}(k) = -\frac{V}{(2\pi\hbar)^2}\frac{eH}{c}\frac{1}{i\pi k}\int_0^\infty d\varepsilon\, n_F(\varepsilon)\int_0^{\sqrt{2m\varepsilon}} dp_z\, \exp\left(\frac{i2\pi k}{\hbar\omega_c}\left(\varepsilon - \frac{p_z^2}{2m}\right) - i\pi k\right)$$

$$= -\frac{V}{(2\pi\hbar)^2}\frac{eH}{c}\frac{1}{i\pi k}\int_0^\infty d\varepsilon\, n_F(\varepsilon)\int_0^{\sqrt{2m\varepsilon}} dp_z\, \exp\left(\frac{i2\pi k}{\hbar\omega_c}\left(\varepsilon - \frac{p_z^2}{2m}\right)\right)(-1)^k. \qquad (22.46)$$

It is important, that the exponent in (22.46) has a large phase since:

$$\varepsilon_F \gg \hbar\omega_c. \qquad (22.47)$$

Now we have to determine effective integration region for the first integral over dp_z in (22.46):

$$\int_0^{\sqrt{2m\varepsilon}} dp_z\, \exp\left(\frac{i2\pi k}{\hbar\omega_c}\left(\varepsilon - \frac{p_z^2}{2m}\right)\right). \qquad (22.48)$$

If we had the right to expand the upper limit of the integration on ∞, then the integral acquired the Gaussian form and could be taken elementary:

$$\int_0^\infty dp_z\, \exp\left(\frac{i2\pi k}{\hbar\omega_c}\left(\varepsilon - \frac{p_z^2}{2m}\right)\right) = \sqrt{\frac{\pi}{2}\frac{\hbar\omega_c m}{i\pi k}} = \sqrt{\frac{\hbar\omega_c m}{i2k}}. \qquad (22.49)$$

Note that at the upper limit we have $p_z \sim p_F$, and moreover $\frac{p_F^2}{2m} \gg \hbar\omega_c$ according to Eq. (22.47). Hence in the integral (22.48) over dp_z we can really set the upper limit $\sqrt{2m\varepsilon} \to \infty$, and thus get Eq. (22.49).

Then the thermodynamic potential $\Omega_{\text{mag.}}(k)$:

$$\Omega_{\text{mag.}}(k) = -\frac{V}{(2\pi\hbar)^2}\frac{eH}{c}\frac{1}{i\pi k}\sqrt{\frac{\hbar\omega_c m}{i2k}}(-1)^k\int_0^\infty d\varepsilon\, n_F(\varepsilon)\exp\left(\frac{i2\pi k\varepsilon}{\hbar\omega_c}\right)$$

$$= -\frac{V}{(2\pi\hbar)^2}\frac{eH}{c}\frac{\exp\left(-\frac{i\pi}{4}\right)}{i\pi k}\sqrt{\frac{\hbar\omega_c m}{2k}}(-1)^k\int_0^\infty d\varepsilon\, n_F(\varepsilon)\exp\left(\frac{i2\pi k\varepsilon}{\hbar\omega_c}\right), \qquad (22.50)$$

where $\sqrt{i} = \exp\left(\frac{i\pi}{4}\right)$.

Further on it is convenient to introduce the variable:

$$y = \varepsilon - \mu, \qquad (22.51)$$

and expand the integration over this variable from $-\infty$ to ∞, using the fact, that $-\mu$ is large and negative.

On top of that we can introduce the variable:

$$\gamma_k = \frac{2\pi k}{\hbar\omega_c}. \tag{22.52}$$

As a result, the thermodynamic potential $\Omega_{\text{mag.}}(k)$ equals to:

$$\Omega_{\text{mag.}}(k) = -\frac{V}{(2\pi\hbar)^2}\frac{eH}{c}\frac{\exp\left(-\frac{i\pi}{4}\right)}{i\pi k}\sqrt{\frac{\hbar\omega_c m}{2k}}(-1)^k\exp\left(i\gamma_k\mu\right)\int_{-\infty}^{\infty}dy\,\frac{\exp(i\gamma_k y)}{\exp\left(\frac{y}{T}\right)+1}. \tag{22.53}$$

The remaining integral in the infinite limits can be closed by an infinite-distant Watson contour C in the complex plane (see Fig. 22.6b). After that we can evaluate

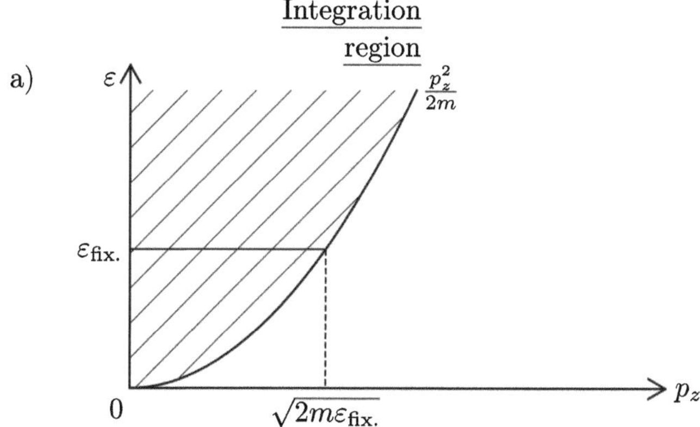

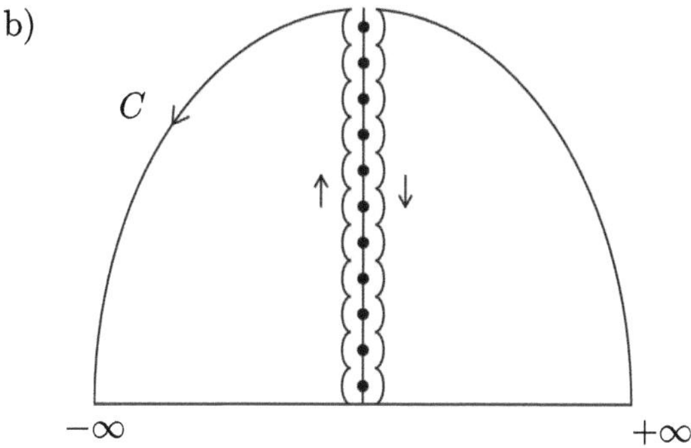

Fig. 22.6 **a** Integration region of the momentum projection on z-axis parallel to the magnetic field **b** Complex Watson contour and the poles on the imaginary axis [1]

the integral by the method of residues [1,2] since it has the poles on the complex axis in the points:

$$y = (2l + 1)i\pi T. \tag{22.54}$$

Let us stress that in these points $\exp\left(\frac{y}{T}\right) = \exp\left(i\pi(2l + 1)\right) = -1$, and Fermi-Dirac distribution function diverges.

Moreover, the integral over the closed contour (which equals to the sum of the integral from $-\infty$ to ∞ and the integral over the infinite-distant contour) is zero:

$$\oint = \int_{-\infty}^{\infty} + \int_C = 0. \tag{22.55}$$

As a result, initial integral is given by the sum of the residues which lie on the complex axis inside the closed contour:

$$\int_{-\infty}^{\infty} = -\int_C = 2\pi i \sum_{l=0}^{\infty} \text{Res}. \tag{22.56}$$

Then the integral:

$$J = \int_{-\infty}^{\infty} dy \frac{\exp(i\gamma_k y)}{\exp\left(\frac{y}{T}\right) + 1} = 2\pi i \sum_{l=0}^{\infty} (-T)\exp(-\pi T \gamma_k(2l + 1)). \tag{22.57}$$

Summation of the infinite geometric series in (22.57) can be performed easily and yields:

$$J = -2\pi i T \frac{\exp(-\pi T \gamma_k)}{1 - \exp(-2\pi T \gamma_k)} = -\pi i T \frac{2}{\exp\left(\pi T \gamma_k\right) - \exp\left(-\pi T \gamma_k\right)}$$

$$= -\pi i T \frac{1}{\sinh(\pi T \gamma_k)} = -\frac{i}{\gamma_k} \frac{\pi T \gamma_k}{\sinh(\pi T \gamma_k)} = -\frac{i}{\gamma_k} \Psi(T, k), \tag{22.58}$$

where we introduced the dimensionless function:

$$\Psi(T, k) = \frac{\pi T \gamma_k}{\sinh(\pi T \gamma_k)}. \tag{22.59}$$

Hence the thermodynamic potential:

$$\Omega_{\text{mag.}}(k) = \frac{V}{(2\pi\hbar)^2} \frac{eH}{c} \frac{\exp\left(-\frac{i\pi}{4}\right)}{i\pi k} \sqrt{\frac{\hbar\omega_c m}{2k}} (-1)^k \exp\left(i\gamma_k \mu\right) \left(\frac{i}{\gamma_k}\right) \Psi(T, k)$$

$$= \frac{V}{(2\pi\hbar)^2} \frac{eH}{c} \frac{1}{\pi k \gamma_k} \sqrt{\frac{\hbar\omega_c m}{2k}} (-1)^k \exp\left(i\gamma_k \mu - \frac{i\pi}{4}\right) \Psi(T, k). \tag{22.60}$$

Correspondingly the real part of the thermodynamic potential oscillates:

$$\operatorname{Re} \Omega_{\text{mag.}}(k) = \frac{V}{(2\pi\hbar)^2} \frac{eH}{c} \frac{1}{\pi k \gamma_k} \sqrt{\frac{\hbar \omega_c m}{2k}} (-1)^k \Psi(T,k) \cos\left(\gamma_k \mu - \frac{\pi}{4}\right).$$

(22.61)

Note that the phase of the oscillations is large:

$$\Phi = \gamma_k \mu = \frac{2\pi k}{\hbar \omega_c} \mu \gg 1.$$

(22.62)

The phase variation on 2π arises under the variation of the magnetic field:

$$\Delta\Phi = \frac{2\pi k}{\hbar \omega_c} \mu \frac{\Delta H}{H} = 2\pi, \text{ and } \frac{k\mu}{\hbar \omega_c} \frac{\Delta H}{H} = 1,$$

(22.63)

where ΔH is the variation of the magnetic field.

Hence, sufficiently small variation of field leads to the variation of phase on 2π and thus to significant oscillations:

$$\frac{\Delta H}{H} = \frac{\hbar \omega_c}{k\mu} \ll 1.$$

(22.64)

Using the fact that the cyclotron frequency is proportional to the field $\omega_c = \frac{eH}{mc} \sim H$, we can represent (22.64) in the form:

$$\left(\frac{\Delta H}{H^2} \sim \left|\Delta \frac{1}{H}\right|\right) = \frac{\hbar e}{mc} \frac{1}{k\mu}.$$

(22.65)

Let us emphasize that in the temperature factor entering in Eq. (22.61):

$$\Psi(T,k) = \frac{\pi T \gamma_k}{\sinh(\pi T \gamma_k)},$$

the value of the argument:

$$\pi T \gamma_k = \pi T \frac{2\pi k}{\hbar \omega_c} = \frac{2\pi^2 T k}{\hbar \omega_c} \approx 20k \frac{T}{\hbar \omega_c}$$

(22.66)

becomes large at not very low temperatures.

In this case the function $\sinh(\pi T \gamma_k) \approx \frac{1}{2} \exp(\pi T \gamma_k)$, and the temperature factor reads:

$$\Psi(T,k) \approx 2\pi T \gamma_k \exp\left(-\pi T \gamma_k\right).$$

(22.67)

Then the thermodynamic potential:

$$\mathrm{Re}\,\Omega_{\mathrm{mag.}}(k) = \frac{V}{(2\pi\hbar)^2}\frac{eH}{c}\sqrt{\frac{\hbar\omega_c m}{2k}}(-1)^k\left(\frac{2\pi T\gamma_k}{\pi k\gamma_k}\right)\exp\left(-\pi T\gamma_k\right)\cos\left(\gamma_k\mu - \frac{\pi}{4}\right)$$

$$= \frac{V}{(2\pi\hbar)^2}\frac{eH}{c}\sqrt{\frac{\hbar\omega_c m}{2k}}(-1)^k\frac{2T}{k}\exp\left(-\pi T\gamma_k\right)\cos\left(\gamma_k\mu - \frac{\pi}{4}\right). \tag{22.68}$$

Moreover, in the sum over k in expression (22.34) for the total thermodynamic potential the main role belongs to the first harmonic with $k = 1$. As a result the total potential:

$$\Omega \approx \Omega_{\mathrm{nonmag.}} + 2\,\mathrm{Re}\,\Omega_{\mathrm{mag.}}(k = 1), \tag{22.69}$$

where:

$$\mathrm{Re}\,\Omega_{\mathrm{mag.}}(k = 1) = -\frac{2TV}{(2\pi\hbar)^2}\frac{eH}{c}\sqrt{\frac{\hbar\omega_c m}{2}}\exp(-\frac{2\pi^2 T}{\hbar\omega_c})\cos\left(\frac{2\pi\mu}{\hbar\omega_c} - \frac{\pi}{4}\right). \tag{22.70}$$

Let us stress once more that to have quantum oscillations we should fulfill the condition $\frac{T}{\hbar\omega_c} \ll 1$. In this case the comb in the density of states in Fig. 22.4a is not smeared by the temperature. The Dirac step function is blurred close to the Fermi surface in the region of the order of the temperature T, and has a form in this case presented on Fig. 22.7. Thus the intermediate temperatures for which the factor $\Psi(T, k)$ in Eq. (22.67) is exponentially decreasing correspond to the interval:

$$\frac{1}{20} \leq \frac{T}{\hbar\omega_c} \ll 1. \tag{22.71}$$

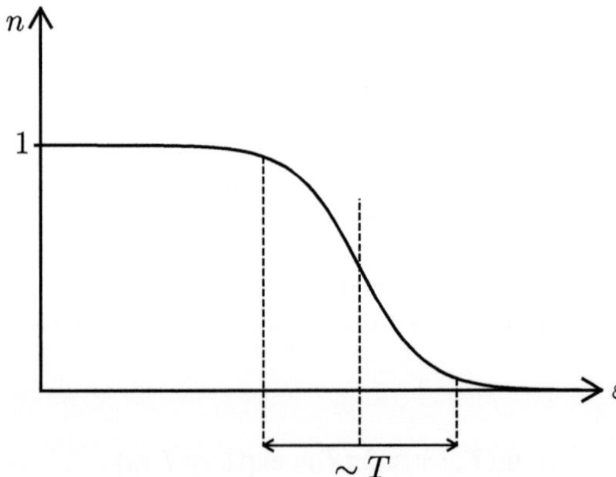

Fig. 22.7 Temperature blurring of the Dirac step function for the detection of the quantum oscillations in bismuth [1]

If, however, the temperatures are very low, and $\frac{T}{\hbar\omega_c} \ll \frac{1}{20}$, then:

$$\sinh\left(\pi T \gamma_k\right) \approx \pi T \gamma_k,$$

and as a result:

$$\Psi(T,k) = \frac{\pi T \gamma_k}{\sinh(\pi T \gamma_k)} = 1. \tag{22.72}$$

In this case magnetic contribution to the thermodynamic potential:

$$\mathrm{Re}\,\Omega_{\mathrm{mag.}}(k=1) = -\frac{V}{(2\pi\hbar)^2}\frac{eH}{c}\sqrt{\frac{\hbar\omega_c m}{2}}\,\frac{1}{\pi\gamma_{k=1}}\cos\left(\frac{2\pi\mu}{\hbar\omega_c} - \frac{\pi}{4}\right)$$

$$= -\frac{Vm}{(2\pi\hbar)^2}\frac{eH}{mc}\sqrt{\frac{\hbar\omega_c m}{2}}\frac{\hbar\omega_c}{2\pi^2}\cos\left(\frac{2\pi\mu}{\hbar\omega_c} - \frac{\pi}{4}\right) = -\frac{V(\frac{m}{2})^{\frac{3}{2}}}{(2\pi\hbar)^2}\frac{1}{\pi^2\hbar}(\hbar\omega_c)^{\frac{5}{2}}\cos\left(\frac{2\pi\mu}{\hbar\omega_c} - \frac{\pi}{4}\right). \tag{22.73}$$

Note that quantum oscillations were firstly discovered in bismuth. Bismuth is a transition metal and there is small number of electrons on elementary cell in it. Hence the chemical potential μ is small in bismuth. On the other hand, electron effective mass in bismuth is small [1,4]:

$$m^* \approx 10^{-2}\,m. \tag{22.74}$$

That is why the cyclotron frequency in bismuth is large:

$$\omega_c = \frac{|e|H}{m^*c}. \tag{22.75}$$

Correspondingly, the factor $\frac{\mu}{\hbar\omega_c}$ decreases, but still is much larger than 1. As a result, there is enough resolution in the experiment to measure the variation of magnetic field ΔH in bismuth against the background of large fields H (the ratio $\frac{\Delta H}{H} = \frac{\hbar\omega_c}{\mu}$ increases to some extent, but is still smaller than 1).

If we know the behavior of the magnetic part of the thermodynamic potential, then we can find the behavior of the oscillating part of the magnetization as well:

$$M_{\mathrm{osc.}} = -\frac{\partial\Omega}{\partial H} = -\frac{2\partial\,\mathrm{Re}\,\Omega_{\mathrm{mag.}}(k=1)}{\partial H}. \tag{22.76}$$

Substituting expression (22.73) for $\mathrm{Re}\,\Omega_{\mathrm{mag.}}(k=1)$ in (22.76) we obtain at very low temperatures:

$$M_{\text{osc.}} \sim \frac{\partial}{\partial H}((\hbar\omega_c)^{\frac{5}{2}}\cos\left(\frac{2\pi\mu}{\hbar\omega_c} - \frac{\pi}{4}\right)) \approx -(\hbar\omega_c)^{\frac{5}{2}}\frac{2\pi\mu}{\hbar}\sin\left(\frac{2\pi\mu}{\hbar\omega_c} - \frac{\pi}{4}\right)\frac{\partial}{\partial H}(\frac{1}{\omega_c})$$

$$= (\hbar\omega_c)^{\frac{5}{2}}\frac{2\pi\mu}{\hbar}\frac{mc}{|e|}\frac{1}{H^2}\sin\left(\frac{2\pi\mu}{\hbar\omega_c} - \frac{\pi}{4}\right) = (\hbar\omega_c)^{\frac{5}{2}}\frac{2\pi\mu}{\hbar\omega_c}\frac{1}{H}\sin\left(\frac{2\pi\mu}{\hbar\omega_c} - \frac{\pi}{4}\right). \qquad (22.77)$$

Note that deriving Eq. (22.77) we differentiated only rapidly oscillating function. Thus, we can see, that if the thermodynamic potential oscillates, then all the other thermodynamic quantities are also oscillating.

Note that usually experimentalists are measuring magnetic susceptibility. For $\frac{\mu}{\hbar\omega_c} \gg 1$ the oscillating part of the susceptibility is proportional to the oscillating component of the thermodynamic potential:

$$\chi_{\text{osc.}} = \frac{\partial M_{\text{osc.}}}{\partial H} \sim (\hbar\omega_c)^{\frac{5}{2}}\frac{2\pi\mu}{\hbar\omega_c}\frac{1}{H}\frac{2\pi\mu}{\hbar}\frac{\partial}{\partial H}\left(\frac{1}{\omega_c}\right)\cos\left(\frac{2\pi\mu}{\hbar\omega_c} - \frac{\pi}{4}\right)$$

$$= -(\hbar\omega_c)^{\frac{5}{2}}\frac{2\pi\mu}{\hbar\omega_c}\frac{1}{H^2}\frac{mc}{|e|H}\frac{2\pi\mu}{\hbar}\cos\left(\frac{2\pi\mu}{\hbar\omega_c} - \frac{\pi}{4}\right)$$

$$= -(\hbar\omega_c)^{\frac{5}{2}}(\frac{2\pi\mu}{\hbar\omega_c})^2\frac{1}{H^2}\cos\left(\frac{2\pi\mu}{\hbar\omega_c} - \frac{\pi}{4}\right)$$

$$\sim \text{Re}\,\Omega_{\text{mag.}}(k=1)(\frac{2\pi\mu}{\hbar\omega_c})^2\frac{1}{H^2} \sim \chi_{\text{osc.}}(k=1). \qquad (22.78)$$

The first qualitative considerations on the magnetic moment oscillations in strong fields and at low temperatures belong to L.D. Landau [7].

I.M. Lifshitz and Yu. A. Kosevich [13,14,19,20] constructed consistent one-particle theory of De Haas–Van Alphen effect [12]. The full theory of De Haas–Van Alphen effect with an account of many-body interactions is still far from being complete [21–23].

De Haas–Van Alphen effect plays very substantial role in the modern solid-state physics and in particular in the physics of strongly correlated systems [24–28] with heavy fermions [28–41].

In these systems many-body effects connected with electron–electron interaction are very important. They lead in particular to the emergence of the heavy electron masses [21–23,28–41] $m^* \sim 200\,$m, and, correspondingly, of the narrow electron bands with the width of the order of $W \sim 50\,$K for f-electrons of the rare-earth elements [21–23,28–41].

The bands are often half-filled and thus $n_f \to 1$ (we have practically one f-electron on site). In this case an important role belongs to the Mott–Hubbard localization effects [24–27].

Experiments on the measurement of the electron specific heat indicate that in strong fields $H \sim (6 \div 10)\,$T effective mass decreases strongly with the field increase. At the same time, in De Haas–Van Alphen effect the decrease of the heavy electron mass does not always manifest itself.

Consistent many-body theory of De Haas–Van Alphen effect should eliminate this contradiction in the interpretation of different experiments.

22.5 Landau Diamagnetism

In the previous section we found strongly oscillating part of the thermodynamic potential. It occurs, however, that there is also a static contribution to the potential $\Omega_{\text{mag.}}(k)$ related to Landau diamagnetism [1,2,7].

To find this contribution let us return back to the expression (22.46) for $\Omega_{\text{mag.}}(k)$:

$$\Omega_{\text{mag.}}(k)$$

$$= -\frac{V}{(2\pi\hbar)^2}\frac{eH}{c}\frac{1}{i\pi k}\int_0^\infty d\varepsilon\, n_F(\varepsilon)\int_0^{\sqrt{2m\varepsilon}} dp_z \exp\left(\frac{i2\pi k}{\hbar\omega_c}\left(\varepsilon - \frac{p_z^2}{2m}\right)\right)(-1)^k.$$

Let us analyze the integral over dp_z in (22.46)

$$I = \int_0^{\sqrt{2m\varepsilon}} dp_z \exp\left(\frac{i2\pi k}{\hbar\omega_c}\left(\varepsilon - \frac{p_z^2}{2m}\right)\right)$$

more thoroughly. Then we can notice that effective integration region for this integral is connected not only with the points $\varepsilon = \varepsilon_F$, $p_z \to 0$ (as we discussed in the previous section deriving oscillating part of the potential), but also with the points where the phase is small and correspondingly:

$$p_z = \sqrt{2m\varepsilon} - y, \tag{22.79}$$

where y is small. Hence:
$p_z^2 \approx 2m\varepsilon - 2y\sqrt{2m\varepsilon}$, and

$$\frac{p_z^2}{2m} \approx \varepsilon - \frac{y}{m}\sqrt{2m\varepsilon}. \tag{22.80}$$

As a result: $\varepsilon - \frac{p_z^2}{2m} \approx \frac{y}{m}\sqrt{2m\varepsilon}$. At the same time $dp_z = -dy$, and the integral becomes equal:

$$I = \int_0^{\sqrt{2m\varepsilon}} dy \exp\left(\frac{i2\pi k}{\hbar\omega_c}\frac{y}{m}\sqrt{2m\varepsilon}\right) = \int_0^{\sqrt{2m\varepsilon}} dy \exp\left(i\gamma_k y\sqrt{\frac{2\varepsilon}{m}}\right), \tag{22.81}$$

where we introduced again the parameter: $\gamma_k = \frac{2\pi k}{\hbar\omega_c}$.

This integral is taken elementary and yields:

$$I = \frac{1}{i\gamma_k}\sqrt{\frac{m}{2\varepsilon}}\left(-1 + \exp\left(i\gamma_k\sqrt{2m\varepsilon}\sqrt{\frac{2\varepsilon}{m}}\right)\right) = \frac{1}{i\gamma_k}\sqrt{\frac{m}{2\varepsilon}}(-1 + \exp(2i\gamma_k\varepsilon)).$$

$$(22.82)$$

Further on we should calculate the integral over $d\varepsilon$ in the expression for the thermodynamic potential:

$$\int_0^\infty d\varepsilon\, n_F(\varepsilon)\frac{1}{i\gamma_k}\sqrt{\frac{m}{2\varepsilon}}(-1 + \exp(2i\gamma_k\varepsilon)). \qquad (22.83)$$

Let us utilize the fact that for low temperatures the Femi-Dirac distribution function is almost the step function $n_F(\varepsilon) \approx 1$ for $\varepsilon < \varepsilon_F$ and rapidly decreases practically to zero for $\varepsilon > \varepsilon_F$. Then the integral (22.83) approximately equals to:

$$\int_0^{\varepsilon_F} d\varepsilon \frac{1}{i\gamma_k}\sqrt{\frac{m}{2\varepsilon}}(-1 + \exp(2i\gamma_k\varepsilon)) = \frac{1}{i\gamma_k}\sqrt{\frac{m}{2}}\int_0^{\varepsilon_F} d\varepsilon \frac{1}{\sqrt{\varepsilon}}(-1 + \exp(2i\gamma_k\varepsilon))$$

$$= \frac{1}{i\gamma_k}\sqrt{\frac{m}{2}}(I_1 + I_2). \qquad (22.84)$$

The first integral in (22.84) can be easily calculated and reads:

$$I_1 = -\int_0^{\varepsilon_F} d\varepsilon \frac{1}{\sqrt{\varepsilon}} = -2\sqrt{\varepsilon_F}. \qquad (22.85)$$

The second integral can be reduced to the Gaussian form and is given by:

$$I_2 = \int_0^{\varepsilon_F} d\varepsilon \frac{1}{\sqrt{\varepsilon}}\exp(2i\gamma_k\varepsilon) = 2\int_0^{\varepsilon_F} d(\sqrt{\varepsilon})\exp(2i\gamma_k\varepsilon) = 2\int_0^{\sqrt{\varepsilon_F}} dz\, \exp(2i\gamma_k z^2)$$

$$\approx 2\int_0^\infty dz\, \exp(2i\gamma_k z^2) = 2\frac{1}{\sqrt{2i\gamma_k}}\frac{\sqrt{\pi}}{2} = \frac{\sqrt{\pi}}{\sqrt{2i\gamma_k}}. \qquad (22.86)$$

But we work in the limit of large Fermi energy:

$$\varepsilon_F\gamma_k \gg 1, \text{ and } |I_1| \gg I_2. \qquad (22.87)$$

Hence the second term with an exponent $\exp(2i\gamma_k\varepsilon)$ in the r.h.s. of Eq. (22.83) does not work at integration by energy.

Thus:

$$\Omega_{\text{mag.}}(k) = \frac{V}{(2\pi\hbar)^2}\frac{eH}{c}\frac{1}{i\pi k}\frac{1}{i\gamma_k}\sqrt{\frac{m}{2}}2\sqrt{\varepsilon_F}(-1)^k$$

$$= \frac{V}{(2\pi\hbar)^2}\frac{eH}{c}\frac{1}{\pi}\frac{2}{2\pi}\frac{1}{i^2}\sqrt{\frac{m\varepsilon_F}{2}}\frac{(-1)^k}{k^2}\hbar\omega_c = -\frac{Vm}{(2\pi\hbar)^2}\frac{eH}{mc}\frac{1}{\pi^2}\frac{p_F}{2}\hbar\omega_c\frac{(-1)^k}{k^2}$$

$$= -\frac{Vmk_F}{(2\pi\hbar)^2}\frac{1}{2\pi^2}(\hbar\omega_c)^2\frac{(-1)^k}{k^2} = \text{Re}\Omega_{\text{mag.}}(k) \propto H^2. \tag{22.88}$$

We can see, that the contribution to the components of the magnetic part of the thermodynamic potential $\text{Re}\Omega_{\text{mag.}}(k)$ related to the small phase from the integration region over dp_z in (22.46) is proportional to magnetic field squared and has an alternating sign depending on k. At the same time, this contribution to the potential is not oscillating (in contrast to the regions close to the points $\varepsilon = \varepsilon_F$, $p_z \to 0$).

We can sum up all the components of $\text{Re}\Omega_{\text{mag.}}(k)$ in (22.34) and get for the non-oscillating magnetic part of the total potential [1,2]:

$$\Omega_{\text{mag.}} = 2\,\text{Re}\sum_{k=1}^{\infty}\Omega_{\text{mag.}}(k) = -\frac{Vmk_F}{(2\pi\hbar)^2}\frac{1}{\pi^2}(\hbar\omega_c)^2\sum_{k=1}^{\infty}\frac{(-1)^k}{k^2}. \tag{22.89}$$

But the infinite series in (22.89) is tabulated and equals to [1,2]:

$$\sum_{k=1}^{\infty}\frac{(-1)^k}{k^2} = -\frac{\pi^2}{12}. \tag{22.90}$$

Hence quadratic in field non-oscillating contribution to the thermodynamic potential is given by:

$$\Omega_{\text{mag.}} = -\frac{Vmk_F}{(2\pi\hbar)^2}\frac{1}{\pi^2}(\hbar\omega_c)^2\left(-\frac{\pi^2}{12}\right) = \frac{Vmk_F}{(2\pi\hbar)^2}\frac{1}{12}(\hbar\omega_c)^2. \tag{22.91}$$

Correspondingly, magnetic moment is linear in field and negative (as it should be for the diamagnetic contribution):

$$M_{\text{diam.}} = -\frac{\partial\Omega}{\partial H} = -\frac{Vmk_F}{(2\pi\hbar)^2}\frac{2}{12}\hbar\omega_c\frac{\hbar e}{mc} = -\frac{Vmk_F}{(2\pi\hbar)^2}\frac{1}{6}\hbar\omega_c\frac{\hbar e}{mc} < 0. \tag{22.92}$$

As a result magnetic susceptibility:

$$\chi_{\text{diam.}} = \frac{\partial M_{\text{diam.}}}{\partial H} = -\frac{1}{6}\frac{Vmk_F}{(2\pi\hbar)^2}\left(\frac{\hbar e}{mc}\right)^2 < 0. \tag{22.93}$$

22.6   Problem to Lecture 22. Difference in Factor 3 Between Landau Diamagnetism and Pauli Paramagnetism

In the beginning of this Lecture, we determined already the paramagnetic Pauli contribution to susceptibility:

$$\chi_{\text{para.}} = \frac{\partial M}{\partial H} = 2\mu_B^2 g(\varepsilon_F) V,$$

where $g(\varepsilon_F)$ is the density of states on one spin projection, and μ_B is electron Bohr magneton for spin $S = \frac{1}{2}$:

$$\mu_B = \frac{\hbar e}{2mc}. \tag{22.94}$$

For quadratic spectrum, as we already discussed in Lecture 21, the density of states on one spin projection:

$$g(\varepsilon_F) = \frac{1}{2} \frac{mk_F}{2(\pi\hbar)^2} = \frac{mk_F}{4(\pi\hbar)^2}. \tag{22.95}$$

Substituting the definitions of Bohr magneton from Eq. (22.94) and density of states on one spin projection from (22.95) in the susceptibility (22.93), we get [1,2]:

$$\chi_{\text{diam.}} = -\frac{1}{6} \frac{Vmk_F}{(2\pi\hbar)^2} \left(\frac{\hbar e}{mc}\right)^2 = -\frac{1}{6} V g(\varepsilon_F) (2\mu_B)^2 = -\frac{2}{3} V g(\varepsilon_F) \mu_B^2 = -\frac{1}{3} \chi_{\text{para.}} \tag{22.96}$$

Thus, the paramagnetic Pauli contribution exceeds in 3 times by absolute value the diamagnetic Landau contribution ($\chi_{\text{diam.}} = -\frac{1}{3}\chi_{\text{para.}}$) [1,2].

Note, that this relation is obtained for the bare electron mass. Effective electron mass can be substantially larger than the bare one (as it is often the case in many strongly correlated electron systems). Then the ratio of the susceptibilities can be different from 3 [42].

22.7   Bohr–Sommerfeld Quasiclassical Quantization

For high Landau levels $n \sim \frac{\mu}{\hbar\omega_c} \gg 1$ the quasiclassical approximation works [15], and we can use Bohr–Sommerfeld formulas for action quantization.

Let us remind, that in the presence of the magnetic field the quasiclassical projections of the generalized momentum in Landau gauge read [15]:

$$p_x = P_x - \frac{e}{c} A_x = P_x + \frac{e}{c} Hy, \tag{22.97}$$

$$p_y = P_y, \, p_z = P_z. \tag{22.98}$$

Then the adiabatic invariant [15,43] for y-component of the generalized momentum is quantized according to Bohr–Sommerfeld theory:

$$\oint P_y d Q_y = \oint p_y dy = \frac{c}{eH} \oint p_y d p_x = \frac{c}{eH} S(\varepsilon, p_z) = 2\pi \hbar (n + \frac{1}{2}),$$
(22.99)

where $Q_y = y$ is the generalized coordinate, $S(\varepsilon, p_z)$ is the orbital area in (x, y)-plane for fixed ε and p_z.

Note that deriving (22.99) we used the relation (22.97) according to which:

$$P_x = \text{const, and } dy = \frac{c}{eH} d p_x.$$
(22.100)

It is important, that from relation (22.99) we can extract the spectrum $\varepsilon(p_z, n)$ measuring in experiment the orbital area $S(\varepsilon, p_z)$.

Let us vary on 1 the number of Landau levels: $\Delta n = 1$. Then the orbital area in (22.99) varies as:

$$\frac{c}{eH} \frac{\partial S}{\partial \varepsilon} \Delta \varepsilon = 2\pi \hbar.$$

Hence the variation of the energy

$$\Delta \varepsilon = \frac{\hbar e H}{c} 2\pi (\frac{\partial S}{\partial \varepsilon})^{-1}.$$
(22.101)

But for fixed p_z:

$$\Delta \varepsilon = \hbar \omega_c = \frac{\hbar e H}{m^* c}.$$
(22.102)

Then the inverse effective mass of electron in magnetic field:

$$\frac{1}{m^*} = 2\pi (\frac{\partial S}{\partial \varepsilon})^{-1}.$$
(22.103)

Let us recollect now the continuous parameter x which can be defined as:

$$x = \frac{c}{eH} \frac{1}{2\pi \hbar} S(\varepsilon, p_z) - \frac{1}{2} = n.$$

We introduced this parameter in the preceding sections instead of the discrete variable n with the help of the Poisson transformation [1,2]. Then in the components $\exp(i 2\pi k x)$ of the lattice sum in the Poisson formula

$$\exp(i2\pi kx) \rightarrow \exp\left(i2\pi k \frac{c}{eH}\frac{1}{2\pi\hbar}S(\varepsilon, p_z)\right). \tag{22.104}$$

The phase in (22.104) effectively sits on the energies $\varepsilon \sim \varepsilon_F$ and momentum projection along the field direction $p_z \sim p_z^*$. Moreover, the optimal value of p_z^* corresponds to the condition of the stationary phase:

$$\left(\frac{\partial S}{\partial p_z}\right)|_{p_z^*} = 0. \tag{22.105}$$

The orbital area close to the extremum can be expanded as:

$$S = S^* + \frac{1}{2}\frac{d^2 S}{dp_z^2}|_{p_z^*}(p_z - p_z^*)^2, \tag{22.106}$$

where:

$$S^* = S(\mu, p_z^*) \tag{22.107}$$

is the area of the extreme cross section (see Fig. 22.8).

As a result, the Gaussian integral over dp_z similar to the integral (22.46) appears again for the magnetic contribution to k-component of the thermodynamic potential $\Omega_{\mathrm{mag.}}(k)$.

For extreme cross sections we can find the relative variation of the magnetic field $\frac{\Delta H}{H}$ in formula (22.99), which leads to the variation of the adiabatic invariant (of the dimensionless action $I = \frac{S_{\mathrm{eff.}}}{\hbar}$) on one period:

$$\frac{c}{eH}S\left(\mu, p_z^*\right)\frac{1}{\hbar}\frac{\Delta H}{H} = 2\pi. \tag{22.108}$$

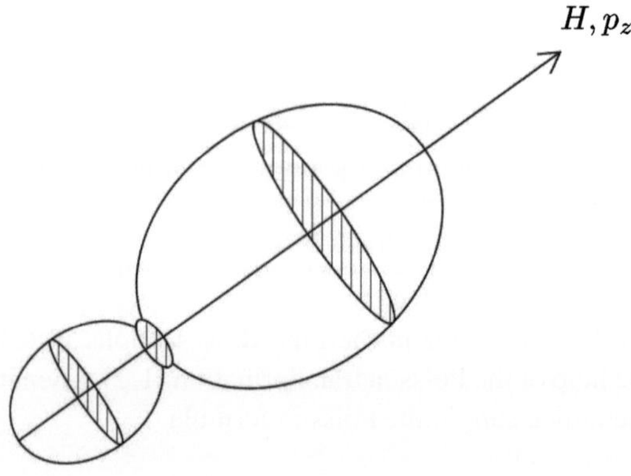

Fig. 22.8 Extreme cross sections for the quasiclassical description of electron motion in magnetic field [1]

Then the relative variation of magnetic field is given by:

$$\frac{\Delta H}{H} = \frac{2\pi \hbar e H}{c S^*}.$$ (22.109)

Thus, from the rotation period we can find extreme cross sections S^*. Moreover, rotating magnetic field, we can determine the shape of the Fermi surface [1,4].

References

1. Yu. Kagan, *Lectures on the Solid-state Theoretical Physics*, Moscow, MEPHI, 1981–1982. Part II. Electrons - unpublished
2. L.D. Landau, E.M. Lifshitz, *Statistical Physics, Part I* (Butterworth-Heinemann, 1980)
3. A.A. Abrikosov, *Fundamentals of the Theory of Metals* (Elsevier Science Publishers, North-Holland, Amsterdam, 1988)
4. I.M. Lifshitz, M.Y. Azbel, M.I. Kaganov, *Electron Theory of Metals* (Consultants Bureau, New York, 1973)
5. C. Kittel, *Quantum Theory of Solids* (Wiley, 1963)
6. N.W. Ashcroft, N.D. Mermin, *Solid State Physics* (Holt, Rinehart and Winston, 1976)
7. L.D. Landau, Diamagnetismus der Metalle. Z. Phys. **64**, 629 (1930)
8. L. Schubnikow, W.J. De Haas, Magnetic resistance increase in single crystals of bismuth at low temperatures. Proc. R. Neth. Acad. Arts Sci. (in German) **33**, 130 (1930)
9. L. Schubnikow, W.J. De Haas, New phenomena in the change in resistance of bismuth crystals in a magnetic field at the temperature of liquid hydrogen. Proc. R. Neth. Acad. Arts Sci. (in German) **33**, 363–378 (1930)
10. L. Schubnikow, W.J. De Haas, New phenomena in the change in resistance of bismuth crystals in a magnetic field at the temperature of liquid hydrogen II. Proc. R. Neth. Acad. Arts Sci. **33**, 418 (1930). (in German)
11. L. Schubnikow, W.J. De Haas, The change in resistance of bismuth crystals in a magnetic field at the temperature of liquid nitrogen. Proc. R. Neth. Acad. Arts Sci. **33**, 433 (1930) (in German)
12. W. De Haas, P.M. Van Alphen, The dependence of the susceptibility of diamagnetic metals upon the field. Proc. Acad. Sci. Amsterdam **33**, 1106 (1930). http://www.dwc.knaw.nl/DL/publications/PU00015989.pdf
13. I.M. Lifschitz, A.M. Kosevich, On the theory of the De Haas–Van Alphen effect for particles with an arbitrary dispersion law. Dokl. Akad. Nauk, USSR **96**, 963 (1954)
14. I.M. Lifschitz, A.M. Kosevich, Sov. Phys. JETP **2**, 636 (1956)
15. L.D. Landau, E.M. Lifshitz, *Quantum Mechanics, Non-relativistic Theory* (Pergamon Press, New York, 1977)
16. D. Shoenberg, *Publisher, Magnetic Oscillations in Metals* (Cambridge University Press, 2009)
17. M. Suzuki, I.S. Suzuki, Lecture notes on Solid State Physics: De Haas–Van Alphen effect. State University of New York at Binghamton (2006). https://web.archive.org/web/20110718064004/, https://www2.binghamton.edu/physics/docs/note-dhva.pdf, https://en.wikipedia.org/wiki/Binghamton_University
18. J. Kübler, *Theory of Itinerant Electron Magnetism* (Oxford University Press, Oxford, 2000), https://books.google.com/books?id=M70tL9xaOxoC&q=lifshitz+kosevich+equation
19. A.Yu. Grosberg, Personalia: I. M. Lifshitz-100th birthday anniversary. Physics Uspekhi **61**, 84 (2018)
20. V.G. Peschanskii, Yu.A. Kolesnichenko, On the 60th anniversary of the Lifshitz–Kosevich theory. Low Temp. Phys. **40**, 267 (2014). http://aip.scitation.org/doi/10.1063/1.4871744

21. G. Lonzarich, Publisher, Lifetime in Superconductivity and Magnetism. A tribute to Professor David Shoenberg (Cambridge Scientific Publishers, 2011)
22. L. Taillefer, J. Flouquet, G.G. Lonzarich, Normal and superconducting phases of heavy fermions. Physica B **169**, 257 (1991)
23. M. Springford, Heavy fermion compounds, studied using de Haas–van Alphen effect. Physica B **171**, 151 (1991)
24. J. Hubbard, Proc. R. Soc. Lond. **A276**, 238 (1963)
25. J. Hubbard, Proc. R. Soc. Lond. **A277**, 237 (1964)
26. N.F. Mott, *Metal–Insulator Transitions* (Taylor and Francis LTD, London, 1974)
27. N.F. Mott, E.A. Davis, *Electronic processes in Non-crystalline Materials* (Oxford University Press, 1979)
28. P. Fulde, *Electron Correlations in Molecules and Solids*, Springer Series in Solid-State Sciences (1993)
29. A.C. Hewson, *The Kondo Problem in Heavy Fermions*. Cambridge Studies in Magnetism (Cambridge University Press, 1993)
30. C.M. Varma, Y. Yafet, Phys. Rev. B **13**, 2950 (1976)
31. K.G. Wilson, Rev. Mod. Phys. **47**, 773 (1975)
32. P.W. Anderson, G. Yuval, D.R. Hamann, Phys. Rev. B **1**, 4464 (1970)
33. P. Nozieres, JLTP **17**, 31 (1974)
34. H. Keiter, N. Grewe, in *Valence Fluctuations in Solids*, ed. by L.M. Falicov, W. Hanke, M.B. Maple (North-Holland, Amsterdam, 1981)
35. D.M. News, N. Read, Adv. Phys. **36**, 799 (1987)
36. P. Coleman, Phys. Rev. B **35**, 5072 (1987)
37. Yu. Kagan, N.V. Prokof'ev, Electron polaron effect and quantum diffusion of heavy particle in metals. Sov. Phys. JETP **63**, 1276 (1986)
38. Yu. Kagan, N.V. Prokof'ev, Quantum tunneling in metals and the heavy fermion problems. Sov. Phys. JETP **66**, 211 (1987)
39. M.Yu. Kagan, A.G. Aronov, Search for non-Fermi liquid behavior in the two-band model with one narrow band. Czechoslovak. J. Phys. **46**, 965 (1996)
40. M.Yu. Kagan, V.V. Val'kov, Anomalous resistivity and the origin of a heavy mass in the two-band model with one narrow band. JETP **140**, 196 (2011)
41. M.Yu. Kagan, V.V. Val'kov, Anomalous resistivity and Electron polaron effect in the two-band Hubbard model with one narrow band. J. Supercond. Nov. Magn. **25**, 1379 (2012)
42. E.G. Batyev, Pauli paramagnetism and Landau diamagnetism. Physics Uspekhi **179**, 1374 (2009)
43. L.D. Landau, E.M. Lifshitz, *Mechanics* (Butterworth-Heinemann, 1976)

Lecture 23. The Cooper Problem

23

Abstract

We start this Lecture with the evaluation of the sign and magnitude of electron–phonon interaction close to the Fermi surface and deep inside Fermi sphere. On the basis of these results, we solve the famous Cooper problem in the BCS theory and find the binding energy of the Cooper pair. We determine the critical temperature of the superconducting transition and the value of the superconducting gap as well. We analyze the role of strong electron–electron repulsion which is always present in metals and show that the superconductivity is not destroyed due to the presence of Tolmachev logarithm in the effective coupling constant. In conclusion we provide a brief introduction to the unconventional superconductivity.

23.1 Amplitude of Electron–Phonon Interaction

As we already discussed in first Lecture of the second part of our Course, the secondly quantized Hamiltonian of electron–phonon interaction has a form [1–4]:

$$H = \sum_{k\sigma q\alpha} A_{q\alpha} a_{k\sigma}^{\dagger} a_{k-q\sigma} \left(b_{q\alpha} + b_{-q\alpha}^{\dagger} \right), \tag{23.1}$$

where α is the phonon branch of the spectrum, $a_{k\sigma}^{\dagger}$ and $a_{k-q\sigma}$ are the operators of electron creation and annihilation, $b_{q\alpha}$ and $b_{-q\alpha}^{\dagger}$ are the operators of annihilation and creation of phonons, $A_{q\alpha}$ is the matrix element of electron–phonon interaction:

$$A_{q\alpha} = i \sqrt{\frac{\hbar N}{2 M \omega_{\alpha}(f)}} (q e_{\alpha}(f)) \frac{V_{ei}(q)}{\varepsilon(q, 0)} \Delta(f - q). \tag{23.2}$$

(We consider this matrix element in detail in Lectures 15 and 19).

Let us remind, that in the expression for matrix element (23.2) $e_\alpha(f)$ are the polarization vectors for the phonon spectrum, N is the number of the elementary cells, M is the nuclear mass.

In Eq. (23.2) $V_{ei}(q)$ is the Fourier component of electron–ion interaction, $\varepsilon(q, 0)$ is static dielectric permitivity, $\omega_\alpha(f)$ are the frequencies of the phonon spectrum, while the symbol $\Delta(f - q)$ denotes equality of the wave vectors f and q (with the accuracy of the inverse lattice vector K).

For the acoustic branches of the phonon spectrum and small wave vectors $f = q$, and the matrix element reads:

$$A_{q\alpha} = i\sqrt{\frac{\hbar N}{2M\omega_\alpha(q)}}\,(q e_\alpha(q))\,\frac{V_{ei}(q)}{\varepsilon(q, 0)}. \tag{23.3}$$

Hamiltonian (23.1) conserves the number of electrons and does not conserve the number of phonons. The vertex of electron–phonon interaction for two possible processes in Eq. (23.1) is presented in Fig. 23.1. It corresponds to phonon emission by electron and phonon picking up processes.

Let us find the amplitude of the interaction between electrons via the phonon exchange (see Fig. 23.2). In the first two orders of the perturbation theory the amplitude of the interaction acquires the form:

$$f = -\frac{m}{4\pi\hbar^2}\left[V_{fi} + \sum_s \frac{V_{is}V_{sf}}{E_i - E_s}\right], \tag{23.4}$$

where the initial state in the second term of Eq. (23.4) and in Fig. 23.2 is connected with two electrons with the wave vectors and spin projections given by:

$$i = |k_1\sigma_1.k_2\sigma_2\rangle. \tag{23.5}$$

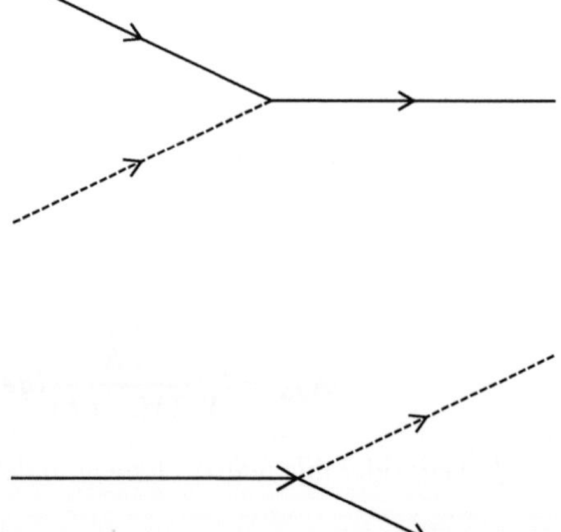

Fig. 23.1 The vertex of electron–phonon interaction can be related to phonon emission by electron and phonon picking up processes [1]

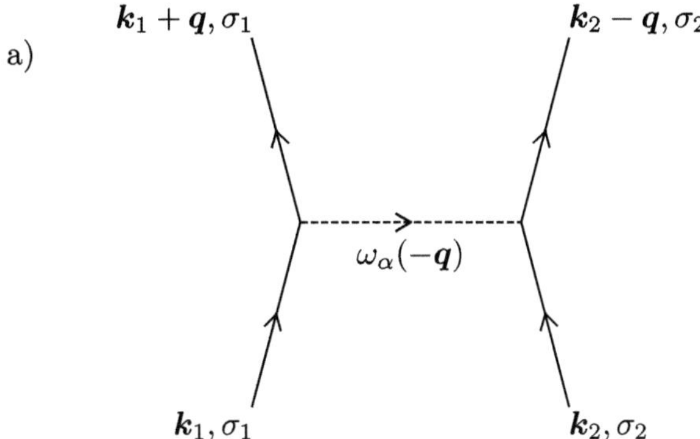

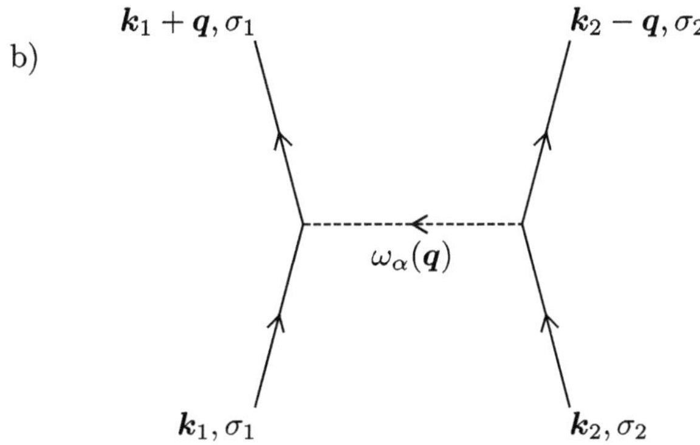

Fig. 23.2 Amplitude of electron–phonon interaction in the second order of the perturbation theory corresponds to the interaction between electrons via the phonon exchange. In this case two processes are possible. On **a** we present the process when the first electron emits the phonon with momentum $-q$, while the second electron is absorbing it. On **b** we describe the second possible process when the second electron emits the phonon with momentum q, while the first electron is absorbing it. At the same time, initial and final states of two electrons are the same for these two diagrams [1,12]

At the same time, the final state (after phonon emission by the first electron and it's picking up by the second electron in Fig. 23.2a and in inverse process in Fig. 23.2b) corresponds to two electrons with the wave vectors and spin projections:

$$f = |k_1 + q\sigma_1, k_2 - q\sigma_2\rangle . \tag{23.6}$$

Note, that the energy, momentum and spin conservation laws are fulfilled for final and initial state. Thus, the energies of the initial and finite state are equal:

$$E_i = E_f, \text{ and } \varepsilon_{k_1} + \varepsilon_{k_2} = \varepsilon_{k_1+q} + \varepsilon_{k_2-q}. \tag{23.7}$$

Let us stress, that for superconductivity problem it is just the second order of the perturbation theory that matters [1,3–9]. The point is that it conserves not only the number of electrons, but the number of phonons as well.

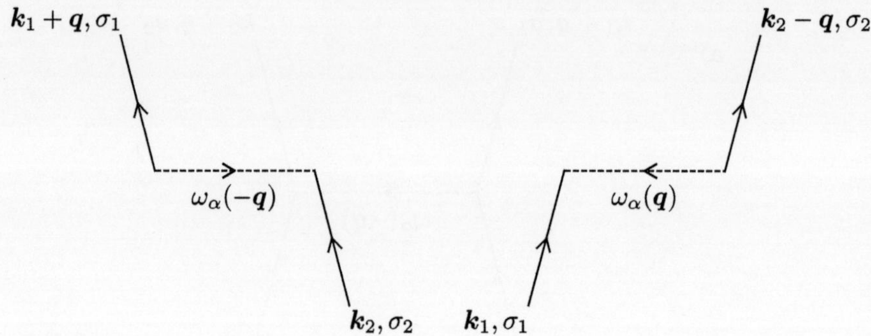

Fig. 23.3 Two possible intermediate states in the second order of the perturbation theory. For the first process **a** in the intermediate states there are: first electron after phonon emission, second electron before phonon absorption, and the phonon with momentum $-q$. For the second process **b** in the intermediate state we get second electron after phonon emission, first electron before phonon absorption and the phonon with momentum q [1,12,13]

The processes which do not conserve the number of phonons also are important for other problems such as the investigation of the conductivity or other kinetic effects in metals [1,10,11].

In the second order of the perturbation theory according to Fig. 23.3a, b, we have two intermediate states, related to the emission processes, namely of phonon with momentum $-q$ emitted by the first electron (Fig. 23.3a), and of phonon with momentum q emitted by the second electron (Fig. 23.3b) [1,12,13].

For the first process we get in the intermediate state: <u>first electron after phonon emission, second electron before phonon absorption, and the phonon with momentum $- q$</u> [12,13]:

$$s_1 = |k_2\sigma_2, k_1 + q\sigma_1, \omega_\alpha(-q)\rangle. \tag{23.8}$$

For the second process we obtain in the intermediate state: <u>second electron after phonon emission, first electron before phonon absorption and the phonon with momentum q</u> [12,13]:

$$s_2 = |k_1\sigma_1, k_2 - q\sigma_2, \omega_\alpha(q)\rangle. \tag{23.9}$$

The energies of these two intermediate states are equal to:

$$E_{s_1} = \varepsilon_{k_2} + \varepsilon_{k_1+q} + \hbar\omega_\alpha(-q),$$
$$E_{s_2} = \varepsilon_{k_1} + \varepsilon_{k_2-q} + \hbar\omega_\alpha(q). \tag{23.10}$$

Moreover, for each of the two diagrams in Fig. 23.2 initial and final energies of two electrons:

$$E_i = \varepsilon_{k_1} + \varepsilon_{k_2}, \tag{23.11}$$

and:

$$E_f = \varepsilon_{k_1+q} + \varepsilon_{k_2-q} \tag{23.12}$$

equal to each other.

Correspondingly, for the second order contribution to the interaction amplitude in the denominator of the second term in Eq. (23.4) enters the following energy difference [12, 13] from the first intermediate state:

$$E_i - E_{s_1} = (\varepsilon_{k_1} + \varepsilon_{k_2}) - (\varepsilon_{k_2} + \varepsilon_{k_1+q} + \hbar\omega_\alpha(-q)) = \varepsilon_{k_1} - \varepsilon_{k_1+q} - \hbar\omega_\alpha(q), \tag{23.13}$$

where we took advantage of the parity of phonon frequencies:

$$\hbar\omega_\alpha(-q) = \hbar\omega_\alpha(q). \tag{23.14}$$

At the same time, the energy difference for the second intermediate state is given by:

$$E_i - E_{s_2} = (\varepsilon_{k_1} + \varepsilon_{k_2}) - (\varepsilon_{k_1} + \varepsilon_{k_2-q} + \hbar\omega_\alpha(q)) = \varepsilon_{k_2} - \varepsilon_{k_2-q} - \hbar\omega_\alpha(q). \tag{23.15}$$

If we take into account now the energy conservation law (23.7), then this energy difference can be written as:

$$E_i - E_{s_2} = \varepsilon_{k_1+q} - \varepsilon_{k_1} - \hbar\omega_\alpha(q). \tag{23.16}$$

Thus, the contribution to the interaction amplitude in (23.4) from the first intermediate state s_1 equals in the second order of the perturbation theory:

$$\sum_\alpha \frac{A_{-q\alpha} A_{q\alpha}}{E_i - E_{s_1}} = \sum_\alpha \frac{|A_{q\alpha}|^2}{\varepsilon_{k_1} - \varepsilon_{k_1+q} - \hbar\omega_\alpha(q)}, \tag{23.17}$$

In Eq. (23.17) we accounted that $e_\alpha(-f) = e_\alpha^*(f)$ for the polarization vectors, and hence $A_{-q\alpha} = A_{q\alpha}^*$ for the matrix elements of electron–phonon interaction. As a result:

$$A_{-q\alpha} A_{q\alpha} = |A_{q\alpha}|^2. \tag{23.18}$$

Similarly, the contribution to the interaction amplitude from the second intermediate state yields:

$$\sum_\alpha \frac{A_{q\alpha} A_{-q\alpha}}{E_i - E_{s_2}} = \sum_\alpha \frac{|A_{q\alpha}|^2}{\varepsilon_{k_1+q} - \varepsilon_{k_1} - \hbar\omega_\alpha(q)}. \tag{23.19}$$

Then, the sum of the contributions to the interaction amplitude from two intermediate states reads [1, 12, 13]:

$$\sum_\alpha \frac{|A_{q\alpha}|^2}{\varepsilon_{k_1} - \varepsilon_{k_1+q} - \hbar\omega_\alpha(q)} + \sum_\alpha \frac{|A_{q\alpha}|^2}{\varepsilon_{k_1+q} - \varepsilon_{k_1} - \hbar\omega_\alpha(q)}$$

$$= 2\sum_\alpha \frac{|A_{q\alpha}|^2 \hbar\omega_\alpha(q)}{(\varepsilon_{k_1} - \varepsilon_{k_1+q})^2 - (\hbar\omega_\alpha(q))^2}. \tag{23.20}$$

Let us assume now that the initial energies of two electrons with the wave vectors k_1 and k_2 lie above the Fermi surface in the narrow energy shell much smaller than the Debye temperature. Thus:

$$\varepsilon_{k_1} - \varepsilon_F \ll \hbar\omega_D, \text{ and } \varepsilon_{k_2} - \varepsilon_F \ll \hbar\omega_D. \tag{23.21}$$

We know that electrons cannot fall into the filled fermionic background in Fig. 23.4. Correspondingly their final energies (after emission and absorption) also lie in the Debye shell. As a result, the absolute value of the energy difference in (23.20) is much less than Debye temperature:

$$|\varepsilon_{k_1} - \varepsilon_{k_1+q}| \ll \hbar\omega_D. \tag{23.22}$$

At the same time, the frequencies of the phonon spectrum are only less, or of the order of the Debye frequency:

$$\hbar\omega(q) \leq \hbar\omega_D. \tag{23.23}$$

Hence in the denominator of Eq. (23.20):

$$(\varepsilon_{k_1} - \varepsilon_{k_1+q})^2 \ll (\hbar\omega_\alpha(q))^2, \tag{23.24}$$

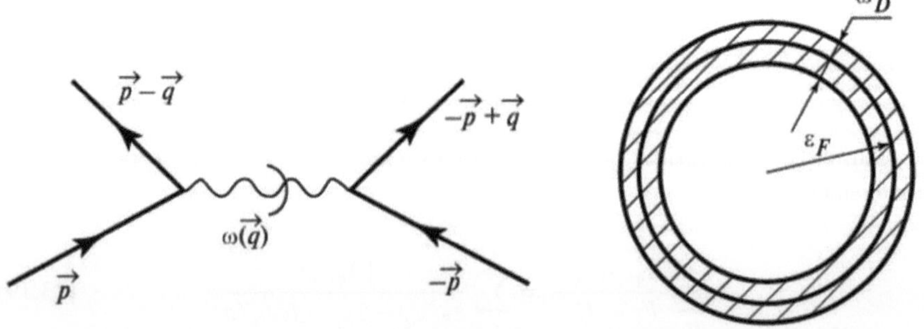

Fig. 23.4 Two electrons living above the filled Fermi surface (as in the early solution of the Cooper problem [6]) or above and below Fermi surface (as in the exact BCS theory [5, 7]). Electrons attract each other via the phonon exchange when their energies are lying in the narrow energy shell which is much less than Debye temperature [1, 12]

and the second order contribution to the effective electron–phonon interaction yields:

$$V_{\text{eff.}} = (-2) \sum_{\alpha} \frac{|A_{q\alpha}|^2}{\hbar \omega_{\alpha}(q)} < 0. \tag{23.25}$$

Note that the amplitude is related to the effective interaction entering in (23.25) as:

$$f = -\frac{m}{4\pi\hbar^2} V_{\text{eff.}} \tag{23.26}$$

For the acoustic phonons with small wave vectors q effective interaction is given by:

$$\begin{aligned}
V_{\text{eff.}} &= (-2) \sum_{\alpha} \frac{|A_{q\alpha}|^2}{\hbar \omega_{\alpha}(q)} = (-2) \sum_{\alpha} \frac{\hbar N}{2M\omega_{\alpha}(q)} \frac{1}{\hbar \omega_{\alpha}(q)} |q e_{\alpha}(q)|^2 \left(\frac{V_{ei}(q)}{\varepsilon(q,0)} \right)^2 \\
&= -\sum_{\alpha} \frac{N}{M(\omega_{\alpha}(q))^2} |q e_{\alpha}(q)|^2 \left(\frac{V_{ei}(q)}{\varepsilon(q,0)} \right)^2.
\end{aligned} \tag{23.27}$$

Thus, two electrons lying close to the Fermi surface always attract each other (when we neglect the Coulomb interaction between them).

Moreover, the screened electron–ion interaction on small wave vectors does not depend on q:

$$\frac{V_{ei}(q)}{\varepsilon(q,0)} = -\frac{4\pi Z e^2}{V(q^2 + \kappa^2)} \approx -\frac{4\pi Z e^2}{V\kappa^2} = \text{const}. \tag{23.28}$$

That is why effective interaction in (23.27):

$$V_{\text{eff.}} = -\frac{N}{M} \left(\frac{4\pi Z e^2}{V\kappa^2} \right)^2 \sum_{\alpha} \frac{1}{(\omega_{\alpha}(q))^2} |q e_{\alpha}(q)|^2 \sim -\frac{N}{Mc^2} \left(\frac{4\pi Z e^2}{V\kappa^2} \right)^2 = \text{const} \tag{23.29}$$

for the linear spectrum of acoustic phonons:

$$\omega_{\alpha}(q) = c_{\alpha} q \tag{23.30}$$

does not depend on q as well.

Note, that if we lift electrons higher above the Fermi surface, then they will repel each other [1,3–9].

It is important that electrons while interacting with each other are moving practically in a two-dimensional way in momentum space. They cannot go neither down in energy (since all the states inside the Fermi sphere are occupied) nor up in energy (the temperatures are low enough).

We know from the Quantum mechanics [2] that for the two-dimensional motion of the attracting particles the bound state always arises (even at very weak attraction). But the bound state of two electrons corresponds to composed boson. Hence superconductivity is the superfluidity of bi-electron (Cooper) pairs.

23.2   Evaluation of the Magnitude of Electron–Phonon Interaction

Let us demonstrate that attraction between electrons due to electron–phonon interaction is not small. We will show that it is at least comparable with electron–ion interaction.

Indeed, let us evaluate the dimensionless constant of electron–phonon interaction λ, which as we will see in the next sections determines the critical temperature in the BCS model [5–7]. This constant (if the contribution of acoustic phonons is a dominant one) is given by the product of the absolute value of effective interaction in (23.29) and the density of (electron) states (multiplied by the system volume):

$$\lambda = V |V_{\text{eff.}}| g(\varepsilon_F). \tag{23.31}$$

For the quadratic spectrum of electrons the density of states [1,3]:

$$g(\varepsilon_F) = N(0) = \frac{m k_F}{2\pi^2 \hbar^2} \sim \frac{n_{\text{el.}}}{\varepsilon_F}. \tag{23.32}$$

Then:

$$V g(\varepsilon_F) \sim V \frac{n_{\text{el.}}}{\varepsilon_F} = \frac{N_{\text{el.}}}{\varepsilon_F}. \tag{23.33}$$

Note that the number of atoms for simple crystal (with one atom on elementary cell) equals to the number of lattice sites, From the electroneutrality condition it reads:

$$ZN = N_{\text{el.}}, \tag{23.34}$$

where Z is the ion charge.

Hence the density of electron states multiplied by the system volume:

$$V g(\varepsilon_F) \sim \frac{ZN}{\varepsilon_F}. \tag{23.35}$$

At the same time, the absolute value of the effective interaction by order of magnitude is given by:

$$|V_{\text{eff.}}| \sim \frac{N}{M \overline{c^2}} \left(\frac{4\pi Z e^2}{V \kappa^2} \right)^2, \tag{23.36}$$

where $\overline{c^2}$ is average sound velocity squared.

That is why the dimensionless constant:

$$\lambda = V|V_{\text{eff.}}|g(\varepsilon_F) \sim \frac{ZN}{\varepsilon_F} \frac{N}{Mc^2} \left(\frac{4\pi Ze^2}{V\kappa^2}\right)^2 = \frac{Z}{\varepsilon_F} \frac{1}{Mc^2} \left(\frac{4\pi Ze^2}{\kappa^2}\right)^2 \frac{N^2}{V^2}$$

$$= \frac{Z}{\varepsilon_F} \frac{1}{Mc^2} \left(\frac{4\pi Ze^2}{\kappa^2}\right)^2 \frac{1}{\Omega_0^2} = \frac{Z\varepsilon_F}{Mc^2} \left(\frac{4\pi Ze^2}{\varepsilon_F \kappa^2 \Omega_0}\right)^2, \tag{23.37}$$

where the volume of the elementary cell:

$$\Omega_0 = \frac{V}{N}.$$

But zeroth Fourier harmonics of the screened electron–ion interaction for the quadratic electron spectrum:

$$\frac{4\pi Ze^2}{\kappa^2} = \frac{4\pi Ze^2 \varepsilon_F}{6\pi e^2 n_{\text{el.}}} = \frac{2Z\varepsilon_F}{3n_{\text{el.}}}, \tag{23.38}$$

where we used expression for the square of the Thomas–Fermi wave vector [14,15]:

$$\kappa^2 = \frac{6\pi e^2 n_{\text{el.}}}{\varepsilon_F}.$$

Correspondingly:

$$\frac{4\pi Ze^2}{\varepsilon_F \kappa^2 \Omega_0} = \frac{2Z\varepsilon_F}{3n_{\text{el.}} \varepsilon_F \Omega_0} = \frac{2ZN}{3n_{\text{el.}} V} = \frac{2ZN}{3N_{\text{el.}}} = \frac{2}{3}. \tag{23.39}$$

As a result, we get finally:

$$\lambda \sim \frac{Z\varepsilon_F}{Mc^2} \sim \frac{Z\varepsilon_F}{M\omega_D^2 a^2} = \frac{Z\varepsilon_F^2}{\hbar^2 \omega_D^2} \frac{\hbar^2}{Ma^2 \varepsilon_F}, \tag{23.40}$$

where a is interatomic distance.

But the Fermi energy:

$$\varepsilon_F = \frac{p_F^2}{2m} \sim \frac{\hbar^2}{ma^2}. \tag{23.41}$$

That is why:

$$\frac{\hbar^2}{Ma^2 \varepsilon_F} \sim \frac{m}{M}, \tag{23.42}$$

and hence:

$$\lambda \sim \frac{Z\varepsilon_F^2}{\hbar^2 \omega_D^2} \frac{m}{M}. \tag{23.43}$$

However, we know that in metals the ratio:

$$\frac{\varepsilon_F}{\hbar\omega_D} \sim \sqrt{\frac{M}{m}}.$$
(23.44)

Then from Eq. (23.43) we can see that the dimensionless coupling constant $\lambda \sim 1$. In most of the conventional superconductors (Hg, Al, Nb) this constant equals to:

$$\lambda \sim \left(\frac{1}{3} \div \frac{1}{4}\right)$$
(23.45)

(see [1, 12, 13, 16–22]). It is larger in Pb and metallic hydrogen.

It is important to note, that the screening does not contribute significantly in the amplitude of electron–ion interaction on large wave vectors $q \sim k_F \gg \kappa$.

Hence the smallness of this interaction compared to the Fermi energy is conserved. Namely:

$$\frac{4\pi Z e^2}{\varepsilon_F k_F^2 \Omega_0} = \frac{4\pi Z e^2}{\varepsilon_F \kappa^2 \Omega_0}\frac{\kappa^2}{k_F^2} \sim r_S \ll 1$$
(23.46)

in dense electron plasma for $r_S \ll 1$.

23.3   Cooper Problem

Let us show that the bound state exists for quasi-two-dimensional motion of two electrons with opposite momenta (and spins in case of conventional superconductors) in momentum space (see Fig. 23.5).

Let us denote the coordinate part of the two-electron wave function as: $\Psi(\mathbf{r}_1, \mathbf{r}_2)$. If spins of two electrons are opposite and thus the total spin of the pair S = 0, then

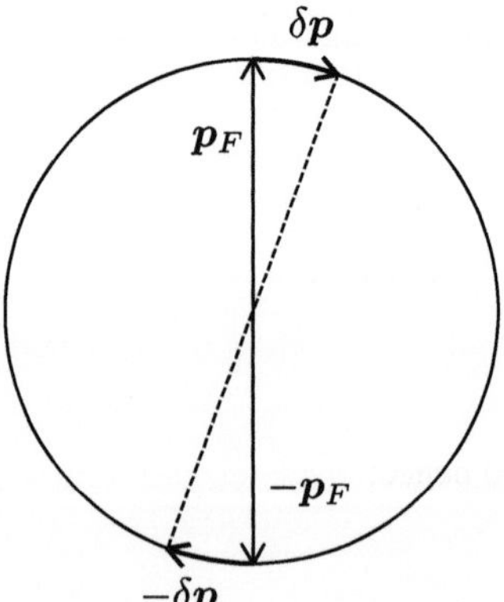

Fig. 23.5 Bound state formation for two electrons with opposite momenta $\mathbf{p}_F$ and $-\mathbf{p}_F$ for quasi-two-dimensional coherent motion along the Fermi surface in momentum space [1, 12]

the spin part of the two-electron wave function is antisymmetric (it corresponds to spin singlet state). In this case coordinate part is symmetric.

Let us expand coordinate part of the wave function $\Psi(\mathbf{r}_1, \mathbf{r}_2)$ in series over the states of two electrons. In the Cooper problem these one-electron states are plane waves:

$$\exp(i\mathbf{k}_1\mathbf{r}_1)\exp(i\mathbf{k}_2\mathbf{r}_2),$$

where we assume that the normalized volume $V = 1$.

As always in the two-particle problem, let us introduce the coordinates of the center of mass and relative motion. The center of mass coordinate for two particles with equal masses reads:

$$\mathbf{R} = \frac{\mathbf{r}_1 + \mathbf{r}_2}{2}, \tag{23.47}$$

At the same time, the coordinate of the relative motion:

$$\boldsymbol{\rho} = \mathbf{r}_1 - \mathbf{r}_2. \tag{23.48}$$

Correspondingly, coordinates of the first and second particle are given by:

$$\mathbf{r}_1 = \mathbf{R} + \frac{\boldsymbol{\rho}}{2}, \mathbf{r}_2 = \mathbf{R} - \frac{\boldsymbol{\rho}}{2}. \tag{23.49}$$

It is convenient also to introduce the center of mass momentum:

$$\mathbf{K} = \mathbf{k}_1 + \mathbf{k}_2, \tag{23.50}$$

and the momentum of the relative motion:

$$\mathbf{q} = \frac{\mathbf{k}_1 - \mathbf{k}_2}{2}. \tag{23.51}$$

Then:

$$\exp(i\mathbf{k}_1\mathbf{r}_1 + i\mathbf{k}_2\mathbf{r}_2) = \exp(i\mathbf{K}\mathbf{R} + i\mathbf{q}\boldsymbol{\rho}). \tag{23.52}$$

Let us stress, that the interaction in our case depends only on the coordinate of the relative motion $V\left(|\mathbf{r}_1 - \mathbf{r}_2|\right) = V(\rho)$.

We can expand two-particle wave function in series over plane waves as follows:

$$\Psi(\mathbf{R}, \boldsymbol{\rho}) = \sum_{q K} c_q(\mathbf{K}) \exp(i\mathbf{K}\mathbf{R}) \exp(i\mathbf{q}\boldsymbol{\rho}). \tag{23.53}$$

Let us solve the Schrodinger equation for the two-particle problem. Schrodinger equation in momentum space has the form ($\hbar = 1$):

$$\left(\frac{K^2}{4m} + \frac{q^2}{m} - E \right) c_q + \sum_{q_1} V_{qq_1} c_{q_1} = 0, \tag{23.54}$$

where kinetic energy of the center of mass $\frac{K^2}{4m}$ is related to the total mass $2m$ of two particles with mass m, while kinetic energy of the relative motion $\frac{q^2}{m}$ is connected with the reduced mass $\frac{m}{2}$.

Hence:

$$c_q = -\frac{1}{\frac{K^2}{4m} + \frac{q^2}{m} - E} \sum_{q_1} V_{qq_1} c_{q_1}. \tag{23.55}$$

As we showed in the beginning of the Lecture, interaction of two electrons close to the Fermi surface corresponds to attraction due to phonon exchange. In other words, if two electrons are living in the narrow energy shell where ($\hbar = 1$):

$$|\varepsilon_1 - \varepsilon_F| \ll \omega_D, |\varepsilon_2 - \varepsilon_F| \ll \omega_D, \text{ and } |\varepsilon_1 - \varepsilon_2| \ll \omega_D, \tag{23.56}$$

then the interaction between them can be described by attractive (negative) constant:

$$V_0 = -|V_0| < 0.$$

In this case Eq. (23.55) is simplified and can be written as:

$$c_q = \frac{|V_0|}{\frac{K^2}{4m} + \frac{q^2}{m} - E} \sum_{q' \neq q}' c_{q'}, \tag{23.57}$$

where uppercase stroke at sum sign means that in Eq. (23.57) we sum up over q' only in the narrow energy shell close to the Fermi surface.

Let us sum up the l.h.s. and r.h.s. of Eq. (23.57) over $q \neq q'$:

$$\left(\sum_q' c_q \right) = \left(\sum_q' \frac{|V_0|}{\frac{K^2}{4m} + \frac{q^2}{m} - E} \right) \left(\sum_{q'}' c_{q'} \right). \tag{23.58}$$

Further on, replacing summation over q by q' in the r.h.s. of Eq. (23.58) and reducing left and right by $\left(\sum_q' c_q \right)$, (assuming that for nontrivial solution $\left(\sum_q' c_q \right) \neq 0$) we obtain:

$$1 = \sum_q' \frac{|V_0|}{\frac{K^2}{4m} + \frac{q^2}{m} - E}. \tag{23.59}$$

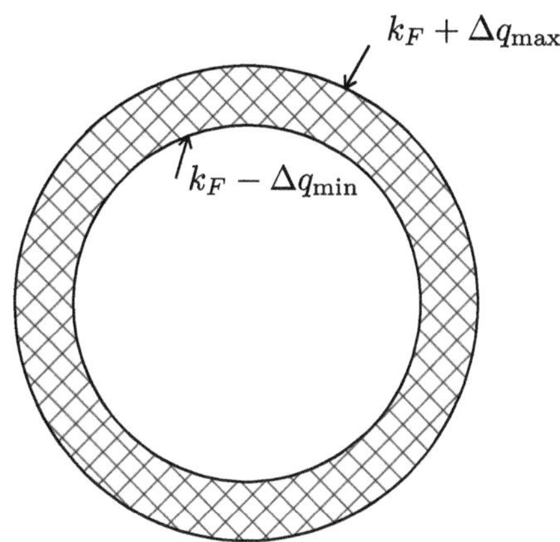

Fig. 23.6 The Cooper pairing of two electrons in momentum space in the narrow energy shell close to the Fermi surface for wave vector of the relative motion $q = k_F + \Delta q$. We show in the figure the upper and lower boundaries of the shell $k_F + \Delta q_{\mathrm{max}}$ and $k_F - \Delta q_{\mathrm{min}}$ [1]

Let us solve the problem for the total momentum of two electrons:

$$K = |\boldsymbol{k}_1 + \boldsymbol{k}_2| \to 0. \tag{23.60}$$

Then the momentum of the first particle: $\boldsymbol{k}_1 \to -\boldsymbol{k}_2$, and momentum of the relative motion: $\boldsymbol{q} \approx \frac{2\boldsymbol{k}_1}{2} \approx \boldsymbol{k}_1$. But $|\,\boldsymbol{k}_1| \sim k_F$. Hence, $q \sim k_F$.

Let us denote the momentum of the relative motion of two electrons as:

$$q = k_F + \Delta q, \tag{23.61}$$

where Δq is small (see Fig. 23.6).

Using Eq.'s (23.50 and 23.51) we can write initial wave vectors of two electrons as:

$$\boldsymbol{k}_1 = \boldsymbol{q} + \frac{\boldsymbol{K}}{2}, \, \boldsymbol{k}_2 = -\boldsymbol{q} + \frac{\boldsymbol{K}}{2}. \tag{23.62}$$

Hence for small values of the total momentum $\boldsymbol{K}$ we get from the cosine theorem:

$$k_1^2 = q^2 + qK \cos\theta + \frac{K^2}{4} \approx q^2 + qK \cos\theta, \tag{23.63}$$

and:

$$k_2^2 \approx q^2 - qK \cos\theta, \tag{23.64}$$

where we neglected the term $\frac{K^2}{4}$ and introduced angle θ between wave vectors $\boldsymbol{q}$ and $\boldsymbol{K}$.

23.4　Limitations on the Integration Regions

Firstly, we should satisfy two conditions on the integration regions, connected with the filling of the Fermi sphere:

$$k_1^2 > k_F^2, \text{ and } k_2^2 > k_F^2. \tag{23.65}$$

More tough condition from these two can be written as follows:

$$q^2 - qK|\cos\theta| > k_F^2. \tag{23.66}$$

Further on let us substitute the definition of vector q in (23.61) in (23.66). As a result, for small values of Δq we obtain:

$$k_F^2 + 2\Delta q k_F - qK|\cos\theta| > k_F^2. \tag{23.67}$$

Reducing left and right by k_F^2, we get the first of the limitations:

$$2\Delta q k_F > qK|\cos\theta|, \tag{23.68}$$

and, correspondingly (see Fig. 23.6):

$$\Delta q_{\min} = \frac{qK|\cos\theta|}{2k_F} \approx \frac{k_F K|\cos\theta|}{2k_F} = \frac{K|\cos\theta|}{2}. \tag{23.69}$$

The second limitation can be derived from the conditions:

$$k_1^2 \le k_F^2 + 2m\omega_0, \text{ and } k_2^2 \le k_F^2 + 2m\omega_0, \tag{23.70}$$

where ω_0 is a typical phonon frequency.

The toughest of these two conditions can be written in the form:

$$q^2 + qK|\cos\theta| \le k_F^2 + 2m\omega_0. \tag{23.71}$$

Substituting again the definition (23.61), we obtain:

$$k_F^2 + 2\Delta q k_F + qK|\cos\theta| \le k_F^2 + 2m\omega_0, \tag{23.72}$$

or reducing left and right by k_F^2:

$$2\Delta q k_F + qK|\cos\theta| \le 2m\omega_0. \tag{23.73}$$

Hence:

$$\Delta q_{\max} \approx \frac{2m\omega_0}{2k_F} - \frac{k_F K|\cos\theta|}{2k_F} = \frac{m\omega_0}{k_F} - \frac{K|\cos\theta|}{2}. \tag{23.74}$$

So, we can proceed now from the summation to the integration in (23.59) accounting for two limitations on $\Delta q_{\min}$ and $\Delta q_{\max}$. Then:

$$1 = |V_0| \int \frac{d^3\boldsymbol{q}}{(2\pi)^3} \frac{1}{\frac{K^2}{4m} + \frac{q^2}{m} - E} \approx |V_0| \int \frac{d^3\boldsymbol{q}}{(2\pi)^3} \frac{1}{\frac{k_F^2}{m} + \frac{2\Delta q k_F}{m} - E}, \qquad (23.75)$$

where we neglect the small term $\frac{K^2}{4m}$ in denominator and write $\frac{q^2}{m}$ in terms of k_F and Δq using definition (23.61).

Further on it is convenient to count the binding energy of the two particles from their doubled chemical potential $2\varepsilon_F$ and introduce the notation:

$$E = 2\varepsilon_F - \varepsilon, \qquad (23.76)$$

where $-\varepsilon$ is the negative binding energy. After that we can rewrite the measure of integration in the spherical coordinates as:

$$\int \frac{d^3\boldsymbol{q}}{(2\pi)^3} = \frac{4\pi}{(2\pi)^3} \int_0^1 d\cos\theta \int_{\Delta q_{\min}}^{\Delta q_{\max}} k_F^2 d\Delta q. \qquad (23.77)$$

As a result, Eq. (23.75) acquires the form:

$$
\begin{aligned}
1 &= |V_0| \frac{4\pi}{(2\pi)^3} \int_0^1 d\cos\theta \int_{\Delta q_{\min}}^{\Delta q_{\max}} k_F^2 d\Delta q \, \frac{1}{\frac{k_F^2}{m} + \frac{2\Delta q k_F}{m} - 2\varepsilon_F + \varepsilon} \\
&= |V_0| \frac{4\pi}{(2\pi)^3} \int_0^1 d\cos\theta \int_{\Delta q_{\min}}^{\Delta q_{\max}} k_F^2 d\Delta q \, \frac{1}{\frac{2\Delta q k_F}{m} + \varepsilon} \\
&= |V_0| \frac{4\pi k_F^2 m}{(2\pi)^3} \int_0^1 d\cos\theta \int_{\Delta q_{\min}}^{\Delta q_{\max}} \frac{d\Delta q}{2\Delta q k_F + m\varepsilon} \\
&= |V_0| \frac{4\pi k_F^2 m}{2(2\pi)^3 k_F} \int_0^1 d\cos\theta \ln\left(2\Delta q k_F + m\varepsilon\right) \big|_{\Delta q_{\min}}^{\Delta q_{\max}} \\
&= |V_0| \frac{m k_F}{(2\pi)^2} \int_0^1 d\cos\theta \ln \frac{2 k_F \Delta q_{\max} + m\varepsilon}{2 k_F \Delta q_{\min} + m\varepsilon} \\
&= |V_0| \, g\left(\varepsilon_F\right) \int_0^1 d\cos\theta \ln \frac{2m\omega_0 - k_F K |\cos\theta| + m\varepsilon}{k_F K |\cos\theta| + m\varepsilon}. \qquad (23.78)
\end{aligned}
$$

In (23.78) we used expression (23.69) and (23.74) for $\Delta q_{\min}$ and $\Delta q_{\max}$, and introduced again the notation:

$$g\left(\varepsilon_F\right) = N(0) = \frac{m k_F}{(2\pi)^2}$$

for the density of states at the Fermi level for one spin projection in case of quadratic spectrum of electrons (assuming $\hbar = 1$).

Further on in the nominator of the expression for the logarithm we can neglect the binding energy ε in comparison with the phonon frequenct ω_0 (we would like to fulfill the condition $\varepsilon \ll \omega_0$). Besides that we assume that $k_F K \ll m\omega_0$. Then lifting the modular brackets for $\cos\theta$, we get:

$$1 = |V_0| g\left(\varepsilon_F\right) \int_0^1 d\cos\theta \ln \frac{2m\omega_0}{k_F K \cos\theta + m\varepsilon}, \tag{23.79}$$

where $\varepsilon > 0$.

After that we can take the integral in (23.81) by parts. Hence:

$$1 = |V_0| g\left(\varepsilon_F\right) \left\{ \cos\theta \ln \frac{2m\omega_0}{k_F K \cos\theta + m\varepsilon} \Big|_0^1 + \int_0^1 d\cos\theta \frac{k_F K \cos\theta}{k_F K \cos\theta + m\varepsilon} \right\}$$

$$= |V_0| g\left(\varepsilon_F\right) \left\{ \ln \frac{2m\omega_0}{k_F K + m\varepsilon} + \int_0^1 d\cos\theta \left(1 - \frac{m\varepsilon}{k_F K \cos\theta + m\varepsilon} \right) \right\}$$

$$= |V_0| g\left(\varepsilon_F\right) \left\{ \ln \frac{2m\omega_0}{k_F K + m\varepsilon} + 1 - \frac{m\varepsilon}{k_F K} \ln \left(k_F K \cos\theta + m\varepsilon\right) \Big|_0^1 \right\}$$

$$= |V_0| g\left(\varepsilon_F\right) \left\{ \ln \frac{2m\omega_0}{k_F K + m\varepsilon} + 1 - \frac{m\varepsilon}{k_F K} \ln \frac{k_F K + m\varepsilon}{m\varepsilon} \right\}$$

$$= |V_0| g\left(\varepsilon_F\right) \left\{ \ln \frac{2m\omega_0}{m\varepsilon + k_F K} + 1 - \frac{m\varepsilon}{k_F K} \ln \left(1 + \frac{k_F K}{m\varepsilon} \right) \right\}. \tag{23.80}$$

Let us consider first physically most important case when two electrons have antiparallel momenta, and the total momentum $K = 0$. In this case expanding the second logarithm in the last line in Eq. (23.80) in series up to linear terms we obtain:

$$\ln \left(1 + \frac{k_F K}{m\varepsilon} \right) \approx \frac{k_F K}{m\varepsilon}. \tag{23.81}$$

Now we can rewrite the equation for the binding energy of the Cooper pair in the form:

$$1 = |V_0| g\left(\varepsilon_F\right) \left\{ \ln \frac{2m\omega_0}{m\varepsilon} + 1 - \frac{m\varepsilon}{k_F K} \frac{k_F K}{m\varepsilon} \right\} = |V_0| g\left(\varepsilon_F\right) \left\{ \ln \frac{2m\omega_0}{m\varepsilon} + 1 - 1 \right\}$$

$$= |V_0| g\left(\varepsilon_F\right) \ln \frac{2m\omega_0}{m\varepsilon} = |V_0| g\left(\varepsilon_F\right) \ln \frac{2\omega_0}{\varepsilon}. \tag{23.82}$$

From Eq. (23.82) we can find the binding energy:

$$\varepsilon_b = 2\omega_0 \exp\left(-\frac{1}{|V_0| g\left(\varepsilon_F\right)} \right), \tag{23.83}$$

and correspondingly:

$$E_b = 2\varepsilon_F - \varepsilon_b < 2\varepsilon_F. \tag{23.84}$$

Let us stress, that the point where the interaction is absent ($V_0 = 0$) in (23.82) and (23.83) is the point of the substantial singularity, since the energy cannot be expanded in series in the neighborhood of $V_0 = 0$.

That is why we cannot obtain the result for the binding energy summing up the finite number of terms of the perturbation theory.

On the diagrammatic language [4, 12, 13, 23, 24], we should sum up the infinite geometric series of the so-called ladder diagrams in the Cooper (particle-particle) channel to get the pole of the two-particle vertex in the form of Eq. (23.82):

$$1 = |V_0|\, g\,(\varepsilon_F) \ln \frac{2\omega_0}{\varepsilon}. \tag{23.85}$$

Here we have an analogy between superconductivity theory [5–7] and Quantum field theory [25]. Since the binding energy E_b in (23.84) is less than $2\varepsilon_F$ or two rest energies of the particles in Quantum electrodynamics:

$$E_b < 2mc^2, \tag{23.86}$$

then we should count the binding energy of the Cooper pair from the background (from the initial level).

Comment 1. In exact BCS theory [5,7] in the denominator of the exponent in (23.83) for the pair binding energy enters two times larger quantity $2\,|V_0|\, g\,(\varepsilon_F)$, that is two times larger density of states $2\, g\,(\varepsilon_F)$, since deriving (23.83) we account only electrons living above the Fermi surface, but do not take into account electrons below Fermi surface. In more exact theory we should account them as well.

In the next Lecture we will see that the binding energy of the Cooper pair ε_b by order of magnitude equals to the critical temperature of the superconducting transition. Note, that for conventional superconductors (Hg, Al, Nb) the critical temperature equals to $T_C \sim (1 \div 10)$ K [1, 12, 13, 16–24, 26–35].

This estimate follows from formula (23.83) for the binding energy under the substitution of the typical phonon frequency (of the order of Debye frequency):

$$\hbar\omega_0 \sim \theta_D \sim (100 \div 200) \text{ K}, \tag{23.87}$$

and attractive coupling constant of electron–phonon interaction:

$$\lambda = 2\,|V_0|\, g\,(\varepsilon_F) \sim \left(\frac{1}{3} \div \frac{1}{4}\right). \tag{23.88}$$

In this case the binding energy and the critical temperature for $\lambda = \frac{1}{3}$:

$$\varepsilon_b \sim T_C \sim \theta_D \exp(-\frac{1}{\lambda}) \sim \theta_D \exp(-3) \sim 10 \text{ K}. \tag{23.89}$$

Comment 2. We consider symmetric coordinate wave function of two electrons $\Psi(r_1, r_2)$. Hence, the spin part of the two-electron wave function $\Phi(\sigma_1, \sigma_2)$ should be antisymmetric with respect to the permutation of electron spins.

Comment 3. We found fundamental restructuring of electron spectrum (and not just the renormalization of the spectrum). In fact, we separate the bound state of two electrons solving the Cooper problem [6].

This fundamental reconstruction of the spectrum should be taken into account in the many-body problem [5,7] (on the level of the Bogoliubov $u - v$ transformation for electron systems [8,9]) to get the energy gap in the spectrum of one-particle excitations in superconductor.

Note, that in magnetic field we also have a radical reconstruction of electron spectrum leading to the emergence of Landau levels [2,36] and to the finite motion of electron in the plane perpendicular to the field direction.

23.5   Critical Velocity in Superconductor

Let us find the critical velocity for the center of mass of the Cooper pair for which the bound state disappears:

$$\varepsilon_b \left(K_{\text{crit.}} \neq 0 \right) = 0.$$

If in Eq. (23.79) we put $\varepsilon = 0$, then in this case:

$$\begin{aligned}
1 &= |V_0|\, g\,(\varepsilon_F) \int_0^1 d\cos\theta \ln \frac{2m\omega_0}{k_F K \cos\theta} \\
&= |V_0|\, g\,(\varepsilon_F) \left[\ln \frac{2m\omega_0}{k_F K} - \int_0^1 d\cos\theta \ln\cos\theta \right].
\end{aligned} \tag{23.90}$$

But the integral:

$$\int_0^1 d\cos\theta \ln\cos\theta = -1. \tag{23.91}$$

Hence we obtain:

$$\begin{aligned}
1 &= |V_0|\, g\,(\varepsilon_F) \left[\ln \frac{2m\omega_0}{k_F K} + 1 \right] \\
&= |V_0|\, g\,(\varepsilon_F) \ln \frac{2m\omega_0 e}{k_F K}.
\end{aligned} \tag{23.92}$$

Correspondingly:

$$\ln \frac{2m\omega_0 e}{k_F K} = \frac{1}{|V_0|\, g\,(\varepsilon_F)}, \tag{23.93}$$

and thus:

$$\frac{k_F K}{m} = 2\omega_0 e \exp\left\{ -\frac{1}{|V_0|\, g\,(\varepsilon_F)} \right\} = e\varepsilon_b(K = 0). \tag{23.94}$$

As a result, the critical velocity of the center of mass of the pair:

$$V_{\text{crit.}} = \frac{K_{\text{crit.}}}{2m} = \frac{e\varepsilon_b(K=0)}{2k_F} \sim \frac{\varepsilon_b}{k_F}. \tag{23.95}$$

Note, that in many-body BCS theory [5,7] according to Landau criterium of superfluidity [3]:

$$V_{\text{crit.}} \sim \frac{\Delta}{k_F} \sim \frac{T_C}{k_F}, \tag{23.96}$$

where:

$$\Delta \sim \varepsilon_b \sim T_C \tag{23.97}$$

is the superconducting gap (in the spectrum of elementary excitations).

More detailed derivation of the expression (23.97) for electron excitations in superconductor will be presented in the next Lecture.

So, the binding energy decreases as a function of the center of mass momentum K. Moreover, the critical velocity exists for which the Cooper pairs have zero binding energy ε_b ($V_{\text{crit.}}$) $= 0$ (and the composed boson consisting of two electrons is destroyed).

This is the main difference between the destruction of superconductivity in metal and superfluidity in helium [36].

In superfluid He-4 for velocities higher than the critical one, the counterflow of spontaneously generated excitations destroys the superfluid flow [36,37]. As a result, the superfluid helium goes to the normal state. Nevertheless, it remains Bose liquid even in the normal state [1].

In superconductor for the velocities larger than the critical one we get the destruction of the Cooper pairs, and superconducting Bose liquid of Cooper pairs in metal transforms to the normal Fermi liquid of unpaired electrons.

The same happens at higher temperatures. For temperatures higher than the critical one $T > T_C$ superfluid He-4 becomes normal Bose liquid. At the same time, superconductor becomes normal fermionic metal [1] (Fig. 23.7).

23.6 Coherence Length of the Cooper Pair

Coherence length (or the size of the spatial localization of the pair) can be found from the estimate for the pair binding energy.

The binding energy by order of magnitude equals to the difference between electron energies lying close to the Fermi surface and on the Fermi surface itself [1]:

$$\varepsilon_b \sim \frac{(p_F + \delta p)^2}{2m} - \frac{p_F^2}{2m} \approx \frac{p_F \delta p}{m} = v_F \delta p, \tag{23.98}$$

where δp is uncertainty (smearing over momentum of the Cooper pair), and $v_F = \frac{p_F}{m}$ is the Fermi velocity.

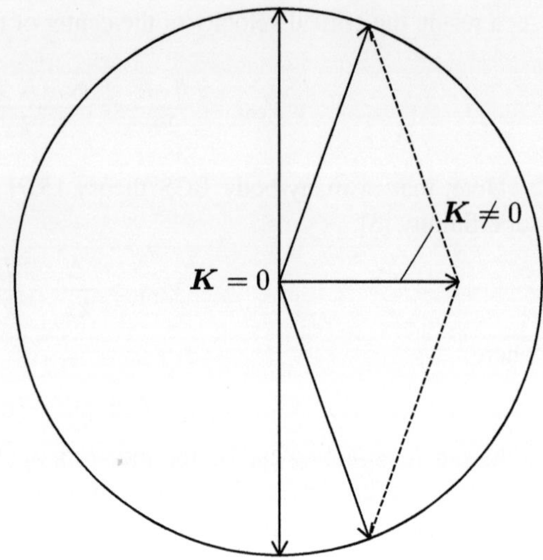

Fig. 23.7 Cooper pairs with zero total momentum $K = 0$ and total momentum different from zero ($K \neq 0$), when we get certain angle between the momenta of two electrons of the pair [1]

Then, from Heisenberg uncertainty principle [2], coherence length (or the smearing of the pair over coordinate):

$$\xi_0 \sim \frac{\hbar}{\delta p} \sim \frac{\hbar v_F}{\varepsilon_b} \sim \frac{\hbar v_F a}{\varepsilon_b a} \sim a\frac{v_F p_F}{\varepsilon_b} \sim a\frac{\varepsilon_F}{\varepsilon_b} \gg a, \tag{23.99}$$

where a is interatomic distance, and $\frac{\hbar}{a} \sim p_F$.

But the typical value of the Fermi energy $\varepsilon_F \sim 10^4$ K, whereas the binding energy $\varepsilon_b \sim 10$ K. Hence the coherence length in conventional superconductors: $\xi_0 \sim 10^3 a \sim (3000 \div 4000)$ Å strongly exceeds the interatomic distance [1, 3–10, 12, 13].

Hence the Cooper pairs in conventional superconductors are very extended and strongly overlapping with each other. As a result, inside each pair there are a lot of electrons belonging to other Cooper pairs (see Fig. 23.8). This is the case of weak superconductivity (superconductivity in momentum space).

Let us emphasize, that another limiting case is possible corresponding to strong superconductivity (superconductivity in coordinate space) [38–41]. In this case the local (compact pairs) or bi-polarons (bi-electron molecules) arise in the system. This situation is realized for example in superconducting bismuth oxides (BaKBiO) [23, 24, 42–46] and quantum Fermi gases in magnetic traps in the regime of Feshbach resonance [47–52].

We can say that the limit of strong superconductivity corresponds to Bose–Einstein condensation (BEC) [53–58] of preformed local pairs [12, 23, 38–46]. The crossover between these two limiting cases in the modern terminology is called the BCS-BEC crossover (between the extended and local pairs). The more details on the BCS-BEC crossover in different systems can be found in [12, 23, 59–63].

Note that in many high-temperature superconductors [64–79] and superconductors with heavy fermions [80–83] we have just the intermediate situation with rather

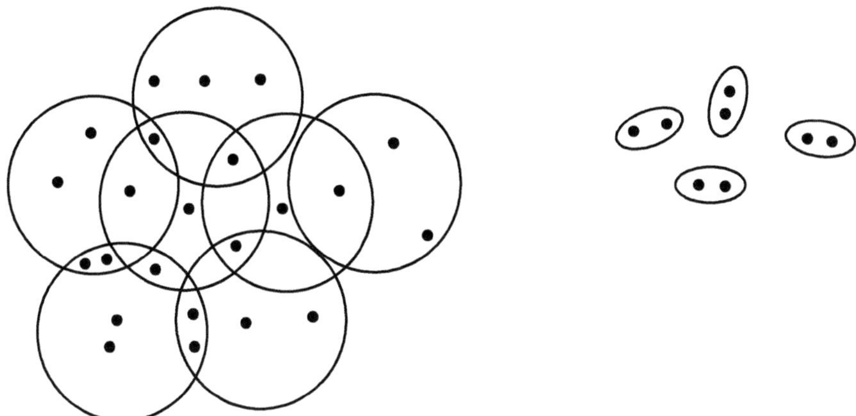

Fig. 23.8 Two limiting cases of the BCS-BEC crossover between extended and local pairs. On the left side of the figure, we show extended Cooper pairs in conventional superconductors (Hg, Al, Nb, Pb). On the right side we illustrate the compact pairs in superconducting bismuth oxide (BaKBiO) and in quantum Fermi gases Li-6 and K-40 in magnetic traps in the regime of Feshbach resonance [1,12]

short coherence length $\xi_0 \sim 50\,\text{Å}$. This length nevertheless is much larger than interatomic distance. Thus the pairs are still overlapping (we are still on BCS side of the crossover), but not so strongly as in conventional superconductors [12,23].

23.7 The Tolmachev Logarithm

Let us understand what happens with the Cooper pairs when we are switching on direct Coulomb repulsion between electrons. Let us find the condition for creation of the Cooper pair in this case.

As we know, the amplitude of electron–phonon interaction corresponds to attraction between electrons in the narrow energy shell close to the Fermi surface in momentum space. The Coulomb repulsion acts in the whole unlimited (free) momentum space.

Let us solve again the Schrodinger equation for the two-particle electron wave function in momentum space for the total interaction:

$$V(\rho) = V'_{\text{el.-ph.}}(\rho) + V''_{\text{Coul.}}(\rho) \tag{23.100}$$

for zero total momentum $\boldsymbol{K} = 0$.

In this case, initial Schrodinger equation in momentum space (23.55) on the expansion coefficients c_q of the wave function $\Psi(\boldsymbol{R}, \rho)$ by the plane waves $\exp(i\boldsymbol{K}\boldsymbol{R} + i\boldsymbol{q}\rho)$ acquires the form for $\boldsymbol{K} = 0$:

$$c_q = -\frac{1}{\frac{q^2}{m} - E} \sum_{q_1} V_{qq_1} c_{q_1} = -\frac{1}{\frac{q^2}{m} - E} \left\{ \sum_{q_1} V'_{qq_1} c_{q_1} + \sum_{q_1} V''_{qq_1} c_{q_1} \right\},$$

$$(23.101)$$

where V'_{qq_1} corresponds to electron–phonon interaction and V''_{qq_1}-to electron–electron interaction.

Let us write down the system of two equations. The first equation we can get summing up over the region, where electron–phonon interaction is absent (but the Coulomb interaction is present), while the second one we can derive by the summation over the region where both interactions are present.

Let us mark these regions in momentum space with one uppercase stroke at sum sign (for narrow region) or two strokes (for broad region).

Then the Fourier component for screened Coulomb interaction:

$$V''_0 = \frac{4\pi e^2}{\kappa^2 + q^2} \approx \frac{4\pi e^2}{\kappa^2} = \text{const} > 0 \qquad (23.102)$$

is the positive constant for $\kappa \gg q$.

Note, that in real metal for the values of the correlation radius $r_S \leq 1$ this inequality is violated (and thus the Coulomb interaction V''_0 becomes substantially smaller than $\frac{4\pi e^2}{\kappa^2}$) only for sufficiently large wave vectors $q \sim k_F$.

This fact determines the broad range of action of the Coulomb interaction which is of the order of the Fermi energy ε_F in the energy space.

As a result, Eq. (23.102) reads:

$$c_q = -\frac{V'_0}{\frac{q^2}{m} - E} \sum_{q_1}{}' c_{q_1} - \frac{V''_0}{\frac{q^2}{m} - E} \sum_{q_1}{}'' c_{q_1}, \qquad (23.103)$$

where $V'_0 = -\left|V'_0\right| < 0$ is the negative constant of electron–phonon interaction.

This constant is different from zero in the narrow energy region of the order of the typical phonon frequency: $\hbar\omega_0 \ll \varepsilon_F$.

Correspondingly, after the summation over the second broad region Eq. (23.102) can be rewritten as:

$$\sum_q{}'' c_q = \sum_q{}'' \frac{\left|V'_0\right|}{\frac{q^2}{m} - E} \sum_{q_1}{}' c_{q_1} - \sum_q{}'' \frac{V''_0}{\frac{q^2}{m} - E} \sum_{q_1}{}'' c_{q_1}. \qquad (23.104)$$

But we should remember, that initial electron–phonon interaction V'_{qq_1} is different from zero only in the narrow shell close to the Fermi surface with respect to both indices q and q_1. That is why in the first term in the r.h.s. of Eq. (23.104) works only narrow region related to the sum with one stroke.

Hence:

$$\sum_{q}{}'' c_q = \sum_{q}{}' \frac{|V_0'|}{\frac{q^2}{m} - E} \sum_{q_1}{}' c_{q_1} - \sum_{q}{}'' \frac{V_0''}{\frac{q^2}{m} - E} \sum_{q_1}{}'' c_{q_1}. \tag{23.105}$$

At the same time, after summation of Eq. (23.103) over the first narrow region:

$$\sum_{q}{}' c_q = \sum_{q}{}' \frac{|V_0'|}{\frac{q^2}{m} - E} \sum_{q_1}{}' c_{q_1} - \sum_{q}{}' \frac{V_0''}{\frac{q^2}{m} - E} \sum_{q_1}{}'' c_{q_1}. \tag{23.106}$$

Note, that in the second term in the r.h.s. the Coulomb interaction works well both in narrow and broad regions.

Further on let us introduce convenient notations for the sum of the expansion coefficients of the wave function:

$$\sum_{q}{}'' c_q = c'', \text{ and } \sum_{q}{}' c_q = c'; \tag{23.107}$$

as well as for the sums with the kernels (with the denominators) of the Schrodinger integral equations:

$$\sum_{q}{}' \frac{1}{\frac{q^2}{m} - E} = \left(\sum{}'\right)_q, \text{ and } \sum_{q}{}'' \frac{1}{\frac{q^2}{m} - E} = \left(\sum{}''\right)_q. \tag{23.108}$$

Then in notations (23.107)–(23.108) the system of Eq.'s (23.105)–(23.106) can be written in the form:

$$c'' - |V_0'| \left(\sum{}'\right)_q c' + V_0'' \left(\sum{}''\right)_q c'' = 0, \tag{23.109}$$

$$c' - |V_0'| \left(\sum{}'\right)_q c' + V_0'' \left(\sum{}'\right)_q c'' = 0. \tag{23.110}$$

Reducing the similar terms, we can rewrite the system of Eq.'s (23.109)–(23.110) as:

$$c'' \left[1 + V_0'' \left(\sum{}''\right)_q\right] = c' |V_0'| \left(\sum{}'\right)_q, \tag{23.111}$$

$$c' \left[1 - |V_0'| \left(\sum{}'\right)_q\right] = -c'' V_0'' \left(\sum{}'\right)_q. \tag{23.112}$$

After that we can express c'' via c' from Eq. (23.111):

$$c'' = \frac{c' \, |V_0'| \left(\sum_q'\right)}{1 + V_0'' \left(\sum_q''\right)}.$$

(23.113)

Substituting (23.113) in (23.112) we finally get for c':

$$c' \left\{ 1 - |V_0'| \left(\sum_q{}'\right) + \frac{|V_0'| \, V_0'' \left(\sum_q'\right)^2}{1 + V_0'' \left(\sum_q''\right)} \right\} = 0.$$

(23.114)

This equation has nontrivial solution (for $c' = \sum_q' c_q \neq 0$) when:

$$
\begin{aligned}
1 &= |V_0'| \left(\sum_q{}'\right) \left(1 - \frac{V_0'' \left(\sum_q'\right)}{1 + V_0'' \left(\sum_q''\right)} \right) \\
&= \frac{|V_0'| \left(\sum_q'\right)}{1 + V_0'' \left(\sum_q''\right)} \left(1 + V_0'' \left(\sum_q{}'' - \sum_q{}'\right) \right).
\end{aligned}
$$

(23.115)

Hence multiplying the l.h.s. and r.h.s. of Eq. (23.115) on the denominator $1 + V_0'' \left(\sum_q''\right)$ we obtain:

$$1 + V_0'' \left(\sum_q{}''\right) = |V_0'| \left(\sum_q{}'\right) \left(1 + V_0'' \left(\sum_q{}'' - \sum_q{}'\right) \right).$$

(23.116)

Further on let us divide the l.h.s. and the r.h.s. of Eq. (23.116) on:

$$1 + V_0'' \left(\sum_q{}'' - \sum_q{}'\right).$$

Then:

$$\frac{1 + V_0'' \left(\sum_q''\right)}{1 + V_0'' \left(\sum_q'' - \sum_q'\right)} = |V_0'| \left(\sum_q{}'\right).$$

(23.117)

Now subtracting and adding $V_0'' \left(\Sigma_q'\right)$ to the nominator in the l.h.s. of Eq. (23.117) (23.117) we get:

$$\left(\frac{1 + V_0'' \left(\Sigma_q'' - \Sigma_q'\right)}{1 + V_0'' \left(\Sigma_q'' - \Sigma_q'\right)} + \frac{V_0'' \left(\Sigma_q'\right)}{1 + V_0'' \left(\Sigma_q'' - \Sigma_q'\right)} \equiv 1 + \frac{V_0'' \left(\Sigma_q'\right)}{1 + V_0'' \left(\Sigma_q'' - \Sigma_q'\right)}\right)$$

$$= |V_0'| \left(\sum_q{}'\right). \tag{23.118}$$

Correspondingly, transferring the second term $\dfrac{V_0'' \left(\Sigma_q'\right)}{1 + V_0'' \left(\Sigma_q'' - \Sigma_q'\right)}$ from the left side to the right side of (23.118), we derive the final expression:

$$1 = \left(\sum_q{}'\right)\left(|V_0'| - \frac{V_0''}{1 + V_0'' \left(\Sigma_q'' - \Sigma_q'\right)}\right). \tag{23.119}$$

After that we can use the solution of the Cooper problem according to which for the narrow region:

$$\sum_q{}' \equiv \sum_q{}' \frac{1}{\frac{q^2}{m} - E} = \int \frac{d^3 q}{(2\pi)^3} \frac{1}{\frac{q^2}{m} - E} = g\left(\varepsilon_F\right) \ln \frac{2\omega_0}{\varepsilon}. \tag{23.120}$$

Note that in Eq. (23.120) the binding energy of the Cooper pair $E = 2\varepsilon_F - \varepsilon$, while ω_0 is typical frequency of the phonon spectrum, and $g\left(\varepsilon_F\right)$ is again the density of states on one spin projection (which is 2 times smaller than the correct density of states entering in more exact many-body BCS theory [5,7]).

Analogously, for the broad region (since anyway the singularity is related to the energy $\varepsilon \to 0$) we get:

$$\sum_q{}'' = \int \frac{d^3 q}{(2\pi)^3} \frac{1}{\frac{q^2}{m} - E} = g\left(\varepsilon_F\right) \ln \frac{2\varepsilon_F}{\varepsilon}. \tag{23.121}$$

Then for the sums difference in the denominator of the second term in the r.h.s. of Eq. (23.119):

$$\sum_q{}'' - \sum_q{}' = g\left(\varepsilon_F\right) \ln \frac{\varepsilon_F}{\omega_0}, \tag{23.122}$$

and we derive the famous result of Tolmachev [84]:

$$1 = \ln \frac{2\omega_0}{\varepsilon}\left(|V_0'| g\left(\varepsilon_F\right) - \frac{V_0'' g\left(\varepsilon_F\right)}{1 + V_0'' g\left(\varepsilon_F\right) \ln \frac{\varepsilon_F}{\omega_0}}\right) = \ln \frac{2\omega_0}{\varepsilon}\left(\lambda - \mu^*\right). \tag{23.123}$$

Let us stress that the dimensionless coupling constant of electron–phonon interaction still (as in the case when the Coulomb interaction is absent) in the rigorous theory (with an account of factor 2 for electrons above and below the Fermi surface) is given by:

$$\lambda = 2\left|V_0'\right| g\left(\varepsilon_F\right). \tag{23.124}$$

At the same time, the Coulomb pseudopotential (also with an account of factor 2) yields:

$$\mu^* = \frac{2V_0'' g\left(\varepsilon_F\right)}{1 + 2V_0'' g\left(\varepsilon_F\right)\ln\frac{\varepsilon_F}{\omega_0}} = \frac{\mu_{\text{Coul.}}}{1 + \mu_{\text{Coul.}}\ln\frac{\varepsilon_F}{\omega_0}}. \tag{23.125}$$

The pseudopotential μ^* in comparison with the bare Coulomb potential:

$$\mu_{\text{Coul.}} = 2V_0'' g\left(\varepsilon_F\right) \tag{23.126}$$

is reduced by large Tolmachev logarithm [84]:

$$\ln\frac{\varepsilon_F}{\omega_0} \sim 5 \text{ for } \frac{\varepsilon_F}{\omega_0} \sim 100. \tag{23.127}$$

Let us emphasize, that the bare Coulomb potential for simple metals and total density of states (for both spin projections) as we have shown in Lecture 18 for quadratic electron spectrum equals to:

$$\mu_{\text{Coul.}} = 2V_0'' g\left(\varepsilon_F\right) = \frac{4\pi e^2}{\kappa^2}N(0) = \frac{4\pi e^2}{8\pi e^2 N(0)}N(0) = \frac{1}{2}. \tag{23.128}$$

Thus, the typical values of the Coulomb pseudopotential [1,12,23]:

$$\mu^* \sim \frac{\frac{1}{2}}{1 + \frac{1}{2}\cdot 5} \sim 0.1 \div 0.15. \tag{23.129}$$

At the same time, electron–phonon coupling constant in conventional superconductors as we discussed already:

$$\lambda \sim \frac{1}{3} \div \frac{1}{4}.$$

Thus, effective coupling constant in Eq. (23.123):

$$\lambda - \mu^* > 0 \tag{23.130}$$

for conventional superconductors corresponds to attraction even if we switch on the Coulomb repulsion between electrons. As a result, in this more realistic case the Cooper pairs are emerging again. The binding energy of the pair reads now:

$$\varepsilon_b \sim 2\omega_0 \exp\left(-\frac{1}{\lambda - \mu^*}\right). \tag{23.131}$$

The estimate for the critical temperature of the superconducting transition proves to be the same by order of magnitude in these systems:

$$T_C \sim \theta_D \exp\left(-\frac{1}{\lambda - \mu^*}\right). \tag{23.132}$$

We can say, that electron–phonon interaction (acting in narrow region) is more effective in conventional superconductors than the broad-range Coulomb interaction.

That is why the region of energies arises in Schrodinger equation where the Coulomb interaction is strongly decreased, and $\lambda > \mu^*$.

It is interesting, that it is practically impossible to fulfill much stronger condition $\lambda > \mu_{Coul.}$. As it was shown by Ginzburg and Kizhnitz [22], the point is, that this condition:

$$\lambda - \mu_{Coul.} > 0 \tag{23.133}$$

corresponds to the instability of the ionic lattice in metal.

In the end of this Lecture note, that in certain systems another situation is possible, namely, when not only bare Coulomb potential $\mu_{Coul.} > \lambda$, but also the pseudopotential is larger than the coupling constant of electron–phonon interaction $\mu^* > \lambda$.

This situation takes place in simple and noble metals [12,23] (where the coupling constant of electron–phonon attraction λ is small) and in unconventional superconductors (such as high temperature cuprate superconductor La$_{2-x}$Ba$_x$CuO$_4$ [64], heavy-fermion superconductors UPt$_3$ [82], UBe$_{13}$ [81] and others).

Usually in unconventional superconductors the Tolmachev logarithm $\ln\frac{\varepsilon_F}{\omega_0}$ [84] is small due to the smallness of the Fermi energy ε_F. At the same time, the smallness of the Fermi energy itself is connected either with large effective mass (as in heavy fermions), or with small carrier density (as in semimetals), or with the combination of these two factors (as in cuprate superconductors).

In this case (for $\mu^* > \lambda$) conventional spherically symmetric singlet s-wave pairing (with total spin of the Cooper pair $S = 0$ and orbital momentum of the pair $l = 0$) is impossible [12,23].

However, even in this case the generally anisotropic spin-triplet ($S = 1$) p—wave pairing (with the orbital momentum of the pair $l = 1$) [85,86] is possible. The triplet p—wave pairing is realized for example in superfluid He-3 [87–89], in Sr$_2$RuO$_4$ [90,91], UBe$_{13}$ [81] and other systems. Another example of unconventional superconductivity is the singlet d—wave pairing with the orbital momentum $l = 2$ and total spin of the pair $S = 0$, as in UPt$_3$ [82] and La$_{2-x}$Ba$_x$CuO$_4$ [92–96].

In principle, as for example, in bilayer graphene the more exotic situation with triplet f—wave pairing [97] ($l = 3$) can be also realized. Finally we should mention the pairing with total rotational momentum of the pair $J = |l + S| = 2$ which is related to the case of strong spin-orbital coupling. This type of pairing (with bi-neutron and bi-proton pairs) emerges in the neutron stars [12,23,58,88,89].

Note that one of the basic mechanisms leading to unconventional superconductivity is Kohn–Luttinger mechanism [98] developed in [12,23,24,85,86,92,97]. We analyzed this mechanism in Lecture 18. Another important mechanism is a mechanism of the spin exchange (see for example [12,23,78,93,94,99,100]). It probably plays a dominant role in high-T_C cuprates.

References

1. Yu. Kagan, *Lectures on the Solid-State Theoretical Physics* (Moscow, MEPHI, 1981–1982. Part II. Electrons—unpublished)
2. L.D. Landau, E.M. Lifshitz, *Quantum Mechanics* (Pergamon Press, New York, Non-Relativistic Theory, 1977)
3. E.M. Lifshitz, L.P. Pitaevskii, *Statistical Physics Part II, Theory of the Condensed State* (Pergamon, Elsevier, 2013)
4. A.A. Abrikosov, L.P. Gor'kov, I.E. Dzyaloshinskii, *Methods of Quantum Field Theory in Statistical Physics* (Prentice Hall, Englewood Cliffs, New Jersey, 1963)
5. J. Bardeen, L.N. Cooper, J.R. Schrieffer, Phys. Rev. **108**, 1175 (1957)
6. L.N. Cooper, Phys. Rev. **104**, 1189 (1956)
7. J.R. Schrieffer, *Theory of Superconductivity* (W.A. Benjamin Inc., New York, 1964)
8. N.N. Bogoliubov, V.V. Tolmachev, D.V. Shirkov, *New Method in the Theory of Superconductivity* (Consultants Bureau, New York, London, Chapman and Hall, 1959)
9. N.N. Bogoliubov, *The Theory of Superconductivity* (Gordon and Breach, 1962)
10. A.A. Abrikosov, *Fundamentals of the Theory of Metals* (Elsevier Science Publishers, North-Holland, Amsterdam, 1988)
11. L.D. Landau, E.M. Lifshitz, *Physical Kinetics* (Pergamon, 1981)
12. M.Y. Kagan, *Physics of the Macroscopic Quantum Systems* (Lectures and Seminars, Moscow, 2014). ((in Russian))
13. M.Y. Kagan, A.V. Ozharovskii, *Mathematical methods of the Theory of Superconductivity and Strongly correlated electron systems* (Lecture Course, Part I, Moscow, MEPHI, 1999) (in Russian)
14. L.H. Thomas, The calculation of atomic fields. Math. Proc. Cambridge Philos. Soc. **23**, 542 (1927)
15. E. Fermi, Un Metodo Statistico per la Determinazione di Alcune Proprieta dell' Atomo. Rend. Accad. Naz. Lincei **6**, 602 (1927). ((in Italian))
16. E.A. Linton, *Superconductivity* (Methuen and Co., Ltd., London, 1969)
17. S.V. Vonsovskii, Yu. A. Izumov, E.Z. Kurmaev, *Superconductivity of Transition Metals, Their Alloys and Compounds* (Moscow, 1977) (in Russian)
18. A.C. Rose-Innes, E.H. Rhoderick, *Introduction to Superconductivity* (Pergamon Press, Oxford, London, 1969)
19. V.V. Schmidt, *The Physics of Superconductors* (Springer, Introduction to Fundamentals and Applications, 1997)
20. E. Parks (ed.), *Superconductivity (in two volumes)* (Marcel Dekker Inc., New York, 1969)
21. P.G. De Gennes, *Superconductivity of Metals and Alloys* (W.A. Benjamin Inc., New York-Amsterdam, 1966)

22. V.L. Ginzburg, D.A. Kirzhnits, *High Temperature Superconductivity* (Springer, 2014)
23. M.Y. Kagan, *Modern Trends in Superconductivity and Superfluidity*, Lecture Notes in Physics, vol. 874. (Springer, Dordrecht, 2013)
24. M. Yu. Kagan, K. I. Kugel, A.L. Rakhmanov, A.O. Sboychakov, Electronic phase separation in magnetic and superconducting materials. *Recent Advances*. Springer Series in Solid-State Sciences, vol. 201 (2024)
25. V.B. Berestetskii, E.M. Lifshitz, L.P. Pitaevskii, *Quantum Electrodynamics, Landau–Lifshitz Course on Theoretical Physics*, vol. 4 (Pergamon Press, 1984)
26. D.R. Tiley, J. Tiley, *Superfluidity and Superconductivity* (Van Nostrand Reinhold company, New York-Cincinnati-Toronto-London, Melburn, 1974)
27. A.V. Svidzinskii, Problems of the theory of superconductivity which involve spatial inhomogeneity (Moscow, 1982) (in Russian)
28. B.D. Josephson, Rev. Mod. Phys. **46**, 251 (1974)
29. J. Clarke, F.K. Wilhelm, Nature **453**, 1031 (2008)
30. M. Tinkham, *Introduction to Superconductivity* (Dover Publications, 2004)
31. W. Buckel, R. Kleiner, *Superconductivity* (Wiley-VCH Verlag, An Introduction, 2016)
32. J.B. Ketterson, *Superconductivity* (Cambridge University Press, 1999)
33. P. Fulde, *Electron Correlations in Molecules and Solids*. Springer Series in Solid-State Sciences (1993)
34. J.F. Annett, *Superconductivity, Super-Fluids and Condensates*. Oxford Master Series in Physics (2004)
35. V.L. Pokrovskii, Works by A.I. Larkin on the theory of phase transitions. JETP **117**, 387 (2013)
36. L.D. Landau, E.M. Lifshitz, *Statistical Physics* (Part I, Butterworth-Heinemann, Oxford, 1999)
37. I.M. Khalatnikov, *An Introduction to the Theory of superfluidity* (Perseus Publishing, Cambridge, 2000)
38. M.R. Shafroth, Phys. Rev. B **100**, 463 (1955)
39. M.R. Shafroth, Solid St. Physics **10**, 422 (1960)
40. A.S. Alexandrov, J. Ranninger, Phys. Rev. B **23**, 1796 (1981)
41. M.Y. Kagan, R. Fresard, M. Capezzali, H. Beck, Phys. Rev. B **57**, 5995 (1998)
42. P.W. Anderson, Phys. Rev. Lett. **34**, 953 (1975)
43. C.M. Varma, **61**, 2713 (1988)
44. T.M. Rice, L. Sneddon, Phys. Rev. Lett. **47**, 689 (1981)
45. A.P. Menushenkov, A.V. Kuznetsov, K.V. Klementiev, M.Y. Kagan, JETP **93**, 615 (2001)
46. A.P. Menushenkov, A.V. Kuznetsov, K.V. Klementiev, M.Y. Kagan, J. Supercond. Nov. Magn. **29**, 701 (2016)
47. U. Fano, Nuovo Cimento **12**, 156 (1935)
48. U. Fano, Phys. Rev. **124**, 1866 (1961)
49. H. Feshbach, Ann. Phys. **5**, 357 (1958)
50. H. Feshbach, Ann. Phys. **19**, 287 (1962)
51. C.A. Regal, C. Ticknor, J.L. Bohn, D.S. Jin, Nature **424**, 47 (2003)
52. M.W. Zwierlein, C.A. Stan, C.H. Schunk, S.M.F. Raupach, S. Gupta, Z. Hadzibabic, W. Ketterle, Phys. Rev. Lett. **91**, 250401 (2003)
53. A. Griffin, D.W. Snoke, S. Stringari, *Bose–Einstein Condensation* (Cambridge University Press, 1995)
54. Yu. Kagan, G.V. Shlyapnikov, N.A. Glukhov, JETP Lett. **41**, 238 (1995)
55. Yu. Kagan, G.V. Shlyapnikov, J.T.M. Walraven, Phys. Rev. Lett. **76**, 2670 (1996)
56. L.P. Pitaevskii, S. Stringari, *Bose–Einstein Condensation* (Clarendon Press, Oxford, 2003)
57. C.J. Pethick, S. Smith, *Bose–Einstein Condensation in Dilute Gases* (Cambridge University Press, UK, 2008)
58. B. Svistunov, E. Babaev, N. Prokof'ev, *Superfluid States of Matter* (CRC Press, Taylor and Francis Group, 2015)
59. A.J. Leggett, J. Phys. (Paris), Colloq. **41**, C7 (1980)

60. A.J. Leggett, Modern trends in the theory of condensed matter, Lecture course of the XVI Karpacz winter school of theoretical physics, ed. by A. Pekalski, J. Przystawa (Springer, Berlin, 1980)
61. P. Nozieres, S. Schmitt-Rink, J. Low Temp. Phys. **59**, 195 (1985)
62. R. Combescot, X. Leyronas, M.Y. Kagan, Phys. Rev. A **73**, 023618 (2006)
63. R. Combescot, M. Yu, Kagan, S. Stringari, Phys. Rev. A **74**, 042717 (2006)
64. J.G. Bednorz, K.A. Muller, Rev. Mod. Phys. **60**, 585 (1988)
65. J.C. Philips, *Physics of High-T_c Superconductors* (Academic Press, Boston, 1989)
66. S.A. Wolf, V.Z. Kresin, *Novel Superconductivity* (Plenum Press, New York, 1987)
67. K.A. Muller, G. Benedek, *Phase Separation in Cuprate Superconductors* (World Scientific, Singapore, 1993)
68. D.M. Ginsberg (ed.), *Physical Properties of High Temperature Superconductors* (World Scientific, Singapore, 1994)
69. P.W. Anderson, *The Theory of Superconductivity in the High-T_c Cuprate Superconductors* (Princeton University Press, 1997)
70. J.R. Schrieffer, J.S. Brooks (eds.), *Handbook of High Temperature Superconductivity* (Springer, Theory and Experiment, 2007)
71. S.-I. Uchida, *High Temperature Superconductivity, The Road to Higher Critical Temperature* (Springer Series in Material Sciences, 2015)
72. D.J. Scalapino, Rev. Mod. Phys. **84**, 1383 (2012)
73. C.M. Varma, Rev. Mod. Phys. **92**, 031001 (2020)
74. X.G. Wen, P.A. Lee, Theory of underdoped Cuprates. Phys. Rev. Lett. **76**, 503 (1996)
75. R.B. Laughlin, Science **242**, 525 (1988)
76. V.J. Emery, S.A. Kivelson, Nature **374**, 434 (1995)
77. F.C. Zhang, T.M. Rice, Phys. Rev. B **37**, 3759 (1988)
78. D. Pines, J. Phys. Chem. B **117**, 131145 (2013)
79. E. Abrahams, Present status of the theory of high-temperature superconductivity, in *Advances in Superconductivity* (Springer, 1990)
80. F. Steglich, J. Arts, C.D. Bredl et al., Phys. Rev. Lett. **43**, 1892 (1979)
81. H.R. Ott, H. Rudiger, Z. Fisk, J.L. Smith, Phys. Rev. Lett. **50**, 1595 (1983)
82. G.R. Stewart, Z. Fisk, J.O. Willis, J.L. Smith, Phys. Rev. Lett. **52**, 679 (1984)
83. P.B. Littlewood, G. Varelogianis, in *Proceedings of the First Euro-conference on Anomalous Complex Superconductors* (North-Holland, 1999)
84. V.V. Tolmachev, Doklady of Acad. Sci. USSR **140**, 563 (1961) (in Russian)
85. D. Fay, A. Layzer, Phys. Rev. Lett. **20**, 187 (1968)
86. M.Y. Kagan, A.V. Chubukov, JETP Lett. **47**, 614 (1988)
87. A.J. Leggett, Rev. Mod. Phys. **47**, 331 (1975)
88. D. Vollhardt, P. Wolfle, *The Superfluid Phases of Helium 3* (Taylor and Francis, 1990)
89. G.E. Volovik, *Exotic Properties of Superfluid ^3He* (World Scientific, Singapore, 1992)
90. Y. Maeno, T.M. Rice, M. Sigrist, Phys. Today **54**, 42 (2001)
91. T.M. Rice, M. Sigrist, Jour. Phys.: Condens. Matter **7**, 1643 (1995)
92. M.A. Baranov, M.Y. Kagan, Zeit. Phys. B: Condens. Matter **86**, 237 (1992)
93. M.Y. Kagan, T.M. Rice, J. Phys.: Condens. Matt. **6**, 3771 (1994)
94. J.R. Schrieffer, X.G. Xen, S.C. Zhang, Phys. Rev. B **39**, 11663 (1989)
95. P.A. Lee, N. Nagaosa, X.G. Wen, Rev. Mod. Phys. **78**, 17 (2006)
96. P.A. Lee, Rep. Prog. Phys. **v.71**, 012501 (2008)
97. M.Y. Kagan, M.M. Korovushkin, V.A. Mitskan, Physics Uspekhi **58**, 733 (2015)
98. W. Kohn, J.M. Luttinger, Phys. Rev. Lett. **15**, 524 (1965)
99. D.J. Scalapino, E. Loh, J.E. Hirsch, Phys. Rev. B **34**, 8190 (1986)
100. K. Miyake, S. Schmitt-Rink, C.M. Varma, Phys. Rev. B **34**, 6554 (1986)

Lecture 24. Many-Particle Problem in the Theory of Superconductivity

24

Abstract

In the framework of Bogoliubov $u - v$ transformation for fermions we solve the many-particle problem in the theory of superconductivity and derive the rigorous expression for the superconducting gap. We determine the spectrum of electronic excitations in superconductor and find the critical velocity applying Landau criterion for superfluidity.

24.1 Many-Particle Problem: Bogoliubov $u - v$ Transformation in the Theory of Superconductivity

In the Cooper problem of the previous Lecture, we in fact illustrated an emergence of one Cooper pair [1–20]. In this Lecture, we analyze many-particle problem with large number of electron pairs [1,3,4,8–11]. We will consider this problem in the framework of the Bogoliubov technique [8,9].

In Lecture 3 of the first part of our Course we already applied Bogoliubov $u - v$ transformation to bosonic low-density systems [21], having in mind first of all the diagonalization of the Hamiltonian in weakly non-ideal Bose gas with repulsive interaction between particles.

In this Lecture following the classical papers of Bardeen, Cooper, Schrieffer [2–4] we will now apply Bogoliubov $u - v$ transformation to the fermionic systems [8,9], considering electron gas with attractive interaction between particles.

The Hamiltonian of the many-particle problem in the second quantization acquires the form:

$$H = H_0 + H_{\text{int.}}, \tag{24.1}$$

where the first term H_0 in the Hamiltonian (24.1) describes the bare kinetic energy of electrons:

$$H_0 = \sum_{k\sigma} \varepsilon_k^0 a_{k\sigma}^\dagger a_{k\sigma}. \tag{24.2}$$

Note that operators $a_{k\sigma}^{\dagger}$ and $a_{k\sigma}$ in (24.2) correspond as usual to the creation and annihilation of electrons with the wave vector k and spin projection σ, and satisfy the fermionic anticommutation relations.

Let us remind once more that the energy $\varepsilon_k^0 = \frac{k^2}{2m}$ in (24.2) is the bare kinetic energy of electrons. The momentum and spin are conserved in the first term of the Hamiltonian in Eqs. (24.1)–(24.2).

At the same time, the second term $H_{\text{int.}}$ in the Hamiltonian (24.1) describes two-particle interaction between electrons:

$$H_{\text{int.}} = \frac{1}{2} \sum_{k_1 k_2 q \sigma_1 \sigma_2} V(q) a_{k_1+q\sigma_1}^{\dagger} a_{k_2-q\sigma_2}^{\dagger} a_{k_2\sigma_2} a_{k_1\sigma_1}. \tag{24.3}$$

In the absence of the Cooper pairs, we just have to renormalize electron mass and other Fermi liquid characteristics of electron subsystem (such as specific heat, compressibility, magnetic susceptibility and so on).

But now we have <u>an essential coherent interaction</u>. Let us distinguish this interaction straightforwardly assuming that all other interactions are responsible for the renormalization only.

To do that according to the solution of the Cooper problem let us conserve in Eq. (24.3) only part of the interaction with antiparallel momenta and spins of two electrons:

$$k_1 = -k_2, \sigma_1 = -\sigma_2. \tag{24.4}$$

Then from the total interaction in Eq. (24.3) survives only the following term:

$$H_{\text{eff.int.}} = 2\frac{1}{2} \sum_{k_1 q} V_{\text{eff.}} a_{k_1+q\uparrow}^{\dagger} a_{-k_1-q\downarrow}^{\dagger} a_{-k_1\downarrow} a_{k_1\uparrow}$$

$$= \sum_{k_1 q} V_{\text{eff.}} a_{k_1+q\uparrow}^{\dagger} a_{-k_1-q\downarrow}^{\dagger} a_{-k_1\downarrow} a_{k_1\uparrow}. \tag{24.5}$$

Further on denoting $k_1 + q = k_2$, we get effective Hamiltonian of many-particle problem corresponding to the pair formation:

$$H_{\text{eff.}} = H_0 - |V_0| \sum_{k_1 k_2} a_{k_2\uparrow}^{\dagger} a_{-k_2\downarrow}^{\dagger} a_{-k_1\downarrow} a_{k_1\uparrow}, \tag{24.6}$$

where $V_0 = -|V_0|$ is the total attractive interaction (accounting for electron–phonon attraction and reduced Coulomb repulsion [22]). In Eq. (24.6) we assume that the system volume $V = 1$.

Hamiltonian (24.6) of electron subsystem resembles the Hamiltonian of the Bogoliubov Bose gas. However, there are crucial differences between these Hamiltonians. First of all, due to Pauli principle there are no distinguished states in case of electrons [1,5,6]. Secondly, in our case particles have the spin [1,3,4,10,11].

Let us perform Lejandre transformation [1,6,10,11] for effective Hamiltonian to account for the number of particles conservation and make the Hamiltonian dependent not on the total number of particles N, but on the chemical potential μ:

$$\widehat{H}_{\text{eff.}} \to \widehat{H}'_{\text{eff.}} = \widehat{H}_{\text{eff.}} - \mu \widehat{N}. \tag{24.7}$$

Note that in (24.7) enters the secondly quantized operator of the total number of particles:

$$\widehat{N} = \sum_{k\sigma} a^{\dagger}_{k\sigma} a_{k\sigma}. \tag{24.8}$$

Then kinetic energy of the particles in the transformed Hamiltonian will be counted from the level of the chemical potential:

$$\varepsilon^0_k = \frac{k^2}{2m} \to \varepsilon_k = \varepsilon^0_k - \mu. \tag{24.9}$$

Correspondingly, effective Hamiltonian after Lejandre transformation reads:

$$\begin{aligned}
\widehat{H}'_{\text{eff.}} &= \sum_{k\sigma} (\varepsilon^0_k - \mu) a^{\dagger}_{k\sigma} a_{k\sigma} - |V_0| \sum_{k_1 k_2} a^{\dagger}_{k_2\uparrow} a^{\dagger}_{-k_2\downarrow} a_{-k_1\downarrow} a_{k_1\uparrow} \\
&= \sum_{k\sigma} \varepsilon_k a^{\dagger}_{k\sigma} a_{k\sigma} - |V_0| \sum_{k_1 k_2} a^{\dagger}_{k_2\uparrow} a^{\dagger}_{-k_2\downarrow} a_{-k_1\downarrow} a_{k_1\uparrow}.
\end{aligned} \tag{24.10}$$

Full energy of the system (corresponding before Lejandre transformation to the effective Hamiltonian (24.6)) after the thermodynamic averaging reads:

$$E = E(S, V, N). \tag{24.11}$$

It is the function of entropy S, system volume V, and total number of particles N.

At the same time, after Lejandre transformation and thermodynamic averaging, the functional:

$$\Xi = E - \mu N = \Xi(S, V, \mu) \tag{24.12}$$

depends on entropy S, system volume V, and now on the chemical potential μ.

Let us find fermionic excitations in the system (in other words let us find well-defined, long-living fermionic quasiparticles).

As in the case of the non-ideal Bose gas let us reduce the Hamiltonian (24.7) to the diagonal form with the help of the new ($u - v$ transformed) operators $b^{\dagger}_{k\sigma}$ and $b_{k\sigma}$. New operators are connected with the old operators $a^{\dagger}_{k\sigma}$ and $a_{k\sigma}$ in the following way [1–4, 8–11]:

$$b_{k\downarrow} = u_k a_{k\downarrow} + v_k a^\dagger_{-k\uparrow}, \tag{24.13}$$

$$b_{k\uparrow} = u_k a_{k\uparrow} - v_k a^\dagger_{-k\downarrow}, \tag{24.14}$$

$$b^\dagger_{k\downarrow} = u_k a^\dagger_{k\downarrow} + v_k a_{-k\uparrow}, \tag{24.15}$$

$$b^\dagger_{k\uparrow} = u_k a^\dagger_{k\uparrow} - v_k a_{-k\downarrow}, \tag{24.16}$$

where the Bogoliubov coefficients u_k and v_k are the real functions of the wave vector.

It is important, that the new operators $b^\dagger_{k\sigma}$ and $b_{k\sigma}$ in Eqs. (24.13)–(24.16) should satisfy the same anticommutation relations as the old operators $a^\dagger_{k\sigma}$ and $a_{k\sigma}$. Then, the anticommutator of the new operators:

$$\left\{ b_{k\downarrow} b^\dagger_{k\downarrow} \right\} = b_{k\downarrow} b^\dagger_{k\downarrow} + b^\dagger_{k\downarrow} b_{k\downarrow} = u_k^2 \left(a_{k\downarrow} a^\dagger_{k\downarrow} + a^\dagger_{k\downarrow} a_{k\downarrow} \right)$$

$$+ v_k^2 \left(a^\dagger_{-k\uparrow} a_{-k\uparrow} + a_{-k\uparrow} a^\dagger_{-k\uparrow} \right) + u_k v_k \left(a_{k\downarrow} a_{-k\uparrow} + a_{-k\uparrow} a_{k\downarrow} \right)$$

$$+ u_k v_k \left(a^\dagger_{-k\uparrow} a^\dagger_{k\downarrow} + a^\dagger_{k\downarrow} a^\dagger_{-k\uparrow} \right) = u_k^2 \left\{ a_{k\downarrow} a^\dagger_{k\downarrow} \right\} + v_k^2 \left\{ a^\dagger_{-k\uparrow} a_{-k\uparrow} \right\}$$

$$+ u_k v_k \left\{ a_{k\downarrow} a_{-k\uparrow} \right\} + u_k v_k \left\{ a^\dagger_{-k\uparrow} a^\dagger_{k\downarrow} \right\}$$

$$= u_k^2 \cdot 1 + v_k^2 \cdot 1 + u_k v_k \cdot 0 + u_k v_k \cdot 0 = u_k^2 + v_k^2 = 1. \tag{24.17}$$

Hence u_k and v_k are not only real, but also even functions of the wave vector k. We can convince ourselves in that, when we get explicit expressions for the coefficients u_k and v_k. Note, that in Bose gas with repulsion between particles, as we showed in Lecture 3 of the first part of our Course, the condition on the coefficients u_k and v_k differs from (24.17) by the sign: $u_k^2 - v_k^2 = 1$.

Using the condition $u_k^2 + v_k^2 = 1$ (which is in fact the condition on the determinant of the matrix of the Bogoliubov transformation), we can easily derive inverse transformation (from new to old operators):

$$a_{k\downarrow}(u_k^2 + v_k^2) = u_k b_{k\downarrow} - v_k b^\dagger_{-k\downarrow}, \text{ and } a_{k\downarrow} = u_k b_{k\downarrow} - v_k b^\dagger_{-k\uparrow}. \tag{24.18}$$

Analogously:

$$a_{k\uparrow} = u_k b_{k\uparrow} + v_k b^\dagger_{-k\downarrow}. \tag{24.19}$$

Correspondingly, utilizng the fact that coefficients u_k and v_k are real, we get for old creation operators:

$$a^\dagger_{k\downarrow} = u_k b^\dagger_{k\downarrow} - v_k b_{-k\uparrow}, \tag{24.20}$$

$$a_{k\uparrow}^{\dagger} = u_k b_{k\uparrow}^{\dagger} + v_k b_{-k\downarrow}. \tag{24.21}$$

Then, the first term in the effective Hamiltonian (24.10) (after Lejandre transformation) with an account of the expressions (24.17)–(24.21) written in terms of the new operators $b_{k\sigma}^{\dagger}$ and $b_{k\sigma}$ acquires a form:

$$
\begin{aligned}
\widehat{H}_0' &= \sum_{k\sigma} \varepsilon_k a_{k\sigma}^{\dagger} a_{k\sigma} = \sum_k \varepsilon_k \left(a_{k\uparrow}^{\dagger} a_{k\uparrow} + a_{k\downarrow}^{\dagger} a_{k\downarrow} \right) \\
&= \sum_k \varepsilon_k \left[\left(u_k b_{k\uparrow}^{\dagger} + v_k b_{-k\downarrow} \right) \left(u_k b_{k\uparrow} + v_k b_{-k\downarrow}^{\dagger} \right) \right. \\
&\qquad \left. + \left(u_k b_{k\downarrow}^{\dagger} - v_k b_{-k\uparrow} \right) \left(u_k b_{k\downarrow} - v_k b_{-k\uparrow}^{\dagger} \right) \right] \\
&= \sum_k \varepsilon_k \left\{ \left[u_k^2 b_{k\uparrow}^{\dagger} b_{k\uparrow} + u_k^2 b_{k\downarrow}^{\dagger} b_{k\downarrow} + v_k^2 b_{-k\downarrow} b_{-k\downarrow}^{\dagger} + v_k^2 b_{-k\uparrow} b_{-k\uparrow}^{\dagger} \right] \right. \\
&\qquad \left. + \left[u_k v_k \left(b_{k\uparrow}^{\dagger} b_{-k\downarrow}^{\dagger} + b_{-k\downarrow} b_{k\uparrow} \right) - u_k v_k \left(b_{k\downarrow}^{\dagger} b_{-k\uparrow}^{\dagger} + b_{-k\uparrow} b_{k\downarrow} \right) \right] \right\} \\
&= \sum_k \varepsilon_k \left[u_k^2 \left(b_{k\uparrow}^{\dagger} b_{k\uparrow} + b_{-k\downarrow}^{\dagger} b_{-k\downarrow} \right) + v_k^2 \left(b_{-k\downarrow} b_{-k\downarrow}^{\dagger} + b_{-k\uparrow} b_{-k\uparrow}^{\dagger} \right) \right. \\
&\qquad \left. + u_k v_k \left(b_{k\uparrow}^{\dagger} b_{-k\downarrow}^{\dagger} + b_{-k\downarrow} b_{k\uparrow} - b_{k\downarrow}^{\dagger} b_{-k\uparrow}^{\dagger} - b_{-k\uparrow} b_{k\downarrow} \right) \right].
\end{aligned}
\tag{24.22}
$$

Further on, let us make the replacements $k \to -k$ in terms proportional to v_k^2 and $u_k v_k$ using the fact that the energy:

$$\varepsilon_k = \varepsilon_k^0 - \mu = \varepsilon_{-k} \tag{24.23}$$

as well as the Bogoliubov coefficients:

$$u_k = u_{-k}, \text{ and } v_k = v_{-k}, \tag{24.24}$$

are even functions of the wave vector.

As a result, expression (24.22) can be rewritten as:

$$
\begin{aligned}
\widehat{H}_0' &= \sum_k \varepsilon_k \left[u_k^2 \left(b_{k\uparrow}^{\dagger} b_{k\uparrow} + b_{k\downarrow}^{\dagger} b_{k\downarrow} \right) + v_k^2 \left(b_{k\downarrow} b_{k\downarrow}^{\dagger} + b_{k\uparrow} b_{k\uparrow}^{\dagger} \right) \right. \\
&\qquad \left. + u_k v_k \left(b_{-k\uparrow}^{\dagger} b_{k\downarrow}^{\dagger} + b_{k\downarrow} b_{-k\uparrow} - b_{-k\downarrow}^{\dagger} b_{k\uparrow}^{\dagger} - b_{k\uparrow} b_{-k\downarrow} \right) \right].
\end{aligned}
\tag{24.25}
$$

Now, let us utilize the anticommutation relations according to which we get for the new operators:

$$b_{k\downarrow}b_{k\downarrow}^{\dagger} = 1 - b_{k\downarrow}^{\dagger}b_{k\downarrow}, \text{ and } b_{k\uparrow}b_{k\uparrow}^{\dagger} = 1 - b_{k\downarrow}^{\dagger}b_{k\downarrow}. \tag{24.26}$$

At the same time:

$$b_{-k\downarrow}^{\dagger}b_{k\uparrow}^{\dagger} = -b_{k\uparrow}^{\dagger}b_{-k\downarrow}^{\dagger}, \text{ and } b_{k\uparrow}b_{-k\downarrow} = -b_{-k\downarrow}b_{k\uparrow}. \tag{24.27}$$

Hence Eq. (24.25) is simplified:

$$\widehat{H}_0' = \sum_k \varepsilon_k \left[2v_k^2 + \left(u_k^2 - v_k^2\right)\left(b_{k\uparrow}^{\dagger}b_{k\uparrow} + b_{k\downarrow}^{\dagger}b_{k\downarrow}\right) + 2u_k v_k \left(b_{k\uparrow}^{\dagger}b_{-k\downarrow}^{\dagger} + b_{-k\downarrow}b_{k\uparrow}\right)\right], \tag{24.28}$$

where in the $u_k v_k$-term we used again the transformation $k \to -k$.

Now to rewrite the interaction term

$$H_{\text{eff.int.}} = -|V_0| \sum_{k_1 k_2} a_{k_2\uparrow}^{\dagger} a_{-k_2\downarrow}^{\dagger} a_{-k_1\downarrow} a_{k_1\uparrow} \tag{24.29}$$

in the new operators we should also apply the following expressions:

$$a_{-k\downarrow}^{\dagger} = u_{-k}b_{-k\downarrow}^{\dagger} - v_{-k}b_{k\uparrow} = u_k b_{-k\downarrow}^{\dagger} - v_k b_{k\uparrow}, \tag{24.30}$$

$$a_{-k\downarrow} = u_{-k}b_{-k\downarrow} - v_{-k}b_{k\uparrow}^{\dagger} = u_k b_{-k\downarrow} - v_k b_{k\uparrow}^{\dagger}, \tag{24.31}$$

where we used again the relations:

$$u_k = u_{-k}, \text{ and } v_k = v_{-k}. \tag{24.32}$$

Then the interaction term in new operators reads:

$$H_{\text{eff.int.}} = -|V_0| \sum_{k_1 k_2} \left(u_{k_2}b_{k_2\uparrow}^{\dagger} + v_{k_2}b_{-k_2\downarrow}\right)\left(u_{k_2}b_{-k_2\downarrow}^{\dagger} - v_{k_2}b_{k_2\uparrow}\right)$$

$$\cdot \left(u_{k_1}b_{-k_1\downarrow} - v_{k_1}b_{k_1\uparrow}^{\dagger}\right)\left(u_{k_1}b_{k_1\uparrow} + v_{k_1}b_{-k_1\downarrow}^{\dagger}\right)$$

$$= -|V_0| \sum_{k_1 k_2} \left[u_{k_2}^2 b_{k_2\uparrow}^{\dagger}b_{-k_2\downarrow}^{\dagger} - v_{k_2}^2 b_{-k_2\downarrow}b_{k_2\uparrow} + u_{k_2}v_{k_2}\left(b_{-k_2\downarrow}b_{-k_2\downarrow}^{\dagger} - b_{k_2\uparrow}^{\dagger}b_{k_2\uparrow}\right)\right]$$

$$\cdot \left[u_{k_1}^2 b_{-k_1\downarrow}b_{k_1\uparrow} - v_{k_1}^2 b_{k_1\uparrow}^{\dagger}b_{-k_1\downarrow}^{\dagger} + u_{k_1}v_{k_1}\left(b_{-k_1\downarrow}b_{-k_1\downarrow}^{\dagger} - b_{k_1\uparrow}^{\dagger}b_{k_1\uparrow}\right)\right]$$

$$= -|V_0| \sum_{k_1 k_2} B_{k_2}^{\dagger} B_{k_1}. \tag{24.33}$$

Note that in Eq. (24.33) we introduced convenient notation for the operator:

$$B_{k_1} = u_{k_1}^2 b_{-k_1\downarrow} b_{k_1\uparrow} - v_{k_1}^2 b_{k_1\uparrow}^\dagger b_{-k_1\downarrow}^\dagger + u_{k_1} v_{k_1} \left(b_{-k_1\downarrow} b_{-k_1\downarrow}^\dagger - b_{k_1\uparrow}^\dagger b_{k_1\uparrow} \right).$$

$$(24.34)$$

Correspondingly, Hermitian conjugate operator in Eq. (24.33):

$$B_{k_2}^\dagger = u_{k_2}^2 b_{k_2\uparrow}^\dagger b_{-k_2\downarrow}^\dagger - v_{k_2}^2 b_{-k_2\downarrow} b_{k_2\uparrow} + u_{k_2} v_{k_2} \left(b_{-k_2\downarrow} b_{-k_2\downarrow}^\dagger - b_{k_2\uparrow}^\dagger b_{k_2\uparrow} \right).$$

$$(24.35)$$

Further on let us take advantage again of the anticommutation relation for the terms proportional to $u_{k_1} v_{k_1}$:

$$b_{-k_1\downarrow} b_{-k_1\downarrow}^\dagger = 1 - b_{-k_1\downarrow}^\dagger b_{-k_1\downarrow},$$

$$(24.36)$$

and represent operator B_{k_1} as:

$$B_{k_1} = u_{k_1}^2 b_{-k_1\downarrow} b_{k_1\uparrow} - v_{k_1}^2 b_{k_1\uparrow}^\dagger b_{-k_1\downarrow}^\dagger + u_{k_1} v_{k_1} \left(1 - b_{-k_1\downarrow}^\dagger b_{-k_1\downarrow} - b_{k_1\uparrow}^\dagger b_{k_1\uparrow} \right).$$

$$(24.37)$$

After that from the condition on the minimum of the thermodynamic potential $\Xi(S, V, \mu) = E - \mu N$ at fixed entropy $S = $ const we can determine the coefficients u_k and v_k.

Indeed, the entropy of electron system is given by the well-known combinatorial expression from the Statistical physics [6]:

$$S = - \sum_{k\sigma} [n_{k\sigma} \ln n_{k\sigma} + (1 - n_{k\sigma}) \ln (1 - n_{k\sigma})],$$

$$(24.38)$$

which depends on the occupation numbers $n_{k\sigma}$ of the Fermi gas of electrons.

Hence, if the entropy does not change under the energy minimization, then the occupation numbers also do not change [1]. In new operators (for new reconstructed vacuum) the occupation numbers are equal:

$$n_{k\uparrow} = \left\langle 0 \left| b_{k\uparrow}^\dagger b_{k\uparrow} \right| 0 \right\rangle = \left\langle 0 \left| \widehat{n}_{k\uparrow} \right| 0 \right\rangle, \text{ and } n_{k\downarrow} = \left\langle 0 \left| b_{k\downarrow}^\dagger b_{k\downarrow} \right| 0 \right\rangle = \left\langle 0 \left| \widehat{n}_{k\downarrow} \right| 0 \right\rangle. \quad (24.39)$$

As a result, the operator B_{k_1} in the second term of the effective Hamiltonian reads:

$$B_{k_1} = u_{k_1}^2 b_{-k_1\downarrow} b_{k_1\uparrow} - v_{k_1}^2 b_{k_1\uparrow}^\dagger b_{-k_1\downarrow}^\dagger + u_{k_1} v_{k_1} \left(1 - \widehat{n}_{-k_1\downarrow} - \widehat{n}_{k_1\uparrow} \right). \quad (24.40)$$

Note that the first term in the effective Hamiltonian can be written as:

$$\widehat{H}_0' = \sum_k \varepsilon_k \left[2v_k^2 + (u_k^2 - v_k^2)(\widehat{n}_{k\uparrow} + \widehat{n}_{k\downarrow}) + 2u_k v_k \left(b_{k\uparrow}^\dagger b_{-k\downarrow}^\dagger + b_{-k\downarrow} b_{k\uparrow} \right) \right].$$

$$(24.41)$$

Then the (shifted) energy $E - \mu N$ is determined as an average quantity over the new vacuum:

$$E - \mu N = \left\langle 0 \left| \widehat{H}'_{\text{eff.}} \right| 0 \right\rangle = \left\langle 0 \left| \widehat{H}'_0 \right| 0 \right\rangle + \langle 0 | H_{\text{eff.int.}} | 0 \rangle$$

$$= \left\langle 0 \left| \widehat{H}'_0 \right| 0 \right\rangle - |V_0| \left\langle 0 \left| \sum_{k_1 k_2} B^\dagger_{k_2} B_{k_1} \right| 0 \right\rangle. \tag{24.42}$$

Distinguishing in (24.40) and (24.41) only non-zero diagonal terms proportional to $1, \widehat{n}_{-k\sigma}$, or $\widehat{n}_{k\sigma}$, we get the following expression for the average energy $E - \mu N$:

$$
\begin{aligned}
E - \mu N &= \left\langle 0 \left| \widehat{H}'_0 \right| 0 \right\rangle - |V_0| \left\langle 0 \left| \sum_{k_2} B^\dagger_{k_2} \right| 0 \right\rangle \left\langle 0 \left| \sum_{k_1} B_{k_1} \right| 0 \right\rangle \\
&= \left\langle 0 \left| \sum_{k} \varepsilon_k \left[2v_k^2 + \left(u_k^2 - v_k^2 \right) \left(\widehat{n}_{k\uparrow} + \widehat{n}_{k\downarrow} \right) \right] \right| 0 \right\rangle \\
&\quad - |V_0| \left\langle 0 \left| \sum_{k_2} u_{k_2} v_{k_2} \left(1 - \widehat{n}_{-k_2\downarrow} - \widehat{n}_{k_2\uparrow} \right) \right| 0 \right\rangle \cdot \left\langle 0 \left| \sum_{k_1} u_{k_1} v_{k_1} \left(1 - \widehat{n}_{-k_1\downarrow} - \widehat{n}_{k_1\uparrow} \right) \right| 0 \right\rangle \\
&= \left\langle 0 \left| \sum_{k} \varepsilon_k \left[2v_k^2 + \left(u_k^2 - v_k^2 \right) \left(\widehat{n}_{k\uparrow} + \widehat{n}_{k\downarrow} \right) \right] \right| 0 \right\rangle \\
&\quad - |V_0| \left(\left\langle 0 \left| \sum_{k} u_k v_k \left(1 - \widehat{n}_{-k\downarrow} - \widehat{n}_{k\uparrow} \right) \right| 0 \right\rangle \right)^2 \\
&= \sum_{k} \varepsilon_k \left[2v_k^2 + \left(u_k^2 - v_k^2 \right) \left(n_{k\uparrow} + n_{k\downarrow} \right) \right] \\
&\quad - |V_0| \left[\sum_{k} u_k v_k \left(1 - n_{-k\downarrow} - n_{k\uparrow} \right) \right]^2 .
\end{aligned}
\tag{24.43}
$$

Let us stress that deriving Eq. (24.43) we used the fact, that pairwise crossed convolutions of the new operators $(b^\dagger_{k_2\sigma} b_{k_1\sigma})$ in the interaction term for equal wave vectors $k_2 = k_1$ have macroscopically small statistical weight [1].

Further on the condition of the energy minimization is reduced to the search of the optimal value of one of the coefficients (for example of u_k) under the energy variation (at fixed occupation numbers and fixed $v_k^2 = 1 - u_k^2$):

$$\frac{\delta(E - \mu N)}{\delta u_k} = 0. \tag{24.44}$$

Substituting the relations:

$$2v_k^2 = 2(1 - u_k^2), \text{ and } u_k^2 - v_k^2 = 2u_k^2 - 1, \tag{24.45}$$

we can find the variation of the first term in (24.43):

$$\frac{\delta \left(\sum_k \varepsilon_k \left[2v_k^2 + \left(u_k^2 - v_k^2 \right) \left(n_{k\uparrow} + n_{k\downarrow} \right) \right] \right)}{\delta u_k} = \varepsilon_k \left[-4u_k + 4u_k \left(n_{k\uparrow} + n_{k\downarrow} \right) \right].$$

$$(24.46)$$

At the same time, the variation of the second term:

$$\frac{\delta \left(-\left| V_0 \right| \left[\sum_k u_k v_k \left(1 - n_{-k\downarrow} - n_{k\uparrow} \right) \right]^2 \right)}{\delta u_k}$$

$$= -2 \left| V_0 \right| \left[\sum_k u_k v_k \left(1 - n_{-k\downarrow} - n_{k\uparrow} \right) \right] \frac{\delta (u_k v_k)}{\delta u_k} \left(1 - n_{-k\downarrow} - n_{k\uparrow} \right), \quad (24.47)$$

where the variation of the product of the Bogoliubov coefficients equals to:

$$\frac{\delta \left(u_k v_k \right)}{\delta u_k} = v_k + u_k \frac{\delta v_k}{\delta u_k}. \tag{24.48}$$

After that we can use the variation of the coupling between the coefficients:

$$v_k^2 = 1 - u_k^2, \text{ and } v_k \delta v_k = -u_k \delta u_k. \tag{24.49}$$

Hence:

$$\delta v_k = -\frac{u_k}{v_k} \delta u_k, \text{ and } u_k \frac{\delta v_k}{\delta u_k} = -\frac{u_k^2}{v_k}. \tag{24.50}$$

Then the variation of the product of the coefficients:

$$\frac{\delta \left(u_k v_k \right)}{\delta u_k} = v_k - \frac{u_k^2}{v_k} = \frac{v_k^2 - u_k^2}{v_k}. \tag{24.51}$$

Thus, the variation of the second term equals to:

$$\frac{\delta \left(-\left| V_0 \right| \left[\sum_k u_k v_k \left(1 - n_{-k\downarrow} - n_{k\uparrow} \right) \right]^2 \right)}{\delta u_k} = -2 \left| V_0 \right| \left[\sum_k u_k v_k \left(1 - n_{-k\downarrow} - n_{k\uparrow} \right) \right]$$

$$\cdot \frac{v_k^2 - u_k^2}{v_k} \cdot \left(1 - n_{-k\downarrow} - n_{k\uparrow} \right). \tag{24.52}$$

As a result, condition (24.44) for the minimization of the functional $E - \mu N$ can be written in the form:

$$\varepsilon_k \left[-4u_k + 4u_k \left(n_{k\uparrow} + n_{k\downarrow} \right) \right] = 2 \, |V_0| \left[\sum_k u_k v_k \left(1 - n_{-k\downarrow} - n_{k\uparrow} \right) \right] \cdot \frac{v_k^2 - u_k^2}{v_k}$$
$$\cdot \left(1 - n_{-k\downarrow} - n_{k\uparrow} \right). \tag{24.53}$$

Now we can introduce very important notation for the theory of superconductivity:

$$\Delta = |V_0| \sum_k u_k v_k \left(1 - n_{-k\downarrow} - n_{k\uparrow} \right), \tag{24.54}$$

where Δ is the superconducting gap.

In the BCS theory [2–4] Eq. (24.54) is usually called as self-consistency equation or equation on the superconducting gap.

Then we can rewrite (24.53) as:

$$- 4u_k \varepsilon_k \left(1 - n_{-k\downarrow} - n_{k\uparrow} \right) = 2\Delta \frac{v_k^2 - u_k^2}{v_k} \left(1 - n_{-k\downarrow} - n_{k\uparrow} \right). \tag{24.55}$$

After that reducing the left and right side of (24.55) by the factor $1 - n_{-k\downarrow} - n_{k\uparrow}$ and assuming that in general case this factor is non-zero we get:

$$- 4u_k \varepsilon_k = 2\Delta \frac{v_k^2 - u_k^2}{v_k} = 2\Delta \frac{1 - 2u_k^2}{v_k}. \tag{24.56}$$

Further on multiplying the l.h.s. and r.h.s. of (24.56) on v_k and changing signs, we obtain:

$$4\varepsilon_k u_k v_k = -2\Delta \left(1 - 2u_k^2 \right), \text{ and } 2\varepsilon_k u_k v_k = -\Delta \left(1 - 2u_k^2 \right). \tag{24.57}$$

Now we can square the left and right sides of (24.57):

$$4\varepsilon_k^2 u_k^2 v_k^2 = \Delta^2 \left(1 - 2u_k^2 \right)^2, \tag{24.58}$$

and use again the relation $v_k^2 = 1 - u_k^2$ between the coefficients in the l.h.s. of (24.58). As a result, we derive the biquadratic equation:

$$4\varepsilon_k^2 u_k^2 \left(1 - u_k^2 \right) = \Delta^2 \left(1 - 2u_k^2 \right)^2, \tag{24.59}$$

or equivalently:

$$4u_k^4 \left(\Delta^2 + \varepsilon_k^2 \right) - 4u_k^2 \left(\Delta^2 + \varepsilon_k^2 \right) + \Delta^2 = 0. \tag{24.60}$$

Correspondingly:

$$u_k^4 - u_k^2 + \frac{\Delta^2}{4\left(\Delta^2 + \varepsilon_k^2\right)} = 0, \tag{24.61}$$

and thus:

$$\left(u_k^2 - \frac{1}{2}\right)^2 + \frac{1}{4}\left(\frac{\Delta^2}{\Delta^2 + \varepsilon_k^2} - 1\right) = 0. \tag{24.62}$$

Solution of this equation yields:

$$u_k^2 = \frac{1}{2} \pm \frac{1}{2}\sqrt{1 - \frac{\Delta^2}{\Delta^2 + \varepsilon_k^2}} = \frac{1}{2}\left(1 \pm \sqrt{\frac{\varepsilon_k^2}{\Delta^2 + \varepsilon_k^2}}\right) = \frac{1}{2}\left(1 \pm \frac{|\varepsilon_k|}{\sqrt{\Delta^2 + \varepsilon_k^2}}\right), \tag{24.63}$$

where $\varepsilon_k = \varepsilon_k^0 - \mu$.

Hence:

$$v_k^2 = \frac{1}{2}\left(1 \mp \frac{|\varepsilon_k|}{\sqrt{\Delta^2 + \varepsilon_k^2}}\right). \tag{24.64}$$

Now let us clarify the question about the signs of the Bogoliubov coefficients. From the self-consistency equation (24.54) we can see, that superconducting gap goes to zero only when one of the coefficients u_k or v_k equals to zero.

Besides that, at temperature $T = 0$ and in the absence of the gap the old and new creation and annihilation operators should coincide. But according to the Bogoliubov transformation (24.13):

$$b_{k\downarrow} = u_k a_{k\downarrow} + v_k a_{-k\uparrow}^\dagger.$$

Hence it follows from the condition $b_{k\downarrow} = a_{k\downarrow}$ for $\Delta = 0$ and $T = 0$ that $u_k = 1$ and $v_k = 0$.

Then in the expressions (24.63)–(24.64) we can lift the modular brackets for the energy ε_k, selecting the sign plus for u_k^2 and sign minus for v_k^2.

Thus, finally we can write:

$$u_k^2 = \frac{1}{2}\left(1 + \frac{\varepsilon_k}{\sqrt{\Delta^2 + \varepsilon_k^2}}\right), \tag{24.65}$$

$$v_k^2 = \frac{1}{2}\left(1 - \frac{\varepsilon_k}{\sqrt{\Delta^2 + \varepsilon_k^2}}\right). \tag{24.66}$$

In this case the behavior of the Bogoliubov coefficients squared as functions of energy counted from the Fermi level $\varepsilon_k = \varepsilon_k^0 - \mu$ has the form presented in Fig. 24.1. For $k < k_F$: the energy $\varepsilon_k < 0$, and for the gap $\Delta = 0$: the coefficients squared:

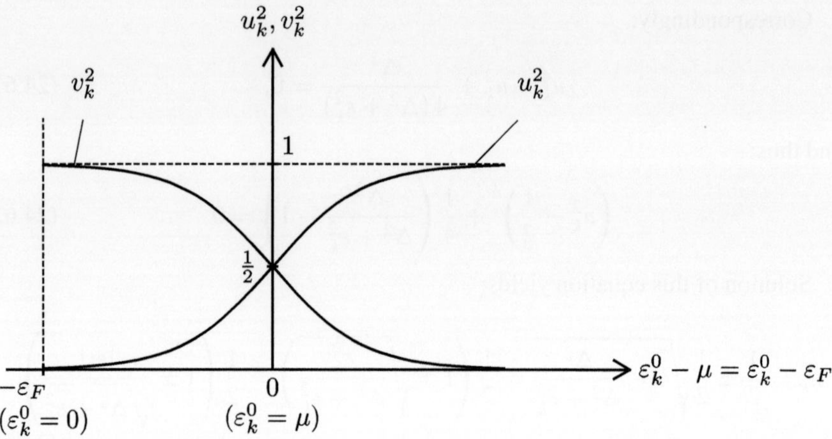

Fig. 24.1 Behavior of the coefficients u_k^2 and v_k^2 as the functions of energy counted from the Fermi level: $\varepsilon_k = \varepsilon_k^0 - \mu = \frac{k^2 - k_F^2}{2m}$ [1]

$u_k^2 = 0$, and $v_k^2 = 1$. At the same time, for $k > k_F$ the energy $\varepsilon_k > 0$, and for the gap $\Delta = 0$: $u_k^2 = 1$, and $v_k^2 = 0$.

If, however, $\Delta \neq 0$, then the transition region exists with the width of the order of Δ around the point $k = k_F$ (on the Fermi surface) where both coefficients squared are positive but less than 1.

That is why, our selection of signs in the Bogoliubov transformation (24.13) for $\Delta = 0$ corresponds to the situation when the new annihilation operator:

$$b_{k\downarrow} = a^{\dagger}_{-k\uparrow} \text{ for } k < k_F, \text{ and } b_{k\downarrow} = a_{k\downarrow} \text{ for } k > k_F.$$

We can say, that our selection of signs in (24.65)–(24.66) can be illustrated as follows: the new operator $b_{k\downarrow}$ is the operator of the hole absorption (or creation of electron with opposite momentum and spin living under the Fermi surface). At the same time, it is the operator of absorption of electron with the same momentum and spin living above the Fermi surface.

Note, that for another selection of signs in (24.65)–(24.66) for $k < k_F$ and $\Delta = 0$ we get for the coefficients squared: $u_k^2 = 1$, and $v_k^2 = 0$. As a result, the new annihilation operator $b_{k\downarrow} = a_{k\downarrow}$ for $k < k_F$ and $b_{k\downarrow} = a^{\dagger}_{-k\uparrow}$ for $k > k_F$.

Let us emphasize that for $\Delta \neq 0$ the product $u_k^2 v_k^2$ equals to:

$$u_k^2 v_k^2 = \frac{1}{4}\left(1 + \frac{\varepsilon_k}{\sqrt{\Delta^2 + \varepsilon_k^2}}\right)\left(1 - \frac{\varepsilon_k}{\sqrt{\Delta^2 + \varepsilon_k^2}}\right)$$

$$= \frac{1}{4}\left(1 - \frac{\varepsilon_k^2}{\Delta^2 + \varepsilon_k^2}\right) = \frac{\Delta^2}{4(\Delta^2 + \varepsilon_k^2)}. \tag{24.67}$$

Assuming that the superconducting gap is not negative $\Delta \geq 0$, we get:

$$u_k v_k = \frac{\Delta}{2\sqrt{\Delta^2 + \varepsilon_k^2}}. \tag{24.68}$$

As a result, the self-consistency equation for the gap (24.54) is given by:

$$\Delta = |V_0| \sum_k \frac{\Delta}{2\sqrt{\Delta^2 + \varepsilon_k^2}} \left(1 - n_{-k\downarrow} - n_{k\uparrow}\right). \tag{24.69}$$

Reducing left and right sides by the superconducting gap, we obtain:

$$1 = \frac{|V_0|}{2} \sum_k \frac{1}{\sqrt{\Delta^2 + \varepsilon_k^2}} \left(1 - n_{-k\downarrow} - n_{k\uparrow}\right). \tag{24.70}$$

Finally, for $\Delta \neq 0$ we should satisfy Eq. (24.57) which contains the minus sign. Substituting expression (24.68) for $u_k v_k$ and (24.65) for u_k^2 in this equation we get identical coincidence of the left and right sides:

$$\left[2\varepsilon_k u_k v_k = \frac{\Delta \varepsilon_k}{\sqrt{\Delta^2 + \varepsilon_k^2}}\right]$$

$$\equiv \left[-\Delta\left(1 - 2u_k^2\right) = -\Delta\left(1 - 1 - \frac{\varepsilon_k}{\sqrt{\Delta^2 + \varepsilon_k^2}}\right) = \frac{\Delta \varepsilon_k}{\sqrt{\Delta^2 + \varepsilon_k^2}}\right]. \tag{24.71}$$

This result also confirms the correct choice of the signs in expressions (24.65) and (24.66).

Let us figure out now what the average occupation numbers $n_{-k\downarrow}$ and $n_{k\uparrow}$ are equal to in the self-consistency equation. We discussed already that these numbers correspond to the new reconstructed vacuum, and, at the same time, are related to the fermionic (Fermi–Dirac) statistics.

It occurs, that the energy of new quasiparticles in the self-consistency equation (24.70) and in the expressions for u_k^2 and v_k^2 contains the superconducting gap $\Delta \neq 0$ and equals to:

$$E_k = \sqrt{\Delta^2 + \varepsilon_k^2}. \tag{24.72}$$

Then the transition region exists in Eqs. (24.65) and (24.66).
Correspondingly, u_k^2 and v_k^2 read:

$$u_k^2 = \frac{1}{2}\left(1 + \frac{\varepsilon_k}{E_k}\right), \; v_k^2 = \frac{1}{2}\left(1 - \frac{\varepsilon_k}{E_k}\right). \tag{24.73}$$

Let us stress that an analogy between Bogoliubov transformation for Bose gas with repulsion [21] and Fermi gas with attraction [8,9] manifests itself again in these expressions.

Note that the self-consistency equation acquires the form:

$$1 = \frac{|V_0|}{2} \sum_k \frac{1}{\sqrt{\Delta^2 + \varepsilon_k^2}} \left(1 - n_{-k\downarrow} - n_{k\uparrow}\right). \tag{24.74}$$

Moreover, the average occupation numbers $n_{-k\downarrow}$ and $n_{k\uparrow}$ for non zero temperatures are equal to each other in the absence of the external magnetic field:

$$n_{-k\downarrow} = n_{k\uparrow} = \frac{1}{\exp\left(\frac{E_k}{T}\right) + 1}. \tag{24.75}$$

Let us emphasize, that the superconducting gap opens on the Fermi surface in Eq. (24.75) for the reconstructed energy:

$$E_k = \Delta \geq 0 \text{ for } k = k_F. \tag{24.76}$$

Note also, that the number of new excitations is not fixed. That is why the chemical potential of the excitation gas $\mu_{\text{exc.}}$ in (24.75) equals zero.

Let us stress, that the chemical potential of the excitation gas $\mu_{\text{exc.}}$ does not have any relation to the chemical potential of the initial electrons μ. The number of excitations is not fixed. They are temperature ones. Hence their number varies with temperature. At $T = 0$ there are no excitations in the system and occupation numbers equal zero:

$$n_{-k\downarrow} = n_{k\uparrow} = 0. \tag{24.77}$$

24.2   Superconducting Gap

Let us proceed from the summation to the integration in the self-consistency equation for the gap (24.74). Then at temperature $T = 0$ (neglecting the Coulomb pseudopotential):

$$1 = \frac{|V_0|}{2} \int_{-\omega_0}^{\omega_0} \frac{g_{\text{tot.}}(\varepsilon)d\varepsilon}{\sqrt{\Delta^2 + \varepsilon^2}} = |V_0| \int_0^{\omega_0} \frac{g_{\text{tot.}}(\varepsilon)d\varepsilon}{\sqrt{\Delta^2 + \varepsilon^2}} = |V_0| \, g_{\text{tot.}}(\varepsilon_F) \int_0^{\omega_0} \frac{d\varepsilon}{\sqrt{\Delta^2 + \varepsilon^2}}, \tag{24.78}$$

where ω_0 is the typical frequency of the phonon spectrum, $g_{\text{tot.}}(\varepsilon_F)$ is the total density of states on the Fermi level (including factor 2, missing in the Cooper problem for the binding energy considered in the previous Lecture).

Deriving this equation, we used the fact that expression under the integral in (24.78) is even. Integration in (24.78) can be performed elementary. As a result, we get the long logarithm:

$$\frac{1}{|V_0|\,g_{\text{tot.}}(\varepsilon_F)} = \ln\left(\varepsilon + \sqrt{\Delta^2 + \varepsilon^2}\right)\Big|_0^{\omega_0}. \tag{24.79}$$

Usually as we already discussed in the previous Lectures, the superconducting gap is small compared with the typical phonon frequency:

$$\Delta \sim T_C \ll \omega_0 \text{ for } \lambda = |V_0|\,g_{\text{tot.}}(\varepsilon_F) \ll 1, \tag{24.80}$$

and hence the long logarithm in the r.h.s. of Eq. (24.79):

$$\ln\left(\varepsilon + \sqrt{\Delta^2 + \varepsilon^2}\right)\Big|_0^{\omega_0} \approx \ln\frac{2\omega_0}{\Delta}. \tag{24.81}$$

Thus, it follows from Eq. (24.79) that:

$$\frac{1}{|V_0|\,g_{\text{tot.}}(\varepsilon_F)} = \ln\frac{2\omega_0}{\Delta}, \tag{24.82}$$

and the superconducting gap in the excitation spectrum yields:

$$\Delta \sim 2\omega_0 \exp\left(-\frac{1}{\lambda}\right). \tag{24.83}$$

Let us remind that in conventional superconductors $\lambda \sim \left(\frac{1}{3} \div \frac{1}{4}\right)$ [18–33].

Note that deriving self-consistency equation (24.74) from the initial equation for the gap (24.69) we assumed, that $\Delta \neq 0$. In fact, formally, solution with $\Delta = 0$ is also possible. We can show, however, that this solution has larger energy and thus to the minimum of energy corresponds only the solution with $\Delta \neq 0$ determined by the formula (24.83).

Moreover, we can prove acting in the spirit of Landau Fermi liquid theory [1,3,4, 7–11] that $E_k = \sqrt{\Delta^2 + \varepsilon_k^2}$ is indeed the energy of the new quasiparticle excitations.

To do that let us find the variation of the thermodynamic functional (24.43) by the occupation number, for example $n_{k\uparrow}$ (at fixed coefficients u_k, v_k, and occupation numbers $n_{-k\downarrow}$ and $n_{k\downarrow}$ for the different spin projection):

$$E_k = \frac{\delta\,(E - \mu N)}{\delta n_{k\uparrow}} = \varepsilon_k\left(u_k^2 - v_k^2\right) + 2\,|V_0|\left[\sum_k u_k v_k\left(1 - n_{-k\downarrow} - n_{k\uparrow}\right)\right] u_k v_k$$

$$= \varepsilon_k\left(2u_k^2 - 1\right) + 2\,|V_0|\left[\sum_k u_k v_k\left(1 - n_{-k\downarrow} - n_{k\uparrow}\right)\right] u_k v_k. \tag{24.84}$$

Substituting expressions for $u_k v_k$ from (24.68) and for u_k^2 from (24.65) in Eq. (24.84), we can rewrite the variation $\frac{\delta(E-\mu N)}{\delta n_k}$ as follows:

$$\frac{\delta(E-\mu N)}{\delta n_k} = \varepsilon_k \frac{\varepsilon_k}{\sqrt{\Delta^2+\varepsilon_k^2}} + 2\Delta \frac{\Delta}{2\sqrt{\Delta^2+\varepsilon_k^2}}$$

$$= \frac{\varepsilon_k^2}{\sqrt{\Delta^2+\varepsilon_k^2}} + \frac{\Delta^2}{\sqrt{\Delta^2+\varepsilon_k^2}} = \sqrt{\Delta^2+\varepsilon_k^2}. \tag{24.85}$$

Thus, the energy of new (reconstructed) quasiparticles $E_k = \sqrt{\Delta^2+\varepsilon_k^2}$ is indeed determined by the formula (24.72).

In Fig. 24.2 we show the quasiparticle (electron) spectrum in normal metal [7, 18, 19] close to the Fermi surface at temperatures above the critical one $T > T_C$:

$$\varepsilon_k = \varepsilon_k^0 - \mu \approx \frac{k^2 - k_F^2}{2m} \approx v_F\,(k - k_F)\,, \tag{24.86}$$

and the spectrum of the superconductor at the temperatures $T < T_C$:

$$E_k = \sqrt{\Delta^2+\varepsilon_k^2} \approx \sqrt{\Delta^2 + v_F^2\,(k-k_F)^2}. \tag{24.87}$$

Quite often in the textbooks the quasiparticle spectrum in the normal metal is denoted by $\xi_k = \varepsilon_k = v_F\,(k-k_F)$ at $\hbar = 1$, or equivalently:

$$\xi_p = v_F\,(p - p_F)\,, \tag{24.88}$$

where: $p = \hbar k$, $p_F = \hbar k_F$.

Correspondingly, the quasiparticle spectrum in superconductor is often written as (see Fig. 24.2):

$$E_p = \sqrt{\Delta^2 + \xi_p^2}. \tag{24.89}$$

It follows from formula (24.87), that the excitation energy is positive both for electron and hole. (We have to spend the energy to create the excitation.)

Moreover, the gap of the order of the pair binding energy appears in the spectrum of the superconductor:

$$\Delta \sim \varepsilon_b. \tag{24.90}$$

Due to the emergence of the gap in the spectrum (24.87), the normal excitations are absent in superconductor at temperature $T = 0$.

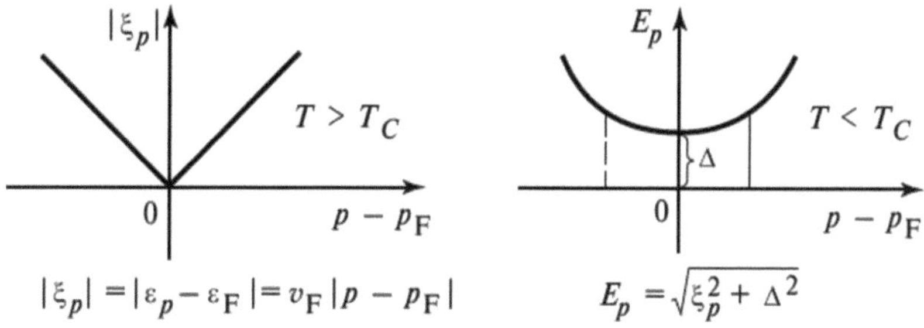

Fig. 24.2 The gapless spectrum of elementary excitations in normal metal $|\xi_p| = v_F |p - p_F|$ at temperatures above the critical one $T > T_C$ and the gapped spectrum in superconductor $E_p = \sqrt{\Delta^2 + \xi_p^2}$ at temperatures $T < T_C$ [1,18]

This fact immediately leads us to the superfluidity since for non-zero Δ we can easily fulfill the Landau criterion [1,10,33].

24.3 Critical Velocity

Let us recollect Lecture 4 on the superfluid helium from the first part of the Course. In this Lecture we discussed already Landau criterion for stopping the superfluid flow [1,10,18–20,33,34] by spontaneous generation of elementary excitations with the momenta antiparallel to the drift velocity of the flow.

Analogously to find the critical velocity in the superconductor let us sit on the superfluid liquid. Then we can show that for small velocity of the flow V it becomes beneficial for us to generate excitations stopping the flow of the Cooper pairs.

Suppose we create an excitation with momentum k and the energy E_k. In this case the total energy of the fluid reads [1,10]:

$$E = \frac{1}{2}MV^2 + kV + E_k, \tag{24.91}$$

where $M = $ const is the total mass of the Cooper pairs superfluid.

Correspondingly, as in superfluid helium, spontaneous generation of excitations is energetically beneficial under the condition [1,10,33,34]:

$$kV + E_k < 0. \tag{24.92}$$

If the generated excitation has momentum antiparallel to the flow, then the scalar product:

$$kV = -kV, \text{ and } -kV + E_k < 0. \tag{24.93}$$

Hence generation of the excitations is energetically beneficial when:

$$V > V_{\text{cr.}} = \min \frac{E_k}{k} = \frac{\Delta}{k_F}. \tag{24.94}$$

We can say that in the absence of the gap Landau criterion for destruction of superfluidity is fulfilled already at very small velocities of the flow. If, however, we have a gap, then the velocity of the flow should be larger than the critical one (determined by the condition (24.94)). At smaller velocities $V < \frac{\Delta}{k_F}$ the superfluid current of the Cooper pairs will flow in the system forever.

24.4 Overlapping (Mixing) of the Bogoliubov Coefficients in Superconductor

In the end of the Lecture let us analyze more detailly the behavior of the Bogoliubov coefficients u_k and v_k in superconductor.

If in the self-consistency equation (24.54) the superconducting gap $\Delta = 0$, then the product of the coefficients $u_k v_k = 0$. At the same time, for $\Delta \neq 0$ the squares of the Bogoliubov coefficients behave as it is shown in Fig. 24.1.

Let us draw separately each of the coefficients u_k and v_k as a function of energy counted from the Fermi level $\varepsilon_k = \varepsilon_k^0 - \mu$ (see Fig. 24.3a, b) and consider more carefully the transition region close to the Fermi surface where the coefficients are mixed and their product $u_k v_k$ is different from zero (see Fig. 24.3c).

Energy dependence of the coefficient v_k resembles the distribution of the Fermi particles (electrons) smeared close to the Fermi surface on the magnitude of the order of the gap Δ.

Dependence of u_k on energy resembles the distribution of holes (also smeared on the magnitude of the order of Δ).

The mixing goes on in the smearing region. In this region the Bogoliubov transformation (24.15) $b_{k\downarrow}^\dagger = u_k a_{k\downarrow}^\dagger + v_k a_{-k\uparrow}$ yields us the possibility to create simultaneously electron with momentum k, and absorb electron with momentum $-k$ (which is equivalent to the creation of hole with momentum k).

Thus, the Bogoliubov transformation in the region of mixing for the coefficients ($u_k v_k \neq 0$) is equivalent to the simultaneous creation of electron and hole.

Actually, the pairing occurs at the coefficients mixing, since the excitations exist in the system which correspond to the mixture of electron and hole.

We can illustrate this fact, starting from the charge conservation law [1]. The charge of electron equals to the sum of the hole charge and the pair charge. This means, that two states with the creation of electron and hole in the Bogoliubov transformation are identical with respect to the gas of temporal excitations (representing the part of the normal component). They are distinguished only by the addition of one pair to the superfluid component (see the classical monograph of J.R. Schrieffer «*Theory of superconductivity*» [4]).

Let us illustrate the importance of the existence of the mixing region. Let us find anomalous average or the probability amplitude of the pair annihilation over the reconstructed vacuum:

$$\langle 0|a_{k\uparrow} a_{-k\downarrow}|0\rangle. \tag{24.95}$$

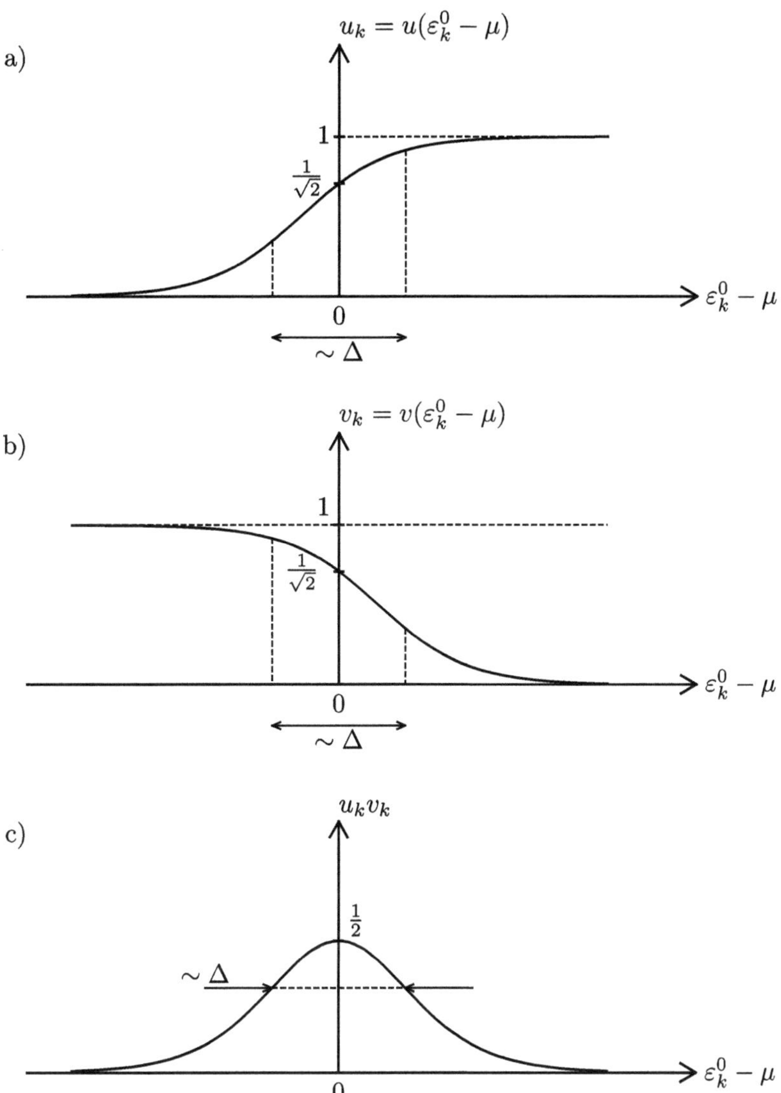

Fig. 24.3 Dependence of the Bogoliubov coefficients u_k (Fig. a), v_k (**b**) and their product (the degree of the coefficients mixing) $u_k v_k$ (**c**) on the energy $\varepsilon_k = \varepsilon_k^0 - \mu$ counted from the Fermi level [1]

Let us emphasize, that this particular combination of old operators enters in the interaction term in the initial effective Hamiltonian (24.10).

Let us transform the old operators to the new ones, utilizing inverse Bogoliubov transformations (24.18) and (24.31). Then:

$$\langle 0|a_{k\uparrow}a_{-k\downarrow}|0\rangle = \left\langle 0\left|\left(u_k b_{k\uparrow} + v_k b^\dagger_{-k\downarrow}\right)\left(u_k b_{-k\downarrow} - v_k b^\dagger_{k\uparrow}\right)\right|0\right\rangle$$

$$= \left\langle 0\left|u_k^2 b_{k\uparrow}b_{-k\downarrow} - v_k^2 b^\dagger_{-k\downarrow}b^\dagger_{k\uparrow} - u_k v_k\left(b_{k\uparrow}b^\dagger_{k\uparrow} - b^\dagger_{-k\downarrow}b_{-k\downarrow}\right)\right|0\right\rangle. \qquad (24.96)$$

Leaving only diagonal terms, we obtain:

$$\langle 0|a_{k\uparrow}a_{-k\downarrow}|0\rangle = u_k v_k\left\langle 0\left|-b_{k\uparrow}b^\dagger_{k\uparrow} + b^\dagger_{-k\downarrow}b_{-k\downarrow}\right|0\right\rangle$$

$$= u_k v_k\left\langle 0\left|\widehat{n}_{-k\downarrow} + \widehat{n}_{k\uparrow} - 1\right|0\right\rangle = u_k v_k\left(2n_k - 1\right), \qquad (24.97)$$

where in the absence of magnetic field $n_{-k\downarrow} = n_{k\uparrow} = n_k$.

At temperature $T = 0$:

$$\langle 0|a_{k\uparrow}a_{-k\downarrow}|0\rangle = -u_k v_k = -\frac{\Delta}{2\sqrt{\Delta^2 + \varepsilon_k^2}}. \qquad (24.98)$$

At finite temperature $T \neq 0$:

$$\langle 0|a_{k\uparrow}a_{-k\downarrow}|0\rangle = -\frac{\Delta}{2\sqrt{\Delta^2 + \varepsilon_k^2}}(1 - 2n_k). \qquad (24.99)$$

Correspondingly, taking the sum over k in (24.99) and multiplying left and right side on $V_0 = -|V_0|$ we get the self-consistency equation for the superconducting gap:

$$-|V_0|\left\langle 0\left|\sum_k a_{k\uparrow}a_{-k\downarrow}\right|0\right\rangle = |V_0|\sum_k \frac{\Delta}{2\sqrt{\Delta^2 + \varepsilon_k^2}}(1 - 2n_k) = \Delta. \qquad (24.100)$$

If $u_k v_k = 0$ (for $\Delta = 0$), then the probability amplitude to create the pair equals zero.

If, however, $u_k v_k \neq 0$, we can introduce the operator of the Cooper pairs $\widehat{\psi}_0$ and the average of this operator over the new vacuum. We will get then the wave function of electron pairs ψ_0. It is proportional to the superconducting gap Δ.

$$\psi_0 \sim \left\langle 0\left|\sum_k a_{k\uparrow}a_{-k\downarrow}\right|0\right\rangle = \frac{\Delta}{V_0}. \qquad (24.101)$$

Concluding the Lecture, let us emphasize, that the many-particle BCS theory [1,3, 4,7,10–20,35–37] based on the Bogoliubov transformation [8,9,21] plays extremely important role not only in the modern theory of superconductivity and strongly correlated electron systems [14,20,35–39] but in the quantum field theory [40–42] and quantum optics [43] as well.

Bogoliubov–De Gennes (BdG) equations [8,9,44,45] generalizing the Bogoliubov transformation on the inhomogeneous case are also very important in the physics of superconductivity and superfluidity [20,46–49].

References

1. Yu. Kagan, *Lectures on the Solid-state Theoretical Physics*, Moscow, MEPHI, 1981–1982. Part II. Electrons - unpublished
2. L.N. Cooper, Phys. Rev. **104**, 1189 (1956)
3. J. Bardeen, L.N. Cooper, J.R. Schrieffer, Phys. Rev. **108**, 1175 (1957)
4. J.R. Schrieffer, *Theory of Superconductivity* (W.A. Benjamin Inc., New York, 1964)
5. L.D. Landau, E.M. Lifshitz, *Quantum Mechanics, Non-relativistic Theory* (Pergamon Press, New York, 1977)
6. L.D. Landau, E.M. Lifshitz, *Statistical Physics, Part I* (Butterworth-Heinemann, Oxford, 1999)
7. A.A. Abrikosov, *Fundamentals of the Theory of Metals* (Elsevier Science Publishers, North-Holland, Amsterdam, 1988)
8. N.N. Bogoliubov, V.V. Tolmachev, D.V. Shirkov, *New Method in the Theory of Superconductivity* (Consultants Bureau, New York, London, Chapman and Hall, 1959)
9. N.N. Bogoliubov (ed.), *The Theory of Superconductivity* (Gordon and Breach, 1962)
10. E.M. Lifshitz, L.P. Pitaevskii, *Statistical Physics, Part II, Theory of the Condensed State* (Pergamon, Elsevier, 2013)
11. A.A. Abrikosov, L.P. Gor'kov, I.E. Dzyaloshinskii, *Methods of Quantum Field Theory in Statistical Physics* (Prentice Hall, Englewood Cliffs, New Jersey, 1963)
12. L.P. Gor'kov, *Selected Papers of Lev P. Gor'kov* (World Scientific, 2014)
13. L.N. Cooper, D. Feldman (eds.), *BCS: 50 years* (World Scientific, 2010)
14. P. Fulde, *Electron Correlations in Molecules and Solids*, Springer Series in Solid-State Sciences (1993)
15. C.P. Enz, *A Course on Many-Body Theory Applied to Solid-State Physics*. Lecture Notes in Physics (World Scientific, Singapore, 1992)
16. M.V. Sadovskii, *Diagrammatics*, Lectures on Selected Problems in Condensed Matter Theory (World Scientific, Singapore, 2006)
17. L.S. Levitov, *Green's Functions in Problems* (Princeton University Press, 2011)
18. M.Yu. Kagan, *Physics of the Macroscopic Quantum Systems*, Lectures and Seminars (Moscow, 2014) (in Russian)
19. M.Yu. Kagan, A.V. Ozharovskii, *Mathematical methods of the Theory of Superconductivity and Strongly correlated electron systems*, Lecture Course, Part I (Moscow, MEPHI, 1999) (in Russian)
20. M.Y. Kagan, *Modern Trends in Superconductivity and Superfluidity*, Lecture Notes in Physics, vol. 874 (Springer, Dordrecht, 2013)
21. N.N. Bogoliubov, J. Phys. USSR **11**, 23 (1947)
22. V.V. Tolmachev, Dokl. Acad. Sci. USSR **140**, 563 (1961) (in Russian)
23. S.V. Vonsovskii, Yu.A. Izumov, E.Z. Kurmaev, *Superconductivity of Transition Metals, Their Alloys and Compounds* (Moscow, 1977) (in Russian)
24. E.A. Linton, *Superconductivity* (Methuen and Co., Ltd., London, 1969)

25. A.C. Rose-Innes, E.H. Rhoderick, *Introduction to Superconductivity* (Pergamon Press, Oxford, London, 1969)
26. V.V. Schmidt, *The Physics of Superconductors, Introduction to Fundamentals and Applications* (Springer, 1997)
27. E. Parks (ed.), *Superconductivity (in Two Volumes)* (Marcel Dekker Inc., New York, 1969)
28. M.L. Cohen, *Superconductivity in Low-Carrier-Density Systems: Degenerate Semiconductors* (W.A. Benjamin Inc., New York-Amsterdam, 1966)
29. V.L. Ginzburg, D.A. Kirzhnits (eds.), *High Temperature Superconductivity* (Springer, 2014)
30. M. Tinkham, *Introduction to Superconductivity* (Dover Publications, 2004)
31. W. Buckel, R. Kleiner, *Superconductivity. An Introduction* (Wiley-VCH Verlag, 2016)
32. J.B. Ketterson, *Superconductivity* (Cambridge University Press, 1999)
33. I.M. Khalatnikov, *An Introduction to the Theory of Superfluidity* (Perseus Publishing, Cambridge, 2000)
34. D.R. Tiley, J. Tiley, *Superfluidity and Superconductivity* (Van Nostrand Reinhold company, New York-Cincinnati-Toronto-London-Melbourne, 1974)
35. J.R. Schrieffer, J.S. Brooks (eds.), *Handbook of High Temperature Superconductivity, Theory and Experiment* (Springer, 2007)
36. A.Y. Kitaev, Physics Uspekhi **44**, 131 (2001)
37. V.V. Val'kov, M.S. Shustin, S.V. Aksenov, A.O. Zlotnikov, A.D. Fedoseev, V.A. Mitskan, M.Y. Kagan, Physics Uspekhi **65**, 1–39 (2022)
38. P.W. Anderson, *The Theory of Superconductivity in the High-T_c Cuprate Superconductors* (Princeton University Press, 1997)
39. P.A. Lee, N. Nagaosa, X.G. Wen, Rev. Mod. Phys. **78**, 17 (2006)
40. S.W. Hawking, Commun. Math. Phys. **43**, 199 (1975)
41. W.G. Unruh, Phys. Rev. D **14**, 870 (1976)
42. S. Katagiri et al., Prog. Theor. Exp. Phys. **12**, 123 (2019)
43. F. Del Anno, S. Del Siena, F. Illuminati, Phys. Rep. **428**, 53 (2006)
44. P.G. De Gennes, *Superconductivity of Metals and Alloys* (W.A. Benjamin Inc., New York-Amsterdam, 1966)
45. A.V. Svidzinskii, Problems of the theory of superconductivity which involve spatial inhomogeneity (Moscow, 1982) (in Russian)
46. G.E. Volovik, *Exotic Properties of Superfluid ^3He* (World Scientific, Singapore, 1992)
47. A.V. Balatsky, G.E. Volovik, V.A. Konyshev, JETP **63**, 1194 (1986)
48. A. Ghosal, M. Randeria, N. Trivedi, Phys. Rev. Lett. **81**, 3940 (1998)
49. M.Y. Kagan, E.A. Mazur, JETP **132**, 596 (2021)

Lecture 25. Thermodynamics of the Superconductor

25

Abstract

In the beginning of the Lecture, we find the temperature behavior of the super-conducting gap and determine specific heat of the superconductor at low and high temperatures. We emphasize the fundamental role of the Meissner effect in the physics of superconductivity and derive expression for the superconducting current. We introduce the concept of the gauge invariance and write down London's equation. In the end of the Lecture, we mention Abrikosov vortices and the critical fields in superconductors of the first and second type.

25.1 Temperature Behavior of the Superconducting Gap

It is intuitively clear that the superconducting gap should disappear at finite temperature T equal to the critical temperature of the superconducting transition $T = T_C$ [1–19].

Indeed, the self-consistency equation for the gap at finite temperature acquires a form:

$$1 = |V_0|\, g_{\text{tot.}}\, (\varepsilon_F) \int_0^{\omega_0} \frac{d\varepsilon\,(1 - 2n(\varepsilon))}{\sqrt{\Delta^2 + \varepsilon^2}}, \tag{25.1}$$

where the average number of excitations

$$n(\varepsilon) = \frac{1}{\exp\left(\frac{\sqrt{\Delta^2 + \varepsilon^2}}{T}\right) + 1} \tag{25.2}$$

grows with the increase of the temperature.

It follows from the structure of Eq. (25.1) that the nominator $1 - 2n(\varepsilon)$ of the expression under the integral decreases for $T \neq 0$.

© The Author(s), under exclusive license to Springer Nature Switzerland AG 2026 495
M. Kagan, *Lecture Notes of Professor Yuri Kagan in Theoretical Solid-State Physics*,
Lecture Notes in Physics 1048, https://doi.org/10.1007/978-3-032-14621-2_25

Hence to fulfill the normalization on unity for the integral in the right side of Eq. (25.1), we should get the decrease of the denominator $\sqrt{\Delta^2 + \varepsilon^2}$. But the denominator can decrease only due to the decrease of the gap Δ. Finally, at temperature which equals to the critical one, the gap should go to zero:

$$\Delta\,(T = T_C) = 0. \tag{25.3}$$

Let us consider more detailly the temperature behavior of the gap both at low $(T \ll T_C)$ and high $(T \to T_C)$ temperatures, using the considerations about the decrease of the gap with the increase of temperature.

25.2 Low Temperatures

Let us denote the magnitude of the gap at temperature $T = 0$ as:

$$\Delta\,(T = 0) = \Delta_0, \tag{25.4}$$

and analyze the temperature behavior of the gap $\Delta\,(T)$ for:

$$T \ll \Delta_0. \tag{25.5}$$

To do that let us rewrite the self-consistency equation (25.1) in the following convenient form ($k_B = \hbar = 1$):

$$
\begin{aligned}
1 = |V_0|\,g_{\text{tot.}}\,(\varepsilon_F) \int_0^{\omega_0} \frac{d\varepsilon\,(1 - 2n\,(\varepsilon))}{\sqrt{\Delta^2 + \varepsilon^2}} &= |V_0|\,g_{\text{tot.}}\,(\varepsilon_F) \int_0^{\omega_0} \frac{d\varepsilon}{\sqrt{\Delta^2 + \varepsilon^2}} \\
- |V_0|\,g_{\text{tot.}}\,(\varepsilon_F) \int_0^{\omega_0} \frac{d\varepsilon\,2n\,(\varepsilon)}{\sqrt{\Delta^2 + \varepsilon^2}} &= |V_0|\,g_{\text{tot.}}\,(\varepsilon_F) \ln \frac{2\omega_0}{\Delta} \\
- |V_0|\,g_{\text{tot.}}\,(\varepsilon_F) \int_0^{\omega_0} &\frac{d\varepsilon\,2n\,(\varepsilon)}{\sqrt{\Delta^2 + \varepsilon^2}}.
\end{aligned}
\tag{25.6}
$$

Deriving (25.6) we used expression (24.83) for the calculation of the integral: $\int_0^{\omega_0} \frac{d\varepsilon}{\sqrt{\Delta^2 + \varepsilon^2}} \approx \ln \frac{2\omega_0}{\Delta}$.

At low temperatures the average number of excitations acquires the Boltzmann form [20]:

$$n(\varepsilon) \approx \exp\left(-\frac{\sqrt{\Delta^2 + \varepsilon^2}}{T}\right). \tag{25.7}$$

Moreover, we can convince ourselves, that at low temperatures in the temperature factor $n(\varepsilon)$ in the second integral term in the self-consistency equation (25.6) contribute only small energies $\varepsilon \ll \Delta$. Then in (25.7) we can expand the square root in series up to the linear terms and get for the average number of excitations:

$$n(\varepsilon) \approx \exp\left(-\frac{\Delta}{T} - \frac{\varepsilon^2}{2\Delta T}\right). \tag{25.8}$$

Correspondingly in the denominator of the integral term in (25.6) we can neglect ε^2 in comparison with Δ^2 under the root. As a result, dividing left and right side of (25.6) on $|V_0|\, g_{\text{tot.}}\,(\varepsilon_F)$, we can write this equation as:

$$\frac{1}{|V_0|\, g_{\text{tot.}}\,(\varepsilon_F)} = \ln\frac{2\omega_0}{\Delta} - \int_0^{\omega_0} \frac{d\varepsilon\, 2n\,(\varepsilon)}{\sqrt{\Delta^2 + \varepsilon^2}} \approx \ln\frac{2\omega_0}{\Delta} - 2\int_0^{\omega_0} \frac{d\varepsilon\, \exp\left(-\frac{\Delta}{T} - \frac{\varepsilon^2}{2\Delta T}\right)}{\Delta}$$

$$= \ln\frac{2\omega_0}{\Delta} - 2\frac{\exp\left(-\frac{\Delta}{T}\right)}{\Delta} \int_0^{\omega_0} d\varepsilon\, \exp\left(-\frac{\varepsilon^2}{2\Delta T}\right). \tag{25.9}$$

But we can see, that the integral $\int_0^{\omega_0} d\varepsilon\, \exp\left(-\frac{\varepsilon^2}{2\Delta T}\right)$ - is almost a Gaussian one and the main role in it play the energies $\varepsilon^2 \sim \Delta T \ll \Delta^2 \ll \omega_0^2$. That is why the upper limit in this integral can be safely expanded on ∞. Hence:

$$\frac{1}{|V_0|\, g_{\text{tot.}}\,(\varepsilon_F)} = \ln\frac{2\omega_0}{\Delta} - 2\frac{\exp\left(-\frac{\Delta}{T}\right)}{\Delta} \int_0^{\infty} d\varepsilon\, \exp\left(-\frac{\varepsilon^2}{2\Delta T}\right)$$

$$= \ln\frac{2\omega_0}{\Delta} - 2\frac{\exp\left(-\frac{\Delta}{T}\right)\sqrt{2\Delta T}}{\Delta} \cdot \int_0^{\infty} dx\, \exp\left(-x^2\right)$$

$$= \ln\frac{2\omega_0}{\Delta} - 2\frac{\exp\left(-\frac{\Delta}{T}\right)\sqrt{2\Delta T}}{\Delta}\frac{1}{2}\sqrt{\pi} = \ln\frac{2\omega_0}{\Delta} - \exp\left(-\frac{\Delta}{T}\right)\sqrt{\frac{2\pi T}{\Delta}}, \tag{25.10}$$

where calculating the Gaussian integral we used dimensionless variable

$$x^2 = \frac{\varepsilon^2}{2\Delta T}.$$

At temperature $T = 0$ the exponential factor $\exp\left(-\frac{\Delta_0}{T}\right) = 0$, and

$$\frac{1}{|V_0|\, g_{\text{tot.}}\,(\varepsilon_F)} = \ln\frac{2\omega_0}{\Delta_0}, \tag{25.11}$$

where $\Delta_0 = \Delta\,(T = 0)$ in agreement with formula (24.83).

At the temperatures different from zero let us replace $\frac{1}{|V_0|g_{\text{tot.}}(\varepsilon_F)}$ in the left side of Eq. (25.10) by $\ln\frac{2\omega_0}{\Delta_0}$ according to Eq. (25.11). At the same time, in the first term in the right side of (25.10) enters $\ln\frac{2\omega_0}{\Delta(T)}$. Let us emphasize, that in the second exponential term in (25.10) with the required accuracy we can replace $\Delta\,(T)$ on Δ_0. Then:

$$\ln \frac{2\omega_0}{\Delta_0} = \ln \frac{2\omega_0}{\Delta\,(T)} - \exp\left(-\frac{\Delta_0}{T}\right)\sqrt{\frac{2\pi T}{\Delta_0}}. \tag{25.12}$$

Hence:

$$\ln \frac{2\omega_0}{\Delta_0} - \ln \frac{2\omega_0}{\Delta(T)} = -\exp\left(-\frac{\Delta_0}{T}\right)\sqrt{\frac{2\pi T}{\Delta_0}}, \text{ and:}$$

$$\ln \frac{\Delta(T)}{\Delta_0} = -\exp\left(-\frac{\Delta_0}{T}\right)\sqrt{\frac{2\pi T}{\Delta_0}}. \tag{25.13}$$

Representing $\Delta(T)$ in (25.13) as:

$$\Delta(T) = \Delta_0 - \delta\Delta(T), \tag{25.14}$$

and using the fact, that:

$$\ln \frac{\Delta(T)}{\Delta_0} = \ln\left(1 - \frac{\delta\Delta(T)}{\Delta_0}\right) \approx -\frac{\delta\Delta(T)}{\Delta_0}, \tag{25.15}$$

we can find the decrease of the gap amplitude at low temperatures:

$$\delta\Delta(T) = \Delta_0 \exp\left(-\frac{\Delta_0}{T}\right)\sqrt{\frac{2\pi T}{\Delta_0}}. \tag{25.16}$$

Thus, at low temperatures the gap is given by:

$$\Delta(T) = \Delta_0\left(1 - \exp\left(-\frac{\Delta_0}{T}\right)\sqrt{\frac{2\pi T}{\Delta_0}}\right). \tag{25.17}$$

At low temperatures it decreases exponentially slow with the increase of temperature.

25.3   High Temperatures

At the temperature equal to the critical one according to Eq. (25.3) the gap $\Delta\,(T = T_C) = 0$ and the square root $\sqrt{\Delta^2 + \varepsilon^2} = \varepsilon$.

In this case the self-consistency equation (25.1) reads:

$$1 = |V_0|\, g_{\text{tot.}}\,(\varepsilon_F) \int_0^{\omega_0} \frac{d\varepsilon\,(1 - 2n\,(\varepsilon))}{\varepsilon}. \tag{25.18}$$

Further on let us use the fact that the nominator in (25.18) equals to:

$$1 - 2n\left(\varepsilon\right) = 1 - \frac{2}{\exp\left(\frac{\varepsilon}{T}\right) + 1} = \frac{-1 + \exp\left(\frac{\varepsilon}{T}\right)}{1 + \exp\left(\frac{\varepsilon}{T}\right)}$$

$$= \frac{\exp\left(\frac{\varepsilon}{2T}\right)\left(\exp\left(\frac{\varepsilon}{2T}\right) - \exp\left(-\frac{\varepsilon}{2T}\right)\right)}{\exp\left(\frac{\varepsilon}{2T}\right)\left(\exp\left(\frac{\varepsilon}{2T}\right) + \exp\left(-\frac{\varepsilon}{2T}\right)\right)} = \frac{2\sinh\frac{\varepsilon}{2T}}{2\cosh\frac{\varepsilon}{2T}} = \tanh\frac{\varepsilon}{2T}. \tag{25.19}$$

As a result, dividing again the left and right side at $T = T_C$ on $|V_0|\, g_{\text{tot.}}\left(\varepsilon_F\right)$, we can write (25.18) as:

$$\frac{1}{|V_0|\, g_{\text{tot.}}\left(\varepsilon_F\right)} = \int_0^{\omega_0} \frac{d\varepsilon \tanh\frac{\varepsilon}{2T}}{\varepsilon}. \tag{25.20}$$

But for small $x = \frac{\varepsilon}{2T} \ll 1$:

$$\tanh x \approx x. \tag{25.21}$$

At the same time, for $x \geq 1$:

$$\tanh x \approx 1. \tag{25.22}$$

Let us evaluate the integral (25.20) with the logarithmic accuracy. Then:

$$\int_0^{\omega_0} \frac{d\varepsilon \tanh\frac{\varepsilon}{2T}}{\varepsilon} \approx \int_{\varepsilon'}^{\omega_0} \frac{d\varepsilon}{\varepsilon} = \ln\frac{\omega_0}{\varepsilon'} = \ln\left(\gamma\frac{\omega_0}{T_C}\right), \tag{25.23}$$

where $\gamma \sim 1$ and we select the lower limit as:

$$\varepsilon' = \alpha T_C = \frac{1}{\gamma} T_C. \tag{25.24}$$

Let us stress, that the typical frequency $\omega_0 \gg T_C$. That is why we get large logarithm in Eq. (25.23):

$$\ln\left(\gamma\frac{\omega_0}{T_C}\right) \gg 1. \tag{25.25}$$

Note that the integral which we neglect in Eq. (25.23) is small in comparison with the conserved one:

$$\int_0^{\varepsilon'} \frac{d\varepsilon \tanh\frac{\varepsilon}{2T}}{\varepsilon} \approx \int_0^{\varepsilon'} \frac{d\varepsilon\frac{\varepsilon}{2T}}{\varepsilon} = \frac{1}{2T_C}\int_0^{\alpha T_C} d\varepsilon = \frac{\alpha}{2} \sim 1 \text{ for } \alpha \sim 1. \tag{25.26}$$

Hence:

$$\frac{1}{|V_0|\, g_{\text{tot.}}\,(\varepsilon_F)} = \ln\left(\gamma \frac{\omega_0}{T_C}\right), \text{ and } T_C = \gamma \omega_0 \exp\left(-\frac{1}{|V_0|\, g_{\text{tot.}}\,(\varepsilon_F)}\right). \qquad (25.27)$$

Exact calculations performed for example in the book of Abrikosov, Gor'kov and Dzyaloshinskii [11] demonstrate that the preexponential factor yields:

$$\gamma \approx 1.14. \qquad (25.28)$$

Comparing expressions for the critical temperature (25.27, 25.28)

$$T_C \approx 1.14\omega_0 \exp\left(-\frac{1}{|V_0|\, g_{\text{tot.}}\,(\varepsilon_F)}\right)$$

and superconducting gap at $T = 0$ (24.84): $\Delta_0 \approx 2\omega_0 \exp\left(-\frac{1}{|V_0|g_{\text{tot.}}(\varepsilon_F)}\right)$ we obtain important relation of the BCS theory for the ratio of the gap amplitude and the critical temperature:

$$T_C \approx 0.57\Delta_0, \text{ or: } \frac{\Delta_0}{T_C} \approx \frac{1}{0.57} \approx 1.75. \qquad (25.29)$$

This relation confirms that for the region of superconducting temperatures:

$$T \leq T_C \sim \Delta_0 \ll \omega_0. \qquad (25.30)$$

At the same time, however, the temperature T is not small compared to the gap amplitude $\Delta(T)$ for $T \sim T_C$.

Correspondingly, for the energy $2\Delta_0$, required to destroy the pair:

$$\frac{2\Delta_0}{T_C} \approx \frac{2}{0.57} \approx 3.5. \qquad (25.31)$$

25.4   Behavior of the Gap Close to T_C

Let us consider the region of temperatures close to the critical one assuming that the following inequality is satisfied:

$$T_C - T \ll T_C. \qquad (25.32)$$

Utilizing expression (25.19), we can write the self-consistency equation in this temperature region as:

$$\frac{1}{|V_0|\, g_{\text{tot.}}\,(\varepsilon_F)} = \int_0^{\omega_0} \frac{d\varepsilon \tanh\frac{\sqrt{\Delta^2+\varepsilon^2}}{2T}}{\sqrt{\Delta^2 + \varepsilon^2}}. \qquad (25.33)$$

After that we can perform the renormalization procedure, subtracting and adding to the right side of Eq. (25.33) the integral $\int_0^{\omega_0} \frac{d\varepsilon \tanh \frac{\varepsilon}{2T}}{\varepsilon}$ which we already used for the calculation of T_C. Then:

$$\frac{1}{|V_0|\, g_{\text{tot.}}(\varepsilon_F)} = \int_0^{\omega_0} \left\{ \frac{\tanh \frac{\sqrt{\Delta^2+\varepsilon^2}}{2T}}{\sqrt{\Delta^2+\varepsilon^2}} - \frac{\tanh \frac{\varepsilon}{2T}}{\varepsilon} \right\} d\varepsilon + \int_0^{\omega_0} \frac{d\varepsilon \tanh \frac{\varepsilon}{2T}}{\varepsilon}$$

$$= \int_0^{\omega_0} \left\{ \frac{\tanh \frac{\sqrt{\Delta^2+\varepsilon^2}}{2T}}{\sqrt{\Delta^2+\varepsilon^2}} - \frac{\tanh \frac{\varepsilon}{2T}}{\varepsilon} \right\} d\varepsilon + \ln\left(\gamma \frac{\omega_0}{T}\right), \qquad (25.34)$$

where we applied the formula $\int_0^{\omega_0} \frac{d\varepsilon \tanh \frac{\varepsilon}{2T}}{\varepsilon} = \ln\left(\gamma \frac{\omega_0}{T}\right)$ from Eq. (25.23) for the second term in the right side of (25.34).

Correspondingly, from (25.27) we get in the left side of (25.34):

$$\frac{1}{|V_0|\, g_{\text{tot.}}(\varepsilon_F)} = \ln\left(\gamma \frac{\omega_0}{T_C}\right).$$

Transferring $\ln\left(\gamma \frac{\omega_0}{T}\right)$ from the right side to the left one and using the relation:

$$\ln\left(\gamma \frac{\omega_0}{T_C}\right) - \ln\left(\gamma \frac{\omega_0}{T}\right) = \ln \frac{T}{T_C}, \qquad (25.35)$$

we can represent Eq. (25.34) in the convenient form:

$$\ln \frac{T}{T_C} = \int_0^{\omega_0} \left\{ \frac{\tanh \frac{\sqrt{\Delta^2+\varepsilon^2}}{2T}}{\sqrt{\Delta^2+\varepsilon^2}} - \frac{\tanh \frac{\varepsilon}{2T}}{\varepsilon} \right\} d\varepsilon. \qquad (25.36)$$

Now we can introduce the dimensionless variable:

$$x = \frac{\varepsilon}{2T} \qquad (25.37)$$

and write Eq. (25.36) as:

$$\ln \frac{T}{T_C} = \int_0^{\frac{\omega_0}{2T}} dx \left\{ \frac{\tanh \sqrt{x^2 + \frac{\Delta^2}{4T^2}}}{\sqrt{x^2 + \frac{\Delta^2}{4T^2}}} - \frac{\tanh x}{x} \right\}$$

$$\approx \int_0^{\infty} dx \left\{ \frac{\tanh \sqrt{x^2 + \frac{\Delta^2}{4T^2}}}{\sqrt{x^2 + \frac{\Delta^2}{4T^2}}} - \frac{\tanh x}{x} \right\}, \qquad (25.38)$$

where the upper limit $\frac{\omega_0}{2T} \gg 1$ can be safely expanded on ∞.

After that, since close to T_C the gap $\Delta(T) \ll T_C$ is small, we can expand the difference of the expressions in the figure brackets in Eq. (25.38) in series over small parameter $\frac{\Delta^2}{4T^2} \ll 1$.

For the transparency of the calculations let us split the integral in the right side of (25.38) on two integrals over dx (from 0 to ~ 1 and from ~ 1 to ∞).

Let us consider first the integral over the region of small x. In this region to escape a mistake we should expand the hyperbolic tangents deeper-up to the quadratic terms inclusively:

$$\tanh x \approx x - \frac{x^3}{3}, \text{ and } \frac{\tanh x}{x} \approx 1 - \frac{x^2}{3} \text{ for } x \ll 1. \tag{25.39}$$

Hence, since the square root $\sqrt{x^2 + \frac{\Delta^2}{4T^2}} \ll 1$ for $x \ll 1$, we obtain for the first term under the integral in the right side of (25.38):

$$\frac{\tanh \sqrt{x^2 + \frac{\Delta^2}{4T^2}}}{\sqrt{x^2 + \frac{\Delta^2}{4T^2}}} \approx 1 - \frac{x^2 + \frac{\Delta^2}{4T^2}}{3}. \tag{25.40}$$

Note, that the difference of two terms in (25.38) is negative for $x \ll 1$:

$$\frac{\tanh \sqrt{x^2 + \frac{\Delta^2}{4T^2}}}{\sqrt{x^2 + \frac{\Delta^2}{4T^2}}} - \frac{\tanh x}{x} \approx 1 - \frac{x^2 + \frac{\Delta^2}{4T^2}}{3} - 1 + \frac{x^2}{3} = -\frac{\Delta^2}{12\,T^2} < 0. \tag{25.41}$$

Thus, the first integral over small x converges and acquires the negative values:

$$\int_0^{\sim 1} dx \left\{ \frac{\tanh \sqrt{x^2 + \frac{\Delta^2}{4T^2}}}{\sqrt{x^2 + \frac{\Delta^2}{4T^2}}} - \frac{\tanh x}{x} \right\} \approx -\frac{\Delta^2}{12\,T^2} \int_0^{\sim 1} dx \sim -\frac{\Delta^2}{T^2} < 0. \tag{25.42}$$

Now we can analyze how the integral in (25.38) behaves at large values of x. For $x > 1$ we can expand the square root as:

$$\sqrt{x^2 + \frac{\Delta^2}{4T^2}} \approx x + \frac{\Delta^2}{8T^2x}. \tag{25.43}$$

As a result, the first term in the figure brackets in Eq. (25.38) reads:

$$\frac{\tanh \sqrt{x^2 + \frac{\Delta^2}{4T^2}}}{\sqrt{x^2 + \frac{\Delta^2}{4T^2}}} \approx \left(1 - \frac{\frac{\Delta^2}{8T^2}}{x^2} \right) \frac{\tanh x}{x} + \frac{\frac{\Delta^2}{8T^2}}{x^2} \frac{1}{\cosh^2 x}. \tag{25.44}$$

Correspondingly, the difference of two terms in (25.38) equals to:

$$\frac{\tanh\sqrt{x^2+\frac{\Delta^2}{4T^2}}}{\sqrt{x^2+\frac{\Delta^2}{4T^2}}}-\frac{\tanh x}{x}=-\frac{\frac{\Delta^2}{8T^2}}{x^2}\frac{\tanh x}{x}+\frac{\frac{\Delta^2}{8T^2}}{x^2}\frac{1}{\cosh^2 x}$$

$$=-\frac{\Delta^2}{8T^2}\frac{1}{x^2}\left(\frac{\tanh x}{x}-\frac{1}{\cosh^2 x}\right). \tag{25.45}$$

At large $x\gg 1$:

$$\frac{1}{\cosh^2 x}\to\frac{4}{\exp(2x)}\to 0,\ \text{and}\ \frac{\tanh x}{x^3}\to 0. \tag{25.46}$$

Thus, the second integral in Eq. (25.38) converges on the upper limit (on infinity) and equals to:

$$-\frac{\Delta^2}{8T^2}\int_{\sim 1}^{\infty}dx\left\{\frac{\tanh\sqrt{x^2+\frac{\Delta^2}{4T^2}}}{\sqrt{x^2+\frac{\Delta^2}{4T^2}}}-\frac{\tanh x}{x}\right\}$$

$$=-\frac{\Delta^2}{8T^2}\int_{\sim 1}^{\infty}dx\frac{1}{x^2}\frac{\tanh x}{x}\sim-\frac{\Delta^2}{T^2}<0. \tag{25.47}$$

That is why, both integrals are convergent (first one on the lower limit, second one on the upper limit), negative and by order of magnitude equal to $-\frac{\Delta^2}{T^2}$.

Correspondingly the sum of these integrals by order of magnitude also equals to $-\frac{\Delta^2}{T^2}$.

Hence the self-consistency equation:

$$\ln\frac{T}{T_C}=-\zeta\frac{\Delta^2}{T^2}, \tag{25.48}$$

where $\zeta\sim 1$.

Expressing the temperature T as:

$$T=T_C-\delta T,\ \text{and:}\ \frac{T}{T_C}=1-\frac{\delta T}{T_C}, \tag{25.49}$$

where the reduced temperature:

$$\frac{\delta T}{T_C}=\frac{T_C-T}{T_C}\ll 1, \tag{25.50}$$

we get for the logarithm in the left side of (25.48):

$$\ln \frac{T}{T_C} = \ln \left(1 - \frac{\delta T}{T_C} \right) \approx -\frac{\delta T}{T_C}. \tag{25.51}$$

As a result:

$$-\frac{\delta T}{T_C} = -\zeta \frac{\Delta^2}{T^2} \approx -\zeta \frac{\Delta^2}{T_C^2}, \text{ and: } \frac{\delta T}{T_C} = \zeta \frac{\Delta^2}{T_C^2}. \tag{25.52}$$

Then finally:

$$\Delta^2 (T) = \frac{T_C}{\zeta} (T_C - T), \text{ and: } \Delta (T) = \frac{1}{\sqrt{\zeta}} \sqrt{T_C (T_C - T)} = \eta T_C \sqrt{\frac{T_C - T}{T_C}}. \tag{25.53}$$

Exact calculations, performed in the book of Abrikosov, Gor'kov and Dzyaloshinskii [11] (25.53) lead to the following result for the constant η in (25.53):

$$\eta \approx 3.06. \tag{25.54}$$

The plot of the dependence of the superconducting gap on temperature in all the temperature range is shown in Fig. 25.1.

It is important to note that if the superconducting gap itself goes to zero as:

$$\Delta(T) \sim \sqrt{T_C - T} \text{ for } T \to T_C, \tag{25.55}$$

then temperature derivative of the gap diverges in the square root fashion:

$$\frac{d\Delta}{dT} \sim \frac{1}{\sqrt{T_C - T}}. \tag{25.56}$$

Note, that in many textbooks the result of Eqs. (25.53) and (25.54) for the temperature dependence of the gap close to T_C is obtained not from the microscopic self-consistency equation for the gap at high temperatures but purely phenomenologically from the minimization of the Ginzburg–Landau (GL) functional [10,17,18,20,21]:

Fig. 25.1 Temperature dependence of the superconducting gap in all the temperature range of the BCS theory [1,17,18]

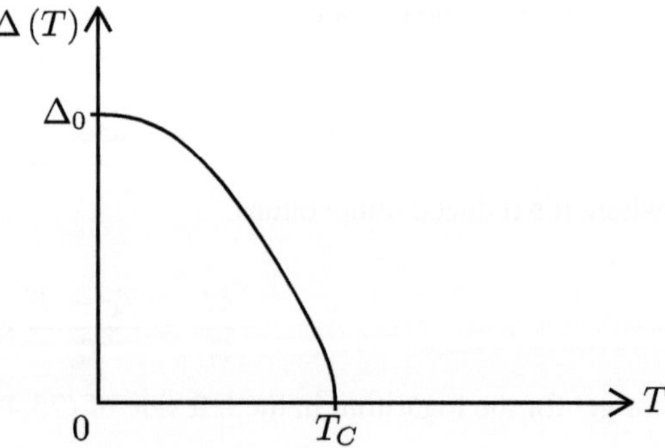

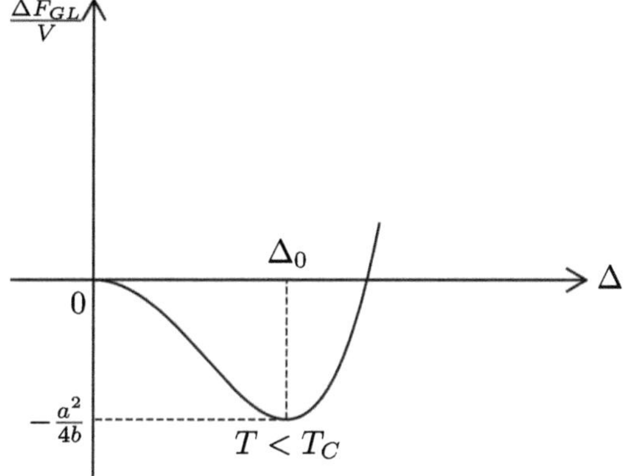

Fig. 25.2 Sombrero hat for the dependence of the GL Free energy $\frac{\Delta F_{GL}}{V}$ from the order parameter (superconducting gap) Δ for temperatures $T < T_C$ [1,17]

$$\frac{\Delta F_{GL}}{V} = a\Delta^2 + b\Delta^4, \text{ and } \min\frac{\Delta F_{GL}}{V} = -\frac{a^2}{4b} \text{ for the optimal gap value:}$$

$$\Delta_0^2 = -\frac{a}{2b} \approx 9.36 T_C(T_C - T), \tag{25.57}$$

where V is the system volume and $\Delta F_{GL} = F_S - F_N$ is Free energies difference of superconducting and normal states.

In Eq. (25.57) as always in Landau theory [22] for the phase transitions of the second order the coefficient at the quadratic term in GL functional $a \propto (T - T_C)/T_C$ linearly depends on the reduced temperature and changes sign in the transition point. At the same time, coefficient at the quartic term is positive: $b \propto \frac{1}{T_C^2} > 0$.

The plot of the dependence of the GL Free energy on the amplitude of the superconducting gap is presented in Fig. 25.2. Behavior of the GL Free energy as a function of the gap resembles the famous <u>sombrero hat</u>. It describes the gap generation in the theory of superconductivity similarly to Higgs mass generation in Quantum field theory [23–25].

Note again that in this Lecture we get expression for the gap at $T \to T_C$ from the solution of the many-particle problem applying Bogoliubov transformation for the effective BCS Hamiltonian leading us to the self-consistency equation for the gap.

Let us remind also that in the previous Lecture we already derived expression for the critical velocity:

$$V_{\text{cr.}} \sim \frac{\Delta(T)}{k_F} \propto \Delta(T). \tag{25.58}$$

Hence, the critical current of the superfluid component reads: $j_{\text{cr.}} \sim V_{\text{cr.}} \sim \frac{\Delta(T)}{k_F} \propto \Delta(T)$. Actually, as it is always the case in superfluid hydrodynamics [1,10,19,26,27]:

$$j_{\text{cr.}} \sim \rho_S V_{\text{cr.}}. \tag{25.59}$$

In Eq. (25.29) enters one more important parameter, namely the superfluid density ρ_S [1,10,19,26,27].

Thus, with the increase of temperature the amplitude of the superconducting gap $\Delta (T)$ decreases together with the critical velocity. Concerning the critical current, we can show (see for example [10,11]) that at temperatures close to T_C both the amplitude of the gap *and the superfluid density* are decreasing.

Let us stress once more that in the theory of the second order phase transitions the gap $\Delta(T)$ plays the role of the order parameter (it is absent at temperatures above the critical one and it emerges at $T = T_C$).

The gap itself is continuous at $T \to T_C$, but its derivative is divergent in the point of the phase transition (similarly to the specific heat which experiences a jump at T_C).

25.5 Specific Heat of the Superconductor

We can determine now the temperature behavior of the specific heat in all the temperature range below T_C. However already at first glance on this problem we can see substantial difficulties [1,10,11].

At low temperatures these difficulties are connected with the fact, that there is small number of excitations due to the presence of the gap in the spectrum. Hence it is difficult to sip energy for our system.

At high temperatures the difficulties are related to the fact that the gap itself and hence the excitation energy are the functions of temperatures.

Let us answer an important question: how we should write expression for the specific heat in the gas of excited particles?

According to Landau Fermi liquid theory [1,10,11], the energy functional depends on the set of the occupation numbers of the new reconstructed quasiparticles. That is why:

$$E_k = \frac{\delta E}{\delta n_k}, \text{ and } \delta E = \sum_k E_k \delta n_k, \tag{25.60}$$

where $E_k = \sqrt{\Delta^2 + \varepsilon_k^2}$ is the excitation energy, ε_k is electron energy in the normal state of metal counted from the Fermi level. At the same time, the fermionic distribution function of the excitation gas:

$$n_k = \frac{1}{\exp\left(\frac{\sqrt{\Delta^2 + \varepsilon_k^2}}{T}\right) + 1}.$$

To evaluate specific heat, we should calculate the derivative of the total energy of the system by temperature $\frac{dE}{dT}$.

It is very important, that according to (25.60) we should calculate only the temperature derivative of the occupation numbers $\frac{dn_k}{dT}$ without differentiation of the excitation energy E_k. Then:

$$\frac{dE}{dT} = \sum_k E_k \frac{dn_k}{dT}.$$ (25.61)

We can understand this result in the following way: the energy functional in Landau Fermi liquid theory depends on the occupation numbers of the excitation gas $E = E(\{n_k\})$. Hence, to calculate the derivative of the total energy by temperature, we differentiate only the occupation numbers:

$$\frac{dE}{dT} = \sum_k \frac{\delta E}{\delta n_k} \frac{dn_k}{dT} = \sum_k E_k \frac{dn_k}{dT},$$ (25.62)

where the derivative of the occupation numbers:

$$\frac{dn_k}{dT} = -\frac{\exp\left(\frac{E_k}{T}\right)}{\left(\exp\left(\frac{E_k}{T}\right) + 1\right)^2} \frac{d}{dT}\left(\frac{E_k}{T}\right).$$ (25.63)

Let us stress, that when we calculate the derivative of the occupation numbers by temperature in (25.63) we already have to account the gap dependence (and thus the dependence of the excitation energy) on temperature. In this case:

$$\frac{d}{dT}\left(\frac{E_k}{T}\right) = -\frac{E_k}{T^2} + \frac{1}{T}\frac{dE_k}{dT} = -\frac{E_k}{T^2} + \frac{1}{T}\frac{1}{2E_k}\frac{d\Delta^2(T)}{dT}.$$ (25.64)

Proceeding in (25.62) from the summation to the integration and taking into account two projections of electron spin, we obtain for the electron specific heat of the superconductor:

$$C_S(T) = 2g(\varepsilon_F) \int d\varepsilon \sqrt{\Delta^2 + \varepsilon^2} \cdot \frac{\exp\left(\frac{\sqrt{\Delta^2+\varepsilon^2}}{T}\right)}{\left(\exp\left(\frac{\sqrt{\Delta^2+\varepsilon^2}}{T}\right) + 1\right)^2}$$
$$\cdot \left[\frac{\sqrt{\Delta^2 + \varepsilon^2}}{T^2} - \frac{\frac{d\Delta^2(T)}{dT}}{2T\sqrt{\Delta^2 + \varepsilon^2}}\right]$$
$$= 2g(\varepsilon_F) \int d\varepsilon \frac{\exp\left(\frac{\sqrt{\Delta^2+\varepsilon^2}}{T}\right)}{\left(\exp\left(\frac{\sqrt{\Delta^2+\varepsilon^2}}{T}\right) + 1\right)^2} \left[\frac{\Delta^2 + \varepsilon^2}{T^2} - \frac{\frac{d\Delta^2(T)}{dT}}{2T}\right],$$ (25.65)

where $g(\varepsilon_F) = \frac{mk_F}{2\pi^2}$ in the formula for the specific heat (25.65) is the density of states on the Fermi surface for one spin projection.

25.6   Low Temperatures $T \ll \Delta_0$

As we know, at low temperatures the gap exponentially slow varies with temperature. According to formula (25.17), the gap at low temperatures $T \ll \Delta_0$ reads:

$$\Delta\left(T\right) = \Delta_0 \left(1 - \exp\left(-\frac{\Delta_0}{T}\right)\sqrt{\frac{2\pi T}{\Delta_0}}\right).$$

Hence, calculating specific heat, we can neglect with required accuracy the temperature dependence of the gap and restrict ourselves with the constant gap in (25.65): $\Delta\left(T\right) = \Delta_0$. Thus, the derivative $\frac{d\Delta^2(T)}{dT} = 0$.

Besides that, similarly to the derivation of expression (25.17), we can consider the Boltzmann distribution function in (25.65) and expand the square root in it, assuming $\Delta^2 \gg \varepsilon^2$. Then the distribution function in (25.65) yields:

$$n\left(\varepsilon\right) \approx \exp\left(-\frac{\Delta_0}{T} - \frac{\varepsilon^2}{2\Delta_0 T}\right).$$

As a result, the specific heat is given by:

$$C_S\left(T\right) \approx 2g\left(\varepsilon_F\right) \int d\varepsilon \left(\Delta_0^2 + \varepsilon^2\right) \frac{1}{T^2} \exp\left(-\frac{\sqrt{\Delta_0^2 + \varepsilon^2}}{T}\right)$$

$$\approx 2g\left(\varepsilon_F\right) \frac{\Delta_0^2}{T^2} \int d\varepsilon \exp\left(-\frac{\Delta_0}{T} - \frac{\varepsilon^2}{2\Delta_0 T}\right)$$

$$\approx 2g\left(\varepsilon_F\right) \frac{\Delta_0^2}{T^2} \exp\left(-\frac{\Delta_0}{T}\right) \int_0^\infty d\varepsilon \exp\left(-\frac{\varepsilon^2}{2\Delta_0 T}\right)$$

$$= 2g\left(\varepsilon_F\right) \frac{\Delta_0^2}{T^2} \exp\left(-\frac{\Delta_0}{T}\right) \sqrt{2\Delta_0 T} \frac{\sqrt{\pi}}{2}$$

$$= \sqrt{2\pi} g\left(\varepsilon_F\right) \frac{\Delta_0^{\frac{5}{2}}}{T^{\frac{3}{2}}} \exp\left(-\frac{\Delta_0}{T}\right) \sim \frac{\Delta_0^{\frac{5}{2}}}{T^{\frac{3}{2}}} \exp\left(-\frac{\Delta_0}{T}\right). \tag{25.66}$$

Qualitatively we can easily understand the result (25.66). To create the excitation, we should spend an energy of the order of Δ_0. But the probability of this process according to Gibbs [22] is proportional to $\exp\left(-\frac{\Delta_0}{T}\right)$. We can say, that at low temperatures we have exponentially small number of excitations in the system.

In other words, specific heat of the superconductor $C_S(T)$ in (25.66) is proportional to the number of broken Cooper pairs, that is to $\exp\left(-\frac{\Delta_0}{T}\right)$.

Correspondingly, if we pump electromagnetic radiation in the system, then its energy ω should be larger than 2Δ to excite electrons. This effect in the physics of superconductivity is called the <u>infrared threshold of the absorption</u> [1,4,7,10,28–34]. It also plays very important role in the physics of ultracold quantum gases [17,19].

25.7 High Temperatures $T \to T_C$

If we put the gap equal zero in all the terms in Eq. (25.65) at high temperatures $T \to T_C$, then we get linear in temperature specific heat of the normal metal C_N [1,7,10,11,20]:

$$C_N(T) = 2g(\varepsilon_F) \int d\varepsilon \frac{\varepsilon^2}{T^2} \frac{\exp\left(\frac{\varepsilon}{T}\right)}{\left(\exp\left(\frac{\varepsilon}{T}\right) + 1\right)^2} = \frac{mk_F}{2\pi^2} \frac{2\pi^2 T}{3} = \frac{mk_F T}{3} \sim T.$$

(25.67)

Let us stress, that the term with the derivative $\frac{d\Delta^2(T)}{dT}$ is important for the calculation of the specific heat at high temperatures $T \sim T_C$.

For $T \to T_C$ according to formulas (25.53, 25.54) the gap squared:

$$\Delta^2(T) \approx (3.06)^2 T_C (T_C - T) \approx 9.36 T_C (T_C - T),$$

and the derivative $\frac{d\Delta^2(T)}{dT}$ is constant negative quantity:

$$\frac{d\Delta^2(T)}{dT} \approx -9.36 T_C < 0. \tag{25.68}$$

Substituting (25.68) in the expression (25.65) for the specific heat at $T \to T_C$, we get:

$$C_S(T) \approx 2g(\varepsilon_F) \int d\varepsilon \frac{\exp\left(\frac{\sqrt{\Delta^2+\varepsilon^2}}{T}\right)}{\left(\exp\left(\frac{\sqrt{\Delta^2+\varepsilon^2}}{T}\right) + 1\right)^2} \left[\frac{\Delta^2+\varepsilon^2}{T^2} + \frac{9.36 T_C}{2T}\right]$$

$$= 2g(\varepsilon_F) \int d\varepsilon \frac{\exp\left(\frac{\sqrt{\Delta^2+\varepsilon^2}}{T}\right)}{\left(\exp\left(\frac{\sqrt{\Delta^2+\varepsilon^2}}{T}\right) + 1\right)^2} \frac{\Delta^2+\varepsilon^2}{T^2}$$

$$+ 2g(\varepsilon_F) \int d\varepsilon \frac{\exp\left(\frac{\sqrt{\Delta^2+\varepsilon^2}}{T}\right)}{\left(\exp\left(\frac{\sqrt{\Delta^2+\varepsilon^2}}{T}\right) + 1\right)^2} \frac{4.68 T_C}{T}. \tag{25.69}$$

Let us subtract from the first term in the right side of (25.69) and add to it expression (25.67) for the specific heat of the normal metal. Then:

$$C_S(T) = C_N(T)$$

$$+2g(\varepsilon_F)\int d\varepsilon \left\{ \frac{\exp\left(\frac{\sqrt{\Delta^2+\varepsilon^2}}{T}\right)}{\left(\exp\left(\frac{\sqrt{\Delta^2+\varepsilon^2}}{T}\right)+1\right)^2} \frac{\Delta^2+\varepsilon^2}{T^2} - \frac{\exp\left(\frac{\varepsilon}{T}\right)}{\left(\exp\left(\frac{\varepsilon}{T}\right)+1\right)^2} \frac{\varepsilon^2}{T^2} \right\}$$

$$+2g(\varepsilon_F)\frac{4.68T_C}{T}\int d\varepsilon \frac{\exp\left(\frac{\sqrt{\Delta^2+\varepsilon^2}}{T}\right)}{\left(\exp\left(\frac{\sqrt{\Delta^2+\varepsilon^2}}{T}\right)+1\right)^2}. \tag{25.70}$$

This expression contains two corrections to the specific heat of the normal metal $C_N(T)$. The first correction to $C_N(T)$ is connected with the expansion of the first term in the figure brackets of (25.70) in series to the linear terms in the small parameter $\frac{\Delta^2(T)}{T^2} \ll 1$. This correction equals to:

$$2g(\varepsilon_F)\frac{\Delta^2(T)}{T^2}\int_0^\infty d\varepsilon\, T^2 \frac{d}{d\varepsilon^2}\left(\frac{\exp\left(\frac{\varepsilon}{T}\right)}{\left(\exp\left(\frac{\varepsilon}{T}\right)+1\right)^2}\frac{\varepsilon^2}{T^2}\right)$$

$$= 2g(\varepsilon_F)\frac{\Delta^2(T)}{T}\int_0^\infty dz\, \frac{d}{dz^2}\left(z^2\frac{\exp(z)}{(\exp(z)+1)^2}\right)$$

$$= 2g(\varepsilon_F)\frac{\Delta^2(T)}{T}\zeta_1, \tag{25.71}$$

where:

$$\zeta_1 = \int_0^\infty dz\, \frac{d}{dz^2}\left(z^2\frac{\exp(z)}{(\exp(z)+1)^2}\right) = \text{const}, \tag{25.72}$$

and we introduced dimensionless variable $z = \frac{\varepsilon}{T}$.

Substituting expression for the gap squared $\Delta^2(T) \approx 9.36 T_C(T_C - T)$ in (25.71) we obtain for the first correction to the specific heat:

$$2g(\varepsilon_F)\frac{\Delta^2(T)}{T}\zeta_1 = 2g(\varepsilon_F)\frac{9.36 T_C(T_C - T)}{T}\zeta_1 \approx 2g(\varepsilon_F)(T_C - T)\zeta_2. \tag{25.73}$$

This correction goes to zero together with the superconducting gap at $T = T_C$.

The second correction in (25.70) is connected with the derivative $\frac{d\Delta^2(T)}{dT}$. This correction with required accuracy can be calculated at $\Delta^2(T) = 0$. In this case it equals to:

$$2g\left(\varepsilon_F\right)\frac{4.68T_C}{T}\int_0^\infty d\varepsilon \frac{\exp\left(\frac{\varepsilon}{T}\right)}{\left(\exp\left(\frac{\varepsilon}{T}\right)+1\right)^2}$$

$$= 9.36T_C g\left(\varepsilon_F\right)\int_0^\infty dz\frac{\exp\left(z\right)}{\left(\exp(z)+1\right)^2} = 9.36g\left(\varepsilon_F\right)T_C\eta_1, \tag{25.74}$$

where:

$$\eta_1 = \int_0^\infty dz\frac{\exp\left(z\right)}{\left(\exp\left(z\right)+1\right)^2} = \text{const}, \tag{25.75}$$

and we introduced again dimensionless variable $z = \frac{\varepsilon}{T}$.

As a result, the second correction to the specific heat $C_N(T)$ reads:

$$9.36g\left(\varepsilon_F\right)T_C\eta_1 = 2g\left(\varepsilon_F\right)T_C\eta_2. \tag{25.76}$$

This correction in contrast to the first one does not go to zero at $T = T_C$.

Thus, the final expression for the specific heat of the superconductor at $T \to T_C$ has the form:

$$C_S\left(T\right) = C_N\left(T\right) + 2g\left(\varepsilon_F\right)\left(T_C - T\right)\zeta_2 + 2g\left(\varepsilon_F\right)T_C\eta_2. \tag{25.77}$$

Note, that the integrals in formulas (25.72) and (25.75) for the corrections to the specific heat are tabulated and can be found in many textbooks on the theory of superconductivity (see for example [10, 11]).

At $T = T_C$ specific heat of the superconducting phase:

$$C_S\left(T_C\right) = C_N\left(T_C\right) + 2g\left(\varepsilon_F\right)T_C\eta_2 = C_N\left(T_C\right) + \frac{4mk_F}{3\zeta\left(3\right)}T_C, \tag{25.78}$$

where $\zeta\left(3\right)$ is the Riemann function $\zeta\left(x\right)$ for $x = 3$ [1, 10, 11, 20].

The final expression for the specific heat of the superconductor $C_S\left(T\right)$ at $T \to T_C$ in (25.77) can be expressed via the specific heat of the normal metal $C_N\left(T_C\right) = \frac{mk_F T_C}{3}$ and can be represented as follows [1, 11]:

$$C_S\left(T\right) = C_N\left(T_C\right)\left[2.43 - 3.77\frac{T_C - T}{T_C}\right]. \tag{25.79}$$

Thus, electron specific heat of the superconductor in the transition point (at $T = T_C$) experiences a jump. This is connected with the fact that the excitation energy is the function of temperature.

The plot for the specific heat $C_S\left(T\right)$ resembles the Greek letter lambda (see Fig. 25.3). This behavior of the specific heat as the function of temperature in superconductor is closely related to the linear temperature dependence of the specific heat of normal metal

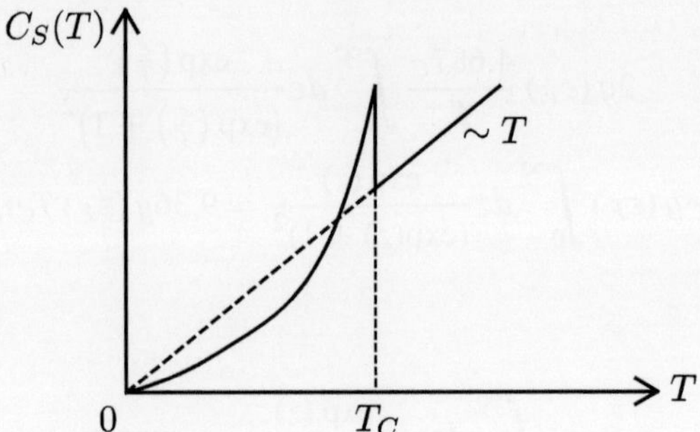

Fig. 25.3 The plot of the electron specific heat of the superconductor as the function of temperature resembles Greek letter lambda. This temperature dependence of the specific heat is closely related to the linear temperature behavior of the specific heat of normal metal $C_N(T) = \frac{mk_FT}{3}$ for $T > T_C$ [1]

$$C_N(T) = \frac{mk_FT}{3} \text{ for } T > T_C.$$

Hence it is customary for the critical temperature to write T_λ instead of T_C and refer to the point of phase transition as to the λ-point. This temperature dependence of the specific heat is typical for the second order phase transitions.

25.8   Superconducting Current and Meissner Effect

It is necessary to emphasize three important facts to derive expression for the superconducting current. Firstly, note the existence of the macroscopic wave function, describing the condensate (the superfluid state).

Secondly, the wave function has the cyclic variable–the phase. Due to this fact we get the quantization of the circulation of the superfluid velocity and the appearance of the quantized vortices [1, 10, 17, 19, 26, 27].

Finally, we should remember that superconductivity is the superfluidity of the Cooper pairs.

In the superfluid helium [1, 10, 17, 19, 26, 27] or in the Bogoliubov Bose gas [35] the existence of the macroscopic wave function is related to the fact that the annihilation operator a_0 of the ground state (of the lowest level of the system occupied by the macroscopically large number of particles $n_0 \gg 1$) can be safely replaced by the c-number, since:

$$a_0 |n_0\rangle = \sqrt{n_0 - 1}\, |n_0 - 1\rangle \approx \sqrt{n_0}\, |n_0\rangle \text{ for } n_0 \gg 1.$$

Analogously, in the theory of superconductivity the operator of the pair creation $B_0 = a_{k\downarrow} a_{-k\uparrow}$ after the averaging over the reconstructed vacuum at $T = 0$ equals to:

$$\langle 0| B_0 |0\rangle = u_k v_k = \frac{\Delta}{2\sqrt{\Delta^2 + \varepsilon_k^2}} = \frac{\Delta}{2E_k}. \tag{25.80}$$

Correspondingly, for $T \neq 0$ the average of the operator of pair creation reads:

$$\langle B\rangle = u_k v_k \left(1 - 2n_k\right) = \frac{\Delta\left(1 - 2n_k\right)}{2\sqrt{\Delta^2 + \varepsilon_k^2}}. \tag{25.81}$$

Thus, $\langle B\rangle \neq 0$ for $\Delta \neq 0$.

In the theory of superfluidity [1, 10, 26, 27] the Ψ-operator of the condensate particles:

$$\widehat{\Psi}_0 = \frac{a_0}{\sqrt{V}}, \tag{25.82}$$

while the Ψ-operator of the excitations in the coordinate space:

$$\widehat{\Psi}\left(r\right) = \frac{1}{\sqrt{V}} \sum_{k \neq 0} \hat{a}_k \exp\left(ikr\right), \tag{25.83}$$

where V is the system volume.

In the normal metal the average of two Ψ-operators with opposite spins occupying the same point r in the coordinate space:

$$\left\langle \widehat{\Psi}_\downarrow\left(r\right) \widehat{\Psi}_\uparrow\left(r\right)\right\rangle = 0, \tag{25.84}$$

where:

$$\widehat{\Psi}_\downarrow\left(r\right) = \frac{1}{\sqrt{V}} \sum_k \hat{a}_{k\downarrow} \exp\left(ikr\right), \text{ and } \widehat{\Psi}_\uparrow\left(r\right) = \frac{1}{\sqrt{V}} \sum_q \hat{a}_{q\uparrow} \exp\left(iqr\right). \tag{25.85}$$

At the same time the average of two Ψ-operators in the superconductor:

$$\left\langle \widehat{\Psi}_\downarrow\left(r\right) \widehat{\Psi}_\uparrow\left(r\right)\right\rangle = \frac{1}{V} \left\langle \sum_k \hat{a}_{k\downarrow} \hat{a}_{-k\uparrow}\right\rangle = \frac{1}{V} \sum_k u_k v_k \left(1 - 2n_k\right)$$

$$= \frac{1}{V} \sum_k \frac{\Delta\left(1 - 2n_k\right)}{2\sqrt{\Delta^2 + \varepsilon_k^2}} = \int \frac{d^3k}{(2\pi)^3} \frac{\Delta\left(1 - 2n_k\right)}{2\sqrt{\Delta^2 + \varepsilon_k^2}} \neq 0 \text{ for } \Delta \neq 0. \tag{25.86}$$

Thus, we can introduce the coordinate-dependent (local) operator for the description of the macroscopic wave function:

$$\hat{\chi}\left(r\right) = \widehat{\Psi}_\downarrow\left(r\right) \widehat{\Psi}_\uparrow\left(r\right). \tag{25.87}$$

The wave function related to this operator can be written as:

$$\chi(r) = \chi_0(r) + \sum_{k \neq 0} c_k \chi_k(r), \qquad (25.88)$$

where $\chi_0(r)$ is the macroscopic wave function describing Bose condensate of pairs with zero momentum, while:

$\sum_{k \neq 0} c_k \chi_k(r)$ is the wave function of the excitations.

The condensate wave function $\chi_0(r)$ separates from the lattice and behaves completely independently. Effectively it behaves as one pair multiplied by the number of pairs.

Nevertheless, certain difficulties arise in this interpretation. They are connected with the fact, that the pairs are excited close to the Fermi surface. Then it seems that their number is proportional to $\frac{\Delta}{\varepsilon_F} N_{\text{el.}}$ and should be small due to the smallness of the Δ-neighborhood of the Fermi surface.

However this is not the case and the total number of pairs at $T = 0$ equals to half the number of electrons. The point is that electrons inside the Fermi surface always behave as the superconducting ones because it is impossible to excite them. Hence it is sufficient for us to make superconducting only electrons living close to the Fermi surface. Remaining electrons can be engaged by them in the superconducting current, and thus the density of superconducting electrons at $T = 0$ equals to the total density.

If we take into account the complex character of the wave function and distinguish slowly varying amplitude and rapidly varying phase in it according to the relation:

$$\chi_0(r) = |\chi_0(r)| \exp(i\Phi(r)), \qquad (25.89)$$

then as Quantum mechanics [36] teaches us, the superconducting current reads:

$$j_S = -\frac{i\hbar 2e}{2(2m)} \left[\chi^* \nabla \chi - \chi \nabla \chi^*\right] = \frac{\hbar 2e |\chi|^2}{2m} \nabla \Phi. \qquad (25.90)$$

Deriving Eq. (25.90) we accounted that the amplitude of the wave function squared is normalized on the density of the superconducting pairs:

$$|\chi|^2 = \frac{n_S}{2}. \qquad (25.91)$$

We keep in mind also that the mass of the pair is $2m$ and the charge of the pair is $2e$.

As a result, the superconducting current:

$$j_S = 2e \frac{n_S}{2} \frac{\hbar}{2m} \nabla \Phi = e n_S \frac{\hbar}{2m} \nabla \Phi = e n_S v_S, \qquad (25.92)$$

where n_S is the number of electrons in the condensate, and $v_S = \frac{\hbar}{2m} \nabla \Phi$ is the superfluid velocity of the pairs.

In magnetic field we should replace the momentum operator $\hat{p} = -i\hbar\nabla$ by the operator of the generalized momentum [36] according to the prescription:

$$\hat{p} \rightarrow \hat{p} - \frac{2e}{c}A, \tag{25.93}$$

where A is the vector potential.

Then the superconducting current besides the standard term $en_S v_S$ contains linear in vector potential London's term $-\frac{e^2 n_S}{mc}A$ [1,10,11,17,37,38]:

$$j_S = \frac{\hbar 2e|\chi|^2}{2m}\left(\nabla\Phi - \frac{2e}{\hbar c}A\right) = en_S v_S - \frac{e^2 n_S}{mc}A. \tag{25.94}$$

Let us prove the gradient (the gauge) invariance of the current j_S. Indeed, the superconducting current proves to be invariant when vector potential is replaced by:

$$A \rightarrow A + \nabla\varphi \tag{25.95}$$

and simultaneously an additional phase is imposed on the wave function:

$$\chi_0 = |\chi_0|\exp\left(i\Phi\right) \rightarrow \chi_0\exp\left(\frac{i2e}{\hbar c}\varphi\right) = |\chi_0|\exp\left(i\Phi + \frac{i2e}{\hbar c}\varphi\right). \tag{25.96}$$

25.9 Quantization of the Magnetic Flux

From Eq. (25.94) for the superconducting current and considerations related to the gradient invariance we can get the condition of the quantization of the magnetic flux.

Let us equate the superconducting current in (25.94) to zero:

$$\frac{\hbar 2e|\chi|^2}{2m}\left(\nabla\Phi - \frac{2e}{\hbar c}A\right) = 0. \tag{25.97}$$

Correspondingly:

$$\nabla\Phi = \frac{2e}{\hbar c}A. \tag{25.98}$$

Let us calculate the circulation from left side and right side of Eq. (25.98) over the closed contour C. Then in the right side we obtain according to the Gauss theorem:

$$\frac{2e}{\hbar c}\oint A dl = \frac{2e}{\hbar c}\iint \operatorname{curl} A dS = \frac{2e}{\hbar c}\iint B dS = \frac{2e}{\hbar c}\Phi_B, \tag{25.99}$$

where $B = \operatorname{curl} A$ is magnetic field, $\Phi_B = \iint B dS$ is magnetic flux.

If we take the circulation over the closed contour C from the left side, then from phase cycling condition we obtain:

$$\oint \nabla \Phi dl = 2\pi n, \qquad (25.100)$$

where n is integer number.

Thus, we derive the condition of the quantization of the magnetic flux:

$$\frac{2e}{\hbar c}\Phi_B = 2\pi n, \text{ and } \Phi_B = \frac{\pi \hbar c}{e}n = n\Phi_0, \qquad (25.101)$$

where:

$$\Phi_0 = \frac{\pi \hbar c}{e} \qquad (25.102)$$

is the superconducting flux quantum.

25.10 London Equation: Meissner Effect

If we take the curl from left side and right side of (25.92) and use the condition of slow variation of superconducting density, then curl $\nabla \Phi = 0$, and we derive London equation [37,38]:

$$\text{curl } j_S = -\frac{e^2 n_S}{mc}\text{curl } A = -\frac{e^2 n_S}{mc}B. \qquad (25.103)$$

Let us solve London equation together with Maxwell equation [39]:

$$\text{curl } B = \frac{4\pi}{c}j_S, \text{ or } j_S = \frac{c}{4\pi}\text{curl } B. \qquad (25.104)$$

Then taking the curl from the left and right side of (25.104), we obtain:

$$\text{curl } j_S = \frac{c}{4\pi}\text{curl curl } B = \frac{c}{4\pi}(\text{grad div } B - \Delta B). \qquad (25.105)$$

Now we can utilize the second Maxwell equation [39], connected with the absence of magnetic charges. According to this equation:

$$\text{div } B = 0, \text{ and: grad div } B - \Delta B = -\Delta B. \qquad (25.106)$$

Hence:

$$\text{curl } j_S = -\frac{c}{4\pi}\Delta B. \qquad (25.107)$$

Substituting Eq. (25.103) in (25.107), we finally get:

$$-\frac{c}{4\pi}\Delta \boldsymbol{B} = -\frac{e^2 n_S}{mc}\boldsymbol{B}, \text{ and: } \Delta \boldsymbol{B} = \frac{4\pi e^2 n_S}{mc^2}\boldsymbol{B}, \tag{25.108}$$

where $\frac{e^2}{mc^2}$ is the typical radius of electron in Quantum electrodynamics [40].

Introducing Meissner penetration depth [41]:

$$\lambda_M^2 = \frac{mc^2}{4\pi e^2 n_S}, \tag{25.109}$$

and selecting the direction and the coordinate dependence of the magnetic field for the superconductor, occupying semi-infinite space (see Fig. 25.4):

$$\boldsymbol{B} = \boldsymbol{e}_z B(x), \tag{25.110}$$

we can rewrite Eq. (25.108) in a simple form:

$$\frac{d^2 B}{dx^2} = \frac{1}{\lambda_M^2} B. \tag{25.111}$$

The solution of this equation corresponds to magnetic field exponentially decaying deep into the superconductor:

$$B(x) = B_0 \exp\left(-\frac{x}{\lambda_M}\right). \tag{25.112}$$

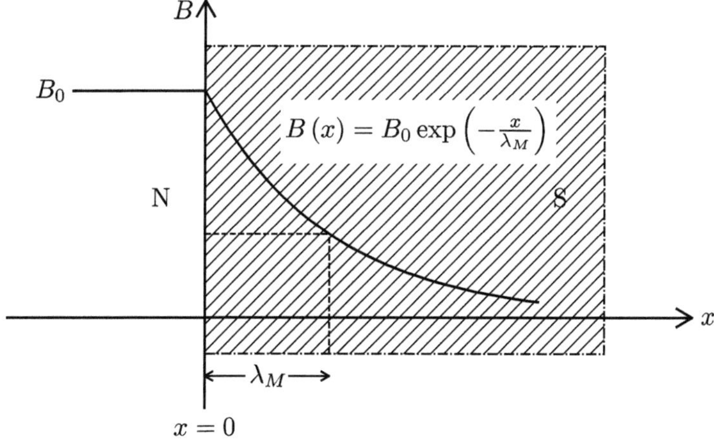

Fig. 25.4 Illustration of the Meissner effect for the case when superconductor occupies semi-infinite space at $x > 0$. We show in the Figure the exponential decay of the magnetic field deep into the superconductor on the distances from the border exceeding the Meissner penetration depth λ_M [1,17,18]

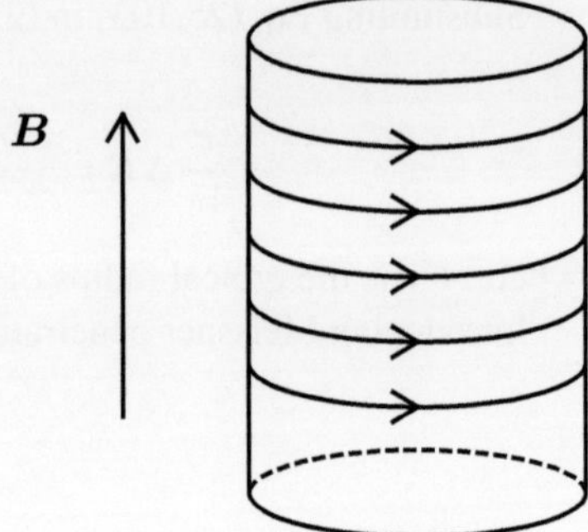

Fig. 25.5 Non-dissipative diamagnetic currents in the surface layer of the width of the order of λ_M arising in the cylindrical superconductor [1]

Thus, λ_M is the penetration depth indeed. In conventional superconductors $\lambda_M \sim 10^{-5}$ cm.

The derivation of the exponential decay in (25.106) assumes firstly that the magnetic field is weak and the number of superconducting electrons does not change $n_S = \text{const}$.

Secondly, the field should decrease slowly on the size of the pair, so that electrons in a pair feel the same field.

In the depth of the superconductor according to the Meissner effect [41] magnetic field is absent:

$$\boldsymbol{B} = \boldsymbol{H} + 4\pi \boldsymbol{M} = 0, \tag{25.113}$$

where $\boldsymbol{H}$ is the external field. That is why magnetic moment emerging in the super-conductor (see Fig. 25.5) equals to:

$$\boldsymbol{M} = -\frac{1}{4\pi}\boldsymbol{H}. \tag{25.114}$$

Thus, superconductor is ideal diamagnet. Magnetic moment appears due to the fact, that according to Maxwell equation curl $\boldsymbol{B} = \frac{4\pi}{c}\boldsymbol{j}_S$, the diamagnetic currents emerge in the system. These non-dissipative currents (arising in the surface layer of the superconductor with the width of the order of λ_M) compensate the external field.

25.11 Other Important Questions in the Theory of Superconductivity

In the end of the Lecture let us briefly mention several other very important questions in the theory of superconductivity:

(1) Two types of superconductors, Ginzburg–Landau parameter distingui-shing type-I and type-II superconductors, London and Pippard limits [1, 10–12, 17–21, 41–51].

(2) Abrikosov vortex lattice [7,42], phase diagram in field, first and second critical fields in type-II superconductors [10,11,20,21,27,42]. Gor'kov estimate [49] for the second critical field.

(3) Abrikosov–Gor'kov theory for superconducting alloys [11,12,52,53]. The dirty limit for Meissner penetration depth and coherence length.

References

1. Yu. Kagan, *Lectures on the Solid-state Theoretical Physics*, Moscow, MEPHI, 1981–1982. Part II. Electrons - unpublished
2. L.N. Cooper, Phys. Rev. **104**, 1189 (1956)
3. J. Bardeen, L.N. Cooper, J.R. Schrieffer, Phys. Rev. **108**, 1175 (1957)
4. J.R. Schrieffer, *Theory of Superconductivity* (W.A. Benjamin Inc., New York, 1964)
5. J.R. Schrieffer, J.S. Brooks (eds.), *Handbook of High Temperature Superconductivity, Theory and Experiment* (Springer, 2007)
6. L.N. Cooper, D. Feldman (eds.), *BCS: 50 Years* (World Scientific, 2010)
7. A.A. Abrikosov, *Fundamentals of the Theory of Metals* (Elsevier Science Publishers, North-Holland, Amsterdam, 1988)
8. N.N. Bogoliubov, V.V. Tolmachev, D.V. Shirkov, *New Method in the Theory of Superconductivity* (Consultants Bureau, New York, London, Chapman and Hall, 1959)
9. N.N. Bogoliubov (ed.), *The Theory of Superconductivity* (Gordon and Breach, 1962)
10. E.M. Lifshitz, L.P. Pitaevskii, *Statistical Physics, Part II, Theory of the Condensed State* (Pergamon, Elsevier, 2013)
11. A.A. Abrikosov, L.P. Gor'kov, I.E. Dzyaloshinskii, *Methods of Quantum Field Theory in Statistical Physics* (Prentice Hall, Englewood Cliffs, New Jersey, 1963)
12. L.P. Gor'kov, *Selected Papers of Lev P. Gor'kov* (World Scientific, 2014)
13. P. Fulde, *Electron Correlations in Molecules and Solids*, Springer Series in Solid-State Sciences (1993)
14. C.P. Enz, *A Course on Many-Body Theory Applied to Solid-State Physics*, Lecture Notes in Physics (World Scientific, Singapore, 1992)
15. M.V. Sadovskii, *Diagrammatics, Lectures on Selected Problems in Condensed Matter Theory* (World Scientific, Singapore, 2006)
16. L.S. Levitov, *Green's Functions in Problems* (Princeton University Press, 2011)
17. M.Yu. Kagan, *Physics of the Macroscopic Quantum Systems*, Lectures and Seminars (Moscow, 2014) (in Russian)
18. M.Yu. Kagan, A.V. Ozharovskii, *Mathematical methods of the Theory of Superconductivity and Strongly Correlated Electron Systems*, Lecture Course, Part I (Moscow, MEPHI, 1999) (in Russian)
19. M.Yu. Kagan, *Modern Trends in Superconductivity and Superfluidity*, Lecture Notes in Physics, vol. 874 (Springer, Dordrecht, 2013)
20. V.L. Ginzburg, L.D. Landau, Zh. Eksp. Teor. Fiz. **20**, 1064 (1950) (in Russian), English translation in: L.D. Landau, Collected papers Pergamon Press, Oxford, 1965), p. 546
21. V.L. Ginzburg, *On Superconductivity and Superfluidity, A Scientific Autobiography* (Springer, 2009)
22. L.D. Landau, E.M. Lifshitz, *Statistical Physics*, Part I (Butterworth-Heinemann, Oxford, 1999)
23. S. Weinberg, Phys. Rev. D **13**, 974 (1976)
24. P.W. Higgs, Phys. Rev. Lett. **13**, 508 (1964). https://doi.org/10.1103%2FPhysRevLett.13.508
25. P.W. Anderson, Phys. Rev. **130**, 439 (1962)
26. I.M. Khalatnikov, *An Introduction to the Theory of Superfluidity* (Perseus Publishing, Cambridge, 2000)

27. D.R. Tiley, J. Tiley, *Superfluidity and Superconductivity* (Van Nostrand Reinhold company, New York-Cincinnati-Toronto-London-Melbourne, 1974)
28. E.A. Linton, *Superconductivity* (Methuen and Co., Ltd., London, 1969)
29. A.C. Rose-Innes, E.H. Rhoderick, *Introduction to Superconductivity* (Pergamon Press, Oxford, London, 1969)
30. V.V. Schmidt, *The Physics of Superconductors, Introduction to Fundamentals and Applications* (Springer, 1997)
31. E. Parks (ed.), *Superconductivity (in Two Volumes)* (Marcel Dekker Inc., New York, 1969)
32. M. Tinkham, *Introduction to Superconductivity* (Dover Publications, 2004)
33. W. Buckel, R. Kleiner, *Superconductivity. An Introduction* (Wiley-VCH Verlag, 2016)
34. J.B. Ketterson, *Superconductivity* (Cambridge University Press, 1999)
35. N.N. Bogoliubov, J. Phys. USSR **11**, 23 (1947)
36. L.D. Landau, E.M. Lifshitz, *Quantum Mechanics, Non-relativistic Theory* (Pergamon Press, New York, 1977)
37. H. London, F. London, Proc. Roy. Soc. A **149**, 71 (1935)
38. H. London, F. London, Physica **2**, 341 (1935)
39. J.C. Maxwell, Phil. Trans. Roy. Soc., London. **155**, 459 (1865). https://en.wikipedia.org/wiki/Philosophical_Transactions_of_the_Royal_Society_of_London
40. V.B. Berestetskii, E.M. Lifshitz, L. P. Pitaevskii, *Quantum Electrodynamics* (Butterworth-Heinemann, 1982)
41. W. Meissner, R. Ochsenfeld, Naturwissenschaften **21**, 787 (1933)
42. A.A. Abrikosov, Sov. Phys. JETP **5**, 1174 (1957)
43. R.P. Feynman, Progress Low Temp. Phys. **1**, Chap. 2, p. 17 (1955)
44. L. Onsager, Nuovo Cimento **6**(Suppl. 2), 249 (1949)
45. J.G. Bednorz, K.A. Muller, Rev. Mod. Phys. **60**, 585 (1988)
46. J.C. Philips, *Physics of High-T_c Superconductors* (Academic Press, Boston, 1989)
47. S.A. Wolf, V.Z. Kresin, *Novel Superconductivity* (Plenum Press, New York, 1987)
48. D.M. Ginsberg (ed.), *Physical Properties of High Temperature superconductors* (World Scientific, Singapore, 1994)
49. L.P. Gor'kov, ZHETP **97**, 833 (1959)
50. L.D. Landau, Z. Angew. Phys. **54**, 629 (1930)
51. A.B. Pippard, Proc. Roy. Soc. A **216**, 547 (1952)
52. A.A. Abrikosov, L.P. Gor'kov, ZHETP **35**, 1158 (1958)
53. A.A. Abrikosov, L.P. Gor'kov, ZHETP **36**, 319 (1959)

Concluding Remarks. Modern Trends of the Theoretical Solid-State Physics

Working on the Lectures of my father I always caught myself thinking that in spite of the years passing after its creation, the Course absolutely did not lose its actuality.

Note, that during last 20 years of the previous century and first 25 years of our century, the modern solid-state theoretical physics was especially booming in two main directions.

The first direction—the theory of mesoscopic physics and localization—is particularly close to the development of modern nanoelectronics and creation of the more and more miniature devices based on quantum principle.

I can recommend to our readers an excellent book of I. Imry on these topics *"Introduction to mesoscopic Physics."*

The second direction was initiated by the discovery of high temperature superconductivity by Alex Muller and George Bednorz in the IBM laboratory in Zurich in 1986. It was connected first of all with the creation and investigation of novel materials promising for the revolution in energetics and computational technique.

Both directions and their intersection are very important also for the various solid-state realizations of the future quantum computers.

Let us emphasize that during his last years in 2016–2019 my father was very interested in the room-temperature superconductivity and lifetime problem of the long-living metastable phases in metallic hydrogen and metallic hydrides at high and normal pressures and underlined the importance of this research.

Working on the Course I paid attention on one more very important detail. Namely, after long and cumbersome calculation leading to the qualitative effect, my father always stopped the calculations and started to analyze and play around the obtained result considering it from all sides and angles. This working style seems to me very useful for young theorists.

Actually, it represents the combination of serious mathematical apparatus of statistical physics and modern solid-state theoretical physics with excellent physical intuition and the background broadness not only in his own field of research but also in related fields.

521

M. Kagan, *Lecture Notes of Professor Yuri Kagan in Theoretical Solid-State Physics*, Lecture Notes in Physics 1048, https://doi.org/10.1007/978-3-032-14621-2

I hope, that the present Course will be useful to young theorists, experimentalists, and material scientists working in scientific laboratories and universities in our country and abroad.

Index

A

Abrikosov
  Abrikosov lattice, 65
  Abrikosov vortex, 368, 495, 519
Abrikosov–Gor'kov
  Abrikosov–Gor'kov theory, 519
Absolute value, 7, 8, 16, 25, 50, 63, 72, 119,
  134, 173, 184, 185, 187, 189–193, 199,
  201, 234, 239, 281, 300, 318, 380, 381,
  438, 448, 450, 505
Absorption
  absorption threshold, 509
  phonon absorption, 198, 446
  photon absorption, 392
Acceleration, 103, 144, 373, 392, 393
Acoustic phonon, 104, 122, 123, 200, 201,
  334, 449, 450
Acoustic spectrum, 105, 199
Adiabaticity
  parameter of adiabaticity, 91
Alkali element, 65, 66
Alloy
  superconducting alloy, xiii, 519
Amorphous system, 4, 7, 8, 177
Amplitude
  amplitude of electron–phonon interaction,
    445
  Bloch amplitude, 3, 386, 389–391
  gap amplitude, 498, 500
  probability amplitude, 171, 173, 177, 184,
    194, 196–198, 490, 492
  scattering amplitude, 8, 9, 31, 41, 42, 277,
    312

Anharmonic
  anharmonic interaction, vii, 93, 203, 206,
    208, 221, 224, 237, 267, 268
  anharmonic term, 93
Anomaly
  Kohn anomaly, xiii, 325, 341, 342, 348,
    349
Anticommutation, 474, 476, 478, 479
Anticommutation relations, 474, 476, 478
Antiferromagnet, 245, 246, 256, 257
Approach
  strong coupling approach, 388
Approximation
  Born approximation, 41, 42, 182, 183,
    188, 189, 228, 233
  Debye approximation, xiii, 107, 111, 113,
    151, 160–162, 410
  Hartree–Fock approximation, 44, 301
Area, 61, 62, 83, 84, 86, 87, 115, 363, 400–
  402, 439, 440
Atom
  hydrogen atom, 8, 9, 42, 43, 247, 249, 251,
    365
Averaging
  temperature averaging, 158, 164, 173, 266

Axis
  main axis, 394
  z-axis, 255

D

De Haas
  De Haas–van Alphen effect, xiii, 417, 424, 434

Debye
  Debye approximation, xiii, 107, 111, 113, 151, 160–162, 410
  Debye frequency, 111, 170, 448, 459
  Debye model, 112, 149, 151, 176
  Debye radius, 325, 339
  Debye temperature, 112, 151, 216, 219, 410, 448

Debye–Waller
  Debye–Waller factor, 167, 184

Delta
  delta function, 20, 65, 183, 185, 194, 198, 212, 213, 219, 220, 379, 382

Density
  normal density, 57, 69, 71, 73, 74
  superconducting density, 516
  superfluid density, 49, 74, 77, 78, 83, 85, 86, 506

Density of states
  density of states on one spin projection, 438, 467
  electron density of states, 397, 406, 407, 413, 422
  phonon density of states, vii, 159, 170, 407, 413
  total density of states, 406, 407, 468, 486

Depth
  penetration depth, 517–519

Derivative
  first derivative, 134, 342, 345
  second derivative, 24, 133, 135, 136, 342, 345
  variational derivative, 373, 378

Diagram
  bubble diagram, 315, 326, 327

Diagrammatic
  diagrammatic expansion, vi, 297
  diagrammatic technique, 315, 320, 325, 326

Diamagnet
  ideal diamagnet, 518

Diamagnetism
  Landau diamagnetism, xiii, 417, 435, 438

Dielectric
  dielectric crystal, xiv, 227, 228, 240

Dielectric function, xiii, 309, 320

Dielectric permittivity, 137, 314, 315, 320, 325, 327, 328, 330–333, 335–342, 344, 356, 357, 361, 375–377, 384

Differential, 155, 170, 182, 183, 237, 265, 367, 401

Dimension
  1D, 5, 124, 163, 166, 349
  2D, 49, 67, 77–79, 81–86, 88, 115, 348, 349
  3D, 20, 72, 78, 83, 84, 111, 118, 151, 349

Dimensionless variable, 17, 22, 48, 110, 151, 299, 323, 331, 346, 497, 501, 510, 511

Dirac
  Dirac step function, 304, 321, 328, 329, 383, 410, 432

Dispersion
  electron dispersion, 406, 407

Displacement
  atom displacement, 147, 149, 157, 203
  mean square displacement, 77, 93, 177, 368

Distance
  interatomic distance, 93, 122, 204, 214, 276, 281, 397, 451, 462, 463

Distribution
  Boltzmann distribution, 508
  Bose–Einstein distribution, 15, 30, 39
  equilibrium distribution, 228, 230, 232–234
  Fermi–Dirac distribution, 298, 383, 408, 410, 427, 430
  Gibbs distribution, 229
  Planck distribution, 21, 149–152, 158, 161, 165, 175, 212, 218, 220, 222, 223, 228, 230

Distribution function
  Boltzmann distribution function, 508
  Bose–Einstein distribution function, 15, 30
  Fermi–Dirac distribution function, 298, 383, 408, 410, 427, 430
  Planck distribution function, 149–152, 158, 161, 165, 175, 212, 218, 220, 223, 228

Drift, 489

Drift velocity, 489

Dulong–Petit
  Dulong–Petit law, 225, 237

E

Effect
  De Haas–van Alphen effect, xiii, 417, 424, 434
  Meissner effect, xiii, 495, 512, 516–518
  Mossbauer effect, vii, xiii, xiv, 167, 176, 177, 183, 184